C0-AVC-642

PLANT COMMUNITY ECOLOGY

Advances in vegetation science 7

Edited by
EDDY VAN DER MAAREL

1985 **DR W. JUNK PUBLISHERS**
a member of the KLUWER ACADEMIC PUBLISHERS GROUP
DORDRECHT / BOSTON / LANCASTER

Plant community ecology:
Papers in honor of
Robert H. Whittaker

Edited by

R.K. PEET

Reprinted from Vegetatio

1985 **DR W. JUNK PUBLISHERS**
a member of the KLUWER ACADEMIC PUBLISHERS GROUP
DORDRECHT / BOSTON / LANCASTER

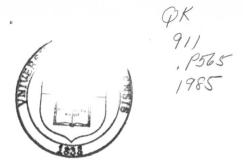

Distributors

for the United States and Canada: Kluwer Academic Publishers, 190 Old Derby Street, Hingham, MA 02043, USA
for the UK and Ireland: Kluwer Academic Publishers, MTP Press Limited, Falcon House, Queen Square, Lancaster LA1 1RN, UK
for all other countries: Kluwer Academic Publishers Group, Distribution Center, P.O. Box 322, 3300 AH Dordrecht, The Netherlands

Library of Congress Cataloging in Publication Data

Main entry under title:

Plant community ecology.

    (Advances in vegetation science ; v. 7)
    1. Plant communities--Addresses, essays, lectures.
2. Botany--Ecology--Addresses, essays, lectures.
3. Whittaker, Robert Harding, 1920-   --Addresses,
essays, lectures. I. Whittaker, Robert Harding,
1920-   . II. Peet, R. K. III. Series.
QK911.P565 1985   581.5'247   85-12543
ISBN 90-6193-532-6

ISBN 90-6193-532-6 (this volume)
ISBN 90-6193-893-7 (series)

Copyright

Table of Contents

1. R. K. Peet. Introduction   1
2. W. E. Westman & R. K. Peet. Robert H. Whittaker 1920–1980: The man and his work. (48: 97–122)   5

I Community analysis

3. M. P. Austin, R. B. Cunningham & P. M. Fleming. New approaches to direct gradient analysis using environmental scalars and statistical curve-fitting procedures. (55: 11–27)   31
4. L. Fresco. An analysis of species response curves and of competition from field data: some results from heath vegetation. (48: 175–185)   49
5. M. J. A. Werger, J. M. W. Louppen & J. H. M. Eppink. Species performances and vegetation boundaries along an environmental gradient. (52: 141–150)   59
6. P. H. Fewster & L. Orlóci. On choosing a resemblance measure for non-linear predictive ordination. (54: 27–35)   69

II Compositional gradients

7. W. E. Westman, Xeric Mediterranean-type shrubland associations of Alta and Baja California and the community/continuum debate. (52: 3–19)   79
8. J. White & D. C. Glenn-Lewin. A numerical analysis of the regional and local variation in North American tall-grass prairie vegetation: structure and composition. (57: 65–78)   97
9. T. R. Wentworth. Distributions of C_4 plants along environmental gradients in southeastern Arizona. (52: 21–34)   111
10. J. M. Bernard, F. K. Seischab & H. G. Gauch Jr. Gradient analysis of the vegetation of the Byron-Bergen swamp, a rich fen in Western New York. (53: 85–91)   125
11. L. Olsvig-Whittaker, M. Shachak & A. Yair. Vegetation patterns related to environmental factors in a Negev Desert watershed. (54: 153–165)   133

III Community dynamics

12. M. E. Harmon, S. P. Bratton & P. S. White. Disturbance and vegetation response in relation to environmental gradients in the Great Smoky Mountains. (55: 129–139)   147
13. W. A. Niering & C. H. Lowe. Vegetation of the Santa Catalina Mountains: Community types and dynamics. (58: 3–28)   159
14. K. Woods. Patterns of tree replacement: canopy effects on understory pattern in hemlock – northern hardwood forests. (56: 87–107)   185

VI

IV Species diversity

15. M. V. Wilson & C. L. Mohler. Measuring compositional change along gradients. (54: 129–141)   207
16. P. Dietvorst, E. van der Maarel & H. van der Putten. A new approach to the minimal area of a plant community. (50: 71–76)   221
17. B. Rice & M. Westoby. Plant species richness at the 0.1 hectare scale in Australian vegetation compared to other continents. (52: 129–140)   237
18. R. M. Cowling. Diversity relations in Cape shrublands and other vegetation in the southeastern Cape, South Africa. (54: 103–127)   249
19. A. Shmida & S. Ellner. Coexistence of plant species with similar niches. (58: 29–55)   275
20. J. Walker & R. K. Peet. Composition and species diversity of pine – wiregrass savannas of the Green Swamp, North Carolina. (55: 163–179)   303
21. H. J. During & J. H. Willems. Diversity models applied to chalk grasslands. (57: 103–114)   321

Introduction

Introduction

R. K. Peet

Dep. of Botany, University of North Carolina, Chapel Hill, N.C. 27514, USA

Robert Whittaker's contributions to ecology were many and remarkably varied. His publication record will long stand as a monument to his greatness, and whatever we do to honor him will likely be rather small in comparison. Less well known were his personal interactions and the impact they had on the development of ecology as well as individual scientists. Over the years he touched many of us and we felt not just a professional but also a deep personal loss in his passing. After his death I was contacted by numerous colleagues who wondered what they might do to honor him. Whittaker had long served on the editorial board of *Vegetatio*, which prompted Eddy van der Maarel to suggest that a series of papers in the journal might be a fitting memorial, and so this project was conceived.

Whittaker was a master of synthesis and during his career he published numerous review papers which showed clearly how his work related to and built on that of others. For this reason it seemed inappropriate and redundant to solicit papers reviewing areas to which Whittaker made important contributions. Instead, I chose to solicit research papers illustrating current applications of approaches Whittaker developed and showing a few of the recent advances which have grown directly from his pioneering work. Past colleagues and students were invited to participate, and a few who had been overlooked also volunteered. To avoid unnecessary delay all papers were published in *Vegetatio* as soon as they were in final form. They are here republished together as a memorial to the man who inspired them. They are arranged in four sections representing areas of plant community ecology which were strongly influenced by Whittaker's work: Methods of community analysis, Analysis of gradients, Community dynamics, and Species diversity.

The first major conceptual contribution Whittaker made to ecology was the introduction of what he called gradient analysis. Henry Gleason had earlier hypothesized species to be distributed individualistically rather than in discrete communities. Starting with Gleason's individualistic model of species distribution, Whittaker visualized species as being distributed along an environmental gradient as a series of overlapping, bellshaped curves from which logically followed a model of continuous vegetation change along the gradient. The four papers in the first section, Community Analysis, represent extensions of the gradient analysis methodology Whittaker developed. One of the shortcomings of Whittaker's gradient analysis studies is that the axes he used cannot readily be compared between studies conducted in different locations. Altitude, a complex gradient combining many covarying factors, does not necessarily mean the same thing in Whittaker's Santa Catalina Mountain study as it does in his Smoky Mountain study. The paper by Austin *et al.* presents statistical methods for identifying less complex and more directly comparable environmental factors for use in gradient analysis studies. The next two papers build directly on Whittaker's model of bell-shaped species distribution curves. The paper by Fresco develops the use of species distribution curves for interpretation of vegetation structure while the paper by Werger *et al.* uses species distribution curves to examine the continuity of community change along a gradient. In addition to developing direct gradient analysis methods, Whittaker recognized the value of indirect (ordination) methods. Together with Gauch & Chase he authored one of the first nonlinear ordination techniques. The paper by Fewster & Orlóci represents continued work toward development of ordination methods consistent with the model of vegetation structure Whittaker developed.

Papers in the second series contain examples of current applications of gradient methods. Westman examines regional vegetation patterns in Californian shrubland vegetation and in addition finds evidence for a possible discontinuity in vegetation composition associated with dominance of a particular shrub species. In an ordination based study of tall grass prairie, White & Glenn-Lewin find that a multidimensional representation improves significantly on the interpretation possible using the traditional single moisture gradient model developed by Curtis in the early days of gradient studies. Wentworth uses a gradient model of vegetation to study the differential distribution of C3 and C4 species in the semiarid mountains of southeastern Arizona. Bernard *et al.* and Olsvig-Whittaker *et al.* both use ordination methods to study small scale vegetation pattern, a topic Whittaker was particularly interested in at the time of his death. In the Olsvig-Whittaker *et al.* paper surface flow of water is found to have considerable potential importance for composition of desert vegetation.

Early in his career Whittaker wrote an important paper on climax concepts. He then turned to other research topics and only late in his career returned to questions of vegetation dynamics. In the intervening period he occasionally suggested that succession might be represented by an added dimension on his typically two dimensional gradient diagrams. While he never personally followed up on this idea, the papers by Harmon *et al.* and Niering & Lowe demonstrate for two of Whittaker's major study areas, the Great Smoky Mountains and the Santa Catalina Mountains, how disturbance can be added as a series of overlays on a gradient representation. In addition, the Niering & Lowe paper completes another aspect of the Santa Catalina vegetation study. Whittaker long viewed community classification and gradient analysis methods as complementary and saw both as contributing unique and important insight. Together with Niering he identified vegetation types for the Santa Catalina study to supplement their gradient representation but had not at the time of his death formally described them. Niering & Lowe present this final component of the work complete with illustrations and photographs Whittaker had helped prepare for this purpose.

A fourth area of plant community ecology to which Whittaker contributed greatly was the study of species diversity. He coined the terms alpha, beta and gamma diversity now in common use and proposed a number of measures for them. Wilson & Mohler present a review of measures of beta diversity and propose several new measures which improve on the originals first proposed by Whittaker in 1960 and which are still in use. The related sampling problem of determining the number of species (alpha diversity) of a community of indeterminant size is discussed in the paper by Dietvorst *et al.* Whittaker himself was more interested in identifying biologically meaningful patterns in species diversity than he was in the actual methods of measurement. The elegance of the patterns of bird diversity identified by MacArthur and colleagues only made the apparent diversity of diversity patterns among plants a greater frustration. He eventually reached the conclusion that to find unifying patterns in plant species diversity would require a massive program of worldwide sampling, something which he hoped to accomplish by the time of his retirement. This is a task now left to others and the papers by Rice & Westoby, Cowling, and Walker & Peet all will contribute to the data base needed to complete it. The identification of factors responsible for the maintenance of different levels of diversity in natural communities remains one of the foremost problems in community ecology. The paper by Shmida & Ellner reviews a number of existing theories and then presents a set of mathematical models to suggest additional mechanisms. The papers by Walker & Peet and During & Willems evaluate possible mechanisms for the maintenance of diversity in two species-rich but very different grassland systems.

We cannot safely project the future directions of science or the long term importance of the contributions made by any one individual. We can, however, assess to some extent the importance of the work of various individuals in the current advancement of science. The papers included in this volume are intended to show how the contributions of R. H. Whittaker have had and continue to have an important guiding influence in ecological research.

Robert H. Whittaker (1920–1980): The man and his work

W. E. Westman[1] & R. K. Peet[2]
[1] *Department of Geography, University of California, Los Angeles, CA 90024, U.S.A.*
[2] *Department of Botany, University of North Carolina, Chapel Hill, NC 27514, U.S.A.*

Keywords: Community ecology, Gradient analysis, History of ecology, Ordination, Species diversity, Vegetation, Whittaker

6

Abstract

R. H. Whittaker enlivened many fields within ecology, systematics and evolution with his insights. Perhaps his most significant contributions to ecology lie in the development of the theories and methods of gradient analysis. Through the verification of the individualistic hypothesis with field data from many regions, and the subsequent development and dissemination of methods for studying species distributions along continua, he helped replace the Clementsian paradigm with a Gleasonian one. His extensive field data on primary production, nutrient cycling patterns and species diversity established new standards for documentation in synecology and helped clarify the basis for site-to-site variation in these variables. Through his broad command of the ecological literature, his writings and his contact with ecologists throughout the world he fostered international understanding of the diversity of approaches to vegetation study.

Introduction

At the time of his death on October 20, 1980, Robert H. Whittaker was recognized as one of the world's leading authorities on plant synecology. During his lifetime Whittaker was a major innovator of methodologies for community analysis and a leader in marshalling field data to document patterns in the composition, productivity and diversity of land plant communities. Through literature review and synthesis, he brought clarity to such disparate fields as classification and ordination of plant communities, plant succession and climax, allelochemistry, evolution and measurement of species diversity, niche theory, and the systematics of kingdoms of organisms. He produced monographs on the vegetation patterns of several montane regions of the United States, and in the last six years of his life extended his work to Mediterranean-climate and arid regions not only of the United States, but of Israël, Australia and South Africa. Whittaker's most often cited work was his undergraduate textbook, *Communities and Ecosystems* (1970c, 1975c), which not only provided an introduction to synecology for thousands of students throughout the world, but also succinctly summarized highly diverse literature in a way that provided new insights for many professional ecologists.

Whittaker's ideas were sometimes the subject of intense controversy during his lifetime (continuum theory, climax pattern concepts, the proposal of five kingdoms of organisms). Critical evaluation of Whittaker's contributions is best left to specialists in his various fields of endeavor, and the ultimate significance of his ideas will have to be evaluated by future historians of ecology. As former students and then colleagues of Whittaker, we lay no special claim to such tasks. Rather, it is our aim in this article to summarize Whittaker's major ideas and contributions, and to describe their development within the perspective of the major events in his training and career. In researching his career, we solicited insights from many of his former associates including doctoral students, research collaborators, and close professional colleagues. The letters we received have been placed in the History of Ecology archives at the University of Georgia.

A biographical sketch

Early life

Robert Harding Whittaker was born the youngest of three children to Clive Charles and Adeline Harding Whittaker on December 27, 1920 in Wichita, Kansas. The family home was in Eureka, Kansas, a town of fewer than 4000 people, located 100 km east of Wichita in a wheat growing, livestock grazing and oil producing area of the state. Clive Whittaker taught zoology at Fairmount College (now University of Wichita), and Adeline taught English at the same institution.[1] Robert's interest and skill in languages was stimulated by his mother and his interest in natural history by his father. Clive Whittaker left his teaching post near the time of Robert's birth and engaged in speculative oil-well drilling in an effort to raise family income to the upper middle-class standard of living to which he and his wife were accustomed as children.[2]

Eureka is in the midst of the gently undulating

[1] See notes at the end of the paper.

topography of the Great Plains. Extensive wheat growing in prairie regions of Kansas with marginally adequate precipitation led to massive topsoil erosion during the dry period of the 1930's when Robert was a teenager. These were also the years of the Great Depression, when concern with economic survival pruned frivolities to their sparest.

The combination of influences from his family and the larger mid-Western society of his day served to embue Whittaker with a belief in the work ethic which he held unswervingly throughout his career. Other values nurtured by that place and era included a stoic acceptance of adversity, a sense of identification with and loyalty to the local community, and a respect for authority, parental and social. In the late 1960's when a number of these values were being questioned, Whittaker felt particularly challenged, and spoke out clearly of his values in final paragraphs of his paper to the Brookhaven Symposium on Diversity and Stability in Ecological Systems (1969c, p. 192):

As ideals comfort, pleasure, passive and commercial entertainment, insulation from adult problems, and relative freedom from real effort and discipline may have singularly unideal consequences. It seems they may impoverish the bases of growth of inner strength, sense of self, respect for work, and sense of community with others and the society and its purposes and cultural heritage. They can produce some youth who have never learned successful management of the angers endemic to the human condition, in whom this anger must direct itself toward destruction and self-destruction. The life of passive pleasure may deny the discipline of growth by which rationality and respect for others are superimposed on the easier ways of irrationality and antagonism. This life leads in an extreme few toward an ideological and emotional solipsism, combined with systematic hatred of restraint and authority, hatred of others' freedom, and antagonism toward understanding of complexity and toward the life of the mind. I describe the extreme, but do so because I believe the United States is, by self-augmenting processes now involving commerce and the communications industry, producing increasing numbers of youth with characteristics toward this extreme.

These views, though rarely expressed in print, were characteristic of Whittaker's value system. The consequences of his own self-discipline, drive for hard work, and respect for 'complexity and . . . the life of the mind,' are seen both in the ecological methods he developed and in the kind of research he produced.

Early in his childhood, Whittaker developed a hobby of collecting butterflies. In teenage years, he would hike through the grazing lands near Eureka, observing butterflies and other remnants of nature. In summers when his family vacationed in Colorado, he would often climb to alpine meadows at dawn, to spend the day alone in the mountains.[2] While Whittaker rationalized his work on mountain vegetation as providing useful vegetation gradients, he also clearly enjoyed montane areas and eagerly sought excuses to study them.[2]

Formal education

In 1938 Whittaker entered Washburn Municipal University in Topeka, Kansas. He received a Bachelor of Arts degree in biology there in 1942, with principal emphasis in zoology and entomology. With the Second World War in full swing, he enlisted in the Army Air Force, and was stationed in England as a weather observer and forecaster until 1946. It was during his spare moments in this period that Whittaker acquainted himself with the classical works of Tansley, Clements, Warming and others. He decided to pursue ecological work upon his return to civilian life, and entered graduate school at the University of Illinois in 1946. He completed his Ph.D. 2½ years later.

Arthur G. Vestal was the botanist and teacher who most influenced Whittaker during his graduate career at Illinois.[3,4] Perhaps Vestal's work on the vegetation of the Colorado Rockies (1914, 1917) where Whittaker spent some of his most pleasant summers, as well as Vestal's scholarly reputation, attracted Whittaker. Although Whittaker applied for graduate standing in the Department of Botany, his application was denied, apparently because of insufficient course background in botany.[3,5] Whittaker was subsequently admitted to the Zoology Department and awarded a fellowship. In March 1946 he initiated graduate studies under the direction of Victor Shelford, who retired from active teaching that summer. Charles Kendeigh replaced

Shelford as Whittaker's adviser in September.[5] Although Whittaker worked with Kendeigh and acknowledged his debt to him, he still called Vestal his 'second adviser.'[4] Whittaker was much taken by Vestal's classroom lectures, which questioned the Clementsian notions of the plant association, and discussed Henry Gleason's alternative views of individualistic species distribution.[4] It is apparent from later conversations that Whittaker was deeply aware of the influence Vestal had in shaping his theoretical attitudes and analytical approach.[3] In his later years at Cornell, Whittaker was obviously pleased to play a similar role for the graduate students of others.

The academic environment Whittaker entered at Illinois has been described in a short autobiographical sketch (1972e). That environment was one of a crossfire of views on the nature of communities. Shelford was a strong Clementsian and Kendeigh's ideas had developed from that tradition. Vestal's sympathies lay with the less popular individualistic views of Gleason. By his own account Whittaker brought with him yet a different viewpoint, the lifezone approach of Charles H. Merriam which fit so well the Colorado mountains that he knew. Recognizing the importance of this unresolved conflict, Whittaker formulated a hypothesis. 'Because of their many interactions, species should tend to evolve toward co-adapted groupings, each representing a favorable, balanced pattern of species interactions adapted to some range of environments. Between these favorable combinations should lie transitions of 'community-level hybridization' with less balanced and more changeable mixtures of species populations' (1972e, p. 690).

Kendeigh had been interested in possible studies in the Great Smoky Mountains and when Whittaker indicated he would like to undertake a study in a mountain area the project was fixed.[4,5] Originally Whittaker set out to test his hypothesis by sampling foliage insects, both in many different kinds of forests, and along an elevational transect from deciduous to spruce-fir forest. It soon became apparent that the test would fail because of the irregularity of insect population levels and because of lack of knowledge about how the underlying plant communities related to each other and the environment. For these reasons Whittaker then undertook a study of the vegetation of the Great

Smoky Mountains. He also attempted to measure microclimatic variation, but abandoned the effort after his weather recording instruments were destroyed by animals[5], and substituted topographic position and aspect as indirect indicators. The field work for both the foliage insect monograph and the Smoky Mountain vegetation monograph was completed in one four month field season.

After returning to Urbana, Whittaker spent the entire academic year analyzing his data on plant population densities and preparing his dissertation on the plant ecology of the Great Smoky Mountains. Having acquired an academic post for the following year, he did not have time to complete analysis of the insect populations before leaving Illinois. After promising to prepare a manuscript on the insects the following year, he successfully defended a dissertation in zoology which contained no reference to animals.

What emerged from Whittaker's work was validation of Gleason's hypothesis and rejection of his own; species were independently distributed along environmental gradients and the hypothesized groups of co-adapted species with parallel distributions could not be found. The significance of the result was obvious to Whittaker and to others like W. H. Camp who wrote Whittaker that his manuscript was 'probably the most important ecological paper of the present century,' and that the method would revolutionize the field.[5] However, because Whittaker's thesis was 'long, wordy . . . speculative and highly theoretical'[5] and in part because it was 'overly aggressive in tearing down opposing theories to make way for his own'[5], publication proved difficult. Whittaker once wrote to Kendeigh in response to a suggestion that the manuscript be shortened, 'I should be appreciative if you would avoid regarding the work as another student's thesis to be edited down to article length for publication. I need not remind you that it is a pioneering work which will mark a turning point in the study of vegetation.'[5] In the intervening years Whittaker's unconventional ideas and approaches were reinforced by the simultaneous, independent work of J. T. Curtis and his colleges and students in Wisconsin. After much negotiating and rewriting H. J. Oosting agreed to publish Whittaker's manuscript in *Ecological Monographs* where it appeared in 1956a, eight years after the dissertation was completed.

Subtler political and cultural values may also have drawn Whittaker's attention to opponents of the Clementsian view. During the 1930's, the community concept became linked with political causes (see Tobey, 1981), as well as being increasingly promoted as an article of faith. Clements and Chaney (1936, pp. 51, 52) drew a parallel between the notions of the plant community as an inter-dependent superorganism, and the desirability of developing interdependence ('cooperation') in human societies to deal with the effects of the Dust Bowl in the Great Plains. This use of holism as a defense of federalism was foreshadowed by the writings of the South African diplomat, Jan Smuts (1926), who used examples of holism in nature to support a variety of political causes. These applications of the community concept were both defended (Phillips, 1935) and attacked (Tansley, 1935, p. 299). In the midst of this imbroglio, Henry Gleason's individualistic notions could be seen as a challenge to those of Clements on both conceptual and political fronts. Significantly, Phillips (1935, p. 227) termed Gleason an 'iconoclast', and further noted (1935, p. 226) that many younger investigators in ecology in the early 1930's favored the Gleasonian viewpoint. Given the political climate of this ecological debate, with its undertones of inter-generational conflict, it is plausible that Whittaker's strong sense of individualism, both political and personal, also led him to sympathize with Vestal's attack on Clements' notions, and to favor the Gleasonian hypothesis.

Whittaker's penchant for immersing himself in new subdisciplines and synthesizing theoretical concepts for testing from the existing literature was a pattern he was to follow throughout his career. While the study of vegetational continua is one such example which he pursued with field data, his first paper (1951) also criticized Clements' regional climatic climax concept. Whittaker proposed that stable vegetation itself responds to site-to-site variations in environment producing gradients of climax vegetation, and that successional patterns leading to this multiplicity of endpoints must therefore also be different. These criticisms, while tied to his Great Smoky Mountains data, were also theoretical criticisms drawn from his analysis of the existing literature. Whittaker subsequently used the literature review and synthesis approach to comment on a variety of topics outside the realm of his own field research, notably the phylogeny of the great groups of organisms (1957b, 1959, 1969b, 1977f, 1978e), and allelochemistry (1970b, f, 1971a, f). Similarly, his major literature reviews on classification of natural communities (1962a) and evolution and measurement of diversity (1972d, 1977g) became definitive works in these fields.

Career and personal development

In 1948 Whittaker was appointed instructor in the Department of Zoology at Washington State College (now University) in Pullman, Washington. While at Washington State, he conducted a comparative study of vegetation on serpentine vs. quartz diorite soils in southern Oregon (1954b, c), and began his field work on the Klamath region (1961c) and the Siskiyou Mountains of Oregon and California in particular (1960b). He also conducted a study of copepod communities of small ponds in the Columbia Basin of Southeastern Washington (1958a). While he rose to the position of Assistant Professor at Washington State by 1950, he was let go in 1951. The reasons for this decision are not known with certainty but the catalyst was a period of fiscal retrenchment. Also, the aggressiveness with which Whittaker attacked the ideas of the established professors and defended his own[5] was not apparently endearing to the faculty.[2] He later admitted that he was at the time 'something of a young turk.'[2]

Having lost his first academic job, Whittaker turned to industry for employment. He was hired in 1951 as a Senior Scientist in the Aquatic Biology Unit of the Department of Radiological Sciences at the Hanford Laboratories of the General Electric Company in Richland, Washington, and remained for three years. With access to laboratory facilities for radio-active tracer studies, Whittaker undertook a detailed microcosm study of the movement of phosphorus in aquaria (1961b). He recognized even at this early date that movement of radionuclides in the environment was an important topic, and that studies of movement of nutrients would be important for understanding ecosystems.

While at the Hanford Laboratories he struck up a romance with Clara Caroline Buehl and the two were married on New Year's day, 1953. Although Clara had an M.A. in biology, her role in the marriage soon became that of wife and mother, rather than of scientific collaborator. In the years of their marriage, they raised three sons: John Charles, Paul Louis, and Carl Robert.

In 1954 after sending some 150 letters of inquiry,[6] Whittaker obtained an academic job as Instructor in the Department of Biology, Brooklyn College, City University of New York. Brooklyn College was primarily an undergraduate college, with some Masters degree students. It did not have the publish-or-perish atmosphere of a major university. During his summers, Whittaker escaped the urban life of New York by returning to the Great Smoky Mountains where he initiated a several-year effort to obtain measurements of the biomass and productivity of the forest communities along an elevational gradient. Since Whittaker's interest was in the entire aboveground primary productivity, he began to develop methods for measuring production of shrubs and herbs (1962b, 1963c), and tree components in addition to trunks (1965b). The main approach he selected was the use of volumetric measurement based on growth rings. He succeeded, through a laborious set of calculations, to obtain productivity estimates for the major plant communities in the mountain range (1966a). This effort laid the ground work for the subsequent development of the dimension analysis methodology. The patience and meticulousness required to carry through such a project virtually single-handedly were enormous. While other ecologists sought alternative routes to production measurement through dynamic techniques like gas exchange, Whittaker's personality was such that the enormous effort and attention to detail required to complete such calculations by the growth-ring and clip-and-weigh methods perhaps even appealed to him. It could be argued that he did not have the funding or training to pursue a more physiological approach. Nevertheless, that he engaged in this type of research at all is, perhaps, a clue to the nature of problems and approaches that he was drawn to.

During this period, Whittaker began collaborative projects related to his quest to characterize productivity gradients. With Neil Cohen and Jerry Olson at Oak Ridge National Laboratories, he applied dimension analysis to some eastern Tennessee trees (1963d). With William Niering of Connecticut College he began a gradient analysis and productivity study of the vegetation of the Santa Catalina Mountains of Arizona resulting in a

series of papers, one of which (1965d) won them the 1966 Mercer Award of the Ecological Society of America for the outstanding ecological paper published in the previous two years.

By 1964, Whittaker was an Associate Professor at Brooklyn College with a substantial reputation primarily as a field-based ecologist producing detailed, data-rich studies, and as a maverick on questions of ecological theory. Among ecologists the latter reputation stemmed primarily from his challenge to three major Clementsian paradigms: the regional climatic climax theory, the notion of unique seres leading to the climatic climax, and the supraroganismal theory of plant association. Among systematists, it derived from his proposed five kingdom system of classification. Although none of these challenges arose entirely without precedent in the literature, the strength of these challenges was due to Whittaker's skill in marshalling exhaustively both the literature and field evidence for his viewpoints, and in writing persuasively.

In 1964 George M. Woodwell persuaded Whittaker to take a year's leave to work with him at Brookhaven National Laboratory. The partnership was electric, in part because the two men shared a commitment to high standards and hard work, and perhaps also because Woodwell's humorful and more expansive style well complemented Whittaker's more intense one. The two developed a profound respect and fondness for each other, and collaborated successfully on eight papers on the Brookhaven oak-pine forest and surrounding vegetation from the point of view of surface area (1967c), biomass and production (1967d, 1968e, g, 1969e), nutrient flow (1967d), and effects of gamma irradiation on structure (1968f, 1973n) and diversity (1969e, 1971b). Their earliest paper (1966b), with W. M. Malcolm, expressed their mutual concern with the hazards posed by technology to ecosystems, and was Whittaker's first published venture into the arena of environmental concerns. During this period, Whittaker and Woodwell formalized and computerized the dimensional analysis system (1969e, 1971c). Whittaker stayed on a second year at Brookhaven, and decided not to return to Brooklyn College. He still thought fondly of the meadows and mountains of his youth and wanted to go West, away from the press of Eastern urban life. He accepted an offer as Professor in the

Department of Population and Environmental Biology at the new University of California campus at Irvine.

Whittaker stayed at Irvine two years, long enough to initiate a study of California vegetation in the San Jacinto Mountains (1976b, 1977b, c) and to begin supervision of a doctoral student, Lawrence McHargue, who studied the vegetation of the Coachella Valley. He also became acquainted with the pygmy forest region of Mendocino County, California through Hans Jenny at Berkeley. Despite the excellent faculty assembled at Irvine, Whittaker was disenchanted at his new location by the dizzying pace of urbanization he observed in southern California at that time.[7] The wildlands he had hoped to find were disappearing from view both by smog and bulldozer.

Disillusioned with California, Whittaker accepted an invitation from Lamont Cole to move to Cornell University in rural Ithaca in September 1968 as Professor of Biology in the Section of Ecology and Systematics. With the recognition that came with his appointment to Cornell, Whittaker began to expand the scope of his career objectives, to consolidate his theoretical positions in the literature, and to become adviser and helper to graduate students and colleagues.

Whittaker's first graduate student at Cornell, Walter Westman, undertook a study of production, nutrient circulation and ordination of the vegetation gradient from pygmy forest through Bishop pine forest to redwoods in the Mendocino County area that Whittaker had earlier secured funding to study. Whittaker spent the summer of 1969 in Mendocino with his family partly working with Westman and his field assistants, but mostly preparing the first edition of his undergraduate text, *Communities and Ecosystems* (1970c). Upon return to Cornell, Hugh G. Gauch, Jr., a Cornell graduate with an M.A. in phycology, was hired to assist Westman with laboratory work. Gauch soon learned computer programming, which he took to with alacrity. The Mendocino vegetation along an edaphic gradient, which was so clear and dramatic in the field, was proving hard to reproduce in ordinations using Bray-Curtis polar ordination, factor analysis or principal components analysis (PCA). After the Mendocino project was completed Whittaker encouraged Gauch to use his mathematical and computer programming talents to

continue testing ordination techniques (1972a, b, 1973i, 1977a, 1978d).

Whittaker recognized in his later papers on gradient analysis the potential usefulness of ordination for revealing the gradient structure of vegetation. Yet, he was concerned lest his work on ordination further alienate European ecologists of the classification tradition, whose friendship he had begun to court. In 1970 he delivered a paper (1972c) to the Rinteln symposium entitled 'Convergences of ordination and classification' in which he took pains to reassure European phytosociologists that ordination techniques, with their characterization of vegetation structure as continua, were not antithetical to the strict classificatory approaches of the Braun-Blanquet school. The paper was both an effort in international diplomacy and in theoretical ecology. Whittaker's standing among European ecologists was solidified, and in 1973 he assumed American editorship of the originally European journal, *Vegetatio*, a post which he held until his death. Soon after joining the editorial board, Whittaker helped formulate a new editorial policy for the journal emphasizing methodology and the conceptual unity of vegetation science in preference to descriptive papers of primarily regional interest. With the encouragement of Eddy van der Maarel, he siphoned a number of American papers on ordination and gradient analysis into the journal, where they appeared interleaved with European phytosociological ones. From Whittaker's perspective, he was seeking now to reunite the camps which had been set asunder by the continuum/community controversy he had helped to flare in the 1950's. He edited and contributed extensively to a book on classification and ordination (1973g, 1978b, c) in which he tried to present with painstaking fairness the values of the various national traditions. These efforts reflected Whittaker's role as conciliator, and increasingly, as respected world authority.

In his years at Cornell, his career reached full bloom, and he embarked on an array of research projects. He continued his work on primary production, working up data for the Hubbard Brook forest (1974c) and the Santa Catalina Mountains (1975k) which he had obtained in the mid-1960's, and compiling summary tables of world primary production in a variety of ecosystems (1970c, 1973l, 1975a, b). The work on primary production permitted calculation of nutrient budgets, and Whittaker conducted studies on nutrient stocks for the Brookhaven forest (1975l), and Hubbard Brook forest (1979i). Whittaker also pursued his early interests in species diversity, stimulated in part by the attention given this topic by G. E. Hutchinson, R. H. MacArthur and their students. General patterns of plant species diversity did not emerge (1977g), and in his later studies of diversity in Mediterranean-climate ecosystems he sought not global generalizations, but local patterns which could be interpreted in light of peculiarities of site history and environment (e.g. 1979a, k, 1980, 1981b).

Although Whittaker was concerned about the environmental problems to which attention was being increasingly drawn in this period, he did not feel comfortable wandering from the scope of his specialization in natural community studies, and his written record on environment and conservation issues is limited to discussions in the final pages of his text (1970c, 1975c), his contributions to the global carbon debate (1973m, 1978f), his 'Coda' to the Brookhaven symposium on diversity and stability (1969c), and short comments on ecological effects of radiation, pesticides (1966b) and weather

modification (1967a), and on the need to preserve natural diversity (e.g. 1960b, 1979f). Whittaker had served on the Governing Board of the Nature Conservancy while at Brooklyn College (1957–62), but resisted temptations to spread his energies in these directions in later years. He clearly preferred to concentrate on his own vegetation studies, and assist others in their efforts in this field.

The growth in Whittaker's reputation from 1965 to the time of his death was exponential. One measure of this is the number of times articles of which he was senior author were cited in the scientific literature. Figure 1 shows the exponential rise in citations, to an average of over 400 per year at the time of his death. During this same period the number of citations of all scientific works also grew exponentially (Fig. 1), whereas the average number of citations per author remained constant at 6 to 8 per year. These trends can be interpreted as indicating that Whittaker had obtained 'authority' status during this period, and was being cited more commonly than other authors in part for this

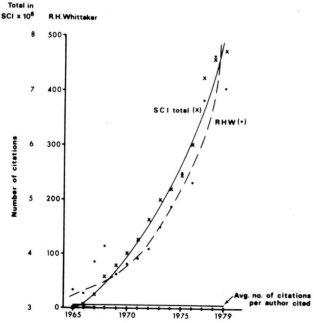

Fig. 1. The number of citations to works of which R. H. Whittaker (RHW) was senior author, the total number of citations for all authors (SCI), and the average number of citations per author cited, during the period 1965–1979.

Table 1. The ten works most often cited between 1965 and 1979, of which R. H. Whittaker was senior author. (Source of data: Science Citation Index).

No. of citations	Communities and Ecosystem
268	R. H. Whittaker. 1st and 2nd editions. MacMillan: New York. 1970, 1975.
251	R. H. Whittaker and P. P. Feeny. Allelochemics: chemical interactions between species. Science 171: 757–770. 1971.
199	R. H. Whittaker. Gradient analysis of vegetation. Biol. Rev. 42: 207–264. 1967.
174	R. H. Whittaker. Dominance and diversity in land plant communities. Science 147: 250–260. 1965.
149	R. H. Whittaker. Vegetation of the Great Smoky Mountains. Ecol. Monogr. 26: 1–80. 1956.
131	R. H. Whittaker. Vegetation of the Siskiyou Mountains, Oregon & California. Ecol. Monogr. 30: 279–338. 1960.
112	R. H. Whittaker. Evolution and measurement of species diversity. Taxon 21: 213–251. 1972.
105	R. H. Whittaker. New concepts of kingdoms of organisms. Science 163: 150–160. 1969.
84	R. H. Whittaker, ed. Handbook of Vegetation Science, Part V. Ordination and Classification of Vegetation. Junk, The Hague. 737 p. 1973.
78	R. H. Whittaker. Forest dimensions and production in the Great Smoky Mountains. Ecology 47: 103–121. 1966.

reason. The fact that his textbook was the most cited work lends credence to this interpretation, as the text was usually cited to document a fact or method, rather than to challenge one. Whittaker's ten most oft-cited works are listed in Table 1.

Whittaker received many inquiries from potential graduate students who sought his guidance, but he limited his acceptance of Ph.D. students for direct supervision to about one per year. During his lifetime, eight students completed their dissertations with him and four additional students completed a major portion of their dissertation work with him (Table 2). Whittaker also hosted several visiting scholars during his Cornell years. Latzi Fresco, Immanuel Noy-Meir, Mike Austin and Mark Hill all worked with Whittaker and Gaugh on ordination problems. The Israeli ecologists Zev Naveh and Avi Shmida worked with Whittaker on sampling and interpretation of diversity. He also supported the professional efforts of a number of ecologists from the Third World, Eastern bloc countries and China. His opposition to what he perceived to be totalitarian governments impelled him to provide particular support for so-called 'dissident' scientists.[8]

Whittaker's years at Cornell were marked both by a significant expansion and solidification of his reputation, and by major events, both joyful and tragic, in his personal life. Probably the most significant event of recognition occurred in 1974, when he was elected to the U.S. National Academy of Sciences. Beyond the Mercer Award, his appointment to Cornell, and his election as Vice-President of the Ecological Society of America in 1971, this was Whittaker's first truly outstanding, tangible evidence of recognition by the scientific establishment. The election marked a stage in Whittaker's career which helped him personally to feel more confident, and ultimately more expansive, in his view of his professional role. Other honors were soon to follow. In 1976, he was appointed to an endowed chair at Cornell, the Charles Alexander Professorship in Biology. During the 1970's he was elected to the American Academy of Arts and Sciences, became an Honorary Member of the British Ecological Society and the Swedish Phytogeographical Society, and was President of the American Society of Naturalists at the time of his death.

In 1974 Clara Whittaker contracted cancer. Her struggle with the disease lasted three years, until she succumbed, at Christmas time, 1977. Although Clara's prolonged illness upset Whittaker greatly he was stoically silent and many of his students and colleagues were not aware of the events that were troubling him. He turned to his traditional values for support, and increased even further the intensity with which he pursued his work. In the period following Clara's death, Whittaker developed a close friendship with his doctoral student, Linda Olsvig. In October 1979, the two were married. Linda later accompanied Whittaker into the field on visits to Israël and South Africa, and took an active interest in his research.

Four months after his second marriage, Whittaker complained of hip pain. X-rays revealed cancer in hip and lungs. 'We were told he had nine months,' writes Linda, 'maybe two years if he was lucky. He wasn't. Bob set himself to complete as much as possible. We had planned to go to Israël, Spain and China – a wonderful trip to complete field work for a variety of projects. We did go to Israël, with Bob on crutches. He conducted himself with great courage and determination, to begin work he would never see completed, but to do his best to see that others would.'[2] His health failed in September and he died October 20, 1980. Shortly before his death, the Ecological Society of America honored him with its highest award, that of Eminent Ecologist.

Although R. H. Whittaker will be remembered by ecologists most for his immensely important research contributions, those who met him found his personality equally arresting. As a person Whittaker was intense and reserved, with an underlying warmth that sparkled intermittently at the surface. His intensity could be intimidating initially, but he was in fact most accessible and supportive to students and colleagues who sought to know him. Those who became closer to him realized that the intensity he radiated was in fact part of a much larger matrix of internal driving forces. These internal tensions, fired by the demands and lessons of his youth, served to forge his formidable intellect with profound sources of physical energy to produce a prodigious scientific output. These same driving forces appear to have been a source of the tenacity with which he was prepared to argue for the values he deeply held. While the weft of these influences on his work must be left to the historian to unravel, it is to the benefit of the scientific community that they helped to produce the masterworks that are his legacy.

Research contributions

Gradient analysis

One of Whittaker's abiding research themes was the examination of patterns of biotic variation along environmental gradients. The methodology of 'gradient analysis' was developed during the initial Smoky Mountains studies and refined in the Siskiyou vegetation study. In both cases Whittaker traversed separate elevation and topographic-moisture gradients, sampling the vegetation at each new exposure, elevation band, or conspicuous vegetation change. The sampling methods used in the Smoky Mountain study were crude and incompletely described, but for the Siskiyou study Whittaker settled on the 0.1 ha sampling units which became standard for his subsequent studies. This standardization allowed Whittaker and others to compare vegetation in geographically distinct areas on a scale hitherto impossible in North America.

Because site-to-site variation in species importance is notoriously large, Whittaker chose to smooth his species/site data by use of weighted averages, permitting underlying patterns to be more readily visible. He juxtaposed the smoothed elevation and topographic-moisture gradients as orthogonal axes, and indicated the location of physiognomic types within the axis space to form community mosaic diagrams. Borrowing from his wartime experience using weather maps with superimposed isobars,[11] he plotted species abundances on the mosaic as isopleths of constant average importance values of a species.

Whittaker's gradient analyses drew from unusually large field data bases. He considered about 300 0.1-ha samples to be an appropriate minimum sample size for a vegetation study of a mountain range.[11] In this way, his work helped establish new standards for sample intensity in ecological field studies. Whittaker used his gradient studies to explore theoretical questions of vegetation structure, and to provide a framework for studies of ecosystem function. In the Great Smokies, the

gradient analysis initially served to test Clementsian hypotheses of community structure, later was used as a basis for production studies (1966a), and still later for studies of understory structure, pursued by Whittaker's graduate student, Susan Bratton.

Whittaker's gradient studies in the Siskiyou Mountains (1960b) appear to have been undertaken for at least three reasons. Firstly, these and the Klamath Mountains were a poorly known but floristically rich 'center' for Arcto-Tertiary forest elements. Elements of this flora also occurred in the Great Smoky Mountains (1961c), providing a basis for crosscontinental comparison. Secondly, the variety of substrates in the Siskiyous provided a new environmental axis for testing gradient structure, and, in particular, to test Clements' notion that different substrates would not support different stable climax communities in the long term (1954c). Thirdly, the area was relatively close to Pullman, where Whittaker worked at the time of these field studies (1948–51). He later encouraged a graduate student, Mark Wilson, to study small scale patterns in serpentine vegetation of the Siskiyous.

Whittaker undertook a third gradient analysis study in the Santa Catalina Mountains of Arizona with William Niering (1964, 1965d, 1968b, c, d, 1975k). The Santa Catalina Mountains, 400 km southeast of the range on which Merriam's (1890) life-zone classification of communities was based, and site of a zonation study by Shreve (1915), provided another opportunity to test alternative hypotheses of species distributions along gradients. Whittaker and Niering found that both species and growth forms tended to vary continuously in abundance across major dominance types, thus supporting Gleason's concepts. Whittaker and Niering further compared vegetation on the north and south sides of the mountain, documented differences between vegetation on limestone and granitic soils, and observed changes with elevation in lifeforms, growthforms, geographic affinities, production, biomass and diversity.

In his final gradient studies, Whittaker sought to examine whether functional attributes varied along environmental gradients. He chose to observe changes in nutrient cycling patterns with changing soil attributes on a single substrate. One such study was located in the San Jacinto and Santa Rosa Mountains of southern California. Here vegetation change could be observed in relation to soil properties along an elevation gradient from salt desert to subalpine forest. The bulk of the study was never prepared for publication. A mosaic diagram constructed by Whittaker and Niering did appear (Hanawalt & Whittaker, 1976b), but only the survey of changes in soil characteristics conducted in collaboration with R. B. Hanawalt was published in detail (Hanawalt & Whittaker, 1976b, 1977b, c). An additional study across an edaphic gradient from pygmy conifer scrub to giant redwood forests in Mendocino County had a similar theme (Westman & Whittaker, 1975b; Westman, 1975, 1978). Several graduate students associated with Whittaker used gradient analysis as part of their dissertation research. Westman (1971, 1975) used these approaches in Mendocino; Peet in the Colorado Front Range (1975, 1978a, b, 1981); Wentworth (1976, 1981) for the Mule Mountains of southeastern Arizona; Lewin (1973, 1975) in the ravines of the Finger Lake Region of New York; and Olsvig in the pine barrens of Long Island (Olsvig *et al.*, 1979b, 1980).

Whittaker also extended gradient analysis methods to applied ecological concerns. In collaboration with George Woodwell (1968f, 1973n) he examined a stress gradient generated by a gamma-emitting cesium source, and discussed Clementsian notions of retrogression in the context of continua. As part of his dissertation research under Whittaker's supervision, Steven Kessell (1979) developed an interactive computer program for forest fire management based on gradient models of vegetation and fuel load observed in Glacier National Park.

Although most of Whittaker's published research was in plant ecology, he retained a lifelong interest in animal ecology. While employed in the Zoology Department at Washington State University he conducted, with C. W. Fairbanks, the study of copepods in small ponds (1958a) and wrote the monograph on Smoky Mountain foliage insects (1952). He later advised two students working on animal communities. Owen Sholes (1979) studied the response of spider communities to the phenological gradient imposed by the *Solidago* inflorescences they inhabited. Steven Sabo (1980) applied gradient analysis methods to the community structure of subalpine birds of the White Mountains of New Hampshire.

The techniques of gradient analysis Whittaker developed provided a battery of methods for the study of vegetation as continua which represented an alternative to the methods of community-oriented phytosociologists. In so doing, Whittaker also developed the means to test hypotheses of gradient structure, and ultimately to replace the Clementsian paradigm with a Gleasonian one.

Classification of communities and organisms

While Whittaker's conceptual style was generally to replace classifications with continua, he clearly recognized the need for categorization as a means to organize information and show relationships. Whittaker further recognized that the tremendous diversity of schemes for classifying vegetation was impeding communication between ecologists in various parts of the world. In 1961 he undertook to review community classification literature. The task was awesome as it required surveying much of the world literature of descriptive ecology for the preceding half century. A full year was spent on this project and the result was a book-sized review (1962a). The monograph won Whittaker the immediate respect of the European community of ecologists, who had not previously encountered a North American who understood so fully what they were trying to accomplish or how their methods related to those of other workers. The essential elements of the review were later presented as a chapter in Whittaker's book *Ordination and Classification of Communities* (1973g, 1978b, c).

Whittaker also had a keen interest in the classification of kingdoms of organisms. He viewed the classical plant/animal dichotomy as highly artificial. Applying ecological criteria, he recognized three groups with different modes of nutrition: primary producers or photosynthetic organisms (plants), consumers (animals), and decomposers (fungi and bacteria) which absorb rather than produce or consume their energy resources (1957b). This nutritional trichotomy provided the basis for his five-kingdom classification (1969d), the procaryotic and eucaryotic single-celled organisms composing the other two kingdoms. His five-kingdom proposal was widely adopted and served to establish Whittaker's reputation among systematists. Whittaker further expressed his interest in the evolution of organisms through his service as

Associate Editor of *Paleobiology* from 1967 to 1979.

Ordination and numerical classification

Whittaker's interest in methods of ordination derived from several sources, as described in his review of gradient analysis (1967b). He was aware of the early efforts by Curtis, Cottam, McIntosh and Bray at the University of Wisconsin to rank samples of Wisconsin vegetation along axes of floristic similarity. The 'leading dominants' technique which Curtis & McIntosh (1951) first attempted consisted of arranging stands to maximize the unimodality of distribution of dominant species. 'Adaptation values,' or weights corresponding to species peaks, were used to calculate weighted stand scores. Since the ultimate arrangement of stands derived from trial arrangement of stands based on their species composition, Whittaker labeled the procedure 'indirect gradient analysis.' In so doing he wished to emphasize that any environmental interpretation of the floristic axis of variation was derived strictly from knowledge of or assumptions about the environmental preferences of the flora, rather than from direct measurement of the habitats in which the plants were growing. Subsequent, more quantitative techniques of ordination developed by Bray & Curtis (1957) and others were included under the category 'indirect gradient analysis' because of their exclusive use of floristic information in axis derivation and interpretation.

Additionally, Whittaker was aware of the potential application to vegetation of the statistical tools of factor analysis and principal components (PCA) advocated by Goodall (1954), Dagnelie (1962) and Orlóci (1966). None of these authors, including Whittaker, recognized at the time both that factor analysis assumed linear responses of species along environmental gradients, and that such linear responses would obscure the intrinsically nonlinear species distributions along environmental gradients. The importance of these limitations was only first publicized by Swan in 1970.

Whittaker was drawn to use and evaluate ordination techniques because he recognized their potential for identifying gradients in species composition and their relationships to the underlying environment. Existing methods of community classification, be they taxonomic (Braun-Blauquet) or nu-

merical (e.g. Williams & Lambert, 1959), did not have this capacity. One of his first efforts in application of ordination methods was to encourage Westman to use factor analysis and polar (Bray-Curtis) ordination on the edaphic gradient from pygmy forest to redwoods in Mendocino County, California. When factor analysis, PCA, polar ordination and various modifications of these failed to reveal the dramatic gradient visible in the field, Westman (1971, 1975) analyzed the statistical assumptions in factor analysis and PCA and noted problems in their application to vegetation data. Whittaker subsequently encouraged Hugh Gauch, Jr., who had been working as a research assistant with him, to construct simulated data based on the notion that species should exhibit the same Gaussian response curves he had obtained with much data averaging and curve-smoothing from his Great Smoky Mountains study (Gauch & Whittaker, 1972a). Gauch & Whittaker (1972b, Whittaker & Gauch, 1973i) proceeded to analyze the ability of polar ordination and PCA to reproduce simulated species gradients or coenoclines of varying species richness (α-diversity) and gradient length (β-diversity) as well as noise levels. This work helped to clarify the distortions which result from use of PCA and factor analysis discussed by Swan (1970), Noy-Meir & Austin (1971) and Beals (1973). Meanwhile, Whittaker edited and contributed extensively to a review of classification and ordination methods in community ecology (1973g).

Unreliability of PCA as a method for identification of important environmental gradients left only polar ordination as a commonly used method. The shortcomings of polar ordination, including subjectivity in endpoint selection, were well known. Whittaker recognized that a logical step in the development of ordination methods for use in ecology would be development of techniques which explicitly assume non-linear species responses. He conceived of an ordination method in which species arrangement would be based on the best simultaneous fit to Gaussian curves. Gauch and Chase developed the algorithm, resulting in Gaussian ordination (Gauch, Chase & Whittaker, 1974a). Because species distributions are rarely perfectly Gaussian, even along strong environmental axes, and because single environmental gradients are rarely strong enough in nature to muffle the influence of secondary habitat factors, Gaussian ordination proved only moderately effective in reproducing accurate floristic axes from either field data or simulated data.

Whittaker was aware of the limitations of the Gaussian assumption, and continued to seek more effective ordination methods (e.g., Kessell & Whittaker, 1976c). He collaborated with other workers (I. Noy-Meir, M. Austin, and later M. O. Hill) who were seeking alternative approaches. After Hill (1973) introduced reciprocal averaging, while pointing out its similarity with Benzécri's (1963–1964) l'analyse factorielle des correspondances, Gauch, Whittaker & Wentworth (1977a) tested it with simulated coenoclines and field data. They found reciprocal averaging far superior in recovering known compositional gradients than any of the methods previously tested, but they also recognized several shortcomings they were unable to resolve. Whittaker invited Hill to Cornell in 1978–79 to work on improving the performance of reciprocal averaging. Hill developed DECORANA (detrended correspondence analysis) during this period (Hill & Gauch, 1980). The final ordination study in which Whittaker participated (Gauch, Whittaker & Singer, 1981a) compared several nonmetric multidimensional scaling techniques to reciprocal averaging and DECORANA. DECORANA proved able to handle data sets of one and several major axes of variation better than the other techniques.

Whittaker's role throughout the development of ordination techniques was to encourage those with whom he worked (especially Hugh Gauch) to explore particular issues arising from his own gradient theories and field data, to contribute data for testing, to review the growing literature, to relate the mathematical results to theories of vegetation structure, and to help write and edit manuscripts on the subject. Whittaker's own limitations in mathematical and computer training inhibited him from becoming directly involved in the details of the programming, but his role as theoretician, critic and supporter of this work was crucial to its progress. With the development of DECORANA and its subsequent successful testing on simulated and field data, and with the production of the Cornell Ecology Program Series and the publication of the second edition of his ordination volume (1978c), Whittaker felt as though his involvement with the development of ordination techniques had reached a natural stopping point.[2]

A second product of Mark Hill's stay at Cornell was the development of TWINSPAN, a new polythetic divisive classification algorithm and program which was a logical outgrowth of Hill's earlier indicator species analysis (Hill, Bunce & Shaw, 1975). The apparent superiority of this technique motivated Whittaker and Gauch to apply the techniques they had developed for evaluation of ordination methods to the analysis of numerical methods of classification. Whittaker's final paper (Gauch & Whittaker, 1981d) in the area of multivariate analysis examined the performance of five hierarchical classification methods in relation to simulated and field data. With the clarity they had brought to earlier analyses, Gauch and Whittaker here carefully established criteria for evaluating classification performance, related the performance criteria to readily-conceptualized features of vegetation structure, and evaluated the performance of a range of classification methods.

The achievements of Gauch, Whittaker, Hill and coworkers in providing community ecologists with numerical methods of analysis of known performance and behavior will undoubtedly stand as signal contributions to the field of community ecology for many years.

Species diversity

Whittaker's interest in species diversity pre-dated the surge of attention given this topic from the late 1950's onward. In his early studies of the Smoky Mountains, Whittaker used his gradient model of vegetation as a framework for describing and interpreting the diversity of both trees (1956a) and foliage insects (1952). Because of sample size variation he selected the index *alpha* of Fisher, Corbet & Williams (1943) to compare diversities. In the Siskiyou study, Whittaker (1960b) similarly included figures showing variation in diversity along the moisture and elevation gradients, and commented on the relation of richness to substrate and continentality.

Whittaker (1960b) studied diversity at several different scales. The diversity of a local community was termed alpha diversity in reference to the commonly used index of Fisher *et al.* As composition of a community at one point on a gradient does not reveal the rate of turnover of species composition along a gradient, he developed measures of this gradient or coenocline differentiation, which he called beta diversity. He termed richness at the regional level gamma diversity and noted that numbers and rates of change in species can be conceptualized along a continuum of scales in the landscape (Whittaker, 1977g).

In 1965(c) Whittaker reviewed diversity measures and models of species abundance using a graph in which the logarithms of importance values were plotted sequentially from the most to the least important species. Whittaker suggested generalizations about the habitat conditions in which different characteristic shapes of these 'dominance-diversity' curves might be found. His discussion helped to show that Preston's lognormal model and MacArthur's broken-stick model of resource partitioning were only two of a wide range of dominance-diversity curves which might occur. The graphical format Whittaker introduced has subsequently been widely used in ecology, archaeology, and other disciplines.

Following his studies of plant species diversity in the Smoky Mountains, Siskiyou's and Santa Catalina's, Whittaker sought the kind of generalizations MacArthur had so successfully described for birds. MacArthur had been able to predict bird species diversity from habitat characteristics, and with E. O. Wilson (1967), had developed a model for predicting species richness at equilibrium in isolated habitats based on simple indices such as habitat patch size and distance from a source for migrants. Such generalizations proved elusive for plants and Whittaker found only numerous, more qualified relationships specific to the taxon, community and area. Whittaker encouraged David Lewin (1975, Glenn-Lewin, 1977) to combine the mountain vegetation data sets with Lewin's data on ravines of upstate New York as part of a massive statistical exercise designed to reveal correlations with species diversity. Few significant correlations were obtained. Whittaker concluded his diversity review in 1977(g) with the observation that 'Although the study of diversity once sought the forms of exact science, it has found instead divergent relationships in different groups and areas, subject much less to prediction than to observation and evolutionary interpretation.' (p. 55).

After failing to find major unifying patterns in species diversity, Whittaker became increasingly interested in understanding what factors were re-

sponsible for the local variability in diversity patterns. He was intrigued by the idea of comparing isolated ecosystems which had developed under similar climatic regimes. The divergent properties of the lizard guilds Pianka studied in California, South Africa and Australia deeply impressed Whittaker.[11] Upon visiting an overgrazed pasture on Mt. Hermon in Israël, he was even more impressed to find a species richness per 0.1 ha almost double the highest he had previously recorded. Literature reports of very high levels of diversity and endemism in the fynbos of South Africa and kwongan of south west Australia contrasted with the low diversity reported for chapparal and mattoral. Intrigued by these observations Whittaker traveled extensively in California, Israël, southern Africa and Australia, comparing diversity in vegetation from regions with Mediterranean climates.[2] Unlike workers who were studying convergences between such regions, Whittaker focused on the causes of the differences in diversity. To him these differences appeared related to community age, substrate fertility and grazing stress. Several papers were published based on this work (Naveh & Whittaker, 1979a, j, 1980; Shmida & Whittaker, 1981b; Whittaker, Niering & Crisp, 1979k; Whittaker, 1977g), but Whittaker's health failed him before he could undertake the planned major synthesis paper.

Community organization and dynamics

Whittaker's contributions to the theory of community structure and function followed several lines, most of which were related to his gradient view of vegetation organization. Although much of this work is interrelated conceptually, it is treated here under several headings.

Niche theory

Whittaker's conceptualization of niche differentiation along environmental gradients (1967b, 1969c, 1970c, 1975c), was his way of explaining the spread of modal peaks of importance value curves for dominant species along environmental gradients. On the one hand, niche differentiation served as a mechanism to explain the coenoclines he was observing along topographic moisture gradients on mountain slopes. On the other hand, the field data provided evidence for the niche hyperspace model presented by G. E. Hutchinson in 1957. Thus Whittaker's concept of an *n*-dimensional space of environmental factors in which Gaussian hypervolumes of species distributions could be found (1967b) corresponded well with Hutchinson's and MacArthur's views of niche.

Because Whittaker's view of niche structure developed from field data and his gradient concept of vegetation variation, he saw clearly the distinction between species distributions along habitat factor axes, and species distributions within a stand in response to the same environmental complex. The second was a concept of niche his gradient studies had not directly characterized, one much closer to Elton's original definition of niche as a species' role in the community. A new term was needed to embrace the concept that the species' environmental tolerances acted both within a site in defining the species niche volume, and between sites along major physical gradients of habitat. With Levin & Root (1973j), he suggested the term 'ecotope' for this combined niche-habitat concept. Whittaker then edited, with Levin, a book of historically important papers on niche theory (1975g).

Small-scale pattern analysis

Unlike R. H. MacArthur, whose inclinations led him further into mathematical conceptualizations of the niche, Whittaker's field orientation led him towards attempts to identify the niche preferences of plants in the landscape. He encouraged his graduate student, Susan Bratton, to identify ways in which understory vegetation in the Great Smoky Mountains might be responding to environmental microheterogeneities. Whittaker recognized that MacArthur and other members of the Princeton school were similarly concerned with environmental heterogeneity or 'grain' in their theoretical studies of species diversity. He sought to make his own contributions at a theoretical level, and collaborated with S. A. Levin (1977h) and D. Goodman (1979h and manuscript) to try to relate population fluctuations due to climate and disturbance at the microscale to the maintenance of alpha diversity, and at a larger scale to beta and gamma diversities. His thinking on these subjects was still in an early stage of formulation at the time of his death, but the larger goal was already clear: to relate individual demographic strategies and microsite heterogeneities to patterns of species diversity at the micro- and ultimately macro-scales.

Concurrent with these efforts, Whittaker began to obtain species richness data using nested quadrats ranging from his initial 0.1 ha (1000 m²) samples through 100, 10, 5 down to 1 m² areas in vegetation from a variety of areas (Naveh & Whittaker, 1979a, j). These data could be used not only to construct species-area curves, but to observe floristic patterns at a range of scales. Borrowing from his ordination work, Whittaker sought to analyze micropatterns by ordinating his 1 m² samples using reciprocal averaging (with Naveh, 1979j, 1980; with Niering & Crisp, 1979k; with Gilbert & Connell, 1979g; with Shmida 1981b; with Morris & Goodman, manuscript). The resulting data could be analyzed to detect microcommunity or synusual groupings, particularly for herb species. He had thus succeeded in characterizing 'patches' in the field, and was in a position to attempt to relate these to theories of population dynamics, disturbance and habitat graininess, and in turn to diversity. This work of synthesis is, of course, the most difficult step, but Whittaker had brought his research to the brink of this phase at the time of his death.

Succession and climax pattern

Whittaker's first publication (1951) was a challenge to the Clementsian notions of a regional monoclimax determined by climate, and to the notion of the seral replacement of one plant association by another. In this paper he suggested that by adopting Gleason's individualistic hypothesis of species distributions, the principle of community continuity (both in space and time) followed. Therefore, individual species, rather than whole communities, must replace each other over time. Site differences will induce differences in floristic composition from place to place, resulting not in a single climax type, or even a polyclimax, but in gradients of steady-state vegetation, a regional 'climax pattern.' The regional climax pattern concept was a direct challenge to the Clementsian paradigm of the day, and set the stage for later analyses of succession on a species-by-species replacement basis pursued by Drury & Nisbet (1973), Connell & Slatyer (1977), Pickett (1976), Horn

(1975) and others in the 1970s.

Whittaker himself returned to his concern with successional theory in the 1970s (1974b, 1975d), and in his paper with Levin (1977h), sought to use microsite heterogeneities and disturbance to understand mosaic successions. He also encouraged two graduate students to study mosaic succession of forest trees. Runkle (1980), a student of Levin, examined tree fall replacement dynamics in the Smoky Mountains and Woods (1979, Woods & Whittaker, 1981c), studied cyclic replacement in northern hardwood forests. Although Whittaker (1974b, 1977h) recognized such processes as direct and cyclic succession and even established 'categories' of successional types, he was not typological in outlook. Instead he viewed successional changes in time and space in relation to a number of axes, and constructed verbally a picture of forces leading toward vegetational stability, countered by forces of disturbance, leading to successional mosaics in the landscape. It was part of Whittaker's writing skill, as well as conceptual talent, to weave together threads (linear axes) representing a variety of environmental and biotic forces into a tapestry which, though complex in design, could evoke almost poetically a picture of the dynamic ecological processes in the mind of the reader.

Biomass, production and nutrient cycling

The earliest approaches to ecosystem function in Whittaker's research were his radiophosphorus tracer studies in aquaria, performed at Hanford Laboratories between 1952–54 (1961b). This early study is remarkable for its effective characterization of phosphorus movement using the compartment model pioneered by E. P. and H. T. Odum only a few years earlier. The aquarium study applied several concepts which were later to find widespread use in nutrient cycling studies: the concentration ratio for elements between trophic levels; turnover rates and decay constants; and the determination of transfer rate constants between compartments.

Whittaker's first major attempt to characterize production in terrestrial ecosystems was his study of aboveground production along the elevational gradient in the Great Smoky Mountains. The novelty of his first approach to production estimation was his extension of production measurement

to all aboveground parts of the vegetation: flowers, fruits, branch wood and bark, and stem wood and bark, of trees and shrubs. He borrowed the concept of volumetric measurement of boles from foresters, but developed his own system for measuring branch biomass and production using both annual xylem ring widths and bud scale scars. His development of logarithmic regressions based on allometric growth assumptions permitted the estimation of forest biomass and production in forests of similar species and physiognomy, but on different sites. With Violet Garfine, a Master's student, he also explored the use of chlorophyll content as an index of productivity (1962c).

In collaboration with G. M. Woodwell, Whittaker expanded and refined his production measurement methods into the technique dubbed 'dimension analysis.' With Woodwell, he developed measures of surface area of forests, which were the first such attempts. During field work in the 1960's, Whittaker applied the dimension analysis approach to vegetation gradients in the Santa Catalina Mountains of Arizona (1975k), the Hubbard Brook watershed in New Hampshire (1974c) and the San Jacinto Mountains of California, although the latter data remain largely unpublished. His discussion of the Santa Catalina data (1975k) was particularly insightful in tracing patterns of primary production and biomass in relation to environmental gradients. Westman and Whittaker (1975b) also applied the technique to Californian coastal conifer forests. Whittaker did not confine use of the method to production measurement, but also began to use dimension analysis techniques, along with tissue nutrient analyses, to develop cation budgets, first for the Brookhaven forest (1975l), and later for the Hubbard Brook forest (1979i).

Whittaker was able to build an extensive base for comparison of biomass and production characteristics in plant communities from his own data. Publication of the massive data tables resulting from the studies was often difficult because of journal policies, but Whittaker argued convincingly that future studies of production depended on the availability of species-specific regression equations, and that recognition of global patterns in production depended on reliable data being available from many sites. Whittaker also compiled a table of primary production and biomass in world ecosystems, first in his text (1970c), later (1973m) refining the estimates for aquatic communities, with G. E. Likens. This table became a basis for major discussions of global primary production totals, and Whittaker & Lieth edited a volume on the subject in 1975(a). The global productivity data gave Whittaker an opportunity to contribute to the debate on vegetation as a net source or sink for atmospheric carbon dioxide (1975i and 1978f).

Whittaker's major contributions to production studies were firstly in his development of a detailed system for static production measurement in temperate forest ecosystems, and secondly in his compilation of extensive sets of data, both from his own work and from the world literature, to reveal major patterns of plant primary production in relation to environmental factors. His extension of this work to nutrient cycling produced an extensive and useful base of empirical information, although as he himself commented in the Hubbard Brook paper (1979i), the degree of effective generalization about patterns of nutrient cycling was somewhat disappointing.

Allelochemistry

The contributions of R. H. Whittaker to allelochemistry derive exclusively from his analysis and synthesis of existing literature on the subject. In 1968, C. H. Muller invited him to be the keynote speaker at a National Academy of Sciences symposium on plant-plant interactions. Muller provided Whittaker with a brief bibliography of recent publications on allelopathy. With characteristic thoroughness, Whittaker researched the literature and provided a lecture entitled 'The Chemistry of Communities' (1971f) which impressed many of the researchers assembled.[3] Whittaker had used both his skill at synthesis of concepts, and his powerful lecturing style, to establish himself in a field that was entirely new to him, and in which he had not done and never subsequently did any original research. It is somewhat ironic that the paper on allelochemics published with Paul Feeny in 1971(a), which was an expansion of the themes in this lecture, became the single most frequently cited article Whittaker ever wrote (Table 1).

Whittaker developed a strong friendship with C. H. Muller while both were at campuses of the University of California in the mid-1960's. They shared a common training from Vestal at Illinois,[3]

and a common cultural and political background from their youths, Muller having been raised in southern Texas. Whittaker funded a chemical analytic facility in Muller's laboratory in the late 1960's. In his review of allelopathy, Whittaker was a strong supporter of Muller's work, although its applicability to field situations was being challenged by others (e.g. Bartholomew, 1970). The ultimate predictive power of this work on allelopathy is still being evaluated. Whittaker's support of it is therefore still a matter for scientific controversy. Whittaker's contribution to the present has lain in his synthesis of evidence from a variety of areas of allelochemistry, and his postulation of a number of evolutionary mechanisms for the origin and selection of such phenomena.

Concluding remarks

R. H. Whittaker's work was characterized on the one hand by a search for patterns and generalization about the structure of biological communities, and on the other by the consistent recognition of the resistance of nature to categorization, classification and simplification. Thus Whittaker emphasized the continuity in the distribution of characters in nature, introducing a calculus to realms of ecological theory hitherto transfixed by categorization. Indeed, it was characteristic of Whittaker to ask of an existing categorical generalization in ecology, How can this generalization be conceptualized in terms of gradients? How can this axis of change be viewed as one of many axes in an n-dimensional space? This same pattern of thought characterized first Whittaker's approach to the distribution of species in relation to environmental factors in space (the gradient theory) and time (the climax pattern), and subsequently to communities and ecosystems.

Whittaker typically selected for study areas with little apparent theoretical cohesion. In both the thoroughness of his reviews and the eclecticism of his approach he was unequalled among ecologists of his time, save perhaps for G. E. Hutchinson. There was not always unanimity of acceptance of the newcomer's interpretations of the literature, and to some Whittaker seemed overly dogmatic once his views were presented. Nonetheless, Whittaker's reviews were consistently thorough and insightful.

Whittaker recognized the importance of hypothesis testing but often preferred to reach generalizations by induction. Typically he would identify a problem, collect the appropriate data, and then interpret the results in what seemed to him the most intuitively satisfying manner. He often smoothed or transformed data to better fit an expected pattern, and deleted 'outliers.' While this ran the risk of overlooking the significant anomaly, it also succeeded in isolating broad patterns which bore well the test of 'replication' from other studies. While Whittaker was prone to draw 'smoothed' Gaussian curves through scatters of points with an impunity that amazed and alarmed his graduate students, he nevertheless resisted the temptation to become too entranced with elegance and simplicity. It was perhaps the latter side of him, along with his own limitations in mathematical training, that dissuaded him from pursuing at greater lengths the kind of mathematical model building characteristic of MacArthur and his followers. A passage at the end of Whittaker's article (with S. Levin, 1977h) on mosaic phenomena concisely illustrates his viewpoint (p. 136):

Ecological theory is not precluded by, but should make realistic allowance for, the intrinsic diversity of ecological phenomena; and ecological research must often center more on analysis, interpretation, comparison, and modeling of cases than on widely applicable generalization. Ecologists have sought a theory or master plan of evolution permitting interpretation of communities through a limited number of strongly linked and widely significant relationships. Such a theory is naturally desired by ecologists as scientists; but . . . there may be no master plan except, perhaps, the evolution of such a diversity of relationships as to frustrate that desire.

It is difficult at this point in the history of the discipline to assess with certainty which of Whittaker's contributions to the science of ecology will be most profound or longlasting. Nevertheless, certain hallmarks of his contributions are already clear: the challenge posed to classificatory approaches to vegetation structure led to the development of the set of theoretical and methodological developments now known as gradient analysis. Although he gave credit to Ramensky and Gleason

for the origins of this tradition and to Curtis and McIntosh for inspiring indirect gradient analysis approaches, it was Whittaker who, more than anyone, solidified the theoretical, methodological and empirical bases for this approach. More than any other unifying notion, it was the conceptualization and demonstration of the continuity of species' response to environmental gradients that characterized Whittaker's work, and which seems likely to be his most enduring contribution.

Through his personal diplomacy, Whittaker built bridges between American and European ecologists over waters which he had troubled with his challenges to phytosociological theories and methods of classification. His reviews of classification and ordination studies, and his global studies of diversity and productivity helped inspire some North American ecologists to increase contacts and collaboration with ecologists beyond their borders.

Whittaker played a major role in generating and disseminating methodologies for the study of plant communities (diversity measurement, gradient analysis, ordination), which he furthered by encouraging Hugh Gauch, Jr. to write computer programs for widescale distribution (Cornell Ecology Programs series). Further, Whittaker's authority status gave his methods a credibility that led to some degree of international standardization of methods. A particularly clear example of this function is the widespread use of the 0.1 hectare samples by workers measuring species richness in vegetation. The loss of this authority in a field short on internationally acknowledged authorities can have a major impact in loosening the cohesion that he helped foster.

Finally, many of Whittaker's contributions remain unpublished in the marginalia of countless manuscripts, and in the correspondence files of myriad colleagues. In addition to his formal editorial duties, Whittaker corresponded with and encouraged an immense number of students and colleagues, many of whom he never personally met. He also delivered a large number of lectures and seminars, characteristically with a sharpness of delivery and density of information that impressed even skeptical audiences. In this way, his influence spread far beyond his published work.

The breadth of Whittaker's influence is a reflection of his own enormous grasp of the fields of ecology and evolutionary biology. Whittaker combined this immensity of vision with a passionate attention to detail and with a personal commitment to collect accurate and abundant supporting data. In his analysis and synthesis of this material he used an intensity of written and oral delivery that reflected his own intellectual ferment. To those who knew him, and are yet to know him, through his writings and scientific influence, he is destined to remain one of the most important figures in community ecology to have lived in this century.

Acknowledgements

We thank the following for responding to our requests for information: B. Chabot, H. Gauch, Jr., D. C. Glenn-Lewin, M. O. Hill, S. C. Kendeigh, S. Levin, H. Lieth, E. van der Maarel, L. T. McHargue, C. H. Muller, W. A. Niering, L. Olsvig-Whittaker, R. Tobey, W. H. Schlesinger, O. Sholes, A. W. Smith, T. R. Wentworth, G. M. Woodwell. The following people provided comments on earlier drafts of the article: E. van der Maarel, P. L. Marks, R. P. McIntosh, W. A. Niering, L. Olsvig-Whittaker, W. H. Schlesinger, R. S. Westman.

Notes

The letters cited were addressed to the authors, and are on file in the History of Ecology Archives, University of Georgia, under 'R. H. Whittaker'.

1. Alice Whittaker Smith, personal communication to W. E. Westman, September 29, 1981.
2. Letter; Linda Olsvig-Whittaker, December 31, 1980.
3. Letter; Cornelius H. Muller, January 12, 1981.
4. R. H. Whittaker (1972e).
5. Letter; S. C. Kendeigh, February 3, 1981.
6. Letter; David C. Glenn-Lewin, April 7, 1981.
7. R. H. Whittaker, personal communication to W. E. Westman, 1969.
8. Letter; Simon Levin, March 18, 1981.
9. Letter; Owen Sholes, January 15, 1981.
10. Letter; Helmut Lieth, February 23, 1981.
11. R. H. Whittaker, personal communication to R. K. Peet, 1972.

Appendix: Publications of Robert H. Whittaker

1948 Whittaker, R. H. A vegetation analysis of the Great Smoky Mountains. Dissertation, University of Illinois, Department of Zoology.

1951 Whittaker, R. H. A criticism of the plant association and climatic climax concepts. Northwest Sci. 26: 17–31.

1952 Whittaker, R. H. A study of summer foliage insect communities in the Great Smoky Mountains. Ecol. Monogr. 22: 1–44.

1953 Whittaker, R. H. A consideration of climax theory: The climax as a population and pattern. Ecol. Monogr. 23: 41–78.

1954a. Whittaker, R. H. Plant populations and the basis of plant indication. (German summ.) Angew. Pflanzensoziol. (Wien), Festschr. Aichinger 1: 183–206.

b. Whittaker, R. H. The ecology of serpentine soils. I. Introduction. Ecology 35: 258–259.

c. Whittaker, R. H. The ecology of serpentine soils. IV. The vegetational response to serpentine soils. Ecology 35: 275–288.

1956a. Whittaker, R. H. Vegetation of the Great Smoky Mountains. Ecol. Monogr. 26: 1–80.

b. Whittaker, R. H. In honor of Erwin Aichinger. Review of Festschrift für Erwin Aichinger zum 60 Geburtstag. 1954. Ecology 37: 396–397.

c. Whittaker, R. H. A new Indian Ecological Journal. Review of Bulletin of the Indian Council of Ecological Research, Vol. 1. Ecology 37: 628.

1957a. Whittaker, R. H. Recent evolution of ecological concepts in relation to the eastern forests of North America. Am. J. Bot. 44: 197–206. Reprinted in, Fifty Years of Botany: Golden Jubilee Volume of the Botanical Society of America, W. C. Steere, ed., p. 340–358. McGraw-Hill, New York, 1958.

b. Whittaker, R. H. The Kingdoms of the living world. Ecology 38: 536–538.

c. Whittaker, R. H. Gradient analysis in agricultural ecology. Review of H. Ellenberg. 1950–54. Landwirtschaftliche Pflanzensoziologie. Ecology 38: 363–364.

d. Whittaker, R. H. Two ecological glossaries and a proposal on nomenclature. Ecology 38: 371.

1958a. Whittaker, R. H. & C. W. Fairbanks. A study of plankton copepod communities in the Columbia Basin, southeastern Washington. Ecology 39: 46–65. Reprinted in, Readings in Population and Community Ecology, W. E. Hazen, ed., p. 369–388.

24

Saunders, Philadelphia, 1964.

b. Whittaker, R. H. A manual of phytosociology. Review of Bharucha, F. R. and W. C. de Leeuw. 1957. A practical guide to plant sociology for foresters and agriculturalists. Ecology 39: 182.

c. Whittaker, R. H. The Pergamon Institute and Russian journals of ecology. Ecology 39: 182–183.

1959 Whittaker, R. H. On the broad classification of organisms. Q. Rev. Biol. 34: 210–226.

1960a. Whittaker, R. H. Ecosystem. *In* McGraw-Hill Encyclopedia of Science and Technology, p. 404–408. McGraw-Hill, New York.

b. Whittaker, R. H. Vegetation of the Siskiyou Mountains, Oregon and California. Ecol. Monogr. 30: 279–338.

c. Whittaker, R. H. A vegetation bibliography for the northeastern states. Review of F. E. Egler. 1959. A cartographic guide to selected regional vegetation literature – where plant communities have been described. Ecology 41: 245–246.

1961a. Whittaker, R. H. Estimation of net primary production of forest and shrub communities. Ecology 42: 177–180.

b. Whittaker, R. H. Experiments with radiophosphorus tracer in aquarium microcosms. Ecol. Monogr. 31: 157–188.

c. Whittaker, R. H. Vegetation history of the Pacific Coast states and the 'central' significance of the Klamath Region. Madroño 16: 5–23.

d. Whittaker, R. H. New Serials. Ecology 42: 616.

1962a. Whittaker, R. H. Classification of natural communities. Bot. Rev. 28: 1–239. Reprinted by Arno Press, New York, 1977.

b. Whittaker, R. H. Net production relations of shrubs in the Great Smoky Mountains. Ecology 43: 357–377.

c. Whittaker, R. H. & V. Garfine. Leaf characteristics and chlorophyll in relation to exposure and production in Rhododendron maximum. Ecology 43: 120–125.

d. Whittaker, R. H. The pine-oak woodland community. Review of J. T. Marshall. 1957. Birds of pine-oak woodland in southern Arizona and adjacent Mexico. Ecology 43: 180–181.

1963a. Niering, W. A., R. H. Whittaker & C. H. Lowe. The saguaro: A population in relation to environment. Science 142(3588): 15–23.

b. Whittaker, R. H. Essays on enchanted islands. Review of The Enchanted Voyage and Other Studies, by G. E. Hutchinson. Ecology 44: 425.

c. Whittaker, R. H. Net production of heath balds and forest heaths in the Great Smoky Mountains. Ecology 44: 176–182.

d. Whittaker, R. H., N. Cohen & J. S. Olson. Net production relations of three tree species at Oak Ridge, Tennessee. Ecology 44: 806–810.

1964 Whittaker, R. H. & W. A. Niering. Vegetation of the Santa Catalina Mountains, Arizona. I. Ecological classification and distribution of species. J. Ariz. Acad. Sci. 3: 9–34.

1965a. Niering, W. A. & R. H. Whittaker. The saguaro problem and grazing in southwestern national monuments. Natl. Parks Mag. 39(213): 4–9.

b. Whittaker, R. H. Branch dimensions and estimation of branch production. Ecology 46: 365–370.

c. Whittaker, R. H. Dominance and diversity in land plant communities. Science 147 (3655): 250–260.

d. Whittaker, R. H. & W. A. Niering. Vegetation of the Santa Catalina Mountains, Arizona: A gradient analysis of the south slope. Ecology 46: 429–452.

1966a. Whittaker, R. H. Forest dimensions and production in the Great Smoky Mountains. Ecology 47: 103–121.

b. Woodwell, G. M., W. M. Malcolm & R. H. Whittaker. A-bombs, bugbombs, and us. NAS-NRC Symposium on 'The Scientific Aspects of Pest Control', by the Brookhaven National Laboratory, U.S. Atomic Energy Commission.

1967a. Whittaker, R. H. Ecological implications of weather modification. *In* Ground Level Climatology, R. H. Shaw, ed., p. 367–384. American Association for the Advancement of Science, Washington, Publ. 86.

b. Whittaker, R. H. Gradient analysis of vegetation. Biol. Rev. 42: 207–264.

c. Whittaker, R. H. & G. M. Woodwell. Surface area relations of woody plants and forest communities. Am. J. Bot. 54: 931–939.

d. Woodwell, G. M. & R. H. Whittaker. Primary production and the cation budget of the Brookhaven forest. Symposium on Primary Productivity and Mineral Cycling in Natural Ecosystems, H. E. Young, ed., p. 151–166. University of Maine Press, Orono.

1968a. Frydman, I. & R. H. Whittaker. Forest associations of southeast Lublin Province, Poland. (German summ.) Ecology 49: 896–908.

b. Whittaker, R. H., S. W. Buol, W. A. Niering & Y. H. Havens. A soil and vegetation pattern in the Santa Catalina Mountains, Arizona. Soil Sci. 105: 440–450.

c. Whittaker, R. H. & W. A. Niering. Vegetation of the Santa Catalina Mountains, Arizona. III. Species distribution and floristic relations on the north slope. J. Ariz. Acad. Sci. 5: 3–21.

d. Whittaker, R. H. & W. A. Niering. Vegetation of the Santa Catalina Mountains, Arizona. IV. Limestone and acid soils. J. Ecol. 56: 523–544.

e. Whittaker, R. H. & G. M. Woodwell. Dimension and production relations of trees and shrubs in the Brookhaven forest, New York. J. Ecol. 56: 1–25.

f. Woodwell, G. M. & R. H. Whittaker. Effects of chronic gamma irradiation on plant communities. Q. Rev. Biol. 43: 42–55.

g. Woodwell, G. M. & R. H. Whittaker. Primary production in terrestrial ecosystems. Am. Zool. 8: 19–30. Reprinted in Energy Flow and Ecological Systems, F. B. Turner, ed.

1969a. Whittaker, R. H. A view toward a National Institute of Ecology. Ecology 50: 169–170.

b. Whittaker, R. H. Een nieuwe indeling van de organismen. Natuur en Techniek 37: 124–132.

c. Whittaker, R. H. Evolution of diversity in plant communities. In Diversity and Stability in Ecological Systems, Brookhaven Symposia in Biology, No. 22, p. 178–195.

d. Whittaker, R. H. New concepts of kingdoms of organisms. Science 163(3863): 150–160.

e. Whittaker, R. H. & G. M. Woodwell. Structure, production, and diversity of the oak-pine forest at Brookhaven, New York. J. Ecol. 57: 155–174.

1970a. Bormann, F. H., T. G. Siccama, G. E. Likens & R. H. Whittaker. The Hubbard Brook ecosystem study: Composition and dynamics of the tree stratum. Ecol. Monogr. 40: 373–388.

b. Brown, W. L., Jr., T. Eisner & R. H. Whittaker. Allomones and kairomones: Transspecific chemical messengers. BioScience 20: 21–22.

c. Whittaker, R. H. Communities and Ecosystems. Macmillan, New York. xi + 162 pp. Reprinted in Japanese edition, Tokyo, 1974.

d. Whittaker, R. H. Neue Einteilung der Organismenreiche. Umschau 16: 514–515.

e. Whittaker, R. H. Taxonomy. In McGraw-Hill Yearbook of Science and Technology 1970, p. 365–369. McGraw-Hill, New-York.

f. Whittaker, R. H. The biochemical ecology of higher plants. In Chemical Ecology, E. Sondheimer and J. B. Simeone, eds., p. 43–70. Academic Press, New York.

g. Whittaker, R. H. The population structure of vegetation. In Gesellschaftsmorphologie (Strukturforschung), (German summ.), R. Tüxen, ed., p. 39–62. Ber. Symp. Int. Ver. Vegetationskunde, Rinteln, 1966. Junk, The Hague.

h. Woodwell, G. M. & R. H. Whittaker. Ionizing radiation and the structure and functions of forests. In Gesellschaftsmorphologie (Strukturforschung), (German summ.), R. Tüxen (ed.), p. 334–339. Ber. Symp. Int. Ver. Vegetationskunde, Rinteln, 1966. Junk, The Hague.

1971a. Whittaker, R. H. & P. P. Feeny. Allelochemics: Chemical interactions between species. Science 171 (3473): 757–770.

b. Whittaker, R. H. & G. M. Woodwell. Evolution of natural communities. In Ecosystem Structure and Function, Proceedings of the 31st Annual Biology Colloquium, J. A. Wiens, ed., p. 137–156. Oregon State University Press, Corvallis.

c. Whittaker, R. H. & G. M. Woodwell. Measurement of net primary production of forests. Reprinted in Productivity of Forest Ecosystems (French summ.), Proceedings of the Brussels Symposium, 1969, P. Duvigneaud, ed., p. 159–175. Unesco, Paris.

d. Brussard, P. F., S. A. Levin, L. N. Miller & R. H. Whittaker. Redwoods: a population model debunked. Science 175: 435–436.

e. Whittaker, R. H. Dry weight, surface area, and other data for individuals of three tree species at Oak Ridge, Tennessee. *In* P. Sollins & R. M. Anderson (eds.), Dry-weight and Other Data for Trees and Woody Shrubs of the Southeastern United States. Oak Ridge National Lab. Pub. ORNL-IBP-71-6. p. 37–38.

f. Whittaker, R. H. The chemistry of communities. *In* Biochemical Interactions Among Plants. National Academy of Sciences. p. 10–18.

1972a. Gauch, H. G., Jr. & R. H. Whittaker. Coenocline simulation. Ecology 53: 446-451.

b. Gauch, H. G., Jr. & R. H. Whittaker. Comparison of ordination techniques. Ecology 53: 868-875.

c. Whittaker, R. H. Convergences of ordination and classification. Reprinted *In* Basic Problems and Methods in Phytosociology (German summ.), Ber. Symp. Int. Ver. Vegetationskunde, Rinteln, 1970, R. Tüxen, ed., p. 39–55. Junk, The Hague.

d. Whittaker, R. H. Evolution and measurement of species diversity. Taxon 21: 213–251.

e. Whittaker, R. H. A hypothesis rejected: the natural distribution of vegetation. *In* W. A. Jensen & F. B. Salisbury (eds.), 1972, Botany: An Ecological Approach. Wadsworth, Belmont, California, p. 689–691. Reprinted *in* W. A. Jensen *et al.* (eds.), 1979, Biology. Wadsworth, Belmont, California, p. 474–476.

1973a. Cottam, G., F. G. Goff & R. H. Whittaker. Wisconsin comparative ordination. *In* Ordination and classification of communities, R. H. Whittaker, ed., p. 193–221. Junk, The Hague.

b. Whittaker, R. H. Approaches to classifying vegetation. *In* Handbook of Vegetation Science, Part V: Ordination and Classification of Vegetation, R. H. Whittaker, ed., p. 325–354. Junk, The Hague.

c. Whittaker, R. H. Community, biological. *In* Encyclopaedia Britannica, p. 1027–1035. 15th edition.

d. Whittaker, R. H. Direct gradient analysis: Results. *In* Handbook of Vegetation Science, Part V: Ordination and Classification of Vegetation, R. H. Whittaker, ed., p. 35–51. Junk, The Hague.

e. Whittaker, R. H. Direct gradient analysis: Techniques. *In* Handbook of Vegetation Science, Part V: Ordination and Classification of Vegetation, R. H. Whittaker, ed., p. 9–31. Junk, The Hague.

f. Whittaker, R. H. Dominance-types. *In* Handbook of Vegetation Science, Part V: Ordination and Classification of Vegetation, R. H. Whittaker, ed., p. 389–402. Junk, The Hague.

g. Whittaker, R. H., ed. Handbook of Vegetation Science, Part V: Ordination and Classification of Vegetation. Junk, The Hague. 737 pp.

h. Whittaker, R. H. Introduction. *In* Handbook of Vegetation Science, Part V: Ordination and Classification of Vegetation, R. H. Whittaker, ed., p. 1–6. Junk, The Hague.

i. Whittaker, R. H. & H. G. Gauch, Jr. Evaluation of ordination techniques. *In* Handbook of Vegetation Science, Part V: Ordination and Classification of Vegetation, R. H. Whittaker, ed., p. 289–321. Junk, The Hague.

j. Whittaker, R. H., S. A. Levin & R. B. Root. Niche, habitat, and ecotope. Am. Nat. 107: 321–338.

k. Whittaker, R. H. & G. E. Likens. Carbon in the biota. *In* Carbon and the Biosphere, G. M. Woodwell & E. V. Pecan, eds., 281–302. U.S. Atomic Energy Commission, CONF-720510, Springfield, Virginia.

l. Whittaker, R. H. & G. E. Likens. Introduction. *In* The Primary Production of the Biosphere, Symp. given at the Second Congress of the American Institute of Biological Sciences, Miami, 1971. Human Ecol. 1: 301–302.

m. Whittaker, R. H. & G. E. Likens. Primary production: The biosphere and man. *In* The Primary Production of the Biosphere, Symp. given at the Second Congress of the American Institute of Biological Sciences, Miami, 1971. Human Ecol. 1: 357–369.

n. Whittaker, R. H. & G. M. Woodwell.

Retrogression and coenocline distance. *In* Handbook of Vegetation Science, Part V: Ordination and Classification of Vegetation, R. H. Whittaker, ed., p. 55–73. Junk, The Hague.

1974a. Gauch, H. G., Jr., G. B. Chase & R. H. Whittaker. Ordination of vegetation samples by Gaussian species distributions. Ecology 55: 1382–1390.

b. Whittaker, R. H. Climax concepts and recognition. *In* Handbook of Vegetation Science, Part VIII: Vegetation Dynamics, R. Knapp, ed., p. 139–154. Junk, The Hague.

c. Whittaker, R. H., R. H. Bormann, G. E. Likens & T. G. Siccama. The Hubbard Brook ecosystem study: Forest biomass and production. Ecol. Monogr. 44: 233–254.

1975a. Lieth, H. & R. H. Whittaker (eds.) The Primary Productivity of the Biosphere. Springer-Verlag, New York. 339 pp.

b. Westman, W. E. & R. H. Whittaker. The pygmy forest region of northern California: Studies on biomass and primary productivity. J. Ecol. 62: 493–520.

c. Whittaker, R. H. Communities and Ecosystems. 2nd Edition. Macmillan, New York. 385 pp. Reprinted in Japanese edition, Tokyo, 1978.

d. Whittaker, R. H. Functional aspects of succession in deciduous forests. *In* Sukzessionsforschung (German summ.), W. Schmidt, ed., p. 377–405. Ber. Symp. Int. Ver. Vegetationskunde, Rinteln, 1973.

e. Whittaker, R. H. The design and stability of plant communities. *In* Unifying Concepts in Ecology, W. H. van Dobben & R. H. Lowe-McConnell, eds., p. 169–181. Report of Plenary Sessions, 1st International Congress of Ecology, The Hague, 1974. Junk, The Hague & Pudoc, Wageningen.

f. Whittaker, R. H. Vegetation and parent material in the western United States. *In* Vegetation und Substrat, (German summ.), H. Dierschke, ed., p. 443–465. Ber. Symp. Int. Ver. Vegetationskunde, Rinteln, 1969.

g. Whittaker, R. H. & S. A. Levin (eds.) Niche: Theory and Application. Benchmark Papers in Ecology. Dowden, Hutchinson and Ross, Stroudsburg, Pennsylvania. 448 pp.

h. Whittaker, R. H., S. A. Levin & R. B. Root.

On the reasons for distinguishing 'niche, habitat, and ecotope.' Am. Nat. 109: 479–482.

i. Whittaker, R. H. & G. E. Likens. The biosphere and man. *In* Primary Productivity of the Biosphere, H. Lieth and R. H. Whittaker, eds., p. 305–328. Springer Verlag, New York.

j. Whittaker, R. H. & P. L. Marks. Methods of assessing terrestrial productivity. *In* Primary Productivity of the Biosphere, H. Lieth and R. H. Whittaker, eds., p. 55–118. Springer Verlag, New York.

k. Whittaker, R. H. & W. A. Niering. Vegetation of the Santa Catalina Mountains, Arizona. V. Biomass, production, and diversity along the elevation gradient. Ecology 56: 771–790.

l. Woodwell, G. M., R. H. Whittaker & R. A. Houghton. Nutrient concentrations in plants in the Brookhaven oak-pine forest. Ecology 56: 318–332.

1976a. Gauch, H. G., Jr. & R. H. Whittaker. Simulation of community patterns. Vegetatio 33: 13–16.

b. Hanawalt, R. B. & R. H. Whittaker. Altitudinally coordinated patterns of soils and vegetation in the San Jacinto Mountains, California. Soil Sci. 121: 114–124.

c. Kessell, S. R. & R. H. Whittaker. Comparisons of three ordination techniques. Vegetatio 32: 21–29.

1977a. Gauch, H. G., Jr., R. H. Whittaker & T. R. Wentworth, A. comparative study of reciprocal averaging and other ordination techniques. J. Ecol. 65: 157–174.

b. Hanawalt, R. B. & R. H. Whittaker. Altitudinal patterns of Na, K, Ca and Mg in soils and plants in the San Jacinto Mountains, California. Soil Sci. 123: 25–36.

c. Hanawalt, R. B. & R. H. Whittaker. Altitudinal gradients of nutrient supply to plant roots in mountain soils. Soil Sci. 123: 85–96.

d. Noy-Meir, I. & R. H. Whittaker. Continuous multivariate methods in community analysis: Some problems and developments. Vegetatio 33: 79–98.

e. Whittaker, R. H. Animal effects on plant species diversity. *In* Vegetation und Fauna,

28

R. Tüxen, ed., p. 409–425. Ber. Symp. Int. Ver. Vegetationskunde, Rinteln, 1976. Cramer, Vaduz.

f. Whittaker, R. H. Broad classification: The kingdoms and the protozoans. *In* Parasitic protozoa, Vol. 1., J. Krier, ed., p. 1–34. Academic Press, New York.

g. Whittaker, R. H. Evolution of species diversity in land communities. *In* Evolutionary Biology, Vol. 10, M. K. Hecht, W. C. Steere & B. Wallace, eds., pp. 1–67. Plenum Publishing Corporation, New York.

h. Whittaker, R. H. & S. A. Levin. The role of mosaic phenomena in natural communities. Theor. Popul. Biol. 12: 117–139.

1978a. Noy-Meir, I. & R. H. Whittaker. Recent developments in continuous multivariate techniques. *In* Ordination of Plant Communities, R. H. Whittaker, ed., pp. 337–378. Junk, The Hague.

b. Whittaker, R. H., ed. Classification of Plant Communities. Junk, The Hague. 408 pp.

c. Whittaker, R. H., ed. Ordination of Plant Communities. Junk, The Hague. 388 pp.

d. Whittaker, R. H. & H. G. Gauch, Jr. Evaluation of ordination techniques. *In* Ordination of Plant Communities, R. H. Whittaker, ed., p. 277–336. Junk, The Hague.

e. Whittaker, R. H. & L. Margulis. Protist classification and the kingdoms of organisms. BioSystems 10: 3–18.

f. Woodwell, G. M., R. H. Whittaker, W. A. Reiners, G. E. Likens, C. C. Delwiche & D. B. Botkin. The biota and the world carbon budget. Science 199(4325): 141–146.

g. Whittaker, R. H. Review of *Terrestrial Vegetation of California,* M. G. Barbour and J. Major, eds. Vegetatio 38: 124–125.

1979a. Naveh, Z. & R. H. Whittaker. Measurements and relationships of plant species diversity in Mediterranean shrublands and woodlands. *In* Ecological Diversity in Theory and Practice, F. Grassle, G. P. Patil, W. Smith & C. Taillie, eds., p. 219–239. International Co-operative Publishing House, Fairland, Maryland.

b. Olsvig, L. S., J. F. Cryan & R. H. Whittaker. Vegetational gradients of the pine plains and barrens of Long Island. *In* Pine Barrens:

Ecosystem and Landscape, R. T. T. Forman, ed., p. 265–282. Academic Press, New York.

c. Sabo, S. R. & R. H. Whittaker. Bird niches in a subalpine forest: An indirect ordination. Proc. Natl. Acad. Sci. 76: 1338–1342.

d. Shmida, A. & R. H. Whittaker. Convergent evolution of deserts in the old and new world. *In* Werden und Vergehen von Pflanzengesellschaften, O. Wilmanns & R. Tüxen, eds., 437–450. Ber. Symp. Int. Ver. Vegetationskunde, Rinteln, 1978. Cramer, Vaduz.

e. Whittaker, R. H. Appalachian balds and other North American heaths. *In* Heathlands and Related Shrublands of the World, Vol. 9. A. Descriptive Studies, R. L. Specht, ed., p. 427–440. Elsevier, Amsterdam.

f. Whittaker, R. H. Vegetational relationships of the pine barrens. *In* Pine Barrens: Ecosystem and Landscape, R. T. T. Forman, ed., p. 315–331. Academic Press, New York.

g. Whittaker, R. H., L. E. Gilbert & J. H. Connell. Analysis of two-phase pattern in a mesquite grassland, Texas. J. Ecol. 67: 935–952.

h. Whittaker, R. H. & D. Goodman. Classifying species according to their demographic strategy. I. Population fluctuations and environmental heterogeneity. Am. Nat. 113: 185–200.

i. Whittaker, R. H., G. E. Likens, F. H. Bormann, J. S. Eaton & T. G. Siccama. The Hubbard Brook ecosystem study: Forest nutrient cycling and element behavior. Ecology 60: 203–220.

j. Whittaker, R. H. & Z. Naveh. Analysis of two-phase patterns. *In* Contemporary Quantitative Ecology and Related Ecometrics, G. P. Patil and M. Rosenzweig, eds., p. 157–165. International Co-operative Publishing House, Fairland, Maryland.

k. Whittaker, R. H., W. A. Niering & M. D. Crisp. Structure, pattern, and diversity of a mallee community in New South Wales. Vegetatio 39: 65–76.

1980 Naveh, Z. & R. H. Whittaker. Structural and floristic diversity of shrublands and woodlands in northern, Israel and other Mediterranean areas. Vegetatio 41: 171–190.

1981a. Gauch, H. G., Jr., R. H. Whittaker & S. B.

Singer. A comparative study of nonmetric ordinations. J. Ecol. 69: 135–152.

b. Shmida, A. & R. H. Whittaker. Pattern and biological microsite effects in two shrub communities, southern California. Ecology 62: 234–251.

c. Woods, K. D. & R. H. Whittaker. Canopy-understory interaction and the internal dynamics of mature hardwood and hemlock-hardwood forests. In D. West, H. H. Shugart & D. B. Botkin (eds.), Forest Succession: Concepts and Application. Springer-Verlag, New York. p. 305–323.

d. Gauch, H. G., Jr. & R. H. Whittaker. Hierarchical classification of community data. J. Ecol. 69: 537–557.

Manuscripts

a. Goodman, D. & R. H. Whittaker. Classifying species according to their demographic strategy. II. A critique of theories concerning r- and k-selection.

b. Whittaker, R. H., J. Morris & D. Goodman. Pattern analysis in savanna at Nylsvley, South Africa.

Table 2. Doctoral dissertations directed by R. H. Whittaker. Degrees were awarded by Cornell University, except where indicated.

A. Dissertations completed under R. H. Whittaker (all at Cornell University)

W. E. Westman, 'Production, Nutrient Circulation and Vegetation-Soil Relations of the Pygmy Forest Region of Northern California.' 1971.

D. C. (Glenn-)Lewin, 'Diversity in Temperate Forests.' 1973.

T. McHargue, 'A Vegetational Analysis of the Coachella Valley, California.' University of California, Irvine. 1973.

S. P. Bratton, 'The Structure and Diversity of Harbaceous Understory Communities in Temperate Deciduous Forest.' 1975.

R. K. Peet, 'Forest Vegetation of the East Slope of the Northern Colorado Front Range.' 1975.

T. R. Wentworth, 'The Vegetation of Limestone and Granite Soils in the Mountains of Southeastern Arizona.' 1976.

O. Sholes, 'Response of Arthropods to the Phenology of Host-Plant Inflorescences, Concentrating on the Host Genus *Solidago*.' 1980.

S. Sabo, 'Community Ecology of Subalpine Birds of the White Mountains.' 1980.

L. Olsvig, 'A Comparative Study of Northeastern Pine Barrens Vegetation.' 1980.

B. Dissertations not completed under R. H. Whittaker

S. R. Kessel,** 'Gradient Modeling: Resource and Fire Management.'

K. D. Woods,* 'Interstand and Intrastand Pattern in Hemlock-Northern Hardwood Forests.' 1981.

M. Wilson,* 'Niche and Habitat in the Low Elevation Vegetation on Serpentine Soils in the Siskiyou Mountains, Oregon.' 1982.

References***

Bartholomew, B., 1970. Bare zone between California shrub and grassland communities: the role of animals. Science 170: 1210–1212.

Beals, E. W., 1973. Ordination: mathematical elegance and ecological naiveté. J. Ecol. 61: 23–36.

Benzécri, J. P., 1963–1964. Cours de linguistique mathématique: 3 ème et 4 ème leçons; Rennes. Inst. Statist. Univ. Paris, Multigr.

Bratton, S. P., 1976. Resource division in an understory herb community: responses to temporal and microtopographic gradients. Am. Nat. 110: 679–693.

Bray, J. R. & Curtis, J. T., 1957. An ordination of the upland forest communities of southern Wisconsin. Ecol. Monogr. 27: 325–349.

Chabot, B. F. & Bunce, J. A., 1979. Drought-stress effects on leaf carbon balance. In: O. T. Solbrig et al. (eds.), Topics in Plant Population Biology, p. 338–355. Columbia Univ. Press, N.Y.

Clements, F. E. & Chaney, R. W., 1936. Environment and life in the Great Plains. Carnegie Inst. of Washington Suppl. Publ. 24. 54 pp.

Connell, J. H. & Slatyer, R. O., 1977. Mechanisms of succession in natural communities and their role in community stability and organization. Am. Nat. 111: 1119–1144.

Curtis, J. T. & McIntosh, R. P., 1951. An upland forest continuum in the prairie-forest border region of Wisconsin. Ecology 32: 476–496.

* Not formally completed under R. H. Whittaker.
** Not formally awarded.
*** In addition to references listed in Appendix and Table 2.

Dagnelie, P., 1960. Contribution à l'étude des communautés végétales par l'analyse factorielle. Bull. Serv. Carte Phytogeogr. Ser. B. 5: 7–71, 93–195.

Drury, W. H. & Nisbet, I. C. T., 1973. Succession. J. Arnold Arb, 54: 331–368.

Fisher, R. A., Corbet, A. S. & Williams, C. B., 1943. The relation between the number of species and the number of individuals in a random sample of an animal population. J. Anim. Ecol. 12: 42–58.

Glenn-Lewin, D. C., 1977. Species diversity in North American temperate forests. Vegetatio 33: 153–162.

Goodall, D. W., 1954. Objective methods for the classification of vegetation. III. An essay in the use of factor analysis. Aust. J. Bot. 2: 304–324.

Hill, M. O., 1973. Reciprocal averaging: an eigenvector method of ordination. J. Ecol. 61: 237–249.

Hill, M. O., Bunce, R. G. H. & Shaw, M. W., 1975. Indicator species analysis, a divisive polythetic method of classification, and its application to a survey of native pinewoods in Scotland. J. Ecol. 63: 597–613.

Hill, M. O. & Gauch, H. G., Jr., 1980. Detrended correspondence analysis: an improved ordination technique. Vegetatio 42: 47–58.

Horn, H. S., 1975. Markovian properties of forest succession. In: M. L. Cody and J. M. Diamond (eds.), Ecology and Evolution of Communities, p. 196–211. Belknap Press, Cambridge, Mass.

Kessell, S. R., 1979. Gradient Modeling: Resource and Fire Management. Spring-Verlag, N.Y.

Lewin, D.C., 1975. Plant species diversity in ravines of the southern Finger Lakes region, New York. Can. J. Bot. 53: 1465–1472.

MacArthur, R. H. & Wilson, E. O., 1967. The Theory of Island Biogeography. Monogr. in Population Biology 1. Princeton University Press, Princeton, N. J.

Merriam, C. Hart., 1890. Results of a biological survey of the San Francisco Mountain region and desert of the Little Colorado in Arizona. North American Fauna 3: 1–128.

Noy-Meir, I. & Austin, M. P., 1971. Principal component ordination and simulated vegetational data. Ecology 51: 551–552.

Orlóci, L., 1966. Geometric models in ecology. I. The theory and application of some ordination methods. J. Ecol. 54: 193–215.

Peet, R. K., 1978a. Latitudinal variation in southern Rocky Mountain forests. J. Biogeogr. 5: 275–289.

Peet, R. K., 1978b. Forest vegetation of the Colorado Front Range: patterns of species diversity. Vegetatio 37: 65–78.

Peet, R. K., 1981. Forest vegetation of the Colorado Front Range: composition and dynamics. Vegetatio 45: 3–75.

Pickett, S. T. A., 1976. Succession: an evolutionary interpretation. Am. Nat. 110: 107–119.

Phillips, J., 1935. Succession, development, the climax, and the complex organism: an analysis of concepts. Part III: The complex organism: conclusions. J. Ecol. 23: 488–508.

Runkle, J., 1979. Gap phase dyamics in climax mesic forests. Ph. D. dissertation. Cornell Univ., N.Y.

Sabo, S. R., 1980. Niche and habitat relations in subalpine bird communities of the White Mountains of New Hampshire. Ecol. Monogr. 50: 241–259.

Shreve, F., 1915. The vegetation of a desert mountain range as conditioned by climatic factors. Carnegie Inst. Washington Publ. 217: 1–112.

Smuts, J. C., 1926. Holism and Evolution. Macmillan, N.Y.

Swan, J. M. A., 1970. An examination of some ordination problems by use of simulated vegetational data. Ecology 51: 89–102.

Tansley, A. G., 1935. The use and abuse of vegetational concepts and terms. Ecology 16: 284–307.

Tobey, R., 1981. Saving the Prairies: The Founding School of American Plant Ecology, 1895–1955. University of California Press: Berkeley and Los Angeles.

Vestal, A. G., 1914. Prairie vegetation of a mountain front area in Colorado. Bot. Gaz. 58: 377–400.

Vestal, A. G., 1917. Foothills vegetation in the Colorado Front Range. Bot. Gaz. 64: 353–385.

Wentworth, T. R., 1981. Vegetation on limestone and granite in the Mule Mountains, Arizona. Ecology 62: 469–482.

Westman, W. E., 1975. Edaphic climax pattern of the pygmy forest region of California. Ecol. Monogr. 45: 109–135.

Westman, W. E., 1978. Patterns of nutrient flow in the pygmy forest region of northern California. Vegetatio 36: 1–16.

Williams, W. T. & Lambert, J. M., 1959. Multivariate methods in plant ecology. I. Association-analysis in plant communities. J. Ecol. 47: 83–101.

Woods, K., 1979. Reciprocal replacement and the maintenance of codominance in a beech maple forest. Oikos 33: 31–39.

Accepted 30.10.1981.

New approaches to direct gradient analysis using environmental scalars and statistical curve-fitting procedures*

M. P. Austin[1], R. B. Cunningham[2] & P. M. Fleming[1]
[1] Division of Water and Land Resources, CSIRO, P.O. Box 1666, Canberra City, A.C.T. 2601, Australia
[2] Department of Statistics, Australian National University, P.O. Box 4, Canberra City, A.C.T. 2601 Australia

Keywords: Aspect, Australia, Direct gradient analysis (DGA), Environmental scalar, Eucalypt, Generalized Linear Modelling (GLM), Radiation index, Realized niche, Species response curve

Abstract

The conceptual framework of direct gradient analysis (DGA) is discussed in relation to the functional, factorial approach to vegetation. Both approaches use abstract simplified environment gradients with which to correlate vegetation response. Environmental scalars based on physical process models of environment and/or known biological growth processes can be incorporated to make analyses less location specific. An example of an environmental scalar (radiation index) for converting aspect and slope measurements to the more biologically relevant radiation input at a site is given.

The problem of the shape of species response curves to environmental gradients is examined using a sample of 1 286 plots from eucalypt forest in southern New South Wales. An important conclusion is that skewed or bimodal response curves may be due to unsatisfactory distribution of observations and/or unrecognized environmental factors. The use of Generalized Linear Modelling (GLM) as a method for providing a statistical basis for DGA is presented. Analyses using GLM, and presence/absence data are presented for a range of eucalypt species (*Eucalyptus rossii, E. dalrympleana, E. fastigata* etc.). Successful prediction of species distributions (realized niches) can be achieved with mean annual temperature, mean annual rainfall, radiation index and geology. Quadratic terms are required in many cases, indicating bell-shaped response curves. The major variability associated with species niches is shown to be related to a limited number (4) of environmental factors. DGA with biologically relevant scalars and appropriate statistical methods is suitable for studying many problems of species' realized niches and plant community composition.

Preface

Robert Whittaker challenged the orthodoxy of his day with the aid of explicit quantitative methods. Subsequently he accepted the challenge to derive general principles from his results, in particular in Gauch & Whittaker (1972) though not perhaps with due regard for statistical nicety (see Westman & Peet, 1982). It is with pleasure therefore that we are able to offer statistical support for one of his insights, the occurrence of bell-shaped ecological response curves.

* Acknowledgements: We thank R. B. Good, J. Duggin, R. Florence and others for making their data available, K. Christenson & E. M. Adomeit for assistance with data analysis, M. F. Hutchinson & D. N. Body for their help with rainfall estimation and H. A. Nix for making his temperature estimates available, and P. Werner, L. F. M. Fresco, I. Noy-Meir, P. Cochrane, W. E. Westman, R. K. Peet and C. R. Margules for comments on the manuscript.

Introduction

R. H. Whittaker made many contributions to ecology (Westman & Peet, 1982). One of the earliest was the concept of a vegetation continuum expressed in terms of environmental gradients, i.e.

direct gradient analysis (Whittaker, 1951, 1954, 1956, 1960, 1967; Whittaker & Niering, 1965) which may yet prove to be Whittaker's most important legacy to the current generation of community ecologists. The idea of analyzing vegetation and species patterns graphically in relation to explicit environmental gradients is essentially simple but does have weaknesses. Whittaker was aware of many of the potential criticisms (Whittaker, 1978) but felt that the graphic illumination of complex ecological problems achieved by the method made the minor difficulties of interpretation acceptable.

This paper examines first the conceptual framework for direct gradient analysis (DGA); second, methods for converting the commonly used environmental gradients of aspect and altitude to variables more relevant to plant growth are discussed; and third, some results on the shape of species response curves are discussed in relation to previous work. Finally an approach to the use of statistical models with DGA is presented.

Conceptual framework

The conceptual framework of DGA is closely related to that of Jenny (1941) regarding the functional, factorial approach to the differentiation of soils. The idea of applying the approach to plants (Major, 1951; Crocker, 1952) was published shortly after the first work of Robert Whittaker (1951), though the first major study using the functional factorial approach appears to have been that of Perring (1958, 1959, 1960) for NW European chalk grasslands.

The essence of the approach is that vegetation properties (v) can be expressed as a function of five factors (Major, 1951, after Jenny, 1941)

$$v = f(cl, p, r, o, t \ldots)$$

where cl = climate, p = parent material, r = topography, o = biotic factor, t = time . . . and possibly other variables. Such a statement seems something of a truism and non-operational. Jenny using soils (s) showed that if four of the five factors were held constant, relationships between soil properties and the fifth factor, e.g. climate, could be demonstrated and analyzed statistically, i.e.

$$s = f(cl : p, r, o, t)$$

Direct gradient analysis as first used by Whittaker is similar and assumes that vegetation properties are related to mesoscale climate and topography when parent material, the biotic factor and time are held constant. In Jenny's notation

$$v = f(cl, r : p, o, t)$$

where mesoscale climate is represented by altitude, and topography ($\equiv$ moisture index) by aspect and position as the axes of the gradient analysis diagrams. Thus DGA can be said to be a graphical analysis procedure for a two-factor, functional factorial approach to the study of vegetation. Perring (1960) separates two components of topography, aspect and slope for his graphical procedure but also has a third factor, climate, incorporated by producing graphical analyses for four climatically distinct areas. Whittaker (1960) also introduced a third factor, parent material by presenting DGA diagrams for different geologies in the same area. Peet (1978) has incorporated latitude (climate?) and successional status (time), as variables. These graphical approaches allow the complex distribution patterns of both individual species and of vegetation parameters such as species richness and productivity (Whittaker & Niering, 1975) to be detected and described in a way which has provided ecological insight.

Environmental gradients have been described and investigated by a wide spectrum of methods, utilizing a variety of concepts. Austin (1980) and Austin & Cunningham (1981) for example divided environmental gradients into three types – indirect, direct and resource gradients. Indirect or complex environmental gradients, e.g. altitude, are those whose influence on plant growth is indirect; it is the location-specific correlation of altitude to rainfall, wind and temperature which is responsible for changes in vegetation performance with altitude. Direct environmental gradients (single factor gradients in Whittaker's term?), e.g. pH, are those where the factor has a direct physiological effect on plant growth, while a resource gradient, e.g. nitrogen, is one where the factor is directly used as a resource for plant growth. There is no absolute distinction between a direct environmental factor and a resource. A particular resource, e.g. nitrate, may when available in supra-abundant quantity act as a direct environmental gradient influencing plant growth through osmotic and chemical toxicity effects. Light is the energy resource for plants, while temperature is a major direct environmental factor.

The functional factorial approach (Jenny, 1941; Whittaker, 1956; Perring, 1958) has used indirect environmental gradients; however plant performance (p) can be represented as a function of four major environmental 'factors':

$$p = f(n, w, t, l)$$

where n = nutrient, w = water, t = temperature, l = light (see also Ellenberg, 1950; Bakuzis & Hansen, 1965; Mueller-Dombois & Ellenberg, 1974). Topography and climate are environmental factors which provide either the inputs which drive the functions of the ecosystem (e.g. rainfall, and solar radiation) or act as modifiers of those inputs (e.g. aspect, slope or soil texture) which in conjunction determine the availability of resources to plants. Indirect environmental factors act through their intercorrelated component variables influencing the way in which the inputs or forcing functions of rain and radiation determine the availability of resource factors at the plant surface or the ambient values of direct environmental factors such as temperature (Fig. 1). In order to understand vegetation/environment relationships we need to develop generalized expressions which summarize how indirect environmental factors determine the availability of resources.

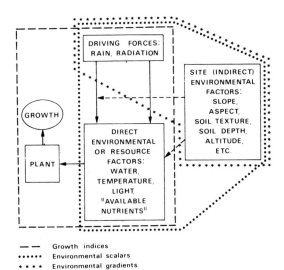

— — Growth indices
•••••• Environmental scalars
• • • • Environmental gradients

Fig. 1. Schematic diagram of the relationship of various environmental variables, types of scalars and the variables they incorporate which have been used to represent their effects on plant growth.

Attempts have been made to develop gradient analysis in this direction. Waring & Major (1964) developed an approach which considered explicitly each of the four resources. They correlated species basal area or frequency with particular environmental variables which gave reasonable estimates of the resource variables, e.g. the environmental moisture gradient was represented by the lowest value of available soil moisture observed during the year, expressed as a percentage of the storage capacity. For each gradient, species were classified by their response type and a species-weighted index calculated; the final environmental gradients were thus indirect (vegetational) ordination axes.

The calculation of scalars (Loucks, 1962; Waring & Major, 1964; see also Waring *et al.,* 1972; Zobel *et al.* 1976) represents an attempt to relate environmental gradients to direct growth factors. The method, however, involved a significant arbitrary element; the scalars are not necessarily based on appropriate physical processes. Waring & Major attempted to use available knowledge of environmental process, but it is arguable whether the minimum observed soil moisture they used is a dependent variable determined by the vegetation or an independent environmental factor. This arbitrary element need not be present; environmental scalars based on physical processes could be developed and used to estimate particular components of resource availability. Fitzpatrick & Nix (1970) devised a growth index for studying Australian grasslands based on three scalars of moisture, temperature and light. Such an index attempts to integrate the effect of several different resource variables.

Thus, three types of measurement have been used for expressing environmental gradients of variables:

(a) Indirect environmental gradients (Austin, 1980) of Jenny (1941), and Whittaker (1951) using simple, measurable factors such as altitude, and aspect. These may be thought of as representational gradients which are readily measured in the field and represent a suite of interconnected variables.

(b) Direct environmental gradients (Austin, 1980), e.g. the environmental scalars of Loucks (1962) and Waring & Major (1964) using nomograms or environmental variables selected on the basis of presumed effects on plants. These may be considered as estimators of individual component variables which determine availability of resources but are not readily measurable.

(c) Growth indices (Fitzpatrick & Nix, 1970) based on the known generalized physiological responses of plants. These are intended to integrate the various resources into a composite measure of how a plant responds to its environment.

The first does not take into account the impact of either environmental or physiological processes on the correlation between environment and vegetation. The second has the potential for incorporating both the complexity of environmental processes and the plant/environment response but only limited progress has so far been made. The growth index approach has so far no means of incorporating other environmental variables, e.g. aspect, into the index.

These environmental gradients or indices are abstractions from the known physiological and environmental complexity. Such abstraction is needed, however, if we are to distinguish between the effects of different environmental factors. We can then test whether further factors and interactions are necessary *for the level of prediction we wish to achieve.* For example, for a specific location (i.e. where the rainfall/altitude relationships remain the same) in mountainous terrain, a DGA based on the indirect environmental gradients of aspect and altitude may be sufficient for the purposes of a particular study (e.g. Whittaker, 1956). For niche studies, direct environmental gradients may be necessary in order to relate species distribution to physiological response, e.g. the substitution of mean annual temperature for altitude, annual incoming solar radiation for aspect, soil pH or soil nutrient concentration for geology.

The simplification implicit in the abstraction of environmental gradients will mean that estimates of the gradients are likely to be noisy and biased. However, provided the relationships between vegetation variables and the environmental gradients are important (i.e. account for a large proportion of the variance) as well as statistically significant, then the results of DGA will continue to contribute usefully to community analysis and theory.

Although the environmental influences on vegetation are often held to be too numerous and complicated to be represented in such a simple way, the evidence of Whittaker is that in some environ-

ments, two or three indirect environmental gradients may summarize much of the observed variation in species distributions. The hypotheses arising from this discussion of DGA are therefore:

(1) Only a limited number (2–6) of environmental factors (gradients) account for a major proportion of the variability in vegetation composition.

(2) If environmental gradients are expressed as resource gradients improved prediction will be achieved.

(3) If a large proportion of the variance is accounted for, then numerous hypotheses regarding species niche can be tested, e.g. what shapes are species niches?; does the degree of overlap between species fit any rule?

To achieve the goal of testing these hypotheses we need: 1) improved environmental models and scalars; 2) explicit analysis procedures capable of handling more than three gradients and providing statistical methods for estimating relationships.

The remainder of this paper addresses these problems of environmental models and analytical procedures in relation to the first hypothesis above, namely that only a limited number of factors need to be considered.

Environmental scalars

Here we briefly indicate how certain environmental gradients may be measured to provide more rational environmental scalars clearly based on physical processes.

Topography and aspect

The use of a topographic or aspect measure assumes that temperature or moisture are important for plant distribution. Apart from soil differences related to topographic position, the major variable related to topography is incoming solar radiation. This may differ by a factor of 10 between protected south-facing gullies and exposed north-facing slopes in winter at 36 °S latitude. If slope, aspect and latitude of a site are known, then the potential solar radiation (in the absence of the atmosphere) can be calculated (see Lee, 1963).

The influence of the atmosphere and of the cutoff in direct radiation due to the surrounding horizon can also be taken into account (Fleming, 1971), together with the diffuse radiation component. For any given site, a radiation index (*RI*) can be calculated expressing the relative amount of total radiation received compared with that received by a flat, exposed plain. The method used is that of Fleming (1971) and details can be found in Fleming & Austin (1983). The index is based on known physical processes and clearly stated assumptions, removing the need for many of the *ad hoc* transformations often applied to aspect scalars.

Radiation indices based on the model have been used in studies of the influence of slope and aspect of chalk grassland species (Austin, 1972), and of shade gradients on lawn dynamics (Austin & Belbin, 1981) in Britain. Several studies of eucalypt distributions have used the radiation index as a measure of aspect (Austin, 1971, 1978; Austin & Cunningham, 1981). In all cases, the index has proved a better predictor of species distribution than simple measures of slope or aspect. The index can also be used to build more complex scalars incorporating rainfall, evapotranspiration and available soil moisture storage. See Austin (1971, 1972) and Austin & Cunningham (1981) for further discussion.

Altitude

Altitude is an indirect environmental gradient which can be partitioned into several direct environmental factors. Two obvious variables which change with altitude are temperature and rainfall. These variables also change in a complex fashion with latitude and longitude. No simple physical process model can estimate them at present. Statistical procedures exist which can provide a prediction for either temperature or rainfall, given the altitude, latitude, and longitude of a site. These procedures are based on fitting a smooth but complex surface to stations with suitable climatic data, and have reached a considerable degree of sophistication. The technique used to obtain estimates for rainfall and temperature for analysis in this paper was a Laplacian smoothing spline function of three variables: altitude, latitude and longitude (Hutchinson, in preparation, Hutchinson & Bischof, 1983).

Both physical process models and statistical methods can provide improved estimates for DGA.

Problems associated with the shape of species response curves to environmental gradients

Gauch & Whittaker (1972) put forward a number of propositions about species response curves, in particular that measures of species performance show a 'Gaussian' or bell-shaped response curve in relation to environmental gradients. Austin (1976) in a review of published curves, suggested that other shapes of curves were as frequent. There are several alternative hypotheses regarding the shape of the response curves (Austin, 1980), but detailed data on which to base tests, are limited.

Austin, Cunningham & Good (1983) collated a data set of 1 286 observations from various sources, of the presence and absence of eucalypt species on plots of approximately 0.1 ha size in sclerophyll forest and alpine habitats in southeastern New South Wales, Australia. The irregular-shaped sample area extends from the western boundary of the Kosciusko National Park to the New South Wales coast between Bermagui and Batemans Bay. (Details of the location of sites can be obtained from the senior author.) Measurements had been made at each plot for altitude, aspect, slope, geology and location (from which estimates for temperature, radiation index and rainfall could be derived (see Austin *et al.*, 1983) for further details). We examine the shape of species qualitative response curves to altitude, the effect of other environmental factors and of converting altitude to an estimate of mean annual temperature.

Altitude

The large number of observations allows examination of the altitudinal distribution of eucalypt species to be undertaken in some detail (Austin *et al.*, 1983). The probability of occurrence is calculated for every 100 m zone based on the proportion of sites within the zone at which the species are found (Fig. 2). Four types of curves can be recognized:
(a) 'Gaussian'. A typical bell-shaped curve, e.g. *Eucalyptus delegatensis.*
(b) Skewed. Examples are *E. radiata* and *E. pauciflora.*
(c) Platykurtic. These are species with low maximum probability and a flattened response curve; examples are *E. fastigata* and *E. muellerana.*

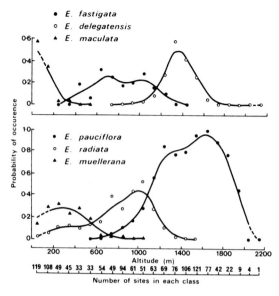

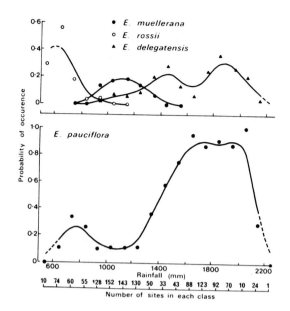

Fig. 2. Probability of occurrence (frequency) of selected eucalypt species in relation to altitude expressed per total number of observations per 100 m altitude class; a one-dimensional direct gradient analysis (DGA). Curves are based on weighted moving average $x'_0 = \frac{1}{4}x_{-1} + \frac{1}{2}x_0 + \frac{1}{4}x_{+1}$, endpoints $x'_E = 2/3x_E + 1/3x_{E\pm1}$.

Fig. 3. Probability of occurrence of selected eucalypt species in relation to mean annual rainfall. Curves as in Figure 2.

(d) Indeterminate. These species are those with occurrences only at low altitudes with a maximum between 0–200 m, e.g. *E. maculata,* and whose response curve is thus truncated at sea level.

Species with high probabilities of occurrence (e.g. *E. pauciflora*) are characteristic of higher altitudes. Those species with curves approaching a bell-shaped response curve are also characteristic of higher altitudes. Species at higher altitudes, i.e. at the extreme of the environmental gradient, appear able to exploit a wider range of other environmental variables, e.g. rainfall or a variety of aspects. In the absence of noise or the influence of other factors, including competition from other species, these species tend to show a bell-shaped distribution. The corollary of this is that species with 'non-Gaussian' response curves are less 'controlled' by altitude and should show more obvious relationships with other environmental factors.

Rainfall

If the occurrences of eucalypt species are plotted against the rainfall estimates for the plots, a similar

variety of response curves is obtained, including a bimodal form, e.g. *E. pauciflora* (Fig. 3). Several species with lower probabilities of non-bell-shaped curves with respect to altitude appear to have better defined curves with respect to rainfall. Most species now show lower maximum probability values, though there are exceptions, e.g. *E. rossii.* In many cases the variability is markedly increased, e.g. *E. delegatensis.*

The bimodal curve for *E. pauciflora* is of particular interest. An explanation for this can be found in terms of a simple technical aspect of studying one-dimensional response curves. The species concerned does not occur below 500 m. Plots occurring below 500 m have, with a few exceptions, a rainfall range of between 850–1 400 mm. If sites below 500 m are removed and the probabilities replotted (Fig. 4), then the bimodal shape disappears and a skewed or 'shouldered' response curve is observed.

The evidence suggests that bimodal curves could be due to sampling and/or data interpretation and analysis problems. The problem can occur in two-dimensional DGA also, if an important variable has been ignored. Bimodal response curves, should not be interpreted to infer ecotypic differentiation or competition, until sampling effects due to incorporating observations beyond the species range have been eliminated. All existing bimodal re-

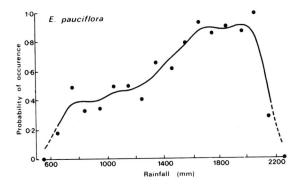

Fig. 4. Probability of occurrence of *E. pauciflora* after removing plots below 500 m. Curves as in Figure 2.

sponse curves need to be re-examined in this light. This is necessary quite apart from problems of estimating the shape of the curve which is considered later.

The question of the shape of species response curves cannot be addressed unless an analysis has been done which demonstrates that the major gradients influencing any particular species have been detected, i.e. the major portion of the species variance in distribution can be ascribed to a limited number of gradients. Questions regarding niche overlap require a similar analysis; as the number of independent environmental gradients is increased, overlap may be expected to decline.

Temperature as a substitute for altitude

When temperature is substituted for altitude as an environmental gradient, the effect of location is also taken into account. Plots at different altitudes may have similar mean annual temperatures depending on their distance from a coast.

Temperature as a more direct environmental gradient might be expected to lead to more bell-shaped or well-defined response curves. Several species curves do become more 'Gaussian' in shape, when plotted against temperature (Fig. 5). For example *E. muellerana* has a coastal distribution. At low altitudes near the coast, temperature changes more rapidly with latitude than altitude. As a consequence, altitude does not reflect temperature differences between sites, and the response curves are truncated (Fig. 2). Using estimated temperature this effect of latitude is allowed for, the response curve now has a maximum probability of occur-

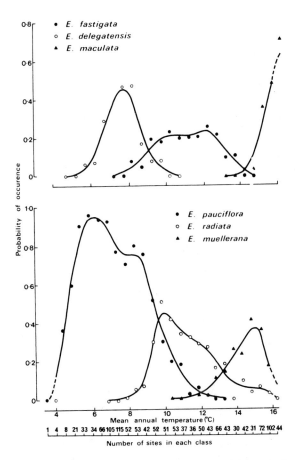

Fig. 5. Probability of occurrence of selected eucalypt species in relation to mean annual temperature. Curves as in Figure 2.

rence suggesting the successful use of a more direct environmental factor.

Other species response curves remain 'flattened' but variance about the moving average curve is reduced, e.g. for *E. fastigata*. The skewness of the response curves of *E. radiata* remains unchanged. Although the results are equivocal regarding the superiority of mean annual temperature over altitude, with respect to smooth response curves, there is nothing to indicate that mean annual temperature is inferior as an environmental gradient. Therefore, mean annual temperature, being more relevant to plant growth is used in further analysis.

Statistical methods for direct gradient analysis

This section presents a brief account of a general framework for the statistical analysis of ecological survey data. The method emphasizes the estimation of relationships between a response and a set of explanatory variables, rather than on hypotheses testing. The methodology is not new and is described in detail elsewhere (see Nelder & Wedderburn, 1972, and Appendix).

Using a general linear model, observed values are predicted as a combination of a systematic component ('explained') and a random component, to which the 'unexplained' variation is assigned. The systematic component can be modelled as a linear combination of explanatory variables, which predicts the observed value. The differences between the observed and predicted values is the random component (or residual) which is described by a probability distribution, usually the normal distribution. Thus if the data consist of a set of observed values represented by y, and $x_1, x_2 \ldots x_k$ are the explanatory variables, the linear model is

$$Y = b_o + b_1x_1 + b_2x_2 + \ldots \qquad + b_kx_k \qquad (1)$$

and the datum is split into

$$y = Y + e$$

where Y is the estimated value and e is the residual.

This formulation is familiar to most readers in the context of regression analysis and analysis of variance models. Recently, a generalization of this formulation has been developed – called the Generalized Linear Modelling (GLM). Before the introduction of the GLM and the availability of computer programs for fitting it, similar analyses were performed by applying variance stabilizing transformations such as $\sqrt{Y_i}$ in the case of counts and $\sin^{-1} \sqrt{Y_{iY}}$ in the case of binomial proportions. These analyses required the assumption of normality of the transformed variables and the necessity of linearity of the explanatory effects. Further, when expected probabilities for binomial data lie outside the range 0.2 to 0.8, as is often the case for ecological data, the arc-sine transformation fails to correct adequately for heterogeneous variances. GLM avoids these assumptions.

The advantages of GLM are that it operates directly on the data and its appropriate statistical distribution, whether it is normal, Poisson, binomial or more complex, and is flexible allowing predictive models involving continuous variables (e.g. rainfall) and/or categorical variables (e.g. different geological substrates) to be combined. The chief

difference for non-statistical users is the use of deviance rather than variance when fitting models with a non-Gaussian error distribution. A brief mathematical description of GLM is given as an Appendix.

The general model used here is that the probability of occurrence of a species can be predicted from mean annual rainfall, mean annual temperature (or altitude), radiation index as a measure of aspect, and the qualitative variable geology.

Austin *et al.* (1983) have shown that GLM using the computer package GLIM (Nelder & Wedderburn, 1972), can demonstrate the statistical validity of curvilinear relationships between the probability of presence of eucalypt species and altitude, rainfall, geology and aspect using the 'log-linear model'. Each independent variable is split into classes and the probability of occurrence in each combination of classes (e.g. granite, rainfall >1 000 mm and radiation <0.95) is modelled using GLIM. Here, the analysis of Austin *et al.* (1983) is extended by incorporating the estimated annual mean temperature in place of altitude and by fitting models with rainfall as a continuous variable rather than as classes. The distribution of five eucalypt species is examined in relation to the environmental variables. Both the log-linear model with factors (qualitative variables), and a mixed model with quantitative variates for temperature, rainfall and radiation indices and a factor for geology, are considered.

Analysis results

E. rossii. An example of a log-linear analysis (similar to Austin *et al.* with mean annual temperature instead of altitude) is given for *E. rossii* in Table 1. The classes were chosen (Table 1) to minimize the number of cells in the contingency table where the probability of occurrence would approach zero and to maximize the number with non-zero probabilities, while operating under the constraint that observations per cell should be as high as possible. Applying strict criteria for accepting a fit to the model ($p < 1\%$), only the main factors are significant but these explain 90% of the deviance.

The coefficients associated with each of the classes (Fig. 6) show the types of behaviour which can be detected. By convention, the coefficient of the first class is given the value zero and the other classes are expressed relative to this, thus the class

Table 1. Binary GLIM model predicting probability of occurrence of *E. rossii,* using a contingency table approach.

Data cells (92 sampled of 108 possible)				No. of classes:	
Factors	Class limits				
Mean annual temperature (T)	10.0, 12.0, 13.5 °C			4	
Mean annual rainfall (R)	780, 1 000 mm			3	
Annual radiation index (RI)	0.95, 1.05			3	
Geology (G)	granite, sediments, volcanics			3	
Total deviance	360.8 df 91				

Main factors model	Residual deviance	df	Change in deviance χ^2	df	p
$T + R + RI + G$	35.09	82			
$-T$	69.44	85	34.35	3	***
$-R$	136.0	84	100.91	2	***
$-RI$	49.66	84	14.57	2	***
$-G$	55.41	84	20.32	2	***

Interaction model					
$T + R + RI + G + 6$ interactions	5.066	52			
$-T.R$	9.92	58	4.85	6	N.S.
$-T.RI$	13.58	58	8.51	6	N.S.
$-T.G$	12.89	58	7.82	6	N.S.
$-R.RI$	12.01	56	6.94	4	N.S.
$-R.G$	14.06	56	8.99	4	<10%
$RI.G$	18.65	56	13.58	4	*

Best model (applying an acceptance level of $p < 1\%$), after backwards elimination.
   Main factors model best. *E. rossii* (p) = $T + R + RI + G$.
      Deviance explained = 90.3%.
      Mean deviance = 0.428.

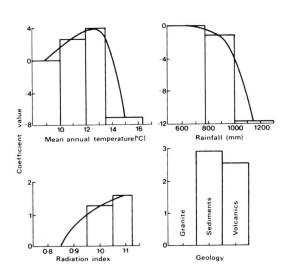

Fig. 6. Graphical presentation of the coefficients associated with the different factors of the contingency table analysis using GLIM and a logit-linear model to predict the probability of occurrence of *E. rossii.*

<10 °C has a value of 0.0 and the class 10–12 °C a value of approximately 3.0. Mean annual temperature coefficients indicate a non-monotonic, possibly skewed response curve. Note the advantage of using classes in that complex curves can be easily fitted. Both rainfall and radiation index coefficients show a slight curve (free hand curves drawn in Fig. 6) rising to maximum at an extreme, at low rainfall (<750 mm) and high radiation index values (i.e. warm northern slopes). Granite is much less likely to have *E. rossii* growing on it than are other rock types. The change in deviance due to each variable gives a measure of its relative importance in predicting the occurrence of *E. rossii.* Rainfall accounts for the highest change in deviance (101.0), temperature is only one third as important (34.4), then geology and finally the radiation index (Table 1).

These model coefficients can be used to predict the probability of occurrence of *E. rossii* for a variety of combinations of environmental gradients. The method is

$$\text{probability} = e^y/(1 + e^y)$$

where $y = a + b_{T(x)} \cdot T_{(x)} + b_{R(x)} \cdot R_{(x)} + b_{RI(x)} \cdot RI_{(x)} + b_{G(x)} \cdot G_{(x)}$; (x) = the class of the observation for each variable; $b_{.(x)}$ = corresponding coefficient for the class; a = constant term.

As the number of classes is increased, the number of cells in the contingency table increases as their product and the number of observations per cell is reduced, leading to greater uncertainty. There is a corresponding increase in interaction terms which give rise to very complex models (e.g. bimodal curves which appear significant), but the very flexibility of the model means that there is high risk of fitting 'significant' but chance differences between cells of the analysis. Reducing the number of classes will, however, reduce the sensitivity of the GLM analysis.

It is possible to fit continuous variables as independent predictors, as with the usual regression analysis, but this requires extensive computation for a moderate sized data set (i.e. 1 286 observations). In order to apply this approach the data were truncated by eliminating observations from those regions where the probability of E. rossii occurring was very low (approx. <1 in 10^4) to reduce the size of the computing problem (Table 2). A continuous model was fitted using the same environmental gradients as variables i.e. temperature, rainfall and radiation index, and geology, with all but geology measured on a continuous scale, assuming quadratic response functions. No attempt has been made to fit more complex models involving interaction terms. Note that it is not valid to estimate total deviance explained from these analysis, only whether there is a significant change (see Appendix for further details).

The analysis (Table 2) tests whether species responses to certain environmental gradients are curvilinear (quadratic), linear or indeterminate, when the direct influence of other factors is taken into account (see Discussion for further comment). The quadratic model is significant for mean annual temperature and radiation index but not for rainfall where only the linear term is significant. Note that removal of either the linear or quadratic term for rainfall makes no significant change in the deviance, though removal of both terms results in a

Table 2. Binary GLIM model predicting probability of occurrence of *E. rossii* using continuous variables as predictors.

Variables	truncated range used				
Mean annual temperature (*T*)	9.5°–13.8 °C				
Mean annual rainfall (*R*)	500 –1 040 mm				
Annual radiation index (*RI*)	0.70–1.12				
Factor					
Geology (*G*)	granite, sediments, volcanics				
Total deviance	399.7 df 603				

Full model	Residual deviance	df	Change in deviance (χ^2)	df	*p*
$T + T^2 + R + R^2 + RI + RI^2 + G(3)$	146.4	595			
$-T$	178.1	596	31.7	1	***
$-T^2$	176.2	596	29.8	1	***
$-(T + T^2)$	182.3	597	35.9	2	***
$-R$	147.4	596	1.0	1	N.S.
$-R^2$	148.9	596	2.5	1	N.S.
$-(R + R^2)$	214.9	597	68.5	2	***
$-RI$	154.5	596	8.1	1	**
$-RI^2$	153.7	596	6.3	1	*
$-(RI + RI^2)$	192.2	597	45.8	2	***
$-G$	161.0	597	14.6	2	***

Best model by backwards elimination

  E. rossii (*p*) = $a + T + T^2 + R + RI + (RI)^2 + G$

    Terms significant at $p < 0.1\%$, coefficient values are not given.

    Residual deviance = 148.9 df 596

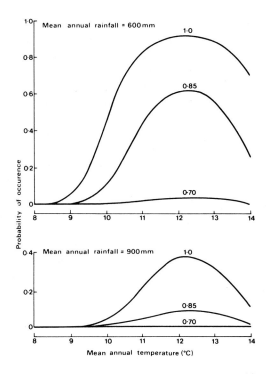

Fig. 7. Predicted response curves of *E. rossii* for probability of occurrence in relation to mean annual temperature at different levels of radiation index (*RI*) and mean annual rainfall. No curve is shown for *RI* = 1.12 as it is indistinguishable from *RI* = 1.0.

large change in deviance. This indicates each term can substitute for the other and only one is necessary to model the influence of rainfall. The model predicts that in the sample area *E. rossii* will have maximum probability of occurrence (Fig. 7) on exposed slopes on sediments, with radiation index values above one, mean annual temperature between 12.0–12.5 °C and mean annual rainfall less than 600 mm. The sample does not contain the full range of environments for the species and hence the optimum environment is not necessarily defined by the model, though the actual geographical distribution is in accord with the predictions made by the model in terms of climatic conditions.

The main center of distribution of *E. rossii* (CSIRO, 1980) is northwest of the area studied here, extending to the Queensland border, and the species is at its southeastern geographical limit within the sample area. This is reflected in the relationships predicted by the model and summarized in Figure 8a.

E. dalrympleana. The sample area constitutes an appreciable portion of this species distribution (Hall *et al.,* 1970). The GLIM analysis (Table 3, Fig. 8b) indicates curvilinear responses to temperature and rainfall, a linear response to radiation index

Table 3. Binary GLIM model predicting probability of occurrence of *E. dalrympleana* using continuous variables as predictors.

Variables	truncated range used
Mean annual temperature (*T*)	5.5°–12.0 °C
Mean annual rainfall (*R*)	600 –2 200 mm
Annual radiation index (*RI*)	0.60–1.12
Factor	
Geology (*G*)	granite, sediments, volcanics
Total deviance	970.5 df 749

Model	Residual deviance	df	Change in deviance (χ^2)	df	p
Linear terms + T^2 + R^2 + RI^2 + $G(3)$	720.2	741			
$-T$	885.1	742	164.9	1	***
$-T^2$	898.5	742	178.3	1	***
$-(T + T^2)$	909.2	743	189.0	2	***
$-R$	727.3	742	7.2	1	**
$-R^2$	729.7	742	9.5	1	**
$-(R + R^2)$	735.2	743	15.0	2	***
$-RI$	720.2	742	<0.1	1	N.S.
$-RI^2$	720.3	742	0.1	1	N.S.
$-(RI + RI^2)$	752.8	743	36.2	2	***
$-G(3)$	728.1	743	7.9	2	**

Best model by backwards elimination

E. dalrympleana (*p*) = $a + T + T^2 + R + R^2 + RI + G$

Terms significant as $p < 1\%$

Residual variance = 720.3 df 742

and increased frequency on volcanics relative to other geologies. The influence of temperature is much greater than that of rainfall (change in deviance for $T = 189.0$, $R = 15.0$) while radiation index is intermediate (change in deviance for $RI = 36.2$). The results accord with a previous contingency analysis using altitude instead of temperature (Austin *et al.*, 1983) and with the discussion of previous work quoted in that paper.

E. fastigata. The sample area encompasses a major part of this species' distribution (Hall *et al.*, 1970). The GLIM analysis (tabular results not presented) indicates curvilinear response curves for temperature and rainfall, and linear for radiation index with higher probability of occurrence on granite (Fig. 8c). The model predicts the highest

probability (0.70 on sediments) of finding *E. fastigata* is under environmental conditions of mean annual temperature approximately 11 °C and mean annual rainfall of 1 200 mm on steep, very protected (gully) southern slopes ($RI = 0.55$). Under similar conditions, but on exposed northern slopes ($RI = 1.12$) the probability falls to 0.22. The relative importance of the environmental variables is temperature (68.3 change in deviance), rainfall (26.1), radiation index (15.3), and geology (7.8). The bell-shaped response curve is a suitable model for this species with respect to temperature and rainfall. In Figure 5, the response curve is flatter and the maximum probability (0.25) is less than suggested by the model (Fig. 8c). This is the result of the obscuring effects of the 'lurking' variable radia-

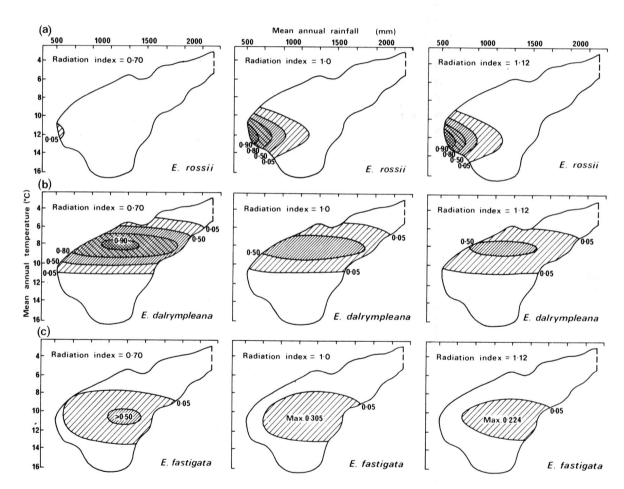

Fig. 8. Predicted response surfaces of probability of occurrence for selected eucalypt species in relation to mean annual temperature and mean annual rainfall at different levels of radiation index on sediments. The predicted surfaces are based on continuous quadratic (or linear) functions of the environmental gradients estimated using GLIM. The outline indicates the distribution of the 1 286 samples in relation to temperature and rainfall.

tion, whose influence is apparent in Fig. 8c.

E. muellerana. This species is approaching its northern geographical limit in the sample area. It is a coastal species, characteristic of southern New South Wales and the East Gippsland region of Victoria, i.e. south of the sample area. The GLIM analysis (tabular results not presented) indicates that a suitable model has temperature, radiation index and geology as important variables. No effect of rainfall is apparent. Temperature is the only environmental factor with a significant quadratic response. Figures 9a and 9b show the predicted response surface. The statistical analysis indicates a 'bell-shaped' response to temperature but provides no statistical evidence for an influence of rainfall (cf. Fig. 3). The distribution of the species beyond the study area would suggest this is not entirely the case, as the species has not been recorded in areas with less than 600 mm mean annual rainfall. The available evidence as indicated by the statistical analysis is insufficient for a positive rainfall response over the range of 800–1 400 mm even though there is an apparent bell-shaped response! The limitations on sample distribution for the data set are such that the response curve in Figure 3 cannot be accepted.

E. maculata. This species is approaching its southern limits in the study area. A significant quadratic function of occurrence with temperature is indicated (GLIM analysis not presented) and a linear relationship with rainfall (Fig. 9c). There is no statistically significant relationship with radiation index. This is unexpected in view of Austin's (1978) subjective interpretation of coastal *E. maculata* communities. In that study it was suggested on the basis of graphical DGA that *E. maculata* dominated communities gave way to an *E. muellerana-E. botryoides* community on protected southern slopes, in the southern part of the study area. The present statistical analysis does not substantiate this interpretation, as there is no term for radiation in the predictive equation.

E. pauciflora. Only one environmental variable has a significant effect on occurrence of *E. pauciflora.* Mean annual temperature has a curvilinear relationship (Fig. 9d). The non-significance of other factors suggests that adaptation to the extremes of an environmental factor may, because of the reduction in number of other species, allow exploitation of a wide range of environmental conditions. Conversely, successful adaptation to an extreme may provide a competitive advantage over other species irrespective of the changes in other environmental factors.

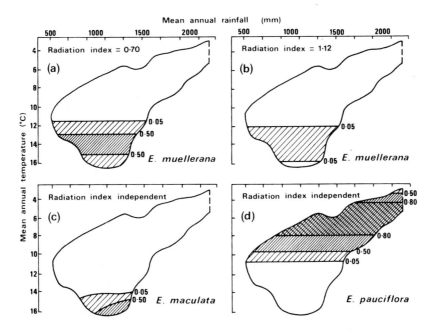

Fig. 9. Predicted response surfaces of probability of occurrence for those selected eucalypts which have simpler or indeterminate (i.e. statistically non-significant) relationships with the gradients temperature, rainfall and radiation (cf. Fig. 8).

Table 4. Significant ($p < 1\%$) terms in predictive model of probability of occurrence for different species.

	Temperature	Rainfall	Radiation Index	Geology
E. rossii	$T + T^2$	R	$RI + RI^2$	( )
E. dalrympleana	$T + T^2$	$R + R^2$	RI	( )
E. fastigata	$T + T^2$	$R + R^2$	RI	( )
E. muellerana	$R + T^2$	N.S.	RI	( )
E. maculata	$T + T^2$	R	N.S.	( )
E. pauciflora	$T + T^2$	N.S.	N.S.	N.S.

Table 4 summarizes the functional predictive models obtained from the GLIM analyses. In all cases, a curvilinear response was significant for temperature. The sample distribution with respect to temperature is apparently sufficient to include the optimum of a number of species. The same does not hold for rainfall; two show a quadratic response, two a linear response, and two are non-significant. With the exception of E. pauciflora, at least three of the four environmental factors are significant predictors of the species occurrence.

Discussion and Conclusion

The functional, factorial approach pioneered by Jenny (1941) and Major (1951) and the direct gradient analysis of Whittaker (1951) have a common conceptual base. They assume that gradients of simple measurable environmental factors are appropriate for analysis of plant communities. Those plant ecologists interested in ecosystem processes, plant competition or population demography may argue that it is inadequate for their purposes. However, DGA provides both a context for their studies and a test of their conclusions. The predictions of physiological ecologists regarding species performance need to be tested against the observational analysis of DGA. The correspondence of their important predictor variables with those variables which best predict species distributions in the 'real world' of vegetation also needs to be examined.

Community ecologists are typically interested in the distal factors (e.g. rainfall) which ultimately influence vegetation composition, while other ecologists concentrate on proximate factors, (e.g. leaf nitrogen content and its influence on grazing pressure). Both often ignore competition between different species.

The section on environmental scalars attempts to show how these approaches can be brought closer together. There is no real conflict. The major requirement is an understanding of environmental processes and how they influence plant growth such that the appropriate level of abstraction can be chosen for the level of prediction required. The level of abstraction is critical. The radiation index is a case where the knowledge and ability to model radiation input is not yet equivalent to the ecologist's observations on species distribution in relation to aspect. It is widely known that in the northern hemisphere those species with a southern distribution in a region, or characteristic of dry habitats, occur preferentially on southwestern slopes (e.g. Perring, 1959, 1960). The scalar, radiation index does not take full account of this. Its estimate of radiation is symmetric about a north–south axis. In the morning, air temperature is lower and it does not reach a maximum until mid afternoon. Thus the combined influence of incoming solar radiation and maximum air temperature on evapotranspiration is greatest in mid afternoon when maximum radiation is received on southwestern slopes. An extension of the process model is required to incorporate this effect. Similarly, estimates of sunshine hours and vapour pressure for the different locations within a study area are needed to allow radiation to be expressed in absolute rather than relative terms.

The graphical analysis of species response to an altitudinal gradient, and the influence of sample distribution in the space defined by the gradient axes serve to indicate that biological interpretations of species response curves must wait on careful assessment of data analysis methods. The evidence for probability of occurrence of the eucalypt species presented in this paper is interpreted to mean that provided the range of variation along a gradient is sufficient, then a bell-shaped or 'Gaussian' response curve is the most frequent type. However, this will be easily detected only when the response is with respect to the dominant environmental factor and sample distribution is adequate. A number of species show skewed curves. The main example is E. pauciflora, the timber-line species. This is consistent with the views of Austin (1976) and van der Maarel (1976) where skewed curves were suggested for species furthest from the optimal conditions for plant growth. The occurrence of subspecies (e.g. E.

pauciflora ssp. *niphophylla* Hall *et al.,* 1970) and the need for confirmatory experimental studies still precludes any final conclusions regarding the occurrence of skewed response curves.

The generalized linear model offers a chance to provide an explicit and rigorous analysis procedure for DGA. By restricting the analysis models to direct effects of different environmental gradients or factors, it has been possible to show statistically that many species have bell-shaped responses without assuming complex interactive models (cf. Austin *et al.,* 1983). Models with greater predictive capability may be developed which incorporate interactions among environmental variables. Given the nature of plant growth responses and the primitive nature of the gradient measures, significant interactions would appear to be very likely.

The hypothesis put forward earlier that only a limited number of environmental factors account for a major proportion of variability in vegetation composition is supported by the statistical analysis. The distribution of the five eucalypt species studied can be reasonably predicted using the four variables examined. Though the second hypothesis, that expression of gradients in terms of resources will improve prediction, has not been examined here, previous papers have provided evidence for this (Austin, 1971; Austin & Cunningham, 1981). The analyses also contribute information on the shape of species response curves. The number of species (of eucalypts in SE New South Wales) with bell-shaped response curves is greater than those with skewed shapes if major environmental factors and sample distribution are taken into account. One limitation which applies to this data set is that while the data set is large (1 286 plots), most species occupy only a small portion of the environmental space sampled. This means that while the absence of a species can be predicted easily, there are (for many species) relatively few positive observations so that estimates of the probability of a species being present are correspondingly limited with respect to statistical confidence. It is hoped that development of a larger data set will allow definition of species ecological response curves or realized niche in relation to four or five environmental factors. This would provide the opportunity to test quantitatively from field data, speculations on niche shape and overlap.

Direct gradient analysis is a correlative proce-dure which describes vegetation–environment relationships and can provide predictive models. However, as J. L. Harper argues in a paper entitled 'After description' (1982) more is required of our ecological explanations than predictions based on correlation. The next step for DGA as a tool for exploratory ecological analysis will be to incorporate plant competition and the influence of higher trophic levels, and then to apply the method at levels of organizations other than the species.

Appendix

Generalized Linear Modelling (GLM)

There are three main components to the GLM.
(1) The data $y = (y_1, y_2, \ldots y_n)$ are assumed to be independently distributed with mean $\mu = (\mu_1, \mu_2, \ldots \mu_n)$.
(2) The linear model is as above and predicts

$$Y = (Y_1, Y_2, \ldots Y_n) = X\beta$$

where β are parameters to be estimated and X is the matrix of explanatory variables.
(3) The link function which connects Y with the mean μ is

$$Y = f(\mu)$$

The three models for the link function, which are subclasses of the GLM, relevant to this study are:
a)   Normally distributed data
     y has a normal distribution with mean μ and variance σ^2, where σ^2 is constant. μ is the mean of the distribution and the predicted Y is related to the mean as follows

$$\mu = Y$$

After fitting the model the total variation can be partitioned into the 'explained' variation and the 'unexplained' variation:

$$\Sigma(y - \bar{y})^2 = \Sigma(\hat{\mu} - \bar{y})^2 + \Sigma(y - \hat{E})^2$$
$$\text{'explained'} \quad \text{'unexplained'}$$

where $\hat{\mu}$ notes the estimated value of μ. This of course is the general linear model described previously.
b)   Poisson distributed data (counts)
     When data are counts, e.g. number of stems, or cell counts, or when data are cross-tabulated in the form of contingency tables, we have the following model:
     The data have a Poisson distribution with mean μ; μ and y are linked by

$$Y = \ln \mu \; ; \; \mu = e^Y$$

Hence the term log-linear model.
c)   Binomial data (proportions)
     The data in this case are counts of the number of 'successes' out of a sample of n. The data have a binomial distribution with mean $\mu = np$, where p is the expected proportion of 'successes'. Several link functions are possible, but in our case the usual one is the logit transformation

$Y = \text{logit}(p) = \ln p/1\text{-}p$

or

$p = e^Y/(1 + e^Y)$

Specification of the linear predictor. Several types of components are possible for the linear predictor i.e. the x_k variables in equation (1). These are

a) Quantitative variates

Such predictors are usual in regression analysis where one parameter is associated with each variate. Occasionally it is necessary to construct transformations of these variates to allow for more complicated functional forms between the response variable and the predictor, e.g. polynomials.

b) Qualitative data

In many applications, explanatory variables arise in the form of nominal or qualitative variates. Sometimes it may help for the purpose of gaining insight into the nature of the relationship between the response and a covariate, to express quantitative variables in the form of grouped variables. Predictor variables of this type are specified as a 'factor' in the linear predictor and a set of parameters is estimated for these variables. This is equivalent to specifying a set of dummy variables having values of 1 when the observation unit occurs in the group and zero otherwise. Because of constraints necessary for the fitting of 'factors' one of the levels will be aliased (i.e. set to zero) with the mean.

c) Mixed terms

The parameter associated with a quantitative variate is allowed to depend on the level of a qualitative variable. Terms of this type allow the specification of more complicated relationships. For example, non-parallel regression lines or a continuous linear relationship to a given value and constancy thereafter, can be fitted readily.

Goodness of fit. The goodness of fit of the model is assessed by the 'deviance'. This is $-2 L_{max}$, where L_{max} is the maximum value of the log-likelihood for the model. For the 'normal errors model' this is the residual sum of squares, which is proportional to a chi-square variable. However the scale factor σ^2 is unknown and has to be estimated. For Poisson and binomial error distributions, the 'deviance' is approximately chi-square since σ^2 is known. In the case of binary data ($n = 1$) or binomial data where n is small, and Poisson data where counts are small the asymptotic distribution theory does not apply and the use of 'deviance' for assessing the adequacy of the model is not valid. Individual terms can however be tested. When fitting a sequence of models the difference between the residual deviances provides a method of testing the significance of additional terms in the model. For binomial or Poisson data these differences are approximately chi-square; for 'normal error models' the usual F statistic has to be constructed.

References

Austin, M. P., 1971. Role of regression analysis in ecology. In: Quantifying Ecology. Proc. Ecol. Soc. Aust. 6: 63–75.

Austin, M. P., 1972. Models and analysis of descriptive vegetation data. In: Mathematical Models in Ecology. Proc. 12th Symp. Br. Ecol. Soc., pp. 61–86.

Austin, M. P., 1976. On non-linear species response models in ordination. Vegetatio 33: 33–41.

Austin, M. P., 1978. Vegetation. In: M. P. Austin & K. D. Cocks, (gen. eds.). Land Use on the South Coast of New South Wales, Vol. 2, pp. 44–67. CSIRO, Melbourne.

Austin, M. P., 1980. Searching for a model for vegetation analysis. Vegetatio 43: 11–21.

Austin, M. P. & Belbin, L., 1981. An analysis of succession along an environmental gradient using data from a lawn. Vegetatio 46: 19–30.

Austin, M. P. & Cunningham, R. B., 1981. Observational analysis of environmental gradients. Proc. Ecol. Soc. Aust. 11: 109–119.

Austin, M. P., Cunningham, R. B. & Good, R. B., 1983. Altitudinal distribution in relation to other environmental factors of several Eucalypt species in southern New South Wales. Aust. J. Ecol. (in press).

Bakuzis, E. V. & Hansen, H. L., 1965. Balsam Fir. Univ. Minnesota Press, Minneapolis.

Crocker, R. L., 1952. Soil genesis and the pedogenic factors. Quart. Rev. Biol. 27: 139.

CSIRO, Division of Forest Research, 1980. Forest Tree Leaflets. 4 vols. CSIRO, Melbourne.

Ellenberg, H., 1950. Unkrautgemeinschaften als Zeiger für Klima und Boden. Ulmer, Stuttgart.

Fitzpatrick, E. & Nix, H. N., 1970. The climatic factor in Australian grassland ecology. In: R. M. Moore. (ed.). Australian Grasslands Ch. 1, pp. 3–26. Australian National University Press, Canberra.

Fleming, P. M., 1971. The calculation of clear day solar radiation on any surface. Paper presented at Aust. Inst. Refrig. Air. Cond. Heating Conference, Perth, May 1971. Mimeo. 24 pp.

Fleming, P. M. & Austin, M. P., 1983. Notes on a radiation index for use in studies of aspect effects on radiation climates. CSIRO Division of Water and Land Resources Tech. Memo (in prep.).

Gauch, H. G. & Whittaker, R. H., 1972. Coenocline simulation. Ecology 53: 446–51.

Hall, N., Johnston, G. M. & Chippendale, G. M., 1970. Forest Trees of Australia. Australian Government Publishing Service Canberra. 334 pp.

Harper, J. L., 1982. After description. In: E. I. Newman (ed.) British Ecological Society, Special Publications Series, 1, pp. 11–25.

Hutchinson, M. F., 1983. Surface fitting and contour drawing programs for small to large data sets (in prep.).

Hutchinson, M. F. & Bischof, R. J., 1983. A new method for estimating the spatial distribution of mean and annual rainfall applied to the Hunter Valley, New South Wales. Aust. Meteorol. Mag. (in press).

Jenny, Hans, 1941. Factors of Soil Formation: a system of quantitative pedology. McGraw Hill, New York. 281 pp.

Lee, R., 1963. Evaluation of solar beam irradiation as a climatic parameter of mountain watersheds. Colo. State Univ. Hydrol. Papers 2: 1–50.

Loucks, O. L., 1962. Ordinating forest communities by means of environmental scalars and phytosociological indices. Ecol. Monogr. 32: 137–66.

Maarel, E. van der, 1976. On the establishment of plant community boundaries. Ber. Deutsch. Bot. Ges. 89: 415–43.

Major, J., 1951. A functional, factorial approach to plant ecology. Ecology 32: 392–412.

Mueller-Dombois, D. & Ellenberg, H., 1974. Aims and methods of Vegetation Ecology. J. Wiley & Sons, New York. 547 pp.

Nelder, J. A. & Wedderburn, R. W. M., 1972. Generalized linear models. J. R. Statist. Soc. A. 135: 370–84.

Peet, R. K., 1978. Latitudinal variation in southern Rocky mountain forests. J. Biogeogr. 5: 275–89.

Perring, F., 1958. A theoretical approach to a study of chalk grassland. J. Ecol. 46: 665–79.

Perring, F., 1959. Topographical gradients of chalk grassland. J. Ecol. 47: 447–81.

Perring, F., 1960. Climatic gradients of chalk grassland. J. Ecol. 48: 415–42.

Waring, R. H. & Major, J., 1964. Some vegetation of the California coastal redwood region in relation to gradients of moisture, nutrients light and temperature. Ecol. Monogr. 34: 167–215.

Waring, R. H., Reed, K. L. & Emmingham, W. H., 1972. An environmental grid for classifying coniferous forest ecosystems. In: J. F. Franklin, L. J. Dempster & R. H. Waring (eds.). Research on coniferous forest ecosystems: First year progress in the coniferous forest biome, US/IBP, pp. 79–92. USDA Forest Service, Pacific Northwest Forest and Range Exp. Station.

Westman, W. E. & Peet, R. K., 1982. Robert H. Whittaker (1920–1980): The man and his work. Vegetatio 48: 97–122.

Whittaker, R. H., 1951. A criticism of the plant association and climatic climax concepts. Northwest Sci. 26: 17–31.

Whittaker, R. H., 1954. Plant populations and the basis of plant indication (German summ.) Angew. Pflanzensoziol. (Wien), Festschr. Aichinger 1: 183–206.

Whittaker, R. H., 1956. Vegetation of the Great Smoky Mountains. Ecol. Monogr. 26: 1–80.

Whittaker, R. H., 1960. Vegetation of the Siskiyon Mountains, Oregon and California. Ecol. Monogr. 30: 279–338.

Whittaker, R. H., 1967. Gradient analysis of vegetation. Biol. Rev. 42: 207–64.

Whittaker, R. H., 1978. Direct gradient analysis. In: R. H. Whittaker, (ed.). Ordination of Plant Communities, pp. 9–50. Handbook of vegetation science 5. Junk, The Hague.

Whittaker, R. H. & Niering, W. A., 1965. Vegetation of the Santa Catalina Mountains, Arizona: A gradient analysis of the south slope. Ecology 46: 429–52.

Whittaker, R. H. & Niering, W. A., 1975. Vegetation of the Santa Catalina Mountains, Arizona. V. Biomass, production and diversity along the elevation gradient. Ecology 56: 771–90.

Zobel, D. B., McKee, A. & Hawk, G. M., 1976. Relationships of environment to composition, structure and diversity of forest communities of the central western Cascades of Oregon. Ecol. Monogr. 46: 135–56.

Accepted 1.7.1983.

An analysis of species response curves and of competition from field data: Some results from heath vegetation*

L. F. M. Fresco**

Department of Plant Ecology, Biological Centre, University of Groningen, P.B. 14, 9750 AA Haren (Gn), The Netherlands

Keywords: Amplitude, Competition, Direct gradient analysis, Ground water, Heathland, Phosphate, Response curve, Simulation

Abstract

The variation in species composition of plant communities has been described by means of the Gaussian distribution (relations between plant populations and environmental variables) and by linear equations (mutual interference between mixed populations). An important presupposition is the existence of a theoretical population, covering 100% of the soil surface. A numerical approximation of the parameters can be obtained by means of an iterative procedure. This technique has been applied to heath vegetation. A small sample has been used to demonstrate the algorithm. The cover of *Calluna vulgaris, Erica tetralix* and *Molinia caerulea* could be explained to a large extent from phosphate and ground water table. A comparison of the results with those of experiments by other authors showed a reasonable similarity. The linear coefficients of interference as means to describe relations between mixed plant populations in a local area have been evaluated by means of dynamic simulation.

Introduction

Many techniques to describe relations between vegetation and environmental variables are available. Recently the problem of conformity between the assumptions on which these techniques are based and properties of plant communities has been discussed extensively (Greig-Smith, 1980; Austin, 1976, 1980). Yet it is often not easy to choose a method which helps the investigator to give a direct answer to his problem and in which the data satisfy the underlying presuppositions.

Whittaker & Gauch (1973) and Gauch, Whittaker & Wentworth (1976) compared a number of techniques with regard to distortion which appears from processing simulated data. Approaches, requiring the least presuppositions, such as Wisconsin Comparative Ordination and Reciprocal Averaging, turned out to be the most useful. A comparative analysis by Prentice (1980) led to the conclusion that Non-metric Multidimensional Scaling is preferable to other methods.

Scott (1974) enumerated the requirements that variables must satisfy when used in a number of techniques. This enumeration made it clear that most requirements (linearity, normality, independence) are untenable in vegetation research. Having discussed the optimum curve, the author concludes: 'Suffice to say there is no general class of mathematical equations which is sufficiently flexible to accomodate (all) these possibilities'.

Others (among which Hill, 1973; Gauch & Chase, 1974; Gauch, Chase & Whittaker, 1974; Ihm & van

* Nomenclature follows Heukels & Van Ooststroom (1977)

** This paper is dedicated to the late professors R. H. Whittaker and D. Bakker, my teachers in ecology. I am very much indebted to M. P. Austin, D. W. Goodall, E. van der Maarel and R. H. Whittaker for critically reading the manuscript, to mrs. A. Severijnse-Waterman for correcting the English text and to mr. Leeuwinga for drawing the graphs. Mrs. F. Engbers and G. de Haan collected the field data.

$$u = (-2 \cdot \ln(O'_{ij}/100))^{1/2} \quad (O'_{ij} > 0) \qquad (6)$$

6. In all stands where $|u| > = 0$, $d(1)_{ij}$, $d(2)_{ij}$ and $d(3)_{ij}$ are computed. (Equations (2), (3) and (4)). $d(1) = d(1)_{ij}$, $d(2) = d(2)_{ij}$, $d(3) = d(3)_{ij}$.

7. In all stands cover percentages are computed by means of the three equations thus obtained: $C(1)_{ij}$, $C(2)_{ij}$ and $C(3)_{ij}$.

8. The three curves are compared to O'_{ij} using any measure of dissimilarity. Good results were obtained with:

$$F_i = \frac{\left(\sum_{j=1}^{n} (C_{ij} - O'_{ij})^2 \right)^{1/2}}{n} \qquad (7)$$

9. The curve with the lowest F_i is considered the appropriate type of local ecological response curve for that species.

Some remarks: (ad 1) if more then one maximum exists, then m_{jk} will be the mean value of the extremes. In that case it is recommended to make $O'_{ij}(max) < 100$. (Ad 6): a statistical estimation of d empirically turned out to be less successful than this approach. Finally it must be kept in mind that the local ecological amplitude has no meaning in itself but is only a proper start for the iterative estimation of the physiological amplitude and of interference. The main goal is to minimize the number of iterative steps.

Next the 'physiological curve' is computed per species (program FYSAM):

1. The procedure is based on a 2f-dimensional space, each axis representing an optimum m_{jk} or a standard deviation d_{jk}. The values of m and d, estimated by means of ECAM represent a point in this space. This is the starting point of the iterative procedure ('screening'). To each point belongs a vector V being the coefficients of interference of the species in question with the other species (equation (5)). The values of m and d, in combination with the assumption that there is a single limiting factor, provide a 'potential percentage coverage' P_{ij} (equation (1)) for each stand.

Since $O_{ij} - P_{ij} = V_{i1}O_{1j} + V_{i2}O_{2j} + ----- + V_{is}O_{sj}$, the vector V can be computed by means of multiple regression. To these vectors P_{ij}, V_i and O_j belongs a computed O^*_{ij}. This O^*_{ij} can be compared with O_{ij} by means of the dissimilarity parameter, presented in equation (7).

2. All values of m are successively multiplied by 1.1 and 0.9, all values d by 1.1. A decrease of d has no ecological sense. E.g. 2 factors need 7 runs with multipliers:

run.	m1	d1	m2	d2
1	1	1	1	1
2	1.1	1	1	1
3	.9	1	1	1
4	1	1.1	1	1
5	1	1	1.1	1
6	1	1	.9	1
7	1	1	1	1.1

For each of the runs the vector V and a dissimilarity value F are computed.

3. The run which delivers the best fit (greatest decrease of F with respect to run 1) provides the starting point for the next series of runs.

4. If there is no decrease of F then at least one more series of runs has to be tried.

5. The point with the best fit is considered the 'physiological amplitude' of the species concerned with respect to all environmental variables.

This method is not a numerical technique, minimizing the dissimilarity analytically. The comparison of great numbers of combinations of values of parameters is not an 'iteration' in a narrow sense of the word. 'Screening' might be a better name. This technique seems to provide solutions which are independent of initial values. In two cases the entire vector space has been scanned. Both times the approach described turned out to provide a correct solution. So far a mathematical proof of the correctness of this method has not been found.

When a species has no clear relations to a factor (the species quantities at both ends of the factor scale are equal to the quantities in the middle), the approximated m is not relevant and d will be great. If the local species-factor relation is linear, m may shift to a value outside the data range. Factors can be correlated within the area studie and so produce factor-complexes. In such a case the decision of which of the factors limits the species is arbitrary for each of the stands. In such cases a factor-complex is represented by one member.

All parameters being approximated, the composition of stands can be 'simulated'. Therefore we consider P and V as known and O as unknown. Equation (5) leads to s equations:

$$P_{ij} = O_j - V_{il}O_l - V_{i2}O_2 \, ----- \, V_{is}O_s(V_{ii} = 0).$$

O can be solved. For preference testing should be executed with other stands than those that were used in the approximation procedure. Extrapolations should be avoided.

Application

An example has been taken from an investigation of a heath vegetation in the north of The Netherlands (Engbers & de Haan, 1977). The data matrix exists of 40 relevés of 2×2 m. They are situated in five different areas, at a maximal distance of 20 km. First a subset containing 10 stands, 3 species and 2 factors is considered. The sequence of calculational steps will be demonstrated for this subset and possibilities and limitations of the method will be clarified. Second a different sample from the same data, containing 10 stands, 6 species and 4 factors will briefly be discussed. Finally the results of a simulation of 6 species using all 40 records will be discussed.

In Table 1 the cover % of the species C (*Calluna vulgaris* (L.) Hull), E (*Erica tetralix* L.) and M (*Molinia caerulea* (L.) Moench) are presented. The environmental variables are g (depth of ground water in cm in May) and p (phosphate in the upper 15 cm of the soil, extracted in ammonium lactate, expressed as mg/1000 g dry soil). The number of remaining species and the sum of their cover percentages are also considered.

The sequence of steps leading to the determination of the local ecological curve of *Calluna* with respect to ground water can easily be followed with the aid of Table 2. Here the smoothening of the curve has been carried out making allowance for the sequence of stands only. The values of the environmental factor can of course be involved. So can more sophisticated techniques of smoothing curves (Austin & Austin, 1980). In practice the final results will not be very much different. The choice of F (equation 7) as a measure of dissimilarity will influence the decision which of the three curves to choose. If e.g. Percentage Similarity would have been applied, the symmetric Gaussian curve would have been selected. Table 3 presents the results of

Table 1. A sample of 10 stands out of 40. These data have been used for the approximation of response curves.

Stand	1	2	3	4	5	6	7	8	9	10
Calluna vulgaris	0	4	0	60	0	0	1	1	40	30
Erica tetralix	70	70	2	4	0	40	50	30	30	10
Molinia caerulea	10	1	1	0	60	10	0	30	0	1
nr. of remaining species	1	3	4	1	3	4	3	5	2	4
total cover of remaining species	1	3	26	10	4	17	25	26	2	33
ground water (cm)	6	48	5	102	52	0	3	4	66	62
phosphate (mg/1000 g)	4	3	26	9	64	3	14	7	13	17

Table 2. The sequence of steps resulting in the 'local ecological response curve' or the 'initial curve' of *Calluna vulgaris* with respect to ground water.

Stand	6	7	8	3	1	2	5	10	9	4	
ground water	0	3	4	5	6	48	52	62	66	100	
o	0	1	1	0	0	4	0	30	40	60	$m_{cg} = 102$
interpol.	0	1	2	3	4	4	17	30	40	60	
o′	0	2	3	5	6	7	29	50	67	100	
u	–	2.80	2.65	2.45	2.37	2.31	1.57	1.18	.89	–	
d1	–	35.4	37.0	39.6	40.5	23.4	31.8	33.9	40.5	–	$d1 = 35.26$
d2	–	1.259	1.222	1.231	1.195	.326	.429	.422	.489	–	$d2 = .822$
d3	–	.242	.254	.273	.280	.184	.254	.280	.340	–	$d3 = .263$
C1	2	2	2	2	2	31	37	53	59	100	$F1 = 2.73$
C2	0	0	0	0	0	66	71	83	87	100	$F2 = 8.25$
C3	3	4	4	4	4	27	32	45	52	100	$F3 = 2.60$

the determination procedure of local ecological curves. Figure 1 gives the curves graphically.

The calculations, leading to the approximation of physiological curves are illustrated by means of the results of the 'screening' for *Calluna* (Table 4). The try-out of combinations is somewhat different from the standard-procedure as described in the previous chapter. This is not indispensable, but if the number of possible combinations is great, it saves time.

The calculated physiological curves are presented in Table 5 together with the coefficients of interference. Table 6 gives the simulated relevés.

The study of the deviations of the separate species may give insight in the limitations of the method. *Calluna* has the largest deviation in stand nr. 2 (obs. 4%, sim. 31%). This is the stand in which a relatively high coverage and a large number of

Table 3. Local ecological curves or initial curves. With regard to the type of distribution, '0' stands for a symmetric (Gaussian) distribution, '+' for a positively skewed (logarithmic Gaussian) distribution and '–' for a negatively skewed (projected logarithmic Gaussian) distribution.

		Calluna	*Erica*	*Molinia*
distrib. (gr. water)		–	0	–
m	"	102.00	27.00	52.00
d	"	.263	23.21	.356
distrib. (phosphate)		0	0	+
m	"	9.00	3.50	64.00
d	"	3.539	6.722	1.653

Table 4. Sequence within the first run of the iterative procedure for the approximation of the physiological response curve and parameters of interference (*Calluna vulgaris*).

Initial values: $m_g = 102.000$ $d_g = .263$ $m_p = 9.000$ $d_p = 3.539$
Multipliers:

m_g	d_g	m_p	d_p	V_{ce}	V_{cm}	F
1	1	1	1	–.0776	–.5144	4.908
1.1	1	1	1	–.0648	–.4282	4.382[2]
.9	1	1	1	–.0780	–.5168	4.200[1]
1	1.1	1	1	–.0975	–.6407	4.886[1]
1	1	1.1	1	–.0584	–.3850	4.230[1]
1	1	.9	1	–.0705	–.4665	4.889[2]
1	1	1	1.1	–.0870	–.5771	4.797[1]
.9	1.1	1	1			4.268
.9	1	1.1	1			4.180
.9	1	1	1.1			4.351
1	1.1	1.1	1	try-out of		4.378
1	1.1	1	1.1	combinations		4.963
1	1	1.1	1.1			4.170
.9	1.1	1.1	1			4.462
.9	1.1	1	1.1			4.390
.9	1	1.1	1.1			4.043[3]
1	1.1	1.1	1.1			4.286
.9	1.1	1.1	1.1			4.371

new values: $m_g = 91.800$; $d_g = .263$; $m_p = 9.900$; $d_p = 3.893$; etc.
final values: $m_g = 82.620$; $d_g = .318$; $m_p = 8.910$; $d_p = 3.893$;
 $V_{ce} = -.1459$; $V_{cm} = -.9543$; $F = 3.128$
'normal run' produces values of F > 3.128 only
'extra run' with multipliers 1.3 and .7 ditto

[1] value of $F <$ 4.908, point for next run.
[2] this value and a '1' value exclude each other.
[3] best combination, initial point for the next run.

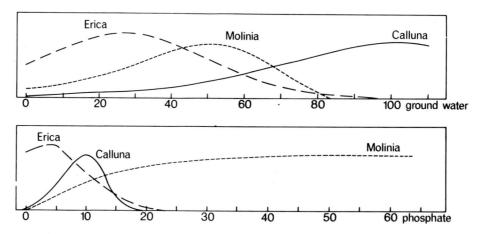

Fig. 1. Smoothed ecological response curves of *Calluna vulgaris, Erica tetralix* and *Molinia caerulea* with respect to ground water (cm below the surface) and phosphate (mg/1000 g). The maximum percentage cover is equal to the maximum observed value.

Table 5. Approximated physiological response curves and parameters of interference.

		Calluna	Erica	Molinia
distribution (gr. water)		–	0	–
m	"	82.620	29.700	42.120
d	"	.318	23.21	.356
distribution (phosphate)		0	0	+
m	"	8.910	3.150	70.40
d	"	3.893	8.052	1.653
V (Calluna)		0	–.0312	–.0601
V (Erica)		–.1459	0	–.1919
V (Molinia)		–.9543	–.1268	0
limiting g/p		5/5	6/4	9/1

crustose lichen species occurs. The next worse simulation was in stand nr. 10. Here the largest coverage of bryophytes could be observed. Stand nr. 9 had a high coverage of the grass *Deschampsia flexuosa*. Stand nr. 8 showed the greatest number of miscellaneous species. The large deviation of *Erica* and *Molinia* in this stand may be caused by this phenomenon. Finally stands nrs. 3 and 5 had the highest cover of *Sphagna*. This can explain the high deviations of *Molinia*.

A comparison of results obtained by others with different methods may elucidate the possibilities of this technique. Gimingham (1972) and Bannister (1969) both found that *Calluna* shows the widest range with regard to water logging, overlapping *Erica,* while *Erica* increases on soils which are wet and poorly drained. This study shows that the range, where both species have a potential cover percentage of at least .5%, occupies the entire gradient studied. A cover percentage of at least 60 is possible for *Calluna* if the ground water level is more than 51 cm below the surface, whereas *Erica* can reach this coverage if the depth of ground water is less than 54 cm. Sheikh (1969) reported that *Erica* shows its greatest vigour where concentration of

major nutrients are low. Comparison of amplitudes with respect to phosphate shows that this study endorses his conclusions. Gimingham mentioned *Erica* not decreasing in performance on wet soils while growing in mixtures with *Calluna*: on relatively dry soils there is a remarkable decrease in vigour. This study leads to the same conclusions as the following table shows.

gr. water depth 10 cm	*Calluna* P = 14	O = 4	rel. decrease =	71%
	Erica 70	70		0%
gr. water depth 80 cm	*Calluna* 100	100		0%
	Erica 10	7		30%

These comparisons show that this approach makes it possible to express the composition of plant communities in terms of species-factor relations and species interference. So far this applies to simply structured and stable vegetation.

A good understanding of spatial patterns can be obtained by comparing the limits of potential presence of the three species with regard to ground water level and phosphate. Without competition with *Calluna* and *Molinia, Erica* is found to be able to grow on quite dry (sandy) soils if the amount of available phosphate is sufficiently low. If the depth of the ground water level is 60 cm, and phosphate is 25 mg/1000 g, *Erica* will not be able to establish a population. If the amount of available phosphate is 1 mg/1000 g only, a coverage of 43% can be predicted. The frequently observed presence of the species near the top of small inland sand dunes may be caused by this phenomenon (within the area of distribution of *Erica cinerea* L. a different equilibrium will be established).

For each of the species the limiting factor can be determined for each combination of the factors within the range studied. In this way the primary effect of a change of an environmental variable can be predicted. E.g. a lowering of the ground water

Table 6. Simulated relevés (to be compared to table 1). *F* is a measure of dissimilarity (equation 7). r is the correlation coefficient.

Stand	1	2	3	4	5	6	7	8	9	10	*F*	*r*
Calluna vulgaris	0	31	0	69	0	0	0	0	57	0	3.95	+.87
Erica tetralix	58	72	0	0	0	43	39	53	28	21	3.09	+.93
Molinia caerulea	11	1	21	1	75	7	11	9	0	16	3.77	+.84
F	4.01	5.71	6.70	2.43	1.29	1.41	5.20	10.39	5.71	11.77		
r	+1.00	+.97	.00	+1.00	+1.00	+1.00	+.96	+.62	+.96	–.86		

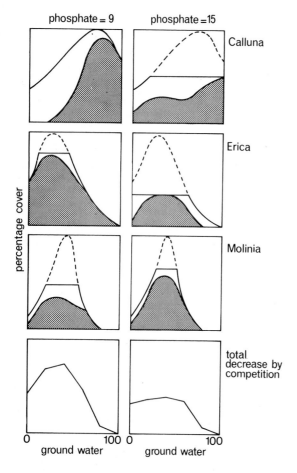

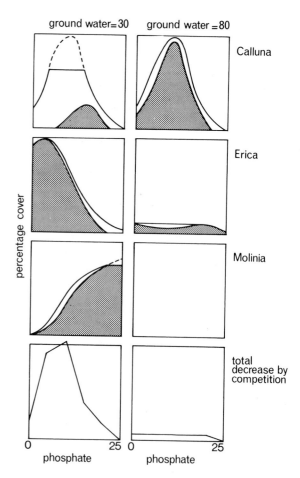

Fig. 2. Computed physiological and ecological response curves with respect to ground water. The dotted parts of the physiological curves indicate limitation by phosphate = 9 mg/1000 g (left column) and phosphate = 15 mg/1000 g (right column). The sum of the distances between the two curves (decrease by competition) is represented in the two bottom graphs.

Fig. 3. Computed physiological and ecological response curves with respect to phosphate. In the left column graphs ground water depth is 30 cm, in the right 80 cm.

table from 60 cm to 100 cm below the surface – the phosphate contents being greater than 30 mg/1000 g – will affect *Molinia* only. All others effects will then be secondary. On the other hand a lowering from 0 cm to 10 cm, while phosphate is 10 mg/1000 g will affect all three species. This indeed might be a step in the direction of understanding the plant community as a system. Such a representation of relations of species to factors may for instance elucidate the 'trigger-factor' principle (Billings, 1952). The sequence of events after a change of an edaphic or biotic factor can hardly be followed and described in the field. This approach makes it possible to formulate hypotheses and to plan experiments.

A study of computed ecological and physiological response curves (Figs. 2 and 3) shows that several shapes are possible. The ecological curve can be similar in form to the physiological curve and almost equal in surface. This is the case if the environment is relatively extreme and competition is almost absent (*Calluna* / phosphate, if the ground water table is low). A shift of the optimum to the right (*Calluna* / phosphate under wet conditions) or to the left (*Molinia* / ground water under poor conditions) is possible. Competition can cause bimodal ecological curves. This can be seen with *Calluna* / ground water / rich and *Erica* / phosphate / dry. This fits in with the types of ecological response, distinguished by Mueller-Dombois &

Ellenberg (1974). The wedge-type response (van der Maarel, 1976) is most similar to truncated curves such as *Erica* / phosphate. Also interesting is the relation between 'competitive decrease' – which is the difference between physiological and ecological response for a given environmental condition – and the environmental gradients. Grime (1973) stated that competition decreases with increasing environmental stress. The results of this simulation are fully conform to this statement. Maximum competition exists if the ground water depth is 30 cm and phosphate is 10 mg/1000 g – the 'physical optimum'.

Next all 40 relevés were simulated by means of the response curves and the parameters of interference obtained with the aid of the sample of 10 relevés. Since our main interest is in the properties of the technique, special attention was paid to the relationship between the deviation of simulated cover percentages to observed values. Dissimilarity was calculated and associated with all environmental factors measured: ground water depth, phosphate, pH-KCl, % organic matter and aluminium (Fig. 4). The three variables not used for the approximation of the curves turned out to have a different relation to dissimilarity. Relatively extreme values of these factors yield simulated relevés with a worse fit. This is not difficult to understand. A low pH for instance can be a limiting factor for one or more of the simulated species. If pH in a relevé is less than the lowest value in the subset, these species can be absent although simulation, based on ground water and phosphate only, will show them present. A pH that is higher than any value in the subset may allow the presence of other species, not involved in this study. Such a species may interfere with the simulated species and so worsen the results. The bad fit of average values of ground water and phosphate will possibly be due to statistical properties of the method.

Finally a new sample (10 records) was taken. Now 4 environmental factors (ground water, phosphate, pH-KCl and percentage organic matter) and 6 species (*Calluna vulgaris, Erica tetralix, Molinia caerulea, Deschampsia flexuosa, Eriophorum angustifolium* and the sum of the 8 *Sphagnum* species present) were processed. As far as *Calluna, Erica, Molinia*, ground water and phosphate were concerned, the final results were not essentially different from those mentioned before. The average fit was slightly better. Interesting is the presence of positive coefficients of interference (e.g. V (*Calluna, Eriophorum*) $= +2.211$; V (*Calluna, Deschampsia*) $= +1.413$). A further consideration leads to the conclusion that this can be caused by an edaphic factor with regard to which one species has a low tolerance, while the other species stores a large amount in its biomass. A study of these results brings about theoretical arguments which will be supported by a simulation of growth of mixed plant populations (see the next chapter).

If species D (*Deschampsia*) is present, it stimulates C (*Calluna*). If D – as a result of competition – has a negative cover percentage (O_D^*-uncorrected < 0; $O_D^* = 0$), this has a negative effect with respect to C. The competitive power of other species not only takes away the positive influence of D but also intensifies negative interference with regard to C.

Between A (*Eriophorum*) and one or more of the other species exists an interaction with respect to C, so that interference with C is different (less negative) if $P_A = 0$. Possibly it concerns E (*Erica*):

$$O_C = P_C - 1.567 \, O_E + 2.211 \, O_A + ----- (P_A > 0)$$
$$O_C = P_C \ 0.797 \, O_E \qquad\qquad + ----- (P_A = 0)$$

This assumption has an ecological meaning. The presence of A stimulates C, but lowers its competi-

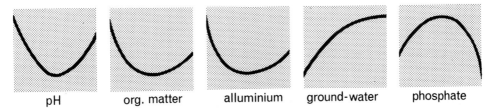

Fig. 4. Five environmental factors related to the fit of computed ('simulated') relevés (40 stands). The vertical axis represents dissimilarity of simulated relevés w.r.t. observed values. Extreme values of factors which are not involved in the simulation cause bad results.

Table 7. Some results of dynamic simulation of three species and one environmental variable.
Parameters: $ax = ay = az = .25$; $kx = ky = kz = 100$; all $l = 0$, except $lzy = .3$; $px + .004$, $py + .040$, $pz + .080$; $tx = 10$, $ty = 15$, $tz = 25$; $q = 10$. Number of steps ('weeks') = 15.

initial			after 15 weeks			
X	Y	Z	X	Y	Z	
20	0	0	20.49	–	–	$P_x = 20.49$
20	20	0	21.11	45.23	–	$V_{xy} = +.014$ (absence of Z)
20	0	20	32.87	–	64.74	$V_{xz} = +.191$
20	20	20	24.06	35.58	42.35	if $V_{xz} = +.191$, then $V_{xy} = -.127$

tive power with respect to E. This will be a matter of horizontal and vertical spatial distribution of individuals (Fig. 5). In the next chapter both types of relationship will be simulated.

Simulation of mixed growth

Several mathematical models underlying dynamic simulation of competition of plant populations have been published (Tilman, 1977; Braakhekke, 1980). None of these models is suited to simulate different growth rates, competition for space, competition for nutrients and toxic levels of minerals present. Such a model should be able to simulate a response curve of a single species and so yield symmetric and skewed curves. Since such a dynamic model is needed to support (or reject) conclusions drawn from the non-dynamic simulation of vegetation, an attempt has been made to develop it.

The applied growth equations for two species are:

$$dx/dt = x.ax.\alpha(x) \cdot \beta(xy) \cdot \gamma(xy) \cdot \delta(xy)$$

$$dy/dt = y.ay.\alpha(y) \cdot \beta(yx) \cdot \gamma(xy) \cdot \delta(yx) \qquad (8)$$

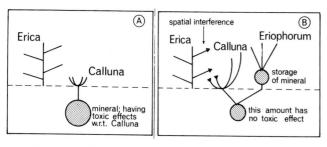

Fig. 5. A schematic representation of interference between *Calluna vulgaris, Erica tetralix* and *Eriophorum angustifolium*. This concerns both the interpretation of vegetation simulation and dynamic simulation.

x is the standing crop of a species X ing/m². If the differential quotient is treated as a difference $\Delta x/\Delta t$, then ax is the maximum possible relative growth within an interval of time Δt.

The third factor reproduces the inhibition of growth as a result of the properties of the individuals of the species. Stretching of cells and distances of transport play a role:

$$\alpha(x) = 1 - x/kx, \qquad (8a)$$

kx being the greatest possible x. The first three factors represent growth by an equation of the logistic type (Fresco, 1973). Competition for space with the other species may be written:

$$\beta(xy) = 1 - lxy.y/ky \qquad (8b)$$

Complete exclusion in space gives $lxy = 1$, spatial independence $lxy = 0$. For the modelling of the scrambling for a nutrient we note:

$$\gamma(xy) = 1 - (px.x + py.y)/q \qquad (8c)$$

q being the amount of the nutrient in the entire system: vegetation plus substrate. px and py represent the fractions of the nutrient in x and y resp. If the nutrient is removed from and supplied to the system, q must be given as a function of t or as a time-series. The equation 8c is symmetric with regard to x and y, so that $\gamma(xy) = \gamma(yx)$. To reproduce the tolerance threshold of a response curve, a growth-stopping amount of the nutrient in the substrate has been assumed:

$$\delta(xy) = 1 - (q - px.x - py.y)/tx \qquad (8d)$$

If cover % has to be simulated in stead of standing crop, the first has to be expressed as a function of the last.

Simulation of the growth curve of a single species with respect to one nutrient gives the 'potential cover' P_x. The curve of a mixture is the 'observed cover' O_x. For each value of q coefficients of interference can be computed.

$$V_{xy} = (O_x - P_x)/O_y$$

A simulation of two species with parameters adapted to *Calluna* and *Erica* and the nutrient to phosphate with variation of initial values (situation at the beginning of the growing season) and of q, showed that linear coefficients of competition are dependent on the amount of nutrient, much less on initial values. This means that a coefficient of interference has to been treated as an average within the range of values of environmental variables measured.

The interaction leading to the behaviour of a species growing in mixtures with two other species, one storing a large amount of a toxic nutrient and the other being a strong competitor for space (Fig. 5) can also be simulated (Table 7).

The fact that an ecologically logical dynamic model supports the assumptions and conclusions from the ordination of vegetation data, described in the previous chapters, provides support for the usefulness of the technique. A final confirmation can be achieved by means of experiments.

References

Austin, M. P., 1976. On non-linear species response models in ordination. Vegetatio 33: 33–41.

Austin, M. P., 1980. Searching for a model for use in vegetation analysis. Vegetatio 42: 11–21.

Austin, M. P. & Austin, B. O., 1980. Behaviour of experimental plant communities along a nutrient gradient. J. Ecol. 68: 891–918.

Bannister, P., 1964. The water relations of certain heath plants with reference to their ecological amplitude. II. Field studies. J. Ecol. 52: 481–497.

Billings, W. D., 1952. The environmental complex in relation to plant growth and distribution. Quart. Rev. of Biol. 27: 251–265.

Billings, W. D., 1974. Environment: concept and reality. In: B. R. Strain & W. D. Billings (ed.). Handbook of vegetation science VI, p. 9–38. Junk, The Hague.

Braakhekke, W. G., 1980. On coexistence: a causal approach to diversity and stability in grassland vegetation. Agricult. Res. Rep. 902, Wageningen, 164 pp. Pudoc.

Ellenberg, H., 1958. Über die Beziehungen zwischen Pflanzengesellschaft, Standort, Bodenprofil und Bodentyp. In: R. Tüxen (ed.). Angewandte Pflanzensoziologie (Stolzenau) 15: 14–18. Junk, The Hague.

Engbers, F. & Haan, G. de, 1977. Analyse van een heidevegetatie. Rep. Dept. of Plant Ecology, Groningen.

Ernst, W., 1978. Discrepancy between ecological and physiological optima of plant species. Oecol. Plant. 13: 178–188.

Fresco, L. F. M., 1973. A model for plant growth. Estimation of the parameters of the logistic function. Acta Bot. Neerl. 22: 486–489.

Gauch, H. G. & Chase, G. B., 1974. Fitting the Gaussian curve to ecological data. Ecology 55: 1377–1381.

Gauch, H. G., Chase, G. B. & Whittaker, R. H., 1974. Ordination of vegetation samples by Gaussian species distribution. Ecology 55: 1382–1390.

Gauch, H. G., Whittaker, R. H. & Wentworth, T. R., 1976. A comparative study of reciprocal averaging and other ordination techniques. J.Ecol. 65: 157–174.

Gimingham, C. H., 1972. Ecology of heathlands. 266 pp. Chapman & Hall, London.

Greig-Smith, P., 1980. The development of numerical classification and ordination. Vegetatio 42: 1–9.

Grime, J. P., 1979. Plant strategies and vegetation processes. Wiley, Chicester, 222 pp.

Groenewoud, H. van, 1976. Theoretical considerations on the covariation of plant species along ecological gradients with regard to multivariate analysis. J. Ecol. 64: 837–847.

Heukels, H. & Ooststroom, S. J. van, 1977. Flora van Nederland. 19th ed. Wolters-Noordhoff, Groningen, 925 pp.

Hill, M. O., 1973. Reciprocal averaging: an eigenvector method of ordination. J. Ecol. 61: 237–249.

Ihm, P. & Groenewoud, H. van, 1975. A multivariate ordening of vegetation data based on Gaussian type gradient response curves. J. Ecol. 63: 767–777.

Maarel, E. van der, 1971. Plant species diversity in relation to management. In: A. S. Watt & E. Duffey (eds.). The scientific management of plant and animal communities for conservation. pp. 45–63. Blackwell, Oxford.

Maarel, E. van der, 1976. On the establishment of plant community boundaries. Ber. Deutsch. Bot. Ges. 89: 415–443.

Mueller-Dombois, D. & Ellenberg, H., 1974. Aims and methods of vegetation ecology. 547 pp. Wiley, New York.

Orlóci, L., 1978. Multivariate analysis in vegetation research. 2nd ed. 276 pp. Junk, The Hague.

Prentice, J. C., 1980. Vegetation analysis and order in variant gradient models. Vegetatio 42: 27–34.

Scott, D., 1974. Description of relationships between plant and environment. In: B. R. Strain & W. D. Billings (eds.). Handbook of vegetation science VI: 49–72. Junk, The Hague.

Sheikh, K. H., 1969. The effects of competition and nutrition on the interrelations of some wet heath plants. J. Ecol. 64: 41–78.

Tilman, D., 1977. Resource competition between planktonic algae: an experimental and theoretical approach. Ecology 58: 334–348.

Whittaker, R. H. & Gauch, H. G., 1973. Evaluation of ordination techniques. In: R. H. Whittaker (ed.). Handbook of vegetation science V: pp. 287–322. Junk, The Hague.

Wuenscher, J. E., 1969: Niche specification and competition modeling. J. Theor. Biol. 25: 436–443.

Wuenscher, J. E., 1974. The ecological niche and vegetation dynamics. In: B. R. Strain & W. D. Billings (eds.). Handbook of vegetation science VI: pp. 39–48. Junk, The Hague.

Accepted 25.9.1981.

Species performances and vegetation boundaries along an environmental gradient*,**

M. J. A. Werger[1], J. M. W. Louppen[2] & J. H. M. Eppink[2]
[1] *Dept. of Plant Ecology, University of Utrecht, Lange Nieuwstraat 106, U.recht, The Netherlands*
[2] *Formerly at the Dept. of Geobotany, University of Nijmegen, Toernooiveld, Nijmegen, The Netherlands*

Keywords: Gradient analysis, Half change, PCA, Species response curve, TABORD, Vegetation boundary

Abstract

A transect in grassland vegetation in the southern part of the Netherlands, showing both a very gradual and a more marked transition in species composition, was sampled as regards species frequency and a number of soil variables. Clustering (TABORD) and Principal Component Analysis revealed the two types of vegetation boundaries, a strong, rigid and a weak, gradual boundary, and the gradient situation in the transect. Species performance curves were tested for Gaussian form, but less than half of the species showed this form even though limits for acceptance were mild. The few species showing a Gaussian response curve clearly illustrate the two types of vegetation boundaries in this transect in agreement with Whittaker's model for such boundaries.

Introduction

A great number of modern methods of vegetation analysis assume that the ecological responses of species to environmental factors follow bell-shaped, or Gaussian, curves. This idea of regular, unimodal species response dates at least from Shelford (1913), but has become increasingly popular since Robert H. Whittaker's (1956) classic study on the Great Smoky Mountains involving a direct gradient analysis of vegetation with altitude as an environmental factor. Despite the considerable popularity of the Gaussian model of species response curves along environmental gradients, there are at least three reasons for a somewhat cautious attitude (cf. Austin, 1980; Austin & Austin, 1980). First, there is only limited direct evidence from field measurements along gradients for the common occurrence of such responses. Though there are examples

available of apparently bell-shaped ecological response curves of some species along obvious vegetation gradients, it has been found that many species in that same vegetation show a variety of more complex responses (see Gauch, 1982 for a review and also Fresco, 1982). Secondly, experimental evidence (Ellenberg, 1953; cf. Knapp, 1967; Mueller-Dombois & Ellenberg, 1974; Austin & Austin, 1980) has shown that varying intensities of competition modify the ecological response curves of species in vegetation in a complex way. The resulting response patterns usually cannot be expressed as Gaussian curves. Thirdly, Austin (1980) has pointed out that considerable confusion has resulted from failure to distinguish between at least three types of environmental gradients, namely indirect environmental gradients (e.g. altitude, aspect), direct environmental gradients (e.g. pH, temperature), and resource gradients (e.g. available nutrients). Species responses to these three types of gradients may differ.

Since plant communities consist of plant populations, knowledge about ecological responses of the

* In memory of R. H. Whittaker
** Species nomenclature follows Heukels & Van Ooststroom (1970).

component species populations is essential for understanding the ecological basis of plant community boundaries. Whittaker (1975, p. 114), possibly motivated by Terborgh's (1971) work, developed a simple model to illustrate that if the ecological responses of co-occurring species to a single or a group of co-varying environmental gradient factors coïncide at least at one end of their ranges, it is likely that clear vegetation boundaries, in the sense of boundaries between adjacent plant communities, are discernible; if they do not coïncide in such a way but overlap in a more or less random or in a regular pattern, transitions between plant communities will be gradual and without identifiable boundaries. It was his belief (see Whittaker, 1967, 1969, 1970) that without disturbance plant communities usually intergrade with plant populations continuously overlapping and replacing each other along the major environmental gradients. Though Whittaker (1967) made explicit that various types of response curves do occur, he suggested that most species would show the bell-shaped form.

Table 1. Soil characteristics along the gradient. Organic values are corrected for carbonate loss through heating. All values represent the average of two determinations. Soil nutrients indicate exchangeable quantities.

plot	clay %	silt %	sand %	organic %	conductivity (μmho)	pH (KCl)	Ca (mg·100 g⁻¹)	K (mg·100 g⁻¹)	Nitrate & nitrite (μg·100 g⁻¹)	P (μg·100 g⁻¹)	Carbonate (%)
1	16	45	39	14	100	3.7	4.1	58	147	48	0.4
4	14	42	44	16	112	3.8	4.8	68	172	76	0.2
7	14	37	49	13	112	3.7	4.4	58	135	68	0.4
10	16	40	44	13	107	3.7	3.6	66	153	52	0.1
13	14	36	50	13	98	3.7	2.0	51	100	56	0.1
16	12	33	55	7	90	3.9	5.8	72	84	43	0.1
19	13	36	51	12	92	3.8	3.7	78	84	40	0.1
22	14	38	48	11	82	3.7	3.0	104	70	40	0.1
25	14	40	46	11	90	3.7	4.0	114	62	48	0.2
28	13	35	52	11	100	3.9	6.3	142	46	36	0.1
31	16	33	51	10	–	4.0	8.5	152	70	60	0.1
34	15	32	53	10	86	4.0	10.4	150	49	72	0.1
37	13	34	53	10	84	4.0	8.7	92	95	84	0.1
40	14	35	51	12	128	5.5	16.0	130	125	168	0.2
43	17	33	50	12	90	4.5	16.8	126	140	112	0.1
46	–	–	–	8	176	6.6	18.2	84	182	96	4.4
49	18	33	49	10	168	6.6	18.4	108	192	132	7.1
52	20	32	48	10	190	6.5	18.4	82	240	104	10.1
55	–	–	–	11	210	6.3	18.4	98	450	141	13.8
58	23	34	43	10	180	6.6	19.2	102	287	160	13.2

It was our purpose to study the patterns of changing importance of plant species along a short, readily identifiable environmental gradient. More precisely, we wanted to know the proportion of species along a gradient which showed bell-shaped ecological response curves, where along the gradient these curves peaked, and how these ecological positions characterized the nature of the different types of vegetation boundaries in the study area. The gradient we studied was basically a nutrient resource gradient, though soil pH and soil texture also varied (see Table 1).

Study area and methods

We selected an area in the southernmost part of the Netherlands, the Bemelerberg, near Maastricht, where on a physiographically homogeneous, SSE-facing grassland slope of 20 to 30° and some 75 m long, both a very gradual and a somewhat steeper change in species composition could be subjectively assessed. The slope consists of nutrient poor Pleistocene alluvial and Oligocene deposits on top of an Upper-Cretaceous chalk sediment (Fig. 1). At the interface of these strata knobs of the chalk bedrock come within about 20 cm of the slope surface, and the alluvial deposit has slid a few meters downslope over the chalk and its weathering product. At the base of the slope the weathered material is covered and partly mixed with a very thin deposit of loess. We sampled this slope with a series of 61 contiguous 1 m² plots starting some meters below the top and

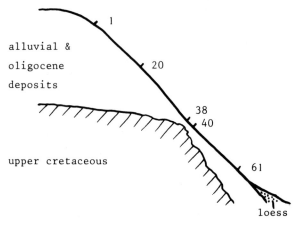

Fig. 1. Positions of plots in the transect showing substrates. Numbers refer to plot numbers.

ending some meters before the base of the slope. The chalk is located near the surface in plots 38 to 40 (Fig. 1). At this section in the transect subjectively recognizable vegetation changes appear to be rather clear, whereas elsewhere these are gradual. Along the transect the vegetation changes from a rather unproductive *Thero-Airion* grassland (plot 1–19) through a *Brachypodio-Sieglingietum* (21 to 28, and perhaps all the way down to plot 38), an open *Mesobromion*-vegetation (40–45), to a dense *Mesobromion*-grassland (at the bottom of the slope) (see Westhoff & Den Held, 1969; Willems, 1982 for notes on these vegetation types). This grassland vegetation is semi-natural in the sense of Westhoff (1971). The climax vegetation would have been *Fago-Quercetum* on the upper slope and *Stellario-Carpinetum* on the lower slope. Peak aboveground standing crop values for the vegetation types encountered were 167, 248, 256 and 299 g · m^{-2} dry weight respectively, and bifacial L.A.I. values were 0.92, 3.73, 4.73 and 7.03 respectively (average of three replicates each).

All 61 plots were subdivided into 100 squares of 1 dm^2 each and the frequency of all vascular species in each plot was determined. Sampling was conducted during the second half of June which is the main flowering season.

In every third plot two soil samples were taken at a depth of about 3 to 5 cm below the A_o. These samples were analysed mechanically and chemically using standard procedures (details are available from the authors). The average values are presented in Table 1.

Vegetation data were clustered using the TABORD clustering program with three fusion limits and offering the original plot data in a random sequence (Van der Maarel *et al.*, 1978). For calculation purposes the frequency values were transformed as follows:

frequency	1–11	12–22	23–33	34–44	45–55
calculation value	1	2	3	4	5

frequency		56–66	67–77	78–88	89–100%
calculation value		6	7	8	9

The total data set as well as a subset of it were also analyzed by Principal Components Analysis (PCA) using Euclidean distance and non-standardized, centred data. Though our use of PCA presupposes linear relationships between species' performance

and the controlling environmental factor and though there are some other restrictive considerations regarding this technique (see e.g. van Groenewoud, 1976; Gauch *et al.*, 1977; Gauch, 1982), we previously had good results with this technique applied in gradient situations with considerably higher β-diversity than in the present data set (Werger, 1978; Werger *et al.*, 1978). A number of other statistical and multivariate techniques were used including Reciprocal Averaging, but since these analyses gave results similar to those obtained by the TABORD- and PCA-analyses, they will not be reported here.

The fit of species responses to Gaussian curves were tested both for parts of the data set and for the entire data set with several plot sequences (Cornell Ecology Program 12: Fit Gaussian curves; Gauch, 1972; Gauch & Chase, 1974).

Results and discussion

Soils

The results of the soil analyses (Table 1) show that the percentage clay increases from plot 40 onwards, and that the silt fraction varies down to plot 28, after which it remains more or less stable. The sand fraction also varies through plot 28, then remains stable till plot 40, and subsequently diminishes proportional to the increasing clay fraction. In general, humus content decreases from about 14% in plot 1 to about 10% around plot 30 after which it remains roughly constant though some plots deviate. Conductivity and pH remain low and stable until plot 37, and increase from there to plot 46 to remain again stable on a higher level. Exchangeable Ca^{2+} values increase steeply between plots 25 to 40 after which there is only a small further increase. K^+ values are low through plot 19, then increase rapidly till plot 31 and subsequently fluctuate in a decreasing trend. Nitrate and nitrite values diminish till the middle of the transect and then rapidly increase again, with an oddly high value at plot 55. Phosphate values show a somewhat similar course as pH and conductivity, and CO_3^{2-} values remain nearly zero till plot 43, followed by a rapid increase till plot 55.

Though the most important changes for many of these variables occur at different points on the transect, it seems clear that there is a marked change in soil conditions following plot 37 and somewhat less marked changes earlier on in the transect. This coïncides with the subjectively assessed change in vegetation in the transect.

Vegetation

The TABORD clustering technique (van der Maarel *et al.*, 1978) was carried out three times with fusion limit values of 50, 60 and 70% of similarity, respectively. The results were basically similar: there is a strong cluster of plots 1 to 19, another strong cluster of plots 39 to 61, and the plots in between are grouped differently according to the similarity level of the fusion limit used. At 50% and 60% plots 20 to 28 are combined with plots 1 to 19 in one cluster, while plots 29 to 38 form a separate third cluster (Table 2). At 70% similarity, a level

given by Westhoff & van der Maarel (1973) as representing the level of similarity between subassociations of a number of communities, plots 1 to 19 and plots 39 to 61 remain separate clusters while plots 20 to 29 and plots 30 to 38 form clusters on their own. Thus, the cluster analysis suggests that there is a strong and stable vegetation boundary at plot 38, while the limit of the cluster containing plots 1 to 19 shifts according to the similarity level for fusion used. The clusters are recognized and listed in a sequence which parallels the field situation of the gradient.

PCA largely confirms this picture. Figure 2 represents the ordination of all plots along the first two axes, accounting for 50.7 and 13.4% of the variance, respectively. The first axis is positively correlated with several of the soil nutrient variables along the gradient, and also with pH (compare Table 1). The second axis does not show full correlation with any of the measured variables though

Table 2. TABORD-clustering at fusion limit 60%. Arrows indicate every tenth plot.

```
                              ↓       ↓       ↓       ↓       ↓       ↓       ↓
Plot number                   1........10........20....... .30....... .40........50........60
Cluster group                 1111111111111111111111111111 2222222222 333333333333333333333333

Hypochaeris radicata          11111.2112121...11...12.1112 2.....1.12 .......................
Jasione montana               7589999998576567989535231322 11.1...... .......................
Rumex acetosella              9999999999999999999999987766 55342234.1 .......................
Agrostis tenuis               7977356763344543854757966666 5555757775 1111111111121111111.1.
Brachypodium pinnatum         1111.45377866689999999998999 9995353212 4463347754578444678999
Festuca ovina                 3312112233112121111111212213 21.1121122 11.....................
Hieracium pilosella           .178212437443111.21436788954 2111126899 86731366245626994.1..11
Hypericum perforatum          .....1121212.11111211.1..... ...12224.. ...22...1..............
Stachys officinalis           ...............1.1.1434243 2159756225 5214121................
Anthoxanthum odoratum         ..................11111111 1334524474 21..1.................
Sieglingia decumbens          ...............1111458744 4112323223 411...................
Hieracium umbellatum          ....................111 23333211.1 ..................1....1
Galium verum                  1.......1.................1 .121..1..1 ..111....123...111....1
Rosa rubiginosa               ......................1 .1.1121..1 .111.3.1..132..........
Rubus caesius                 .............24631. ...................1553
Campanula rotundifolia        ................2.355332223 5131112142 3.112211.1.1...11.1..11
Lotus corniculatus            ...................111122 4542444212 5333333232333552433454 4
Pimpinella saxifraga          .................1111 1111111.11 11.21111131221211221111
Plantago lanceolata           ....................1. 21.1211342 12111311111111.1111111
Centaurea serotina (juv.)     .....................1 1112461222 4345367753333312122444
Achillea millefolium          ..........................413112.11 1...121.......21111.112
Centaurea serotina            ...................1.1111.. 1111.111111111111111.111
Sanguisorba minor             .........................1.33.2212 43353455666678654467454
Carex caryophyllea            ......................122422111 13112132123111111111.1.1
Daucus carota                 .................1112211 13244332232434535354333
Koeleria cristata             ...............1.1111 321312232222.3231......
Arenaria serpyllifolia        ...............11111 68999867999896779747521
Thymus pulegioides            .................1.4 3312444424621.12...111.
```

Table 2. (Cont.)

```
                              ↓       ↓       ↓       ↓       ↓       ↓       ↓
Plot number                   1.......10.......20....... .30....... .40.......50.......60
Cluster group                 1111111111111111111111111111 2222222222 3333333333333333333333333

Potentilla tabernaemontani    .......................... ............1 565342667877478884141.1
Arabis hirsuta                .......................... ..........1 11211134211122212222231
Scabiosa columbaria           ..........................  2111.11211111111.1111.3
Poa pratensis                 ..........................  .112212211224324534797
Trisetum flavescens           ..........................  .1.1111.11111111111111.
Agrimonia eupatoria           ..........................  ....1...1111...12112111
Medicago lupulina             1.........................  ...11121122413312111121
Taraxacum sp.                 ..........................  ...1..11111111..111111.
Carlina vulgaris              .......1..........1111..... .1..1...1.1 25211225444323336685435
Linum catharticum             ..........................  111111.21211.121111....
Unknown sp.                   ..........................  ............11........
Inula conyza                  ..........................  .......1..1.1..1.1.
Prunus spinosa                ..........................  1 344672.................
Sonchus oleraceus             ..........................  .........1.........1..
Capsella bursa-pastoris       ..........................  ...............1.......
Ranunculus bulbosus           ..........................  ...1..........111......
Echium vulgare                ..........................  .......1..1.1.12.11...
Origanum vulgare              ..........................  .......4.........
Plantago media                ..........................  ........11....3.1531.13
Verbascum thapsus             ..........................  ...1..11111111.1....
Leontodon hispidus            ..........................  .1.............
Briza media                   .....................1..1.. 11.........
Papaver rhoeas                ..........................  ...1.1.........
Sarothamnus scoparius         ..........................  .....311..........
Polygala vulgaris             ..........................  ...1... 11...11.............
Arabidopsis thaliana          ..........................  .....1.............
Myosotis arvensis             ..........................  .........1.........
Unknown sp.                   ..................1..11. .1..1.....
Crataegus monogyna            1.........................  3..1. ...........2......
Calluna vulgaris              .....................1.2...
Aira caryophyllea             ....................2....... 1...
Luzula campestris             1.1.1.....................  ...111......
Ulmus sp.                     11........................
```

the nitrogen factor comes fairly close. This failure of the second axis to strongly correlate with any of the measured environmental variables could be interpreted as evidence that this second axis is spurious. It is not uncommon for PCA to bend strong one-dimensional gradients into two-dimensional arches (Gauch, 1982; compare also Werger, 1978; Werger *et al.*, 1978). Clearly plots 1 to 22 are grouped in one cluster, and so are plots 40 to 61, while the other plots are arranged arch-wise in between, with a rather wide gap between plots 22 and 23, and a very wide one between plots 38 and 39. Furthermore, the separation between the first-mentioned cluster and the plots of the arch is along the second axis only, while that between the arch plots and the second-mentioned cluster is along both axes (Fig. 2).

Over the entire transect the number of half changes, based on Whittaker's (1975) coefficient of community, is 4; calculated for sectors this resolves to slightly less then 1 half change in the first 20 plots, just over 2 half changes in plots 21 to 40, and again slightly less than 1 half change in the last 20 plots. This means that the β-diversity over the entire transect, measured as the degree of change in species composition of the plots along the transect and using the graphical method (Whittaker, 1960, 1975), is perhaps somewhat too high to allow an effective ordination of the whole data set at once (Gauch & Whittaker, 1972b; Gauch *et al.*, 1977).

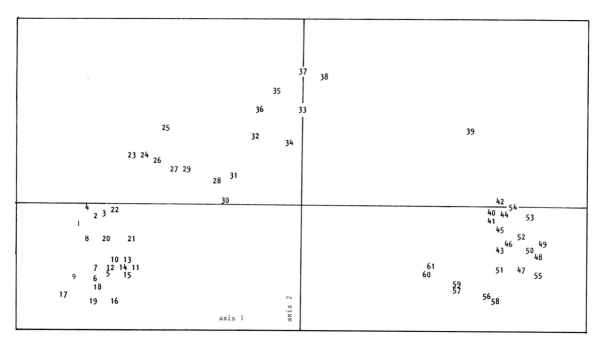

Fig. 2. Ordination of all 61 plots along first and second axes resulting from PCA of the total set of frequency data. Further explanation in text.

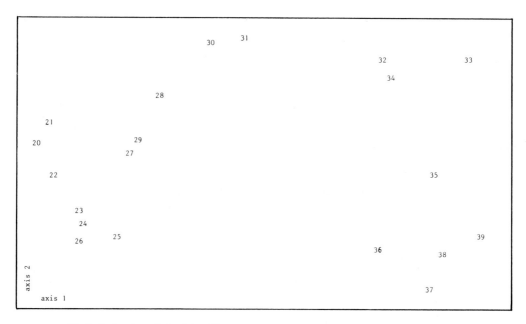

Fig. 3. Ordination of plots 20 to 39 along the first and second axes resulting from PCA.

We therefore decided to do a separate PCA for the plots of the arch between the clusters.

The results of this PCA of plots 20 to 39 are shown in Figure 3. The axes account for 41.5 and 23.6% of the variance, respectively, the total of which is very similar to that in Figure 2. The configuration in Figure 3 shows a slight sinusoid form, with a vague grouping of plots 20 to 26 and 32 to 39. On the third axis (10.3% of variance) plot 39 is distinctly separated from the others. Though axis 1 again can be correlated with several of the soil factors changing along the transect, axis 2 does not appear to relate directly to one of the measured environmental variables. Therefore, the separate analysis of plots 20 to 39 simply confirms the results shown in Figure 2, and suggests that the changes in vegetation reflect a single environmental gradient of one changing variable or a complex of co-varying variables (cf. Werger *et al.*, 1978; Gauch, 1982).

The conclusions from these analyses are that 1) there is a strong boundary between the *Meso-bromion*-vegetation and the other vegetation at plots 38/39, and 2) the boundary between the *Thero-Airion*-vegetation and the *Brachypodio-Sieglingietum* lies somewhere between plots 19 and 22 and is not as pronounced. All plots from 23 to 28 or to 32 apparently are *Brachypodio-Sie-glingietum*-vegetation and somewhat different from plots 32 to 38, though there is a gradual change here without clear boundaries, and it is possible to include this latter group of plots in the *Brachypodio-Sieglingietum*.

Species performance

In the 61 plots of the transect 60 species of vascular plants occurred, but because small juveniles of *Centaurèa serotina* had a strongly different growth form as compared to the adults, they were recorded separately (Table 2). As stated in the introduction, Whittaker (1967) predicted bell-shaped species response curves along a continuously changing environmental gradient with the peaks at different positions. Application of the program for fitting Gaussian curves (Gauch, 1972; Gauch & Chase, 1974) to the frequency data in a sequence as sampled along the transect (1–61) revealed that only 13 out of the total of 42 species which occurred in 10% or more of the plots show a response sufficient-

Table 3. Species identified as having a bell-shaped response curve by Gaussian-curve fit program for various plot sequences giving the percentage of variance accounted for when higher than 50%. N is the number of plots in which the species occurs. For further explanation see text.

Species		% of variance by Gaussian-curve fit			
	N	1–61 transect	1–61 PCA-1	1–61 PCA-3	20–39 PCA
Brachypodium pinnatum	60	–	–	–	90.7
Agrostis tenuis	59	55.2	65.1	54.1	–
Lotus corniculatus	39	–	–	–	71.7
Rumex acetosella	37	91.2	90.2	89.3	80.9
Carlina vulgaris	32	60.9	56.1	70.4	–
Plantago lanceolata	32	–	–	–	61.9
Jasione montana	31	73.9	57.3	82.1	64.2
Sanguisorba minor	30	71.2	64.9	69.1	51.0
Carex caryophyllea	30	–	–	–	71.2
Daucus carota	30	57.8	60.0	–	63.2
Arenaria serpylli-folia	28	65.9	66.7	62.1	–
Hypericum perfo-ratum	24	–	–	–	81.5
Potentilla tabernae-montani	23	51.8	60.3	68.4	–
Sieglingia decum-bens	23	–	–	–	60.2
Poa pratensis	22	80.0	68.0	71.7	–
Anthoxanthum odoratum	21	72.1	67.8	78.4	82.1
Linum catharticum	17	56.6	65.5	70.9	–
Hieracium umbel-latum	14	83.3	67.6	74.6	69.8
Prunus spinosa	7	69.3	78.5	59.7	–

ly similar to a normal curve that 50% or more of the variance is accounted for (Table 3 and Fig. 4). Thus even under these mild standards not fully 31% of the species show the expected pattern. Moreover, comparison of Figure 4 and Table 2 shows that the 'curves' for *Agrostis tenuis, Arenaria serpyllifolia* and *Potentilla tabernaemontani* show stepped responses at plots 38/39 rather than smooth changes, and the first-named species even seems bimodally distributed.

Application of the same programme to the sequence of plots produced in the horseshoe gradient revealed by PCA (Fig. 2) identified exactly the same species as showing a bell-shaped response curve (PCA-1 in Table 3); the variance accounted for differed somewhat but the average was nearly equal. Also, when the frequency data were entered

as a sequence of plots based on a combination of the
sequences from three separate PCA's (plots 1–19,
20–39, 40–61) (PCA-3 in Table 3), the results remain similar, except for *Daucus carota* which now
just misses the 50% of variance level, and the average for the 13 species stays equal.

The soil analyses show that although soil variables are nowhere constant along the transect, a
particularly large portion of the v .riables show major changes in the final segment (plots 40 to 61). The
middle segment (plots 20 to 39) show modest
amounts of change while the first segment is rela-

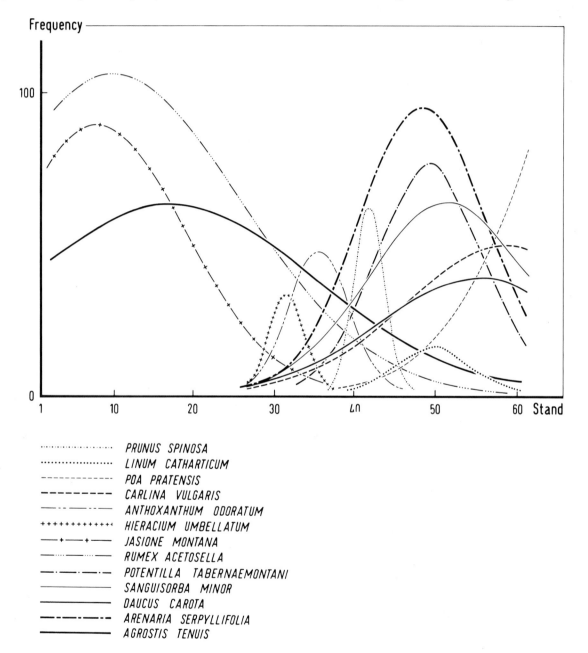

.... *PRUNUS SPINOSA*
.... *LINUM CATHARTICUM*
– – – – – – – – *POA PRATENSIS*
– – – – – – – *CARLINA VULGARIS*
— · — · · — *ANTHOXANTHUM ODORATUM*
+++++++++++ *HIERACIUM UMBELLATUM*
— + — + — *JASIONE MONTANA*
— · · · · — *RUMEX ACETOSELLA*
— · — · — · *POTENTILLA TABERNAEMONTANI*
———————— *SANGUISORBA MINOR*
———————— *DAUCUS CAROTA*
— · — · — · *ARENARIA SERPYLLIFOLIA*
———————— *AGROSTIS TENUIS*

Fig. 4. Bell-shaped response curves of 13 species, accepted as such using the Gaussian-curve fit program, along the transect, showing the
differences in curve-width and numbers of curves in the three parts of the transect. Based on the calculations the fitted curve for *Rumex
acetosella* peaks above 100% frequency. Further explanation in text.

tively homogeneous. Each of the three segments also have their characteristic species combinations as is apparent from Table 2. Figure 4 shows that most of the bell-shaped curves occur on the final segment, less on the middle segment, and few on the first segment, a picture similar to the gradient in soil variables. Thus, bell-shaped response curves appear most common where the change in soil characteristics is most pronounced.

The number and pattern of zero scores clearly influences the results of the Gaussian curve fitting program. For this reason, and because of the clear discontinuities in the transect data resulting in three parts, we applied the curve fitting program to the frequency data of each of the three major segments separately. The plot sequence was determined by the PCA-results and only species occurring at least 7 times in a segment were analyzed. Separate analysis of the three segments meant that approximately bell-shaped response curves within that particular segment could be detected, regardless of species performance along the entire transect, so that, for example, bimodal curves with their optima in different segments could be detected. In contrast, the bell-shaped response of a species with a relatively 'wide' curve over this transect might be much less apparent over a short segment of the transect than if the entire transect were analyzed.

Of 8 species sufficiently common for analysis in the first stretch (plots 1 to 19) only *Brachypodium pinnatum* shows a bell-shaped response curve with over 50% of the variance accounted for, namely 77.2%. In the middle stretch, of the 22 species sufficiently common for analysis, 12 species (or 55%) fit a Gaussian curve with at least 50% of variance accounted for (PCA 20–39 in Table 3). These in part are the same species for which Gaussian response curves could be fit for the whole data set: *Rumex acetosella, Jasione montana* and *Brachypodium pinnatum,* which show their bell-shaped decrease in this part of the transect. In addition, they include the species which characterize this middle part of the transect *(Anthoxanthum odoratum, Hieracium umbellatum* and *Sieglingia decumbens),* a species that strongly increases in frequency in this part of the transect *(Lotus corniculatus),* as well as five species which apparently have small to very small peaks in part of this stretch of the transect (compare Table 2). In the lowest stretch of the transect 32 species occur in more than 6 plots and only 5 of them (16%) show performance curves which are Gaussian with at least 50% of variance accounted for. All five also show Gaussian curves for the whole data set. Thus, in all these tests for Gaussian-curve fit 19 species (or 45%) out of a total of 42 which occurred in at least 10% of all sampled plots show bell-shaped response curves in at least part of the transect.

With reference to the entire data set 31% of the species occurred too rarely in our plots to be tested for bell-shaped response curves, while 21% could be identified as showing such curves. Visual inspection of the frequency data of the remaining species in a sequence as sampled in the transect or determined by PCA suggested that another 10% showed bimodal response curves, 23% strongly skewed unimodal curves, and 15% complex, irregular responses.

Conclusion

The vegetation along the transect appears to change as a result of one major complex of co-varying environmental variables which we suppose to be mainly a gradient of nutrient resources. We draw this conclusion on the basis of 1) the measured changes in soil variables along the transect and 2) the results of the PCA of the vegetation data. The ordination diagram showed basically a horseshoe arrangement of samples which is very similar to those resulting from unidimensional coenoclines of complex gradients ordinated by the same technique (Noy-Meir & Austin, 1970; Gauch & Whittaker, 1972a). This configuration and the relatively high percentages of variance accounted for by the first two axes strongly suggest a unidimensional gradient of a complex of variables (cf. Werger, 1978; Werger *et al.,* 1978; Gauch, 1982).

The soil gradient is not smooth along the entire transect, however. There is rather little change in the top third, and a stronger but gradual change further down, with a rather steep step starting at plots 38/39. This is reflected in both the clustering and the ordination techniques which show the corresponding vegetation changes, and in the shift in β-diversity along the transect.

Though the standards set here for accepting a species performance curve as Gaussian are rather mild, less than one third to one half of the species tested could be regarded as Gaussian. When the

68

variance limit for accepting species as showing a
Gaussian response curve is increased to a more
commonly accepted level very few species remain in
the group of species with bell-shaped curves.

Species accepted as showing Gaussian curves are
primarily those which are characteristic of the lower third of the transect, to a lesser extent those of the
middle part of the transect, and still less those of the
top part. The pictoral presentation of these curves
along the transect (Fig. 4) again reflects the gradual
change in vegetation in the upper middle part of the
transect and a rather strong change around plots
38/39.

The number of species having bell-shaped response curves along gradients of ecologically relevant environmental factors apparently is much
lower than might be expected based on Whittaker's
model. Nonetheless, those species curves identified
as Gaussian in the present study clearly reflect the
two types of boundaries, a strong, rigid boundary at
plots 38/39 and a weak, gradual boundary following plot 19, and seem to confirm the simple model
on the nature of vegetation boundaries presented
by Whittaker (1975).

References

Austin, M. P., 1980. Searching for a model for use in vegetation
analysis. Vegetatio 42: 11–21.
Austin, M. P. & Austin, B. O., 1980. Behaviour of experimental
plant communities along a nutrient gradient. J. Ecol. 68:
891–918.
Ellenberg, H., 1953. Physiologisches und ökologisches Ver-
halten derselben Pflanzenart. Ber. Deutsch. Bot. Ges. 65:
351–362.
Fresco, L. F. M., 1982. An analysis of species response curves
and of competition from field data: Some results from heath
vegetation. Vegetatio 48: 175–185.
Gauch, H. G., 1972. Cornell ecology program 12: fit Gaussian or
normal curve, Cornell Univ., Ithaca.
Gauch, H. G., 1982. Multivariate analysis in community ecology. Cambridge U.P., Cambridge.
Gauch, H. G. & Chase, G. B., 1974. Fitting the Gaussian curve
to ecological data. Ecology 55: 1377–1381.
Gauch, H. G. & Whittaker, R. H., 1972a. Coenocline simulation. Ecology 53: 446–451.
Gauch, H. G. & Whittaker, R. H., 1972b. Comparison of ordination techniques. Ecology 53: 868–875.
Gauch, H. G., Whittaker, R. H. & Wentworth, T. R., 1977. A
comparative study of reciprocal averaging and other ordination techniques. J. Ecol. 65: 157–174.

Groenewoud, H. van, 1976. Theoretical considerations on the
covariation of plant species along ecological gradients with
regard to multivariate analysis. J. Ecol. 64: 837–847.
Heukels, H. & Van Ooststroom, S. J., 1970. Flora van Nederland. Noordhoff, Groningen.
Knapp, R., 1967. Experimentelle Soziologie und gegenseitige
Beeinflussung der Pflanzen. Ulmer, Stuttgart.
Maarel, E. van der, Janssen, J. G. M. & Louppen, J. M. W.,
1978. TABORD, a program for structuring phytosociological tables. Vegetatio 38: 143–156.
Mueller-Dombois, D. & Ellenberg, H., 1974. Aims and methods
of vegetation ecology. Wiley, New York.
Noy-Meir, I. & Austin, M. P., 1970. Principal component ordi-
nation and simulated vegetational data. Ecology 51: 551–552.
Shelford, V. E., 1913. Animal communities in temperate Ameri-
ca. Chicago Univ. Press.
Terborgh, J., 1971. Distribution on environmental gradients:
Theory and a preliminary interpretation of distributional
patterns in the avifauna of the Cordillera Vilcabamba, Peru.
Ecology 52: 23–40.
Werger, M. J. A., 1978. Vegetation structure in the southern
Kalahari. J. Ecol. 66: 933–941.
Werger, M. J. A., Wild, H. & Drummond, B. R., 1978. Vegeta-
tion structure and substrate of the northern part of the Great
Dyke, Rhodesia: gradient analysis and dominance-diversity
relationships. Vegetatio 37: 151–161.
Westhoff, V., 1971. The dynamic structure of plant communities
in relation to the objectives of conservation. In: E. Duffey &
A. S. Watt (eds.), The scientific management of animal and
plant communities for conservation. pp. 3–14. Blackwell,
Oxford.
Westhoff, V. & Held, A. J. den, 1969. Plantengemeenschappen
in Nederland. Thieme, Zutphen.
Westhoff, V. & Maarel, E. van der, 1973. The Braun-Blanquet
approach. In: R. H. Whittaker (ed.), Ordination and classifi-
cation of vegetation. Handb. Veg. Sc. 5: 617–726. Junk, The
Hague.
Willems, J. H., 1982. Het Brachypodio-Sieglingietum Will. &
Blanck. 1975 in Zuid-Limburg. Gorteria 11: 14–21.
Whittaker, R. H., 1956. Vegetation of the Great Smoky Moun-
tains. Ecol. Monogr. 26: 1–80.
Whittaker, R. H., 1960. Vegetation of the Siskiyou Mountains,
Oregon and California. Ecol. Monogr. 30: 279–338.
Whittaker, R. H., 1967. Gradient analysis of vegetation. Biol.
Rev. 42: 207–264.
Whittaker, R. H., 1969. Evolution of diversity in plant commun-
ities. In: diversity and stability in ecological systems. Brook-
haven Symp. Biology 22: 178–196.
Whittaker, R. H., 1970. The population structure of vegetation.
In: R. Tüxen (ed.), Gesellschaftsmorphologie. Ber. Int.
Symp. Rinteln. pp. 39–62. Junk, The Hague.
Whittaker, R. H., 1975. Communities and Ecosystems, 2nd ed.
MacMillan, New York.

Accepted 1.12.1982.

On choosing a resemblance measure for non-linear predictive ordination

P. H. Fewster & L. Orlóci*

Department of Plant Sciences, University of Western Ontario, London, Ontario, Canada N6A 5B7

Keywords: Horseshoe, Linear, Multi-dimensional scaling, Non-linear, Ordination

Abstract

The development of non-linear ordination techniques has stemmed in part from work suggesting that species behave non-linearly to changing environmental factors or gradients. Developments in this area can be seen in two related phases: new algorithms, and the incorporation of new resemblance measures. Emphasis in this paper is placed on resemblance measures incorporated into a method of multi-dimensional scaling. The results show that a resemblance measure which reflects the non-linearities of the data can produce significant improvement in ordination, if the standardizations have not been too 'severe'.

Introduction

The investigation and ordering of vegetation units with respect to known or presumed underlying environmental gradients has long been a major objective of ecological studies. More recent ecological ordinations have evolved from the central idea that the manner in which species respond to environmental influences must be considered. This implies that optimality of the solutions is tied to the method's success in incorporating suitable response models. R. H. Whittaker's contributions were most influential in this area in that they established a theoretical framework incorporating the notions of gradient, response, and utility. It is largely a consequence of his influence that shortcomings in linear ordinations were revealed. This in turn lead to the development of methods which assume non-linear species responses.

The literature (reviewed by Orlóci, 1978) illustrates that early efforts were largely concerned with multi-dimensional configurations where individuals (vegetation plots) occupied positions and the species served as dimensions. An ordination, by contrast, is visualized as a configuration of individuals in a space where major physical (environmental) factors serve as the axes. Hence the problem involves finding the best way to transfer or map individuals in species space into factor space with minimum distortion, and to identify these factors with greatest certainty. In other words, non-linearities need to be unfolded as much as possible so as to obtain a linear ordering.

The complexity of vegetation data can result from a number of factors, such as random variation (noise) and indeterminacy in measurement. More important, however, is the type of species response. This has been demonstrated in both field data and simulation experiments (see van Groenewoud, 1965; Noy-Meir & Austin, 1970; Gauch & Whittaker, 1972). If the response is linear, complexity is not great and efficient ordination algorithms are readily available. In dealing with non-linear data, however, it becomes important that the technique used incorporates devices to handle this non-linearity.

* One of the authors (L. Orlóci) was a recipient of an N.S.E.R.C. grant during the tenure of this project. The authors wish to thank C. Brambilla and G. Salzano for the use of their computer program. A copy of a modified version used here may be obtained from the first author at no charge.

70

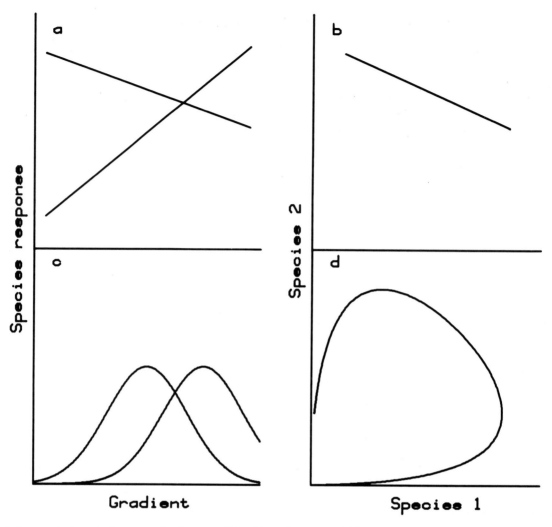

Fig. 1. Response trajectories for two species along a gradient with species response (a) linear and (c) Gaussian with their respective joint scatters (b) and (d) in species space.

Different cases are illustrated in Figure 1. Consider the simplest case where species respond linearly to an environmental gradient (such as changes in elevation up a mountainside), excluding noise. If individuals (vegetation plots) are placed at regular intervals along the gradient, and if two recorded species respond linearly to the gradient, a situation as in Figure 1a would be obtained. If the same information is graphed in species space, where the two species serve as coordinate axes, a straight line is obtained (Fig. 1b). If an ordination seeks to obtain an ordering of individuals, which is meaningful with respect to some environmental gradient (elevation in this case), a transformation is required. In

the linear case this transformation is not complex since the ordering of individuals along the line of joint response in species space (Fig. 1b) is the same as the ordering along the abscissa of Figure 1a. A principal components analysis (PCA) using a Euclidean distance measure will return this line on the first axis. If the example is extended to more than two species, a straight line will still be obtained in species space. Clearly, since a linear species response produces a linear configuration in species space, a linear resemblance measure would be the most meaningful. In fact, since this is a Euclidean space, the Euclidean distance is an appropriate resemblance measure.

If a non-linear species response is assumed, a configuration such as shown in Figure 1c (Gaussian curves) might be obtained. The same information in two-dimensional species space is shown in Figure 1d. Again, the same basic shape (in multi-dimensional space) will be obtained no matter how many species there are. In this case the configuration as represented in species space has a horseshoe shape. Hence the transformation, which takes individual points on this horseshoe and maps them onto a straight line, is necessarily complex. This paper focuses on the development of resemblance measures which can be used in the transformation from a horseshoe to a straight line.

Method

The algorithm used in the analysis accomplishes multi-dimensional scaling (MDS). Lucid descriptions of the method are given by Fasham (1977) and Brambilla & Salzano (1981), after the original outline in Kruskal (1964a, b). Attention is drawn here to a few salient features of the algorithm before remarks concerning the choice of a resemblance measure are made.

MDS works iteratively toward a final solution by comparing distances obtained from the raw data with those from a 'proposed' solution. The choice of a distance measure for the 'proposed' solution therefore determines whether MDS is a linear or non-linear method. In this respect MDS differs from other methods which attempt to handle non-linearity in the data. These include methods which fit curved axes (e.g. Phillips, 1978) or specified response curves (e.g. Johnson, 1973; Gauch, Chase & Whittaker, 1974; Johnson & Goodall, 1979), and those which use regression analysis and scaling to reduce the curvature of an ordination configuration (e.g. Hill & Gauch, 1980).

The version of MDS used here begins either with a random initial point configuration, or one specified using the maximum variance criterion. By this criterion, the p most variable species are used to define an initial configuration, where p is the number of dimensions (D) for which a solution is sought. Kendall (1971) has suggested that the $(p + 1)$D solution of a pD data set is an appropriate strategy, since this reduces the chance of selecting a local minimum as a solution. Since the data sets

tested in this paper all have a single underlying gradient (1D), two-dimensional solutions were sought in all cases.

The choice of an appropriate resemblance measure for data with non-linear species responses is difficult. Numerous possible resemblance measures are conceivable, each specific to a given species response type (cf. Austin, 1979). In any case, the familiar metric resemblance measures are non-optimal when non-linear species responses occur. As a simple example, consider PCA of the data in Figure 1d using Euclidean distance. The result would be a horseshoe, since the algorithm involves a simple geometric rotation in species space. Similarly, when a data set with Gaussian species responses is subjected to MDS analysis incorporating Euclidean distance, the result is again a horseshoe-shaped curve (Fig. 2). A straight line, representing the un-

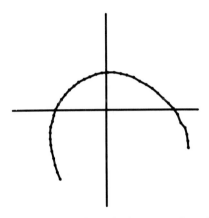

Fig. 2. Results from MDS in ordination space using a Gaussian data set with the linear distance option. Axis 1 is the abscissa and axis 2 the ordinate. Horseshoe-shaped, stress = 0.0079.

derlying gradient to which species are responding, is more desirable. In both cases a linear measure is inappropriate since it does not reflect the response structure of the data. For a curved configuration in species space, the problem amounts to developing an appropriate measure of the distance between two points A and B. A linear resemblance measure will give the distance of line AB, whereas a suitable non-linear measure will give a close approximation to the distance of arc AB.

Orlóci (1978) and Hill & Gauch (1980) have noted that quantitative methods have specific uses and that the user must be careful in applying them. In addition, many authors (e.g. Austin, 1976, 1979, 1980; Werger et al., 1983) have found a variety of species responses in nature such as Gaussian, bimodal, skewed, plateau, and so on. The subject is still very much in the exploratory stage (Feoli & Feoli Chiapella, 1980). However, if a few simplifying assumptions are made, some progress may be made in developing a resemblance measure in which provision is made for the actual curvatures found in nature. An example of this type of approach has been described by Ihm & van Groenewoud (1975) who, assuming Gaussian species responses, applied appropriate transformations to the product moments prior to an eigenanalysis.

In developing appropriate distance measures, the first assumption we make (which will be relaxed later) is that of a single underlying gradient. It is further assumed that all species responses are of the same form, although the actual response type remains open to choice. The objective is that of predicting the ordering of individuals along an unknown gradient based on species scores. The mea-

sured responses Y for a set of p species are assumed to be a function of the levels of an environmental influence to be approximated by a set of ordination scores X (Fig. 3a). Hence $Y = f(X|m)$, where m represents a set of parameters of the response graphs. For any point X on the gradient (abscissa) in Figure 3b, a linear distance (Δ) to a second point $X + \Delta$ is defined. The figure shows that this distance is related to the species response distance $f(X) - f(X + \Delta)$. Although this new distance depends on the position of X along the gradient, a distance which is unique to the type of curve which $f(X|m)$ expresses can be derived. Furthermore, the restriction regarding a single gradient can be relaxed and a similar construction on each of t gradients can be produced, assuming the same type of species response.

Now, let $\Delta = |X_{ij} - X_{ik}|$ be the ith gradient distance between individuals j and k. Then the unique distance between individuals j and k on gradient i (the compositional distance of Orlóci, 1978, 1980) is

$$d^2(j,k|i) = \int_{-\infty}^{\infty} \{f(X|m) - f(X + \Delta|m)\}^2 \, dX \qquad (1)$$

The power 2 was chosen because the integration is possible and because it leads to an interpretable formula. The composite compositional distance,

$$d^2(j,k) = \Sigma \, d^2(j,k|i), \quad i = 1, \ldots, t \qquad (2)$$

gives the distance between individuals j and k. What has been accomplished is a definition of the gradient distance in relation to a distance based on an assumed non-linear species response. Since this distance uses information about the actual species responses, it can be expected to have potential utility when linear species responses cannot be assumed.

Next, a few specific types of species responses are selected, and equation (1) solved to give actual distance (resemblance) measures.

The symmetric Gaussian curve was chosen since responses of this type (bell-shaped) have been reported many times in the literature and are thought to be common (Whittaker, 1956, 1967; van Groenewoud, 1965). The second choice was the skewed Gaussian, since this response type has also been noted (Austin, 1979). Finally, a parabolic curve, which has the basic bell shape but lacks the tails, was chosen. In nature such a response might be expected since a species might be out-competed or

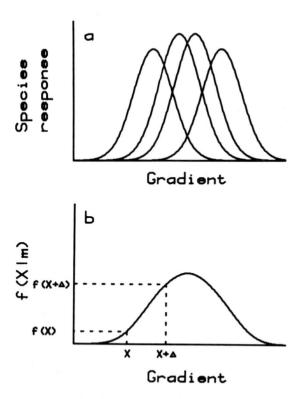

Fig. 3. Response trajectories for several species along a gradient (a) and a standard average trajectory (b).

otherwise selected against at the extremes of its potential range (Forsythe & Loucks, 1972).

Derivation

The integral in equation (1) is now solved. It is noted that standardization of various parameters preceded integration in order to obtain a distance measure independent of these parameters. Note, however, that a substantial loss of general utility may be a consequence of standardizing too many parameters.

For the Gaussian function,

$$y = Be^{-(X-a)^2/2s} \tag{3}$$

where B & s are related to height and width respectively, only standardization to unit height and width $(m = (a,1))$ is needed since the constant a (the level of influence at which response is maximal) drops out following integration. After transformation, the function becomes,

$$y = e^{-(X-a)^2/2} \tag{4}$$

Thus

$$d^2(j,k|i) = \int_{-\infty}^{\infty} \{e^{-(X-a_i)^2/2} - e^{-(X+\Delta-a_i)^2/2}\}^2 \, dX \tag{5}$$

$$\propto 2(1 - s_{ijk})$$

where $s_{ijk} = e^{-\Delta^2/4}$. A similar form was first reported by Gauch (1973).

For the skewed Gaussian function,

$$y = aX^b e^{-cX} \tag{6}$$

the standardization involves setting the parameters a, b and c to unity. Since the mode or abscissa of vertex $X_m = b/c$, an 'extra' or 'more severe' standardization is used compared to the Gaussian. The distance derived following integration is

$$d^2(j,k|i) \propto 1 + (2(\Delta^2 + \Delta - 1)e^{-2\Delta} - 2\Delta e^{-\Delta} \tag{7}$$

For the parabolic function

$$y = -aX^2 + bX + c \tag{8}$$

the standardization involves setting $a = b = 1$ and $c = 0$. This is 'more severe' than that of the skewed Gaussian, since $X_m = -b/2a$. The derived distance is

$$d^2(j,k|i) \propto 3\Delta^4 + \Delta^2 \tag{9}$$

Testing

The next phase is testing the technique (resemblance measure plus method). MDS can incorporate any one of the distance formulae as options for the ordination configuration. The original configuration distances vary, conforming to the ordination configuration distances. In this early stage of development it has been necessary to make some restrictions before generating the test data sets. They are: 1) a single gradient, 2) a few species (10) with the same type of response, and 3) random parameters for the response curves, within certain ranges. The gradient is conceived as being very broad, ranging between two extremes. A range of individual positions was defined between these two extremes where species optima would have an equal (random) chance of occurring. Ranges of constant probability were also chosen for parameters defining the height and width of the curves. The construct simulated the random appearance and disappearance of species along the gradient and implied that individuals (vegetation plots) had fewer species the further they were located from the middle of the gradient.

One data set was generated with Gaussian species responses whereas six sets each were generated for skewed Gaussian and parabolic responses. This was done in order to compare the results for consistency due to the 'extra' standardization. Beta diversity (Whittaker, 1972) ranged from 0.06 hc for the Gaussian set to 3.5 hc for the parabolic sets.

Results

The Gaussian data set produced 2D ordinations as illustrated in Figure 4. The first (Fig. 4a) resulted from the maximum variance option. Repetitions with the random option resulted in the same basic open shape as shown in Figure 4b. This was the only combination of distance measure and option which produced an ordination which is distinctly not a horseshoe. Results from the skewed Gaussian and

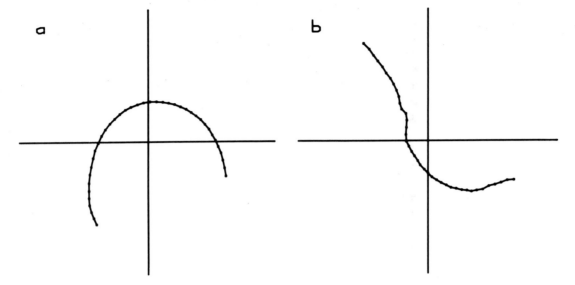

Fig. 4. Results from MDS using a Gaussian data set: (a) horseshoe-shaped, stress = 0.0089 and (b) open-shaped, stress = 0.013.

parabolic data sets, using the maximum variance option, are given respectively in Figures 5 and 6. The curves are asymmetric horseshoe-shaped in both distance options.

Discussion

In general, two aspects of ordination efficiency need to be considered. The first is the possibility that a solution may represent a scrambling of the true ordering, even though the algorithm and resemblance measure used are theoretically appropriate. Problems of this sort may arise, for example, from random variation (noise) in the data. In addition, a complex ordination algorithm like MDS has certain idiosyncrasies (particularly the problem of local minima) which may result in a misordering. The second aspect is the arch or horseshoe effect. Two examples are given (one in PCA and another

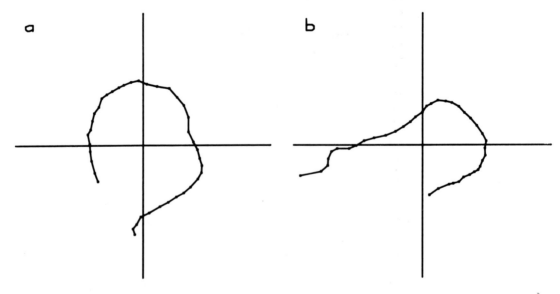

Fig. 5. Results from MDS using skewed Gaussian data sets: (a) involuted, asymmetric horseshoe-shaped, stress = 0.117 and (b) involuted, asymmetric horseshoe-shaped, stress = 0.118.

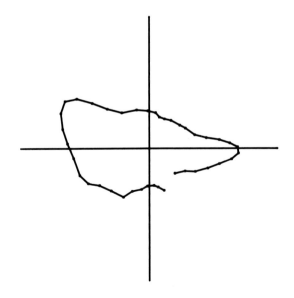

Fig. 6. Result from MDS using a parabolic data set. Asymmetric horseshoe-shaped with much involution, stress = 0.077.

in MDS, Figs. 1c, d and 2) showing that a linear resemblance measure used in ordinating non-linear data produces an involuted, horseshoe-shaped ordination configuration. Kendall (1971) points out that the involution of the horseshoe implies that, without knowing the number of gradients *a priori*, an ordering of individuals may be scrambled. This can be visualized by taking a 2D involuted horseshoe and projecting it onto a single dimension (line). The points at the involuted ends will be mixed in with the middle ones so that the ordering produced is very different from the true one. While some have felt that a linear ordering is necessary and have developed approaches to straightening the ordination configuration (Kendall, 1971; Hill & Gauch, 1980), others (Feoli & Feoli Chiapella, 1980) suggest that the horseshoe effect is revealing rather than detrimental to interpretation of the results.

Our approach in deriving an improved ordination has been to focus on the use of resemblance measures and to illustrate their importance in effectively mapping the data structure from species space into ordination space. Derived measures are combined with the MDS method to produce a non-linear ordination technique. The results using two distance measures derived for the skewed Gaussian and parabolic response are considered first. It has been noted that 'extra' standardization was needed in developing these measures in comparison with the Gaussian. We attribute the relative lack of success of these measures in handling the arc *AB* to these 'extra' standardizations in their derivation.

With respect to the Gaussian measure, MDS performed differently depending on which initialization option was used. Since the algorithm is one of minimization, different solutions may be obtained from different initial configurations: various local minima are conceivable, each returning a different solution. In general the maximum variance option produced less desirable results than the random option. This is because there is no bias in the random option as to the shape of the initial configuration. As such the final result is more a consequence of the distance measure and the method. Other approaches for initializing MDS include using the results of a linear ordination such as PCA. This also tends to give a 2D configuration like Figure 1d, since the initial PCA configuration is a horseshoe. From a developmental and exploratory perspective, the Gaussian measure derived here in conjunction with MDS and the random option has been shown to be workable within the given set of assumptions.

The Gaussian measure was derived by integrating a single response function. Why were not all response curves treated individually so as to obtain a more effective measure? Each curve could be integrated, providing the widths of response curves were first standardized. The results for each species would be the same (cf. equation (5)). Equation (5) could then be used as the collective function relating species to gradient position. Hence the only difference so far is the approach. Another question relates to the standardization of the widths (σ) of the response graphs. If the width for a single species (h) can be made more realistic by incorporating σ^2, integration will give the result $d^2(j,k|i, f(\sigma_h^2))$ which is related to equation (5) but differs by some function of the width. The compositional distance then becomes,

$$d^2(j,k|i) = d^2(j,k|i, f(\sigma_h^2)), \quad h = 1,\ldots,p$$

where p is the number of species. The problem is that since σ is a measure of X, until the gradient is determined, the σ_h^2 remain unknown. This suggests a possible feedback algorithm, in which the distance measure could be made more effective at each step.

Finally, since the perspective of ordination, and indeed that of data analysis, is often exploratory, concern is often directed towards obtaining insights. Attention in this paper has been focused on using resemblance measures which are in some way based on the same type of non-linearity as in the data. However, it is not always possible *a priori* to know much about the data structure. In reference to Figures 4b, 5 & 6, distinctive curves or 'signatures' are produced in the analysis, depending on the measure used, which reflect the underlying data structure. This would also be expected to happen for data with noise. In such a case, a cloud may result showing an overall trend much like one of the known signatures. Hence, it is possible, by trial, to obtain insight into the type of non-linear species response.

The MDS approach to ordination used here is one of many which may give an improvement in the handling of non-linearities in the data. The others include curve fitting, scaling, and *a posteriori* detrending. There are essential differences among these approaches in the conceptualization of the objective and the definition of optimality. With the methods tested here, a solution is regarded as being optimal if the ordination succeeds in unfolding a non-linear configuration. (In this respect, only the Gaussian response measure has been shown to be of utility, since the parabolic and skewed Gaussian measures both returned the horseshoe.) This implies that the type of species response assumed in the derivation of the distance measure is most likely correct. If a horseshoe type ordination configuration is obtained, the original assumption is deemed inappropriate. The actual shape of the ordination configuration may, however, suggest what type of response is depicted by the data. In other words, even if the solution of MDS is not optimal, the ordination still conveys information about properties of the data which relate to non-linearity.

Similar advantages can be seen with curve fitting, such as in the polynomial ordination of Phillips (1978). Here, however, the curvature anticipated by the model before fitting is not concerned directly with the type of response exhibited by the species. Thus as in MDS, non-linear trends are not removed before the user has a chance to detect their presence. By contrast, detrending (Hill & Gauch, 1980) re-

moves trends from the data that the user of non-linear ordinations hopes to detect. This may be completely justifiable and may help greatly in scrutinizing the ordination results, but it cannot be condoned as a general strategy.

References

Austin, M. P., 1976. Performance of four ordination techniques assuming three different non-linear species response models. Vegetatio 33: 43–49.

Austin, M. P., 1979. Current approaches to the non-linearity problem in vegetation analysis. In: Patil, G. P. & Rosenzweig, M. L. (eds.), Contemporary Quantitative Ecology and Related Ecometrics, pp. 197–210. ICPH, Fairland, Maryland.

Austin, M. P., 1980. Searching for a model for use in vegetation analysis. Vegetatio 42: 11–21.

Brambilla, C. & Salzano, G, 1981. A non-metric multi-dimensional scaling method for non-linear dimension reduction. Istituto per le Applicazioni del Calcolo 'Mauro Picone'. Quaderni, Serie III - N. 121. 35 pp.

Fasham, M. J. R., 1977. A comparison of nonmetric multidimensional scaling, principal components and reciprocal averaging for the ordination of simulated coenoclines, and coenoplanes. Ecology 58: 551–561.

Feoli, D. & Feoli Chiapella, L., 1980. Evaluation of ordination methods through simulated coenoclines: some comments. Vegetatio 42: 35–41.

Forsythe, W. L. & Loucks, O. L., 1972. A transformation for species response to habitat factors. Ecology 53: 1112–1119.

Gauch, H. G., 1973. The relationship between sample similarity and ecological distance. Ecology 54: 618–622.

Gauch, H. G., Chase, G. B. & Whittaker, R. H., 1974. Ordination of vegetation samples by Gaussian species distributions. Ecology 55: 1382–1390.

Gauch, H. G. & Whittaker, R. H., 1972. Comparison of ordination techniques. Ecology 53: 868–875.

Groenewoud, H. van, 1965. Ordination and classification of Swiss and Canadian coniferous forests by various biometric and other methods. Ber. Geobot. Inst. ETH Stiftg. Rübel, Zürich 36: 28–102.

Hill, M. O. & Gauch, H. G., 1980. Detrended correspondence analysis: an improved ordination technique. Vegetatio 42: 47–58.

Ihm, P. & van Groenewoud, H., 1975. A multivariate ordering of vegetation data based on Gaussian type gradient response curves. J. Ecol. 63: 767–777.

Johnson, R., 1973. A study of some multivariate methods for the analysis of botanical data. Ph.D. dissertation, Utah State Univ., Logan, Utah.

Johnson, R. W. & Goodall, D. W., 1980. A maximum likelihood approach to non-linear ordination. Vegetatio 41: 133–142.

Kendall, D. G., 1971. Seriation from abundance matrices. In: Hodson, F. R., Kendall, D. G. & Tautu, P. (eds.), Mathematics in the Archaeological and Historical Sciences, pp. 215–252. Edinburgh Univ. Press, Edingburgh.

Kruskal, J. B., 1964a. Multidimensional scaling by optimizing goodness of fit to a nonmetric hypothesis. Psychometrika 29: 1–27.

Kruskal, J. B., 1964b. Nonmetric multidimensional scaling: a numerical method. Psychometrika 29: 115–129.

Noy-Meir, I. & Austin, M. P., 1970. Principal component ordination and simulated vegetational data. Ecology 51: 551–552.

Orlóci, L., 1978. Multivariate Analysis in Vegetation Research. 2nd ed. Junk, The Hague. 451 pp.

Orlóci, L., 1980. An algorithm for predictive ordination. Vegetatio 42: 23–25.

Phillips, D. L., 1978. Polynomial ordination: field and computer simulation testing of a new method. Vegetatio 37: 129–140.

Werger, M. J. A., Louppen, J. M. W. & Eppink, J. H. M., 1983. Species performances and vegetation boundaries along an environmental gradient. Vegetatio (in press).

Whittaker, R. H., 1956. Vegetation of the Great Smoky Mountains. Ecol. Monog. 26: 1–80.

Whittaker, R. H., 1967. Gradient analysis of vegetation. Biol. Rev. 42: 207–264.

Whittaker, R. H., 1972. Evolution and measurement of species diversity. Taxon 21: 213–251.

Accepted 30. 3. 1983.

Xeric Mediterranean-type shrubland associations of Alta and Baja California and the community/continuum debate*

W. E. Westman**

Department of Geography, University of California, Los Angeles, Los Angeles, CA 90024, USA

Keywords: Californian vegetation, Coastal sage scrub, Coastal succulent scrub, Floristic classification, Gradient theory, Mediterranean-type climate, Mexican vegetation

Abstract

A survey of the xeric shrublands of Pacific coastal North America from San Francisco to El Rosario (Mexico), including the inner Channel Islands, was conducted using 99 sample sites of 0.063 ha size. TWINSPAN classification and DECORANA ordination confirmed the existence of two plant formations, distinguishable physiognomically: coastal sage scrub and coastal succulent scrub. Within coastal sage scrub, four floristic associations were recognized: Diablan, Venturan, Riversidian and Diegan. Within coastal succulent scrub, two floristic associations were defined: Martirian and Vizcainan. These associations occur in distinct geographical regions following the coastline, with the Riversidian association occurring in the basin inland from Venturan and Diegan regions. Their locations are strongly correlated with differences in evapotranspirative stress regimes. The Channel Island sites show affinities to several of the mainland associations. The Venturan association can be further subdivided floristically into two subassociations, dominated by *Salvia mellifera* and *S. leucophylla* respectively. These subassociations which are coextensive geographically at a regional scale, typically do not intermingle at a local scale but often meet along sharp boundaries in the landscape. The dominant species segregate by moisture preference, *S. mellifera* preferring coarser-texture soils and more southerly aspects than *S. leucophylla*. Richness and equitability of these sites are depressed relative to other xeric shrubland sites, reflecting the fact that the two subassociations partition the Venturan flora into substantially non-overlapping subsets of species. This segregation of associates between the two *Salvia* dominance types strongly suggests biotic influence of the dominants on subordinate species, perhaps mediated by allelopathy. This biotic interaction, leading to relatively strong floristic subassociations segregating independently in the landscape, would provide an example of the holistic community structure referred to by Clements and his followers, embedded within a larger pattern of continuity in species distributions.

* Nomenclature follows Munz & Keck (1959), Munz (1974) and Wiggins (1980).
** I am grateful to the following for research assistance: S. Coon, E. Hobbs, S. Lavinger, J. F. O'Leary, K. R. Preston, B. Rich and A. Troeger. I also thank the numerous public and private land owners who permitted access to the study sites. This research is based upon work supported by the National Science Foundation under grant DEB 76-81712.

Introduction

The Mediterranean climatic zone of the Pacific coast of North America extends from the Canadian/United States border to northern Baja California, Mexico (50°–30° latitude; UNESCO, 1963). Within this region, from southern Oregon to El Rosario, Mexico (42°–30° latitude), three broad shrubland formations have been recognized. Chaparral is composed of tall ($\geq$2 m), sclerophyllous, evergreen shrubs. Northern coastal scrub consists of shorter (<2 m) sclerophyllous, evergreen shrubs occurring close to the coast from central California northward (Heady *et al.*, 1977). Coastal sage scrub (or 'soft chaparral') is dominated by mesophyllous, partially drought deciduous and seasonally dimorphic shrubs, and occurs primarily from central California southward (Westman, 1981a). Coastal sage scrub occurs in the most xeric habitats of the three, and intergrades with Sonoran (Coloradan and Vizcainan) desert elements to the south and Mohave desert elements to the east.

At the base of the San Pedro Martir in Baja California, Mooney & Harrison (1972) and Mooney (1977) described the xeric shrubland type as 'coastal sage succulent scrub', of which the driest sample contained a more marked succulent element, and no species of sage (*Salvia* spp.; *Artemisia californica*). A xeric shrubland type, often with an increased succulent element and only sometimes predominated by sage species, occurs on the Channel Islands (Power, 1980) and the islands adjacent to northern Baja California. The more succulent of the vegetation types on these islands has been variously termed 'maritime desert shrub' (Dunkle, 1950), 'maritime desert scrub' (Thorne, 1976) and 'maritime cactus scrub' (Philbrick & Haller, 1977).

While coastal sage scrub occurs extensively on the coastal plain from San Francisco (38° N) southward, it occurs in a much narrower zone between chaparral and desert on the eastern side of the Coast Ranges, presumably because the gradient of available moisture is steeper on the continental slopes. Recently several classifications of the major floristic associations within the xeric shrubland formation (coastal sage scrub and adjacent types on desert margin and offshore islands) have been proposed (Kirkpatrick & Hutchinson, 1977; Axelrod, 1978; Westman, 1981a, 1982a). These have left unsettled such questions as the floristic relationship between the mainland and island groups, the basis for distinctions between coastal sage scrub and adjacent desert-margin types, and the more precise location of geographic boundaries for the proposed associations of coastal sage scrub (Axelrod, 1978; Westman, 1982a). In addition, the differences in floristic structure, diversity, composition and habitat among these associations have not been detailed.

A principle objective of this article is to address these questions by analyzing quantitative floristic composition and associated habitat variables for 99 sites from San Francisco to El Rosario, including four Channel islands (Santa Cruz, Santa Catalina, Anacapa, San Clemente). In the process of analyzing these results, using techniques developed in part by R. H. Whittaker, a set of observations has emerged which provide new evidence concerning the community/continuum debate which Whittaker helped flare (Westman & Peet, 1982). Two subassociations of coastal sage scrub appear to exist which are significantly controlled in composition by their dominant species. To the extent that these floristic groupings are influenced by biotic interactions among their components, they provide a supporting example for those who favor a holistic view of community structure, at least in some situations. These groupings, however, can also be seen to be 'noda' within a larger gradient structure (cf. Whittaker, 1967).

Methods

Floristic composition was determined for 99 sites by recording canopy cover of individuals of all plant species along four 25-m line transects within a sample plot 25 m on a side. Cover values of rarer species in the plot, not intercepted by transects, were estimated. Cover values for each plant canopy measured separately were summed to obtain individual overlap foliar cover, which may exceed 100% for the plot. No sampled site had burned less than seven years previously (excepting four sites of minimum ages 5 and 6 yr). Sites were chosen in which shrub cover was not extensively reduced by grazing; they were scored for grazing intensity on a five-point scale (Westman, 1981b).

Variables of vegetation structure sampled were: standing crop of litter mass (from five 0.5-m² quadrats), light penetration to 10 cm aboveground (from 50 systematically-placed points using a quantum meter), and median canopy height. Topographic parameters included latitude, longitude, distance to coast, elevation, slope and aspect. Soil variables were: bulk density, texture, conductivity and pH (saturation extract), available calcium, magnesium, potassium, ammonium and nitrate (sodium acetate extractant, pH 6), base-extractable phosphate (Olsen's method; Jackson, 1958), and total nitrogen (micro-Kjeldahl method). Soil samples were taken at 14–16 cm depth, except for bulk density (8 cm depth). Temperature and precipitation records for California were averaged from the nearest 1–3 weather stations, for the 1950–1970 period (National Climatic Center, Asheville, North Carolina, unpublished; Zabriskie, 1979). Data for Baja California stations were taken from Hastings & Humphrey (1969). Channel Island data were drawn from Dunkle (1950) and unpublished records of D. B. Weismann, D. C. Rentz, and C. Stanton (available at the Santa Cruz Island field station) and the U.S. Weather Service (Federal Building, Los Angeles). Mean annual concentrations of air pollutants were estimated by averaging available values from nearby monitoring stations of State and regional air pollution agencies for the 1963–1977 period. Because of incompleteness of monitoring records, average values for air pollutants presented in Table 3 are drawn from samples of different sizes. Synergistic indices were developed based on enhanced plant damage noted in agronomic literature (Westman, 1981b). Some results from 67 of the samples sites have been reported elsewhere (Westman, 1979a, b, 1980, 1981a, b), and methods of site selection and data collection are described in greater detail in Westman (1981b).

Data reduction was accomplished using programs in the Cornell Ecology Program series: CONDENSE (Singer & Gauch, 1979), DATA-EDIT (Singer, 1980), TWINSPAN (Hill, 1979a) and DECORANA (Hill, 1979b; Hill & Gauch, 1980). Both TWINSPAN and DECORANA have proven robust in tests with simulated data (Gauch & Whittaker, 1981; Gauch et al., 1981).

Two-day indicator species analysis (TWINSPAN) is a polythetic divisive classification method which resembles the Braun-Blanquet tabular sorting methods (Mueller-Dombois & Ellenberg, 1974) in some respects (Hill, 1979a).

Detrended correspondence analysis (DECORANA) is a revised form of reciprocal averaging (RA) ordination (Hill, 1973) in which higher order axes are made independent of the first, and samples are spaced to achieve an even rate of turnover of species along the axes (Hill & Gauch, 1980).

Additional programs in the BMDP series (Dixon & Brown, 1979) were used to calculate correlations and analysis of variance. A diversity program of our own was used to tabulate richness and growth form data, to calculate equitability indices of Simpson (1949), Shannon-Wiener and E_c (Whittaker, 1972), and to plot percentage similarity values along axes of sample separation for the graphical calculation of beta diversity (Whittaker, 1960). The coefficient of community (CC) used is of the form 200 $c/a + b$, where c is the number of species of species in common between stands of richness a and b (Whittaker, 1975), and the percentage similarity (PS) index is that of Bray & Curtis (1957).

Results

Floristic relationships

The 99 sampled sites, extending from Mt. Diablo near San Francisco to slightly S of El Rosario, Mexico, and including the inner Channel Islands, spanned 8° of latitude and 5° of longitude (Fig. 1). They contained 531 species of vascular plants, of which a majority were rare in occurrence or abundance (cf. Westman, 1981b). Thus 275 species occurred in only one or two sites, or attained a maximum cover of less than 3%. The deletion of these rare species from the data set made little difference to the sample ordination, the eigenvalue (EV) for the first DECORANA (DCA) axis increasing from 0.764 to 0.795 as a result of the deletion. Furthermore, differences in total cover from site to site appeared unimportant to floristic distinctions between sites, since relativization of species scores to total 100% had no significant effect on the DCA ordination (first axis EV = 0.762 vs. 0.764 for untransformed data). Although 'outlier' samples to the basic trend in floristic variation occurred, and in the case of certain desert samples were purposely included, their presence in the ordinations did not distort the major axis extracted. When species scores were averaged from adjacent groups of five samples along a RA axis, thereby smoothing outlier effects, the eigenvalue for the subsequent DCA of composite groups changed little (first axis EV = 0.770 vs. EV = 0.764 for ungrouped sites). Transformation of the species data to octave scale ($\log_2$) deemphasized the importance of dominant species, and weakened the effectiveness of DCA (first axis EV = 0.626 vs. EV = 0.764 for untransformed species scores). As a result of these preliminary analyses, effectiveness of the un-

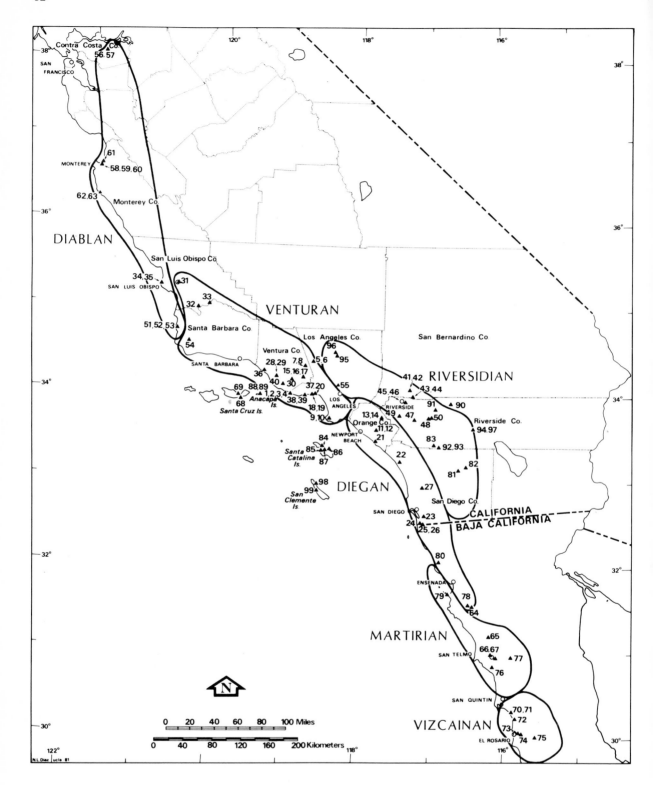

Fig. 1. Location of xeric Mediterranean-climate shrubland associations, based on TWINSPAN classification of 256 species in 99 sites. Sample plot numbers are shown.

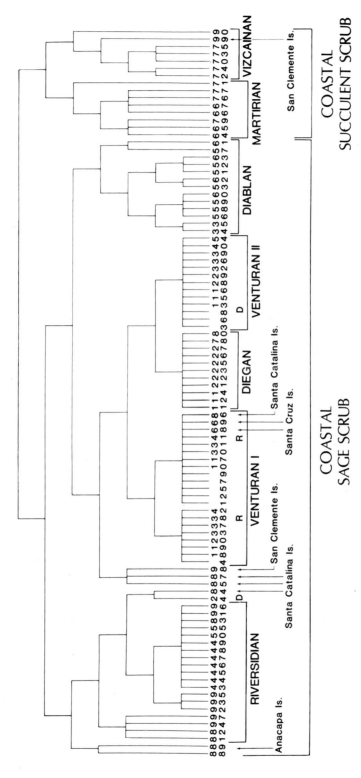

Fig. 2. TWINSPAN classification of 99 sites, using 256 species occurring in at least 3 sites or with at least 3% cover. Site numbers corresponding to Fig. 1 are shown. Outlier sites are shown by letter beneath site number, D = Diegan, R = Riversidian.

relativized species scores for abundant and widespread species in ordinating coastal sage sites was reaffirmed (cf. Westman, 1981a, b). The 256 species occurring in three or more sites or having a cover value of 3% or more were used in all subsequent classification and ordination analyses reported here, unless otherwise stated.

The TWINSPAN classification of samples, when plotted on a map of site locations, revealed relatively circumscribed geographic locations corresponding to each floristic group (Fig. 1). The Californian groups occurred roughly in the regions of the associations previously proposed by Kirkpatrick & Hutchinson (1977), and Axelrod (1978), and were labeled Diablan, Venturan, Diegan and Riversidian accordingly. One major exception to this pattern was the division of Venturan samples into two distinct floristic groups which, although separated by the Diegan group on the sample axis (Fig. 2), were not segregated geographically at a regional scale. These groups (Venturan I and II) differed floristically in that, within Venturan samples, *Salvia mellifera* occurred only in Venturan I and *Salvia leucophylla* only in Venturan II. In addition, *Encelia californica* occurred on 13 of the 18 sites of Venturan I, and was a dominant or codominant (>20% cover) on nine of them; it was present, with <1% cover, on only one site of the Venturan II group.

This distinction between sites dominated by *Salvia mellifera* and *S. leucophylla* had earlier been reported by Westman (1981a) as a result of a three-dimensional RA of a subset of these data. Although the two *Salvia* groups are not segregated geographically at a regional scale, they are commonly segregated spatially at a smaller scale (of the order of tenth hectares), the two groups meeting along sharp ecotones on hillsides (Westman, 1981a, b; 1982c). For example, two pairs of adjacent Venturan I and II sites occurring on the same parent materials within a few tens of meters of each other on a hillside in the Santa Monica Mountains (Sites 3/4; 15/17) had the following coefficients of community (CC) and indices of percentage similarity (PS) respectively: CC = 43, 30%; PS = 6, 12%. By contrast, three pairs of adjacent stands of Riversidian (41/42, 43/44, 45/46) and Diegan (11/12, 13/14, 25/26) sage had mean CC and PS values as follows for Riversidian and Diegan sets respectively: CC = 51,48; PS = 16,42. In situations where the parent material itself varied at a small scale (e.g.

Leo Carrillo State Park; see Fig. 5 in Westman, 1981b), the substrate pattern was not visibly correlated with the shift from Venturan I to II.

Apart from the island samples, four outlier samples can be recognized in the TWINSPAN classification. Sites 41 and 42 in the Riversidian geographic cluster of Figure 1 are classed as Venturan in the hierarchical classification (Fig. 2). These sites both contained *Salvia mellifera* (4% and 93% cover respectively), giving them affinity to Venturan I. The affinity is not complete, however, since Site 41 (dominated by *Encelia farinosa*, a Riversidian and Vizcainan shrub) is ranked among other Riversidian samples in the DCA ordination (discussed below), and Site 42 shows Diablan affinities. The domination of the Diegan Site 13 by *Salvia leucophylla* (55% cover) similarly explains its classification as Venturan II in Figure 2. Site 24, a coastal Diegan site by geographic location, shows a strong Riversidian influence and is thus placed with other sites influenced by both these groups in Figure 1. Its Riversidian and island affinities appear to be partially a reflection of its grazed character, including a high abundance of grasses (*Bromus rubens,* 23% cover; *Agrostis exarata,* 12%).

The Channel Island samples on the whole remain distinct from any one mainland association in the classification (Fig. 2). This fact is even more clearly demonstrated by the arrangement of samples along the first two axes of DCA (Fig. 3). The coastal sage sites on the islands appear to be a composite of elements from several adjacent mainland associations. The island endemics are not a factor in the sample separation observed, since these were eliminated during the initial data screening as occurring in less than three sites. Anacapa samples are seen (Figs. 2 and 3) to be the most distinctive among the island samples, and bear their greatest resemblance to the Riversidian association. Santa Cruz sites bear relation to all four Californian associations (Fig. 3), with particular resemblance to Venturan I (Fig. 2). Santa Catalina island occurs directly offshore from the conjunction of Venturan, Diegan and Riversidian associations, and its samples show relations with all three. In addition, the increased prominence of succulents on a particularly steep, windy coastal bluff (Indian Head) results in its grouping with coastal succulent scrub samples by DCA (Fig. 3). San Clemente is the most southerly of the four islands sampled. The San Clemente

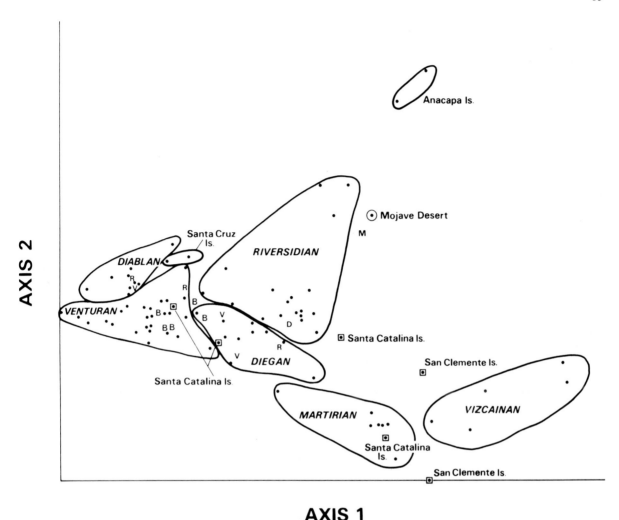

Fig. 3. DECORANA ordination of 99 sites, using unrelativized cover of 256 species occurring in at least 3 sites or with at least 3% cover. Outliers are shown by the letter of the group in which they are classified by TWINSPAN in Fig. 2. B = Diablan; R = Riversidian; V = Venturan; D = Diegan; M = Martirian.

samples, while quite different from each other, both occupy positions (Fig. 3) showing their predominant relationship to the two Baja California associations (Martirian, Vizcainan). The two-dimensional ordination (Fig. 3) is more successful than the one-dimensional classification (Fig. 2) in illustrating the true complexity of floristic origins for the island samples. The prominent role of Riversidian elements in the island floras, despite the substantial separation of the Riverside Basin from the coast and the occurrence of coastal mountain barriers, may be ascribable in part to the heavier influence of grazing animals in both areas, with the resultant

similarity in habitat for exotic herbs of high dispersal ability. The island samples are discussed further in Westman (1982b).

The northern (Diablan) samples occur in a distinct group at the second division of TWINSPAN (Fig. 2), and show closest affinity to Venturan and Diegan associations (Figs. 2 and 3). These northern samples can be further classified as Lucian, Diablan or Franciscan, based on the geographic zones proposed by Axelrod (1978). However, the samples so defined occur intermingled in the TWINSPAN classification, and there was no basis to subdivide them further based on the present sampling information. Only two of the sites, from Mt. Diablo (56, 57), fell into Axelrod's Diablan zone. Additional sampling is therefore needed to determine whether a further division between Lucian and Diablan associations is justified. Axelrod's Franciscan association, although reaching south to include Monterey sites (58–61), also extends into Oregon and contains an abundance of evergreen, sclerophyllous species of northern distribution (*Gaultheria shallon, Vaccinium ovatum, Myrica californica, Rhamnus californica, Baccharis pilularis*). It should therefore be considered more truly part of the northern coastal scrub formation. The Lucian association is located geographically in the coastal region from N Santa Barbara County to San Francisco, in which many of the sample plots classed in the northern group occurred, and might at first seem to be the logical label for the northern group. However, by Axelrod's (1978) definitions, the Lucian association also contains many species of evergreen chaparral (*Arctostaphylos* spp., *Ceanothus* spp.), whereas the Diablan association does not. Like the Franciscan, the Lucian association therefore seems to belong with northern coastal scrub or, in some cases, chaparral. Within the coastal sage scrub formation I have therefore adopted the label Diablan for all samples in the northernmost portion of the range (Fig. 1).

The first division of TWINSPAN separates two groups of Baja Californian sites from the remaining samples (Fig. 2). This division effectively separates associations dominated by mesophyllous, seasonally dimorphic species (coastal sage scrub), from those dominated by succulents and completely deciduous species. The latter two groups are clearly of desert affinity. I have called the northern of the two 'Martirian' because the samples occur on the coastal plain and foothills of the San Pedro Martir, and the southern of the two 'Vizcainan' because of its transitionary character to the Vizcainan (Sonoran) desert of the central Baja peninsula. The Martirian association is essentially synonymous with the 'coastal succulent sage scrub' which Mooney & Harrison (1972) sampled in the foothills of the San Pedro Martir. The Vizcainan association is transi-

tional between coastal succulent scrub and warm desert scrub. Consistent with the usage of 'coastal sage scrub' as a plant formation with four floristic associations (Diablan, Venturan, Riversidian, Diegan), one could consider 'coastal succulent scrub' as a plant formation with two floristic associations (Martirian, Vizcainan). Further sampling south in the Vizcaino desert is needed to determine whether the Vizcainan association would be better classed with the warm desert scrub formation.

Figure 4 presents a hierarchical classification of major species by TWINSPAN, produced in the same process which resulted in the sample classification (Fig. 2). Although 256 species in the 99 sites were classified, only the 100 commonest species are presented. The distinction between the two plant formations is evident, with the Riversidian species occurring in a bridging position. Because the commonest species are by definition among the most widespread, the division of the species into separate associations within formations is not particularly clear, especially at higher levels of the hierarchy. Indeed one of the 'indicator' species for the Riversidian association identified by TWINSPAN is not even listed in Figure 4 (*Eriophyllum wallacei*). Nevertheless, the classification serves as some guide to floristic affinities among widespread species, the separation between the four major *Salvia* species being of particular interest (cf. Fig. 4, Westman, 1981a).

Growth forms, life-cycle types and native status

A number of important physiognomic and floristic differences distinguish the coastal sage scrub and coastal succulent scrub formations. The dominant species of coastal sage scrub are seasonally dimorphic, dropping their large, early-growing season leaves during the dry season, and retaining smaller axillary leaves for much or all of the remaining dry period (Westman, 1981c). The four dominant species of *Salvia* (*mellifera, leucophylla, apiana, munzii*), all of which have this characteristic, are limited to coastal sage scrub, as are a number of codomi-

Fig. 4. Most common cover values for 100 most abundant vascular species and nonvascular groups among those occurring in at least 3 of 99 sites, or with at least 3% cover. 0 = absent from all samples; * = present in a minority of samples; 1 = 0–1% cover value; 2 = 2–4%; 3 = 5–9%; 4 = 10–19%; 5 = 20–100%. Hierarchy at left shows species classification by TWINSPAN. Associations in which species occur are shown with letters.

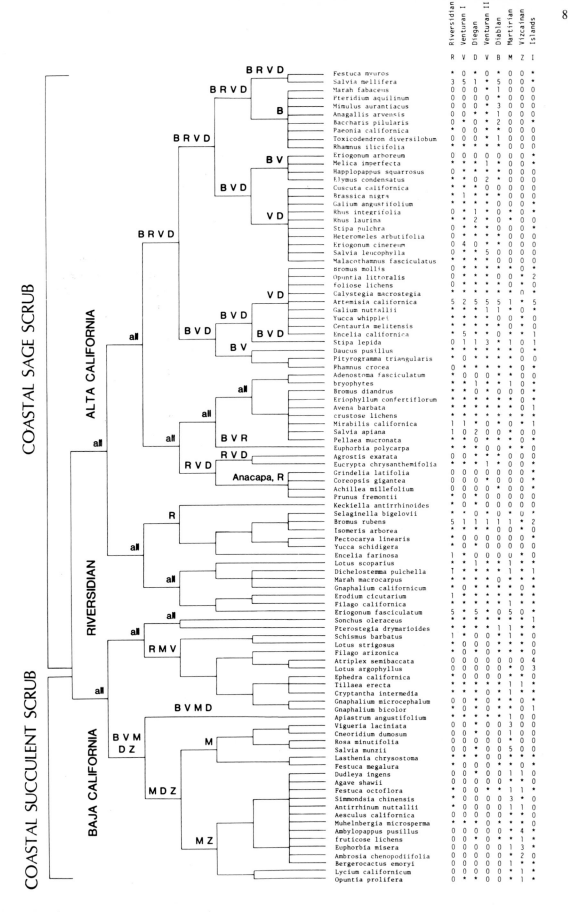

nants (e.g. *Encelia californica, Eriogonum cinereum*). Stem and leaf succulents, which appear in the Diegan association, remain a minor element. In coastal succulent scrub, by contrast, succulents are a dominant element (especially *Cactaceae, Crassulaceae* and *Euphorbiaceae*), and the importance of seasonally dimorphic species is diminished. The deciduous shrubs and small trees which do occur often lose their leaves entirely during the dry season. Indeed, a deciduous small tree element (*Aesculus parryi, Fraxinus trifoliata, Prunus fremontii, Ptelea aptera* and others) is quite common as a minor component of coastal succulent scrub, but occurs in coastal sage scrub only rarely, in the desert-margin sites of San Diego and Riverside Counties. Westman (1981c) has suggested that the encoding of such a photoperiodically-triggered total deciduousness is characteristic of areas with low and unpredictable precipitation and high evapotranspirative stress on the desert margins of the Mediterranean climatic zone.

These broad trends are confirmed by quantitative analysis of the proportions of species in each growth form category (Table 1A) and the percentage foliar cover (unrelativized) of species in each such category (Table 1B). Except for an outsized increase of annual cover by primarily introduced

species in the Riversidian association, the percentage cover of annuals generally increases from the Venturan association southward. The percent cover of introduced species in the Riversidian area (36%) exceeds even that on the heavily grazed island samples (19%); the same is true for annuals (46% vs. 20%). The low shrub cover in the Riversidian group (59%) relative to drier Martirian (73%) and Diegan (85%) groups is also noteworthy. These findings are consistent with the earlier hypothesis (Westman, 1979b, 1981b) based on relativized values and smaller sample size, that factors other than grazing or precipitation, particularly ozone levels, could be causing a reduction in native shrub cover and replacement by annuals in the Riversidian area. However, higher evapotranspirative stress due to higher summer temperatures in the continental Riversidian area may also be responsible for the reduction in shrub cover. Total shrub and tree cover generally increases with decreasing summer evaporative stress (as gauged by mean temperature of the warmest month and annual precipitation), consistent with earlier findings (Westman, 1981a). Venturan II is exceptional in exhibiting a higher shrub and tree cover than expected by this trend (101% vs. 85% for Diablan sites).

Table 1. Growth form and life-cycle types in each floristic association. A. The percentage of all species in the association found in each growth form or life-cycle category. Values do not sum to 100% because species of uncertain or multiple categorization were excluded. B. The average individual overlap foliar cover per site in each growth form or life-cycle category. Values do not sum to 100% because cover values are unrelativized.

Association	% native	% introduced	% annual	% biennial	% perennial	% succulent	% vines	% herbs	% shrubs	% trees
A										
Diablan	85	15	41	1.8	57	1.2	1.8	80	16	0.61
Venturan	83	16	33	1.2	65	3.6	3.0	66	25	1.80
Venturan I	83	16	37	0.8	61	4.8	3.2	65	25	0.81
Venturan II	85	15	25	1.0	74	3.9	2.9	59	31	2.94
Riversidian	91	8	47	0.5	52	4.7	2.6	67	24	0.52
Diegan	88	11	30	0.7	68	6.6	3.7	57	30	1.47
Martirian	93	7	52	0.7	46	7.4	3.4	68	20	0.00
Vizcainan	86	14	46	0.0	54	20.0	1.6	59	18	0.80
B										
Diablan	119	5	13	0.8	110	0.1	1.7	37	85	0.02
Venturan	108	3.5	5.0	0.0	107	1.8	0.4	19	89	1.25
Venturan I	98	4.1	5.5	0.0	96	2.5	0.3	15	84	0.09
Venturan II	125	2.6	4.1	0.0	123	0.8	0.5	25	98	3.13
Riversidian	75	36	46	0.1	65	1.0	0.2	50	59	0.20
Diegan	105	12	13	0.0	103	1.5	6.6	24	85	0.02
Martirian	106	3.7	31	0.1	79	4.1	1.3	34	71	0.00
Vizcainan	74	11	39	0.0	46	9.6	0.1	46	29	0.01

Diversity relations

Diversity variables are summarized in Table 2. The highest richness per 625 m² site occurs in the coastal succulent scrub formation, the Martirian samples being somewhat richer (44 species) than the Vizcainan (33), but also more variable in richness (range: 25–62, s.e. 55; vs. 23–40, s.e. 14). Among the coastal sage groups the Venturan group has a total number of species comparable to the Diablan, and exceeding the Diegan. Nonetheless, the average number of species per site is substantially lower in Venturan sites than in the other three associations. This is because the species are segregated between the two floristic subassociations, with at least 45 species occurring in only one of the two subassociations. Both subassociations, particularly Venturan I (*Salvia mellifera/Encelia californica*), further show strong dominance by their major shrub species. This, combined with the reduced richness, results in lower equitability values (low E_c, high Simpson's index) for Venturan than for any other associations.

Total richness is highest in the Riversidian group, due both to herb and shrub richness. The high richness of native shrubs is particularly significant, since earlier observations (Westman, 1979, 1981b) had indicated a reduction in cover of native species on heavily polluted sites in the Riversidian region, and a question remained about the extent to which the reduction in richness was a result of natural biogeographic factors as opposed to disturbance. The evidence from Tables 1 and 2 is consistent with the hypothesis that native cover and richness are both high in unimpacted portions of the Riversidian region, and that air pollutants may indeed be responsible for a reduction in cover of native species on the most polluted sites. Such evidence is suggestive only, however, and controlled fumigation experiments with ozone and sulfur dioxide on coastal sage scrub shrubs and herbs are currently in progress to test this hypothesis further (K. Preston, L. Weeks, W. Westman, in preparation).

The growth-form diversity of the associations was calculated using the Shannon-Wiener index with cover values for the five growth-form categories listed in Table 1. Growth-form diversity is lowest in the Venturan samples, but otherwise increases toward the warm, arid regions, reaching its peak in the Vizcainan association. This reflects the influx of

Table 2. Diversity variables in the major floristic associations.

	coastal sage scrub						coastal succulent scrub	
	Diablan	Venturan I	Venturan II	Venturan I & II	Riversidian	Diegan	Martirian	Vizcainan
No. of sites	13	21	13	34	19	12	8	7
Species richness								
X Number per 0.0625 ha								
(alpha)	30	18	20	19	29	26	41	33
Std. error	42	19	21	20	17	31	52	14
Total Number in group								
(gamma)	173	127	105	172	198	138	153	129
Total Number of shrub sp.	24	32	33	44	48	42	30	23
Total Number of herb sp.	139	80	59	110	132	75	103	74
Equitability								
Simpson's index	0.282	0.356	0.289	0.330	0.268	0.230	0.147	0.206
s.e.	0.007	0.009	0.001	0.001	0.005	0.004	0.001	0.003
E_c	11.42	6.83	7.53	7.10	9.91	10.25	16.76	12.89
s.e.	6.68	3.12	2.27	2.02	3.59	6.21	9.64	3.28
Species with Cover ⩾20%	1.69	1.62	2.23	1.85	2.05	1.75	1.50	1.57
s.e.	0.16	0.14	0.08	0.08	0.22	0.11	0.51	0.48
Growth-form diversity								
Shannon-Wiener index	0.298	0.246	0.291	0.270	0.340	0.338	0.361	0.398

succulents, and to a lesser extent, small trees, into the flora. The increased growth-form diversity may reflect the larger number of adaptive strategies that have evolved to compete for resources in the increasingly stressed arid environment, although the index itself is sensitive to the ways in which growth forms are categorized.

Beta diversity measured as half-changes in floristic composition using Hill's formula (Whittaker *et al.,* 1979), where half-changes $= (12 \, EV/(1-EV))^{1/2}/ 1.349$ and EV is the eigenvalue of the ordination axis) is 4.6 for the first DCA axis, when all 531 species (untransformed data) are used. When rare species (<3 occurrences, $<3\%$ cover) are deleted, beta diversity predictably decreases somewhat (3.5). The turnover in herbs along the first axis (from Diablan to Vizcainan groups) is greater (5.7 HC) than for shrubs (4.2 HC) (graphical technique). When this trend was partitioned into strict latitudinal vs. longitudinal changes for coastal sage scrub, Westman (1981b) found that turnover in herb composition along a latitudinal gradient was the major contributor to the trend observed, presumably due to the distinct and rich nature of the Riversidian herbaceous flora.

Relations to habitat features

Mean values for major habitat variables in the associations are summarized in Table 3. Differences in levels among these groups, including Channel Island groups not shown, were significant ($P < 0.01$) for all variables except grazing intensity scores and the levels of soil nitrate and salinity (F-test, analysis of variance). All island sites were excluded from the environmental averages, except three sites which were clearly classed with Venturan I in Figure 2 (68,69, Santa Cruz I.; 86, Santa Catalina I.).

The major floristic associations from NW to SE occur along a gradient of increasing evapotranspirative stress, except that the Riversidian association has a higher evapotranspirative stress due to its continental temperature regime. The more southerly associations in this sample set occurred on soils of greater compaction, with lower moisture holding capacity. Vizcainan soils show the heightened salinity, pH and potassium levels characteristic of semi-arid soils. Soil texture averages between associations differed little, except that Riversidian sites

tended to be coarser, and Venturan II samples finer, than the average.

The significantly finer texture on Venturan II than Venturan I sites may help explain the differences in their dominants. *Salvia leucophylla* (the Venturan II dominant) thrives on finer-textured soils than *Salvia mellifera* (Venturan I) (Westman, 1981a). In Mediterranean climates, finer-textured soils dry out near the surface earlier in the growing season than coarser-textured ones (Walter, 1973; Kirkpatrick & Hutchinson, 1980).

In addition, while mean slope values (25°) were identical for the two Venturan groups, aspects differed significantly. 69 percent of the Venturan II sites had aspects in the NE part of the compass (315°–135°), compared to 14% of the Venturan I sites. Poole & Miller (1975) have argued that vegetation on NE aspects in the Mediterranean climatic zone of California develop higher water stress during the dry season because of the higher biomass supported on these aspects as a result of lower evapotranspirative stress early in the growing season. They showed that higher water potential deficits developed on NE than S aspects of inland chaparral by late summer. Venturan II stands do develop higher total cover (127%) than Venturan I stands (104%). As further evidence for the higher drought-tolerance of *S. leucophylla* relative to *S. mellifera,* the former exhibited a significantly higher xylem tension than adjacent individuals of the latter species in an ecotone in the Santa Monica Mountains where the two subassociations met (-36.0 ± 1.6 bars for *S. leucophylla* vs. -25.5 ± 3.1 bars for *S. mellifera;* $\bar{x} \pm 95\%$ confidence interval, n $= 20$; K. Preston, pers. comm.).

Thus where finer-textured soils and poleward-facing aspects co-occur within the Venturan geographic range, one could expect *Salvia leucophylla* and associates (Venturan II) to replace *Salvia mellifera* and associates (Venturan I). In this study, there were only three samples out of 21 in which a Venturan I association occurred on N aspects, and in each case, the soil was coarse-textured ($\leqslant 4.0$ texture index of Westman 1975, where 1 = coarse sand, 10 = clay and 4 = loam). Conversely, there were four cases out of 13 in which Venturan II associations occurred on S aspects, and in each case soils were fine-textured ($\geqslant 4.0$). The loamy texture (index = 4.0) appears to be the boundary point of this distinction, where the two subassociations can

Table 3. Mean values ± standard error for habitat variables associated with major floristic associations.

	Coastal sage scrub						Coastal succulent scrub	
	Diablan	Venturan I	Venturan II	Venturan Combined	Riversidian	Diegan	Martirian	Vizcainan
Number of sites sampled	13	21	13	34	19	12	8	6
Vegetation structure								
Litter mass, g dry mass m²	857 ± 142	2 154 ± 261	1 512 ± 957	1 909 ± 196	473 ± 85	906 ± 125	606 ± 146	648 ± 146
Average canopy height (m)	1.0 ± 0.09	0.9 ± 0.05	1.0 ± 0.09	0.9 ± 0.05	0.9 ± 0.05	1.0 ± 0.07	1.0 ± 0.13	0.7 ± 0.13
% light penetration to 10 cm aboveground	38 ± 4	40 ± 3	39 ± 2	40 ± 2	54 ± 4	46 ± 5	52 ± 4	72 ± 4
Soil variables								
Bulk density (g m⁻³)	1.21 ± 0.08	1.32 ± 0.05	1.24 ± 0.05	1.29 ± 0.04	1.55 ± 0.03	1.47 ± 0.05	1.41 ± 0.05	1.42 ± 0.07
Field capacity (half saturation, %)	26 ± 3	25 ± 2	28 ± 2	26 ± 1	18 ± 1	20 ± 1	18 ± 1	20 ± 1
pH	5.9 ± 0.2	6.8 ± 0.1	6.7 ± 0.1	6.8 ± 0.1	6.8 ± 0.1	6.3 ± 0.2	6.4 ± 0.2	7.4 ± 0.4
Salinity (mS cm⁻³)	0.6 ± 0.3	0.4 ± 0.05	0.3 ± 0.04	0.4 ± 0.04	0.3 ± 0.03	0.3 ± 0.02	0.2 ± 0.03	1.0 ± 0.23
Total nitrogen, %	0.22 ± 0.02	0.23 ± 0.02	0.24 ± 0.02	0.23 ± 0.01	0.08 ± 0.01	0.17 ± 0.01	0.10 ± 0.01	0.06 ± 0.01
Available ammonium-nitrogen (μg g)	3.7 ± 0.7	2.1 ± 0.4	2.0 ± 0.4	2.1 ± 0.3	2.9 ± 0.4	3.8 ± 0.8	3.8 ± 0.5	6.2 ± 1.6
Available nitrate-nitrogen (μg g)	14 ± 1	11 ± 1	10 ± 2	11 ± 1	15 ± 2	12 ± 2	12 ± 1	17 ± 3
Available base-extractable phosphate as P (μg g)	31 ± 8	29 ± 5	25 ± 6	28 ± 4	20 ± 3	16 ± 3	11 ± 2	16 ± 5
Available potassium (μg g)	176 ± 40	126 ± 16	142 ± 29	132 ± 15	108 ± 17	174 ± 26	119 ± 24	269 ± 75
Available calcium (μg g)	1 350 ± 160	3 890 ± 640	3 730 ± 590	3 830 ± 450	1 820 ± 300	1 240 ± 170	1 140 ± 190	2 290 ± 1 050
Available magnesium (μg g)	483 ± 108	724 ± 134	628 ± 133	687 ± 96	216 ± 41	277 ± 59	468 ± 85	238 ± 62
Texture (1 = sand, 10 = clay)	3.7 ± .06	4.1 ± 0.4	6.1 ± 0.4	4.9 ± 0.3	2.7 ± 0.4	3.1 ± .03	3.8 ± 0.8	4.0 ± 1.1
Climate								
Mean temperature, coldest month (°C)	9.1 ± 0.4	11.1 ± 0.4	10.5 ± 0.3	10.9 ± 0.3	9.3 ± 0.4	11.4 ± 0.2	11.9 ± 0.4	13.7 ± 0.4
Mean minimum temperature, coldest month (°C)	3.1 ± 0.4	5.7 ± 0.6	4.5 ± 0.6	5.2 ± 0.4	2.0 ± 0.6	5.3 ± 0.3	*	*
Minimum temperature, coldest month (°C)	2.1 ± 0.4	0.6 ± 0.8	0.4 ± 0.7	0.2 ± 0.6	4.4 ± 0.6	0.3 ± 0.4	*	*
Mean temperature, warmest month (°C)	19.2 ± 0.5	21.8 ± 0.5	21.2 ± 0.7	21.5 ± 0.4	27.0 ± 0.6	21.7 ± 0.4	21.9 ± 0.5	24.1 ± 2.0
Mean maximum temperature, warmest month (°C)	26.8 ± 0.8	27.6 ± 1.1	27.7 ± 1.3	27.6 ± 0.8	36.6 ± 0.5	26.1 ± 1.0	*	*
Maximum temperature, warmest month (°C)	35.5 ± 0.8	33.7 ± 1.2	33.6 ± 1.4	33.7 ± 0.9	41.8 ± 0.5	32.1 ± 1.3	*	*
Mean annual precipitation (cm)	58.5 ± 4.8	36.2 ± 1.8	33.9 ± 1.5	35.3 ± 1.3	25.3 ± 1.9	24.7 ± 1.0	18.6 ± 1.6	11.3 ± 0.9
Mean precipitation driest month (cm)	0.14 ± 0.04	0.0 ± 0.0	0.0 ± 0.0	0.0 ± 0.0	0.02 ± 0.01	0.08 ± 0.07	0.30 ± 0.19	0.0 ± 0.0
Mean precipitation, wettest month	20.3 ± 1.4	15.5 ± 0.6	14.0 ± 0.8	14.9 ± 0.5	9.0 ± 0.8	8.7 ± 0.5	4.3 ± 0.3	3.7 ± 0.6
Air pollution (annual mean concentration)								
Carbon monoxide (μL/L)	1.9 ± 0.3	5.6 ± 0.9	4.5 ± 0.6	5.0 ± 0.5	5.6 ± 0.4	2.3 ± 0.4	*	*
Total hydrocarbons (μL/L)	2.6 ± 0.2	4.6 ± 1.2	2.7 ± 0.0	3.9 ± 0.8	5.1 ± 0.6	1.8 ± 0.3	*	*
Nitric oxide (μL/L)	0.019 ± 0.004	0.037 ± 0.003	0.047 ± 0.013	0.042 ± 0.007	0.044 ± 0.013	0.059 ± 0.017	*	*
Nitrogen oxides (μL/L)	0.042 ± 0.007	0.139 ± 0.046	0.091 ± 0.015	0.121 ± 0.029	0.225 ± 0.029	0.104 ± 0.016	*	*
Nitrogen dioxide (μL/L)	0.028 ± 0.005	0.049 ± 0.009	0.065 ± 0.014	0.054 ± 0.007	0.088 ± 0.016	0.032 ± 0.005	*	*
Ozone (μL/L)	0.04 ± 0.004	0.07 ± 0.009	0.06 ± 0.003	0.07 ± 0.006	0.13 ± 0.013	0.05 ± 0.007	*	*
Particulate (μg/cm³)	61.1 ± 4.7	73.4 ± 4.0	78.6 ± 3.5	75.3 ± 2.8	97.0 ± 4.7	66.0 ± 9.1	*	*
Sulfur dioxide (μL/L)	0.012 ± 0.001	0.032 ± 0.010	0.011 ± 0.003	0.023 ± 0.007	0.025 ± 0.006	0.001 ± 0.001	*	*
Nitrogen dioxide-sulfur dioxide index	1.4 ± 0.0	3.1 ± 0.9	1.2 ± 0.3	2.3 ± 0.6	2.4 ± 0.6	0.3 ± 0.1	*	*
Ozone-sulfur dioxide index	9.8 ± 0.6	27.9 ± 7.9	13.1 ± 1.1	21.6 ± 5.2	26.9 ± 3.6	9.5 ± 1.3	*	*

* Data unavailable.

occur on any aspect. In addition, in this study the Venturan I sites did occur on the average in areas of slightly higher mean annual precipitation (36.2 cm ± 1.8) than Venturan II sites (33.9 ± 1.5). Since the two types can occur on the same hillsides, however, these slight differences in precipitation may simply be artifacts of sample size. Indeed, on a given hillside, unmeasured factors such as runon and runoff, sun angle and shading by adjacent slopes, and substrate depth also contribute to variation in site moisture availability. While the dominant *Salvias* appear to be segregating by moisture preference at a local scale, the factors predominant in influencing moisture stress at a particular site will almost certainly vary.

Encelia californica, a Venturan I codominant, also has a preference for coarse-textured soils. However, it appears to have a moisture preference comparable to *Salvia leucophylla* (Westman, 1981a), so that some other substrate variable, such as its apparent preference for higher phosphate levels (Westman, 1981a; Table 3) may explain its preference for Venturan I. Competitive relations arising from the presence of the dominant *Salvias* (including allelopathy) also may be influencing the occurrence of associated species (see Discussion).

Pearson and rank (Spearman and Kendall's tau) correlations between the 43 environmental variables and the DCA axes of the 99 sites revealed results similar to those from ordinating 67 sites (excluding most of the coastal succulent scrub sites) by reciprocal averaging (Westman, 1981a). The first DCA axis is correlated (Pearson values, all significant at $P < 0.01$) with the NW–SE coastline trend (latitude, $r = -0.72$; longitude, $r = -0.66$), which in turn reflects a gradient of decreasing annual ($r = -0.64$) and monthly maximum ($r = -0.72$) precipitation. As can be seen by comparing Figures 1 and 3, the first axis essentially ranks the samples in their geographic positions relative to each other. In addition, there is a secondary trend toward declining total soil nitrogen levels (total N, $r = -0.47$). Rank correlations produced the same conclusions. The second axis is also somewhat latitudinal ($r = 0.50$), but as can be seen from Figure 3, the major effect is to separate the Riversidian from Diegan, and Diablan from Venturan samples. Because of the predominant offshore winds in summer and the topography of the basin, air pollutants generated in the coastal cities of Los Angeles and Orange Coun-

ties (Venturan, Diegan) are blown inland and accumulate in Riverside and San Bernardino Counties (Riversidian) (Table 3). As a result, the second axis contrast between Riversidian and other associations (except Diablan) results in a strong correlation between this axis and a variety of air pollutants ($r > 0.40$ for ozone, carbon monoxide, sulfur dioxide, hydrocarbons and the NO_2-SO_2 index).

Discussion

The analysis has confirmed the existence of two plant formations of distinct physiognomy within the xeric shrublands of the Mediterranean climatic zone of Pacific North America: coastal sage scrub and coastal succulent scrub. Four associations (Diablan, Venturan (with two subassociations), Riversidian and Diegan) have been recognized for coastal sage scrub on floristic criteria, and two more tentatively identified for coastal succulent scrub (Martirian, Vizcainan). I have used the term 'association' in this study to refer to large floristic groups, consistent with the usage of Axelrod (1950, 1978). While these larger floristic groupings were also recognized by Kirkpatrick & Hutchinson (1977), their use of the term 'association' referred to much smaller groups within the larger 'associations' referred to here. I am using the term 'formation' to refer to groups defined primarily by differences in physiognomy.

The associations show distinct geographic locations following the Pacific coast, with one inland association in the Riversidian region. These relationships are mainly correlated with changes in evapotranspirative stress, although total soil nitrogen is also correlated with the main axis of variation, and air pollutants increase in the Riversidian region. Further sampling of the northern coastal scrub and adjacent chaparral formations in northern California may reveal the need to revise the northern (Diablan) group, but for the present, the Lucian and Franciscan groups of Axelrod (1978) are considered part of one or both of the latter formations, based on their dominance by evergreen, sclerophyllous shrubs.

One of the more interesting results of this study has been the identification of two distinct subassociations within the Venturan group which intermingle at a regional geographical scale, but often meet along sharp boundaries at a local scale. Were subassociations, and the vegetational discontinuities between them, due simply to a change in some habitat feature, one would expect to observe the same phenomenon in the other floristic associations. That one does not (see Fig. 2) suggests that biotic factors, as well as habitat discontinuities, may be causing the vegetation pattern.

The observed suppression of richness and equitability in the Venturan sites is also not well explained by a habitat-discontinuity hypothesis. The latter implies that the 45 associated species not held in common between the two subassociations (Table 2) have evolved to segregate along lines of habitat preference similar to their respective dominants for independent, exclusively physical reasons. It requires postulating concurrent forces of natural selection acting to segregate these 45 species sharply into two groups along a particular habitat factor axis. Furthermore the number of species in fact differentiating the two groups is substantially larger than 45, since there are significant differences in abundance among species which occur in both subassociations (e.g. *Encelia californica;* see lower PS than CC values in Results; also Fig. 4).

An alternative hypothesis to explain the observations is that *Salvia mellifera* and *S. leucophylla* initially segregate in the landscape according to their different moisture preferences, as influenced on the sites by slope position, soil texture and aspect (with vegetative growth further modifying the aspect effect). The respective dominants then influence the associated flora differentially in any of several ways. Possibilities include allelopathic effects of the dominant *Salvias* on their understories (Muller & Muller, 1964; Muller, 1966); partitioning of certain herbivores between the two *Salvias* based on differences in cover protection, or nesting or feeding opportunities, with subsequent impact on the respective understory floras through differences in herbivory; differential effects on understory growth arising from differences in light interception, or moisture or nutrient flow, at the base of *Salvia* shrubs. In the case of allelopathic releases, the influence may be directly on herb species germination or growth as postulated by Muller (1966), or indirect, through influences on soil microorganisms (see Kaminsky, 1981 for an example involving *Adenostoma fasciculatum*). Clearly these alternative mechanisms require further investigation.

The biotic-control hypothesis can explain the three major observations of this study:
(1) the occurrence of relatively sharp boundaries between subassociations at a local scale, but the co-occurrence of these subassociations throughout the Venturan range;
(2) the suppression of species richness and equitability in Venturan sites relative to sites containing other associations in the formation;
(3) the occurrence of a significant number of species co-occurring throughout the Venturan range, but not occurring in both subassociations.
The habitat-discontinuity hypothesis explains neither the observed suppression of species richness, nor the fact that this sharp species segregation by dominants occurs only in the Venturan association. Thus there would appear to be some evidence for the kind of biotic interaction that has given fuel to supporters of the Clementsian community concept over the years.

This set of phenomena occurs only in one association of the six discussed here, and cannot therefore be taken as a predominant trend to counter the widespread application of the principles of species individuality and community continuity fostered by Gleason (1926), Whittaker (1951, 1967) and Curtis (1959). Furthermore, the *Salvias* and *Encelia californica* have been shown to vary continuously in importance along moisture gradients at a large scale (Westman, 1981a). Also the spread of sites seen by ordination of the Venturan association (Fig. 3) indicates that the composition of these two subassociations is variable between sites. Nevertheless, that tendencies for species association exist and are, in some cases, apparently mediated by biotic interactions initiated by the dominant species, is significant. Their existence does suggest that the ultimate basis for the great debate between the proponents of holistic communities and proponents of continua in the last 60 years was the presence in nature of evidence for both points of view.

References

Axelrod, D. I., 1950. Contributions to paleontology. I. Classification of the Madro-Tertiary geoflora. Carnegie Inst. Washington Pub. 590: 1–22.

Axelrod, D. I., 1978. The origin of coastal sage vegetation, Alta and Baja California. Amer. J. Bot. 65: 1117–1131.

94

Bray, J. R. & Curtis, J. T., 1957. An ordination of the upland forest communities of southern Wisconsin. Ecol. Monogr. 27: 325–349.

Curtis, J. T., 1959. The vegetation of Wisconsin: an ordination of plant communities. 657 pp. Univ. Wisc. Press, Madison.

Dixon, W. J. & Brown, M. B. (eds.), 1979. Biomedical Computer Programs, P-series. 880 pp. Univ. Calif. Press, Berkeley-Los Angeles.

Dunkle, M. B., 1950. Plant ecology of the Channel Islands of California. Pub. Allan Hancock Pac. Exped. 13: 247–386.

Gauch, H. G., Jr. & Whittaker, R. H., 1981. Hierarchical classification of community data. J. Ecol. 69: 537–557.

Gauch, H. G., Jr., Whittaker, R. H. & Singer, S. B., 1981. A comparative study of nonmetric ordinations. J. Ecol. 69: 135–152.

Gleason, H. A., 1926. The individualistic concept of the plant association. Bull. Torrey Bot. Club 53: 7–26.

Hastings, J. R. & Humphrey, R. R. (eds.), 1969. Climatological data and statistics for Baja California. Meteorology and Climatology of Arid Regions Tech. Rep. 18, Inst. Atmos. Physics, Univ. Ariz., Phoenix.

Heady, H. F., Foin, T. C., Hektner, M. M., Taylor, D. W., Barbour, M. G. & Barry, W. J., 1977. Coastal prairie and northern coastal scrub. In: Barbour, M. G. & Major, J. (eds.) Terrestrial Vegetation of California. Wiley, N.Y. 733–762.

Hill, M. O., 1973. Reciprocal averaging: an eigenvector method of ordination. J. Ecol. 61: 237–249.

Hill, M. O., 1979. TWINSPAN. A FORTRAN program for arranging multivariate data in an ordered two-way table by classification of the individuals and attributes. 90 pp. Cornell Univ., Ithaca.

Hill, M. O., 1979b. DECORANA. A FORTRAN program for detrended correspondence analysis and reciprocal averaging. 52 pp. Cornell Univ., Ithaca.

Hill, M. O. & Gauch, H. G., Jr., 1980. Detrended correspondence analysis: an improved ordination technique. Vegetatio 42: 47–58.

Jackson, M. L., 1958. Soil chemical analysis. 498 pp. Prentice-Hall, Englewood Cliffs, N.J.

Kaminsky, R. A., 1981. The microbial origin of the allelopathic potential of Adenostoma fasciculatum H & A. Ecol. Monogr. 51: 365–382.

Kirkpatrick, J. B. & Hutchinson, C. F., 1977. The community composition of Californian coastal sage scrub. Vegetatio 35: 21–33.

Kirkpatrick, J. B. & Hutchinson, 1980. The environmental relationships of Californian coastal sage scrub. J. Biogeogr. 7: 23–28.

Mooney, H. A., 1977. Southern coastal scrub. In: Barbour, M. G. & Major, J. (eds.) Terrestrial vegetation of California. pp. 471–490. Wiley, N.Y.

Mooney, H. A. & Harrison, A., 1972. The vegetation gradient on the lower slopes of the Sierra San Pedro Martir in northwest Baja California. Madroño 21: 439–445.

Mueller-Dombois, D. & Ellenberg, H., 1974. Aims and methods of vegetation ecology. 547 pp. Wiley, N.Y.

Muller, C. H., 1966. The role of chemical inhibition (allelopathy) in vegetational composition. Bull. Torrey Bot. Club 93: 332–351.

Muller, W. H. & Muller, C. H., 1964. Volatile growth inhibitors produced by Salvia species. Bull. Torrey Bot. Club 91: 327–330.

Munz, P. A., 1974. A flora of southern California. 1 086 pp. Univ. Calif. Press, Berkeley-Los Angeles-London.

Munz, P. A. & Keck, D. D., 1959. A California flora. 1 681 pp. Univ. Calif. Press, Berkeley-Los Angeles.

Philbrick, R. N. & Haller, J. R., 1977. The southern California islands. In: Barbour, M. G. & Major, J. (eds.) Terrestrial vegetation of California. pp. 893–906. Wiley, N.Y.

Poole, D. K. & Miller, P. C., 1975. Water relations of selected species of chaparral and coastal sage communities. Ecology 56: 1118–1128.

Power, D. M. (ed.), 1980. The California islands: proceedings of a multidisciplinary symposium. 787 pp. Sta. Barbara Museum Nat. Hist., Sta. Barbara, Ca.

Simpson, E. H., 1949. Measurement of diversity. Nature 163: 688.

Singer, S. B., 1980. DATAEDIT. A FORTRAN program for editing data matrices. Cornell Univ., Ithaca.

Singer, S. B. & Gauch, H. G., Jr., 1979. CONDENSE. Convert data matrices from any ORDIFLEX format into a condensed format by samples. 7 pp. Cornell Univ., Ithaca.

Thorne, R. F., 1976. The vascular plant communities of California. In: Latting, J. (ed.) Plant communities of southern California. pp. 1–31. Calif. Native Plant Soc., Berkeley.

UNESCO-FAO, 1963. Bioclimatic map of the Mediterranean zone. Ecological study of the Mediterranean zone. Arid Zone Res. 21. UNESCO, Paris.

Walter, H., 1973. Vegetation of the earth in relation to climate and ecophysiological conditions. English transl. Springer-Verlag, New York-Heidelberg-Berlin.

Westman, W. E., 1979a. A potential role of coastal sage scrub understories in the recovery of chaparral after fire. Madroño 26: 64–68.

Westman, W. E., 1979b. Oxidant effects on Californian coastal sage scrub. Science 205: 1001–1003.

Westman, W. E., 1980. Gaussian analysis: identifying environmental factors influencing bell-shaped species distributions. Ecology 61: 733–739.

Westman, W. E., 1981a. Factors influencing the distribution of species of California coastal sage scrub. Ecology 62: 439–455.

Westman, W. E., 1981b. Diversity relations and succession in Californian coastal sage scrub. Ecology 62: 170–184.

Westman, W. E., 1981c. Seasonal dimorphism of foliage in Californian coastal sage scrub. Oecologia 51: 385–388.

Westman, W. E., 1982a. Coastal sage scrub succession. Proc. Symp. Dynamics and Management of Mediterranean-type Ecosystems. U.S.D.A. Forest Serv., Pac. S.W. For. & Range Expt. Sta., Berkeley. In press.

Westman, W. E., 1982b. Island biogeography: studies on the xeric shrublands of the inner Channel Islands, California. J. Biogeogr. :submitted.

Westman, W. E., 1982c. Plant community structure – spatial partitioning of resources. In: Kruger, F. J., Mitchell, D. T. & Jarvis, J. U. M. (eds.), Mediterranean Type Ecosystems. The Role of Nutrients. Springer-Verlag: Berlin-Heidelberg-New York. In press.

Westman, W. E. & Peet, R. K., 1982. Robert H. Whittaker (1920–1980): the man and his work. Vegetatio 48: 97–122.

Whittaker, R. H., 1951. A criticism of the plant association and climatic climax concepts. Northwest Sci. 26: 17–31.

Whittaker, R. H., 1960. Vegetation of the Siskiyou Mountains, Oregon and California. Ecol. Monogr. 30: 279–338.

Whittaker, R. H., 1967. Gradient analysis of vegetation. Biol. Rev. 42: 207–264.

Whittaker, R. H., 1972. Evolution and measurement of species diversity. Taxon 21: 213–251.

Whittaker, R. H., 1975. Communities and Ecosystems. 2nd ed. 385 pp. Macmillan, N.Y.

Whittaker, R. H., Niering, W. A. & Crisp, M. D., 1979. Structure, pattern, and diversity of a mallee community in New South Wales. Vegetatio 39: 65–76.

Wiggins, I. L., 1980. Flora of Baja California. 1 025 pp. Stanford Univ. Press, Stanford.

Zabriskie, J. G., 1979. Plants of Deep Canyon and the Central Coachella Valley, California. 175 pp. P. L. Boyd Deep Canyon Desert Res. Ctr., Univ. Calif., Riverside.

Accepted 7.7.1982.

Regional and local variation in tallgrass prairie remnants of Iowa and eastern Nebraska*

Jon A. White[1] & David C. Glenn-Lewin**
Department of Botany, Iowa State University, Ames, IA, 50010, U.S.A.
[1] *Present address: Department of Biology, 010A, University of North Carolina, Chapel Hill, NC, 27514, U.S.A.*

Keywords: Gradient analysis, Iowa, Nebraska, Ordination, Quaternary landscapes, Tallgrass prairie, Vegetation

Abstract

Tallgrass prairie vegetation, persisting as numerous small relics scattered across central North America, exhibits a complex pattern of community structure and composition. We sampled vegetation from 11 prairie preserves in Iowa and eastern Nebraska spanning three physiographic regions and a wide variety of upland habitats. Numerical classification and partitioned ordination revealed a complex pattern of both local and geographical variation in the vegetation. The primary coenocline is a complex topographic-moisture gradient. Secondary factors influencing the vegetation include subsoil permeability, chorological differences, and local peculiarities of stand history and dynamics. Local patterns of interaction between the primary and secondary environmental factors vary among Quaternary landscapes resulting in regional vegetation complexity.

Introduction

North American tallgrass prairie vegetation exhibits substantial local as well as geographic variation in composition and structure. This variation in prairie vegetation has been related to topographic position, both in traditional classification studies (e.g. Clements, 1920; Weaver & Fitzpatrick, 1934; Weaver, 1954), and in quantitative gradient analyses (e.g. Curtis, 1955, 1959; Dix & Smeins, 1967; Ayyad & Dix, 1966; Redmann, 1972; Crist & Glenn-Lewin, 1978). Both approaches have emphasized soil moisture as the fundamental environmental factor. This relationship to moisture has led to

* Nomenclature follows Gleason (1952) except for the grasses which follow Pohl (1966).
** Supported by the National Science Foundation (No. GU 2795 and GU 3373), The Nature Conservancy, and the Iowa State Advisory Board for Preserves. We thank T. L. Roberts, S. Taylor, and S. Lundquist for help with field work, and R. K. Peet, A. T. Harrison, and J. K. White for advice, comments, and criticisms.

widespread use of topographic gradients as representative of regional variation in vegetation (Weaver & Fitzpatrick, 1934; Curtis, 1955, 1959; Dix & Smeins, 1967; Ralston, 1968).

It is clearly established that local topographic coenoclines represent complex gradients in which slope position interacts with soil texture and internal drainage (Whitford, 1958; Bliss & Cox, 1964), degree of local relief and external drainage patterns (Whitford, 1958; Brotherson, 1969), soil depth (Dix & Butler, 1960), salinity (Redmann, 1972), and calcium carbonate abundance as influenced by erosion or hydrologic recharge (Brotherson, 1969; Crist & Glenn-Lewin, 1978). Most quantitative analyses of prairie vegetation have used a compositional index to indicate stand or species position on the moisture gradient (Curtis, 1955, 1959; Dix & Smeins, 1967; Ralston, 1968). With this method species most strongly indicative of soil moisture status are used to determine a continuum value (Curtis, 1955, 1959) for each community sample. Though regional studies have acknowledged complexity of the moisture gradient, most have used average composition in coenocline segments to smooth variation. Rather than random fluctuation, however, the noise seen along the continuum often reflects unidentified factors or processes influencing vegetation distribution. These may be lost in the application of the continuum index approach (Peet & Loucks, 1977).

In this study, we employ both locally concentrated and regionally dispersed vegetation samples to simultaneously examine local and regional variation in tallgrass prairie remnants of Iowa and eastern Nebraska. We first use a numerical classification as a basis for describing the vegetation of these prairies, and then an ordination approach to identify important trends. In this way geographic and regional landscape patterns in vegetation demonstrated by broad trends among groups of stands can be identified and compared with patterns among subsets of locally concentrated stands.

Study region

The prairies of Iowa and eastern Nebraska (Fig. 1; ca. 40–44° N 90–98° W) lie in the western portion of the prairie peninsula described by Transeau (1935). Although prairie vegetation blanketed a great majority of the region before European settlement (Dick-Peddie, 1955), only scattered tracts remain. The 11 locations sampled (Fig. 1) include nine of the largest and best preserved in Iowa plus two important tracts in eastern Nebraska.

Climate

The climate of Iowa (Waite, 1978) and Nebraska (Stevens, 1978) is strongly continental with marked seasonal differences in precipitation and temperature. Within the study region, most climatic factors vary from southwest (warmer and drier) to northeast (cooler and wetter). Climate diagrams from the geographical extremes (Fig. 2) illustrate the main climatic features and the degree of climatic difference. In particular, vegetation in the west is likely to experience more frequent and more extreme summer drought than that to the east.

Summers are dominated by warm, moist, Gulf air, and winters by cold, dry Canadian air. Droughts associated with protracted summer incursions of hot, dry air from the southwest occur in cycles of several years (Borchert, 1950; Stevens, 1978); Waite, 1978). During these droughts, rainfall may be virtually nil and daily temperatures high throughout the summer.

Regional landscapes

The land surface of Iowa and eastern Nebraska reflects a history of glacial and interglacial episodes (Wright & Ruhe, 1965; Reed et al., 1965; Ruhe, 1969). Several Quaternary landscapes can be recognized, each distinguished by parent material, relative degree of surface weathering, age, and geographic extent (Fig. 1). The sampled prairies are located on 3 principal landscapes (Ruhe, 1969): the Cary Drift Plain, the Iowan Erosion Surface, and the Kansan Drift Plain.

The Cary Drift Plain is the most recently glaciated landscape (ca. 14 000 yr). It has low relief (3–20 m) and contains numerous shallow, often flooded depressions termed potholes. Terminal and end moraines provide somewhat greater relief on parts of the plain. Drainage patterns, except on the moraines, are poorly defined.

The Iowan Erosion Surface, a gently rolling landscape in the northeastern and northwestern portions of Iowa, was formed during the last glaciation by erosion into glacial drift deposited during earlier glacial episodes. At the end of the last glaciation, this older drift was covered by a thin loess deposit or by a loamy sediment. The underlying drift, being more weathered than the Cary Drift, is high in clay and often forms an impediment to drainage. Scattered ridges of aeolian sand occur on this landscape. Drainage nets are better defined than on the Cary Drift Plain.

The older, loess-mantled Kansan Drift (glaciated 600 000 yr B.P.) is well eroded and thus more hilly than the others. Because of this relief, drainage nets are well defined. During the last glaciation, the hills were covered with loess, which originated from the Missouri River valley. The loess decreases in depth from many meters in western Iowa to a few centimeters in the eastern part of the state (Ruhe, 1969). There was relatively little loess deposition in eastern Nebraska, and the relief there is not as great as in southern and southwestern Iowa. Erosion has removed the loess from slopes, exposing the ancient Kansan drift and paleosols formed in the drift. The remaining loess caps the hills in this landscape.

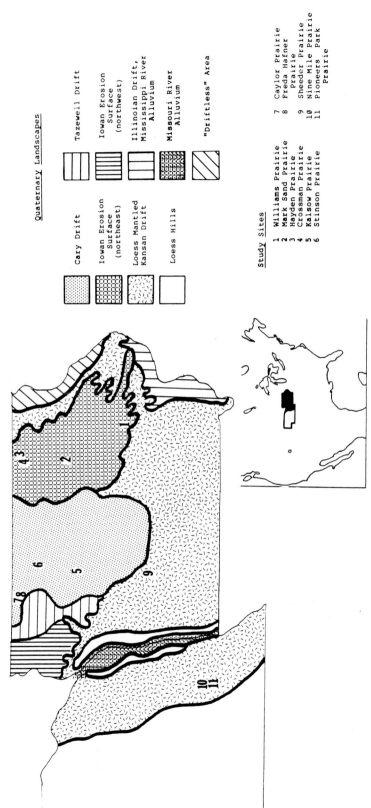

Fig. 1. Regionally distributed Quaternary landscapes in the study region and locations of study sites. Redrawn after Ruhe (1969) and Reed *et al.* (1965).

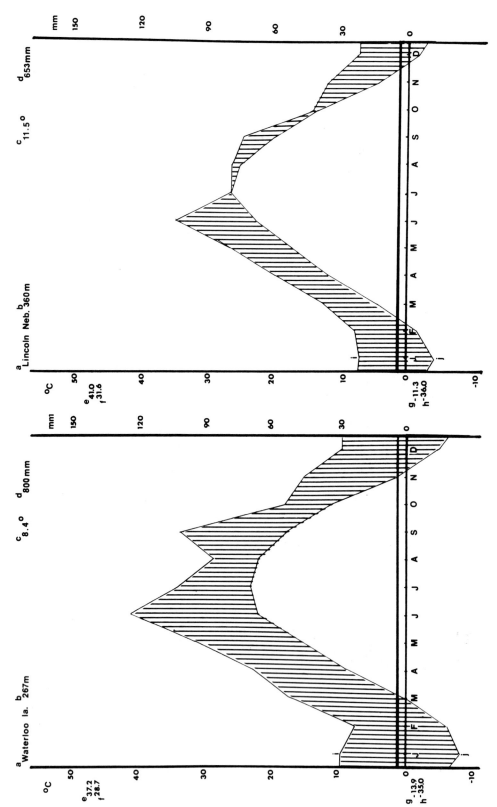

Fig. 2. Representative climate diagrams (after Walter, 1979) for the northeastern and southwestern extremes of the study region (Waterloo, Iowa and Lincoln, Nebraska, resp.), illustrating the principal pattern of geographic variation in climate. a, station; b, elevation; c, mean annual temperature; d, mean annual precipitation; e, maximum recorded temperature; f, mean daily maximum temperature of the warmest month; g, mean daily minimum of the coldest month; h, coldest recorded temperature; i, curve of mean monthly precipitation; j, curve of mean monthly temperature. Months in which daily minimum temperatures fall below freezing are indicated by a black bar along the month axis, while months in which at least one frost occurs are indicated by a diagonally shaded bar.

Methods

Field methods

Within each study site, samples were distributed along transects following topographic gradients (except Crossman and Pioneers' Park Prairies where only 1 and 2 samples, respectively, were recorded). Samples were placed at variable intervals to best represent the range of variation (cf. Whittaker, 1978), and the number of samples per transect varied from 2 to 13 according to the degree of topographic relief.

Although the main object of sampling was upland prairie, some lowland samples were included. The wet prairie samples form a heterogeneous group which does not contain full representation of lowland prairie types. However, taken together they clearly establish the moist end of the topographic vegetation gradient.

Eleven prairies were sampled using a total of 138 samples, each a 2×10 m plot placed with the long axis perpendicular to the topographic gradient (if any) to maximize sample homogeneity. All samples were visited 2 or 3 times during the growing season to record different phenological aspects. Species importance was estimated as the maximum cover observed during the year in the plot (average of estimates from subplots). We recorded 343 plant species but used only the 168 species occurring in 5 or more samples for the analysis.

Analytical methods

We used Two Way Indicator Species Analysis (TWINSPAN: Hill, 1979b; Gauch, 1982) to classify the samples into broad groups based on species composition. TWINSPAN is a divisive, polythetic technique that, like the tabular analysis of European phytosociologists, has the advantage of showing those species most effective in characterizing the types. To examine vegetation–environment relationships, we employed Detrended Correspondence Analysis (DCA or DECORANA: Hill, 1979a; Hill & Gauch, 1980; Gauch, 1982). DCA is especially suited for data with high beta diversity or limited compositional gaps (Gauch & Whittaker, 1981; Gauch et al., 1981).

Soils

Soil profile characteristics vary within and between the landscapes included in our study. The many interrelated physical and chemical characteristics of the soil profile are summarized by soil series. The soil series for each of the 138 plots was identified from small soil cores (Oschwald et al., 1965). U.S. Dept. of Agriculture county soil surveys (Buckner & Highland, 1974; Russell et al., 1974; Brown et al., 1980) integrate data for several moisture-related soil characteristics in each soil series into a few soil permeability classes to which we assigned a rank order scalar of 1 to 6. A topographic-drainage scalar employed by Dix & Smeins (1967) was used to characterize external drainage of each sample. The value of this scalar was determined using their key to drainage conditions, with values ranging from 1 for excessively drained to 6 for permanently incomplete drainage.

Results

Classification

The TWINSPAN classification divided the samples at two scales of environmental variation. The first two TWINSPAN divisions differentiate the wet and dry extremes in the vegetation, with a large intermediate mesic class (Fig. 3). After the two initial divisions, the TWINSPAN groups primarily reflect edaphically distinct sites. On dry sites sand or gravel prairies are distinguished. The mesic prairies divide first by degree of subsoil permeability and then by topographic position. The different substrates (sandy, gravelly, impeded, or unimpeded drainage) are generally associated with different regional landscapes.

Although TWINSPAN clusters are defined by species composition, characterization by 1 or 2 dominant or constant species is inadequate (except perhaps at the highest levels of division). Several graminoids including *Andropogon gerardii, A. scoparius, Sorghastrum nutans, Sporobolus heterolepis, Poa pratensis*, plus several species of *Carex* are dominant in many or all types, though they differ in relative abundance. Differences between types are often best expressed by subtle variations in relative abundances among a large group of species with

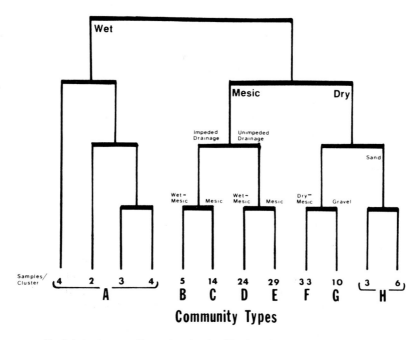

Fig. 3. A dendrogram illustrating the classification of samples by TWINSPAN.

Table 1. Average cover of common plant species by community type (see Fig. 3). Only species attaining at least 5% cover in any one type are included. (Full information is available on request.)

Species	Community type							
		Impeded		Unimpeded				
		Wet		Wet		Dry		
	Wet A	Mesic B	Mesic C	Mesic D	Mesic E	Mesic F	Gravel G	Sand H
Solidago gigantea	8.06	0.42		0.80				
Carex spp.	77.69	3.66	0.51	0.32	0.01			
Helianthus grosseseratus	14.42	3.16	2.96	8.56	1.47	0.20		
Calamagrostis canadensis	9.33	1.12	6.37					
Pycnanthemum virginianum	3.82	6.00	1.41	0.52	1.03			0.22
Carex media		7.02	2.64	1.94	1.00	0.90	0.11	1.20
Muhlenbergia racemosa	3.82	7:48	7.42	0.09	0.15	0.04		
Sporobolus heterolepis		10.12	44.33	4.11	16.22	6.47	2.28	0.22
Sorghastrum nutans		0.60	6.21	4.76	3.35	5.09		0.90
Andropogon gerardii	0.85	20.84	23.84	44.51	31.14	28.24	0.81	1.44
Solidago canadensis	14.69		0.20	19.01	0.23	0.39	0.60	
Poa pratensis	10.49	22.10	3.36	38.13	20.56	14.55	3.44	26.64
Amorpha canescens			4.21	0.23	7.32	3.88	2.81	1.18
Stipa spartea			0.33	0.55	11.68	10.11	9.45	4.20
Zizia aurea	1.15	0.64	7.99	6.44	9.30	1.41		
Andropogon scoparius		16.04	8.77	0.34	3.59	19.93	23.15	3.31
Muhlenbergia cuspidata						0.54	8.15	
Bouteloua hirsuta							8.90	0.64
Sporobolus cryptandrus								8.04
Poa compressa		0.20	1.70			0.49	0.40	15.63
Artemisia ludoviciana		0.20		0.38	0.57	0.19		11.33

intermediate to low importance. Consequently, the TWINSPAN classes do not generally represent dominance types that are clearly identifiable by one or two dominant species, or are identifiable by a few indicator species. To avoid confusion we use geographic and environmental designations for the community types described below. Table 1 presents species importance data for each type.

Wet prairies
Wet prairie type.
Wet prairie stands occupy a variety of low-lying situations including pothole edges, swales, and wet alluvial terraces. All the samples in this type are strongly dominated by *Carex* spp (average cover 78%). The grasses *Poa pratensis, Calamagrostis canadensis, Muhlenbergia racemosa*, and *Andropogon gerardii* (in order of decreasing importance) and various coarse herbs including *Helianthus grosseseratus, Solidago canadensis, S. gigantea*, and *S. graminifolia* occur as codominants in one to a few stands but are totally absent in others.

Mesic prairies
The mesic division is represented by low slope sample sequences from those prairies with relatively great local topographic relief (more than 10 m or slopes of >10%), all upland stands from Iowa prairies with only slight relief or fine textured soils, and the highest positions in otherwise wet local landscapes. Four community types are identified in this division, but they are so compositionally diverse that no simple characterization adequately distinguishes among them. Several important species are widespread reaching their maximum occurrence and cover in these types. Among the dominant grasses, *Andropogon gerardii* and *Sporobolus heterolepis* reach maxima in the mesic prairie.

Wet-mesic, impeded drainage type.
The relief of the Iowan Erosion Surface, although modest, is sufficient to allow recognition of two vegetation types, one on the lower slopes and a second occurring on upper slopes. Stands of this type, occur in low topographic positions on landscapes with gentle relief. Included are stands on alluvial soils of broad, shallow drainage ways, and low-slope positions adjacent to wet prairie communities. Though normally quite moist or even wet, these stands occasionally become dry. The stands share some species with the wet prairie; notably common are *Carex* spp., *Muhlenbergia racemosa*, and *Helianthus grosseseratus*. Many typical mesic prairie species are common, and *Andropogon scoparius*, although most abundant in dry prairies, is important. *Andropogon gerardii* and *Poa pratensis* codominate most stands, with *A. scoparius* of nearly equal importance in many stands. *Carex media, Muhlenbergia racemosa, Sporobolus het-*

erolepis, Helianthus grosseseratus, and *Pycnanthemum virginianum* are of intermediate importance in most stands.

Mesic, impeded drainage type.
These sites occur on upper slopes of the Iowan Erosion Surface. Slowly permeable subsoils derived from weathered Kansan glacial till result in seasonally perched water tables on broad divides and seepage on sideslopes. These heavy subsoils may also contribute to severe dryness during drought. *Sporobolus heterolepis* is the primary dominant, though *Andropogon gerardii* is often codominant and *A. scoparius* is occasionally important. *Amorpha canescens, Aster azureus, Carex media, Fragaria virginiana, Galium obtusum, Heliopsis helianthoides, Poa pratensis, Rosa suffolta, Sorghastrum nutans*, and *Zizia aurea* are common species of intermediate importance.

Wet-mesic, unimpeded drainage type.
A variety of sites on the Cary Drift make up this type, including the slight but broad topographic rises of a gently rolling ground moraine system in the central portion of the drift plain, lower slope positions in the complex junction of rolling drift plain and the terminal moraine of the drift where relief is much greater, and low stands near, but not in, drainage ways and potholes. Also included are sites from the best-drained positions on a wet alluvial terrace above the Iowa River (part of Williams Prairie).

Lowland and moist prairie forbs are very important in this community, often attaining more than 50% of the relative cover. *Andropogon gerardii* and *Poa pratensis* are consistently the most important grasses though *Sorghastrum nutans* and *Sporobolus heterolepis* may be locally abundant in some stands. Abundant forbs include *Aster ericoides, Fragaria virginiana, Galium obtusum, Helianthus grosseseratus, Ratibida pinnata, Solidago canadensis*, and *Zizia aurea*.

Mesic, unimpeded drainage type.
This type is composed of samples from gentle, moderately drained upper slopes throughout the Cary Drift Plain. The crests of small undulations from the pothole-pocked portion of the Cary Drift ground moraine landscapes, and all gentle, middle-to-upper slopes from end moraine systems are included in this type. Finally, a few plots on well-drained alluvium at one Kansas Drift Plain location (Sheeder Prairie), and two atypically well-drained stands from ridge tops of the Iowan Erosion Surface fall in this vegetation type.

The type is dominated by *Andropogon gerardii*. *Poa pratensis* codominates in some sites and *Sporobolus heterolepis* in others; *Stipa spartea* is typical on the crests of gently sloping hills and in other well-drained sites. *Andropogon scoparius* and *Sorghastrum nutans* follow in abundance. *Panicum scribnerianum* is an important small grass. Important forbs include *Amorpha canescens, Aster azureus*, and *Helianthus laetiflorus*, and in the slightly lower stands *Zizia aurea*.

Dry Prairies
Included in the dry prairie division are upper slopes of prairie landscapes with pronounced topographic relief, all sites with rapidly permeable soils on deep, coarse substrates well above the water table, and the western most tallgrass prairie stands sampled (in eastern Nebraska). This is a composi-

tionally heterogeneous group, but is characterized by high importance of *Andropogon scoparius* throughout. Because dry prairies have characteristically more open canopy and low stature, interstitial grasses and forbs (species of low stature) are much more important in species and abundance.

Dry–mesic type.

The well-drained upper slope positions of landscapes with significant local topographic relief (excluding sites with very coarse soils) support dry-mesic prairie vegetation. Portions of terminal moraines, upper slopes in end moraine systems, and slopes on the loess mantled Kansan Drift landscape contribute to this type, as do all the Nebraska plots sampled.

Dominance is variably shared among *Andropogon scoparius, A. gerardii, Stipa spartea, Sporobolus heterolepis, Poa pratensis,* and *Sorghastrum nutans. A. scoparius* often dominates on steep slopes and *S. spartea* on ridge crests and shoulder slopes, or they are co-dominant on sites intermediate between these. Each of *A. gerardii, S. heterolepis, S. nutans,* and *Poa pratensis* is dominant in one or two stands, but is generally subdominant throughout the type. *Bouteloua curtipendula* and *Panicum scribnerianum* are common interstitial grasses. Non-grass species important in the group are *Amorpha canescens, Aster ericoides, Coreopsis palmata, Helianthus laetiflorus, Rosa suffolta,* and *Solidago missouriensis.*

The Nebraska prairie stands are distinct from those in Iowa.

Several species are unique to the two Nebraska sites, the dominance of grasses is more pronounced than in Iowa, and forb cover is less. The two Nebraska sites themselves differ. One is on a xeric valley slope dominated by *Andropogon scoparius* and the other appears more mesic and is co-dominated by *A. gerardii* and *A. scoparius.*

Gravel hill prairie type.

The vegetation of xeric, gravelly, steep upper slopes associated with eskers and kettleholes common along terminal moraines of the Cary Drift is spare and low in stature. *Andropogon scoparius* is dominant on most such slopes, giving way to *Muhlenbergia cuspidata, Bouteloua gracilis,* and *B. hirsuta* on the steepest, driest sites and *Stipa spartea* on the slightly more mesic hilltops. Distinctive forbs include *Ambrosia psilostachya, Aster oblongifolius,* and *Liatris punctata,* but the more important species are *Amorpha canescens, Solidago missouriensis, S. rigida, S. nemoralis,* and *Erigeron canadensis.*

Sand prairie type.

Aeolian sand ridges represent another landform with a distinctive vegetation type. The deep, rapidly permeable, sand tapers off into surrounding wetlands, so that while the surface is very dry, there is ample moisture available at 2–3 m. Vegetation is very spare, a response to the xeric surface but perhaps also to past grazing. *Poa pratensis* and *P. compressa* are dominant grasses which indicate past disturbance, but the true sand prairie flora is represented by *Andropogon scoparius, Paspalum ciliatifolium, Sporobolus cryptandrus,* and several other grasses abundant on

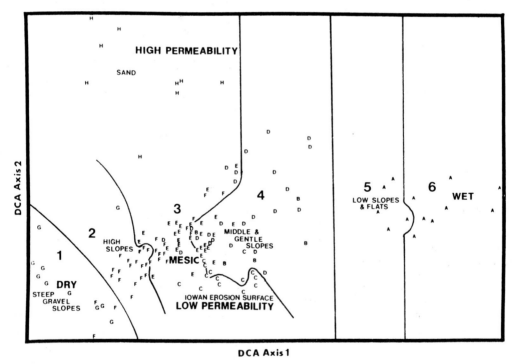

Fig. 4. Ordination of vegetation samples along the first two DCA axes. Isopleths mark the topographic moisture scalar of Dix & Smeins (1967), which ranges from 1, excessively drained, to 6, permanently incomplete drainage. The TWINSPAN groups identified in Fig. 3 are indicated by letters. A, Wet prairie; B, Wet-mesic, impeded drainage; C, Mesic, impeded drainage; D, Wet-mesic, unimpeded drainage; E, Mesic, unimpeded drainage; F, Dry-mesic; G, Gravel; H, Sand.

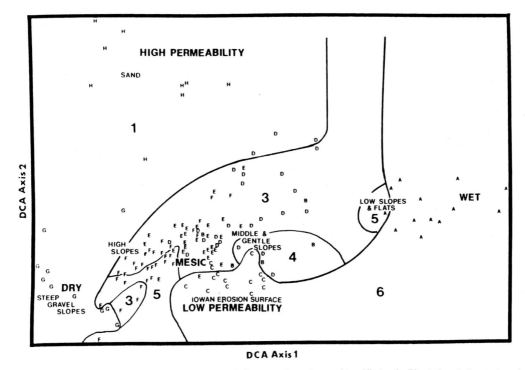

Fig. 5. Ordination of vegetation samples along the first two DCA axes. Samples are identified as in Fig. 4. Isopleths mark an internal drainage scalar based on rates of water percolation (cm/hr) through soil subhorizons. 1, >180 cm/hr; 2, 57–180 cm/hr; 3, 18–57 cm/hr; 4, 2.5–57 cm/hr; 5, 2.5–18 cm/hr; 6, <2.5 cm/hr.

dry prairies. Conspicuous forbs include *Liatris aspera, Ambrosia psilostachya,* and *Oenothera biennis.* A number of weedy species common in sand barrens are restricted to this type of prairie (Curtis, 1959).

Ordination

The ordination of all 138 tallgrass prairie stands on the first two DCA axes is shown in Figures 4 and 5. Individual stands are indicated by symbols representing the TWINSPAN community types (see Table 1). Labels indicate the environmental characteristics of selected regions of the ordination field.

First ordination

Like the first divisions in TWINSPAN, the DCA ordination at the largest scale reconstructs the predominant relationship between soil moisture and community composition in the tallgrass prairie (Figs. 4, 5). The wettest (right) and driest (left) communities are the compositional extremes on the first DCA axis. *Andropogon scoparius, Bouteloua* spp., and *Muhlenbergia cuspidata* (dry prairie spe-

cies) are important in the left portion of Figures 4 and 5, *Andropogon gerardii, Sporobolus heterolepis,* and *Poa pratensis* in the central portion, and *Carex* species toward the right. The distinctive composition of sand prairies (upper left of Fig. 4) accounts for the second dimension. The Nebraska stands (climatologically the most arid) are distinguished from the Iowa stands on the third axis (not shown).

Two important aspects of soil moisture are topographic position and soil permeability. The topographic moisture scalar (Dix & Smeins, 1967) is essentially unidirectional along the first axis (isopleths in Fig. 4), thus showing the relationship between topographic position and community composition. (F of stand position among topographic classes = 148, $p(>F) \ll 0.01$; percent of variation in stand position on the first axis among topographic classes = 89%.) The second ordination axis is strongly related to internal soil drainage with the highly permeable sands located at the top and the poorly permeable soils at the bottom (Fig. 5). A comparison of the two scalars shows them to be

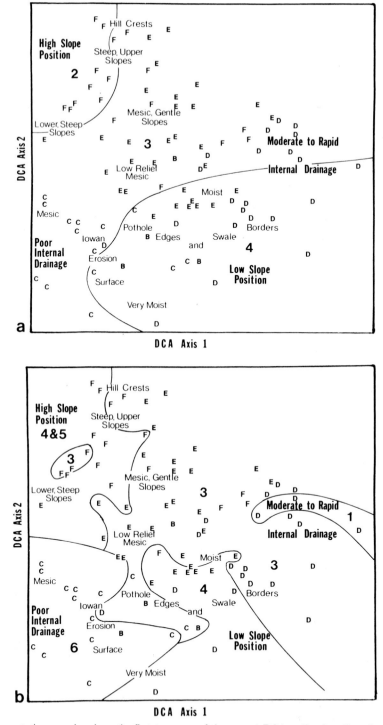

Fig. 6. Distribution of vegetation samples along the first two axes of the second DCA ordination. Samples and descriptors are in identical positions in each figure. Samples are designated by community type as in Fig. 4. Descriptors indicate environmental characteristics associated with ordination positions. Boldface descriptors indicate topographic and soil permeability gradient extremes. In (a) isopleths and boldface numerals mark the topographic gradient scalar as in Fig. 4. In (b) isopleths and boldface numerals delimit soil permeability classes as in Fig. 5.

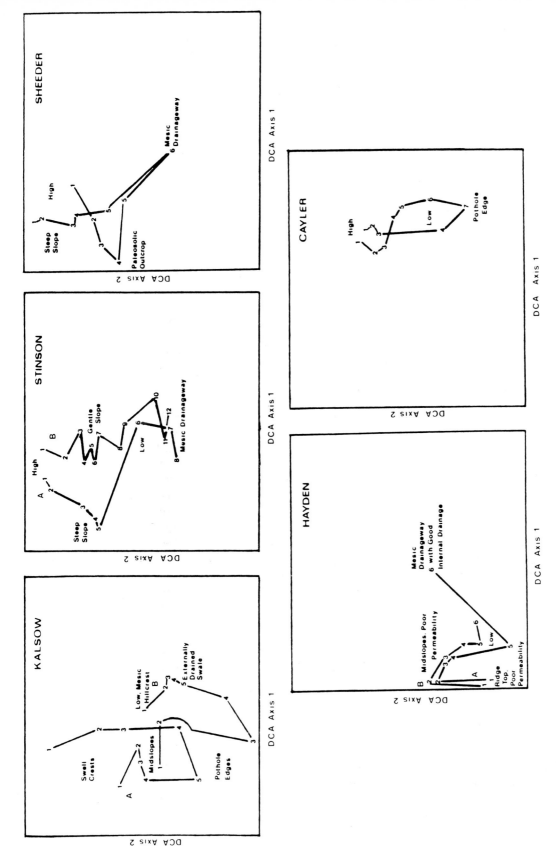

Fig. 7. The distribution of individual transects from five study sites along the first two axes of the second DCA ordination. The positions of individual samples are identical to those in Fig. 6. The numbers identifying samples represent relative topographic position within a transect. Separate transects from the same site are identified A or B.

correlated. Plots excessively drained by topographic position also have coarse, rapidly permeable soils, while plots samples at the lowest topographic positions have poorly permeable soils.

The vegetation pattern along the second axis, however, is not always consistent with this topography–permeability relationship. Among samples in intermediate topographic positions (central portion of Figs. 4, 5) there is a sequence from sites with poor subsoil permeability and moderate external drainage on topographic highs (lower center, *Sporobolus heterolepis* and *Andropogon gerardii* dominant) to topographically lower sites with moderately poor internal drainage (center, more dominance by *A. gerardii* and less by *S. heterolepis*) to topographically higher sites again, but with moderate to rapid internal drainage (upper center, *A. gerardii* and *Poa pratensis* dominant). Thus, the reciprocal reinforcement of topography and permeability found at the extremes of the moisture gradient breaks down in the middle.

Second ordination

Ordination of vegetation data with significant diversity of multiple dimensions of variation may reveal large-scale patterns of compositional and environmental variation, as described above, but it is often less helpful for distinguishing patterns among proximate stands in ordination space (Gauch *et al.*, 1981). Partitioning a complex data set by deleting extremes, and then reordinating the reduced data set is a useful technique for revealing fine-scale patterns not otherwise apparent among samples (Peet, 1980). Consistent with this approach, wet prairie, gravel hill prairie, and sand prairie samples, the compositional extremes of the first two DCA axes, were deleted. In addition, Nebraska samples were deleted because they formed a compositional extreme on the third DCA axis.

The first and second DCA axes for the remaining 87 stands are shown in Fig. 6. Community types and sites (Figs. 6a and b) are generally segregated in the same way as by TWINSPAN, but relationships within and between sites are now apparent. In Fig. 7 specific topographic transects within individual prairies are traced, the relative topographic position of each point being indicated by its number.

This second ordination (Figs. 6 and 7) provides enhanced resolution of the relationship of vegetation to topographic position and subsoil permeability. A topographic gradient (Figs. 6a and 7) runs from the upper left (high slopes) to the lower right (low slopes), and a permeability gradient (Figs. 6b and 7) from the lower left (very poor) to the upper right (moderately fast).

There are three important aspects of this second ordination. First, the transects (Fig. 7) run roughly parallel to one another indicating a generality of vegetation response to topographic position. Second, separation of the transects along the permeability axis indicates that the topographic response differs according to local soil permeability. Slight differences in orientation of the transects reflect differences in the local interaction of topography and permeability. Third, sharp turns or deflections in the topographic sequences correspond to known discontinuities in the local interaction of topography and soil permeability. Soils with poor permeability and seasonally shallow water tables adjacent to potholes at Kalsow Prairie (Cary Drift Plain) support vegetation different from the typical low-slope vegetation of externally drained swales. The vegetation of broad ridge tops at Hayden Prairie (Iowan Erosion Surface), where seasonally perched water tables occur due to impeded internal drainage, shifts abruptly at the slopes where external drainage reduces the effects of the heavy subsoil. Another vegetation shift occurs on low slopes as the soil parent material shifts from weathered glacial till to deep alluvium with good internal drainage. At Sheeder Prairie (loess-mantled Kansan Drift Plain) a paleosolic outcrop represents a discrete mid-slope zone of heavy clay soil with poor permeability.

Thus, the topographic–permeability relationship is important for understanding the vegetation of tallgrass prairies, but this relationship is not a simple one. The different study sites are located on different geomorphological landscapes, and the variations between locations are related to the differences in the topo-edaphic characteristics of the regional landscapes they represent.

Discussion

The tallgrass prairie vegetation of Iowa and eastern Nebraska varies in a complex, multidimensional manner, reflecting vegetational responses at various scales to topographic position, regional and local edaphic characteristics, and geography. Each regional landscape (and to some degree each location) has a distinctive vegetation pattern related to its topographic–environmental complex. Herein lies much of the complexity of prairie vegetation. The unidimensional continuum approach does not suffice to describe differences among localities or regional landscapes.

The primary vegetation pattern corresponds to a soil moisture gradient, which is most evident along topographic gradients. Reconstruction of the full coenocline requires contributions from various locations and landscapes with different topographic characteristics; no one site or landscape encompasses the full range of environments. The most xeric sites are the upper slopes of terminal moraines of the Cary Drift, which are both internally and externally well drained. These grade into the moderately steep slopes of the Cary Drift end moraines and the dissected Kansan Drift landscape. Moderate slopes of the end moraines, moderate slopes of the Kansan Drift Plain in Nebraska, and slope bases of the dissected portions of the Kansan Drift Plain in Iowa represent yet more mesic environments along the gradient. The gentlest slopes of upland prairie come from the ground moraines of the Cary Drift or high points of alluvial terraces. The wettest sites come from the lowest positions in any landscape.

A second pattern related vegetation variation to internal soil drainage (i.e. subsoil permeability). In some situations, internal drainage varies in concert with topographic position thereby reinforcing the extremes of soil moisture conditions: coarse, internally drained soils are found on topographic highs, and heavy soils with poor internal drainage are in low topographic positions. This combination produces the complex soil moisture gradient (sensu Whittaker, 1967) which underlay the continuum employed in previous regional prairie analyses (e.g. Curtis, 1955, 1959; Dix & Smeins, 1967; Ralston, 1968). However, the internal drainage variable is differentiated as a distinct aspect of the moisture regime by cases in which the two aspects of soil drainage vary independently or inversely. These patterns are best illustrated in Fig. 6. The implication is that prairie species are distributed differentially with respect to both topographic position and subsoil permeability.

Geographic separation may contribute to differ-

ences between prairie types. For instance, floristic and structural differences distinguish the dry–mesic Nebraska prairies from those in Iowa. These differences may represent a macroclimatic influence, or perhaps different rates of postglacial migration (Benninghoff, 1968). Finally, size and history of the prairie relics may also be important; populations of prairie species are confined to a few small and widely separated remnants. Local population dynamics since isolation, exacerbated by small population sizes and limited migration, may have contributed to independent extinction or expansion of individual isolated populations.

References

Ayyad, M. A. G. & Dix, R. L., 1964. An analysis of a vegetation-–microenvironmental complex on prairie slopes in Saskatchewan. Ecol. Monogr. 34: 421–442.

Benninghoff, W. S., 1968. Biological consequences of Quaternary glaciations in the Illinois region. University of Illinois, College of Agriculture Special Publication 14: 70–77.

Bliss, L. C. & Cox, G. W., 1964. Plant community and soil variation within a northern Indiana prairie. Am. Midl. Nat. 72: 115–128.

Borchert, J. R., 1950. The climate of the central North American grassland. Ann. Assoc. Am. Geogr. 40: 1–39.

Brotherson, J. D., 1969. Species composition, distribution, and phytosociology of Kalsow Prairie, a mesic tall-grass prairie in Iowa. Ph.D. diss, Iowa State University, Ames, Iowa.

Brown, L. E., Quandt, L., Scheinost, S., Wilson, J. & White, D., 1980. Soil Survey of Lancaster County Nebraska. USDA Soil Conservation Service. 174+ pp.

Buckner, R. L. & Highland, J. D., 1974. Soil survey of Howard County, Iowa. USDA Soil Conservation Service. 131+ pp.

Clements, F. E., 1920. Plant indicators. Carnegie Inst. Wash., Publication 290: 114–139.

Crist, A. & Glenn-Lewin, D. C., 1978. The structure of community and environmental gradients in a northern Iowa prairie. Proc. Fifth Midwest Prairie Conference, pp. 57–64.

Curtis, J. T., 1955. A prairie continuum in Wisconsin. Ecology 36: 558–566.

Curtis, J. T., 1959. The vegetation of Wisconsin. University of Wisconsin Press, Madison. 657 pp.

Dick-Peddie, W. A., 1955. Presettlement forest types in Iowa. Unpubl. MS thesis. Ames, Iowa, Iowa State University.

Dix, R. L. & Butler, J. E., 1960. A phytosociological study of a small prairie in Wisconsin. Ecology 41: 316–327.

Dix, R. L. & Smeins, F. E., 1967. The prairie, meadow, and marsh vegetation of Nelson County, North Dakota. Can. J. Bot. 45: 21–58.

Gauch, H. G., 1982. Multivariate analysis in community ecology. Cambridge University Press. 298 pp.

Gauch, H. G. & Whittaker, R. H., 1981. Hierarchical classification of community data. J. Ecol. 69: 537–557.

110

Gauch, H. G., Whittaker, R. H. & Singer, S. B., 1981. A comparative study of nonmetric ordinations. J. Ecol. 69: 135-152.

Gleason, H. A., 1952. The new Britton and Brown illustrated flora of the northeastern United States and adjacent Canada. N.Y. Botanical Gardens, 3 vols.

Hill, M. O., 1979a. DECORANA – A FORTRAN program for detrended correspondence analysis and reciprocal averaging. Cornell University, Ithaca, N.Y.

Hill, M. O., 1979b. TWINSPAN – A FORTRAN program for arranging multivariate data in an ordered two-way table by classification of the individuals and attributes. Cornell University, Ithaca, N.Y.

Hill, M. O. & Gauch, H. G., 1980. Detrended correspondence analysis, an improved ordination technique. Vegetatio 42: 47-58.

Oschwald, W. R., Riecken, F. F., Dideriksen, R. I., Scholtes, W. H. & Schaller, F. W., 1965. Principal soils of Iowa. Special Rep. 42. Iowa State University Cooperative Extension Service, Ames, Iowa.

Peet, R. K., 1980. Ordination as a tool for analyzing complex data sets. Vegetatio 42: 171-174.

Peet, R. K. & Loucks, O. L., 1977. A gradient analysis of southern Wisconsin forests. Ecology 58: 485-499.

Pohl, R. W., 1966. The grasses of Iowa. Iowa State Journal of Science 40: 341-566.

Ralston, R. D., 1968. The grasslands of the Red River valley. Ph.D. diss., University of Saskatchewan.

Redmann, R. E., 1972. Plant communities and soils of an eastern North Dakota prairie. Bull. Torrey Bot. Club 99: 65-76.

Reed, E. C., Dreeszen, V. H., Bayne, C. K. & Schultz, C. B., 1965. The Pleistocene in Nebraska and northern Kansas. In: H. E. Wright & D. G. Frey (eds.). The Quaternary of the United States. Princeton University Press, Princeton, New Jersey.

Ruhe, R. V., 1969. Quaternary landscapes in Iowa. Iowa State University Press, Ames, Iowa. 255 pp.

Russell, R. C., Dideriksen, R. I. & Fisher, C. S., 1974. Soil survey of Guthrie County, Iowa. USDA, Soil Conservation Service.

Stevens, W. R., 1978. Nebraska. In: The Climates of the States. National Weather Service, National Oceanic and Atmospheric Administration. Gales Research Company. Book Tower, Detroit.

Transeau, E. N., 1935. The prairie peninsula. Ecology 16: 423-437.

Waite, P. J., 1978. Iowa. In: The Climates of the States. National Weather Service, National Oceanic and Atmospheric Administration. Gales Research Company. Book Tower, Detroit.

Walter, H., 1979. Vegetation of the Earth. Springer-Verlag, New York. 274 pp.

Weaver, J. E., 1954. North American Prairie. Johnsen Publ. Co., Lincoln, Nebraska. 348 pp.

Weaver, J. E. & Fitzpatrick, T. J., 1934. The prairie. Ecol. Monogr. 4: 109-295.

Whitford, P. B., 1958. A study of prairie remnants in southeastern Wisconsin. Ecology 39: 727-733.

Whittaker, R. H., 1967. Gradient analysis of vegetation. Biol. Rev. 42: 207-264.

Whittaker, R. H., 1978. Ordination of Plant Communities. Junk, The Hague. 388 pp.

Wright, H. E. & Ruhe, R. V., 1965. Glaciation of Minnesota and Iowa. In: H. E. Wright & D. G. Frey (eds.). The Quaternary of the United States. Princeton University Press, Princeton, New Jersey.

Accepted 19.3.1984.

Distributions of C_4 plants along environmental and compositional gradients in southeastern Arizona*

T. R. Wentworth**

*Department of Botany, North Carolina State University, Raleigh, NC 27650 USA****

Keywords: Arizona, C_3, C_4, Gradient analysis, Ordination, Photosynthetic pathways, Vegetation

Abstract

Distributional patterns of C_4 plants were investigated in 4 study areas located in se Arizona: granite slopes in the Mule Mountains, limestone slopes in the Mule Mountains, calcareous bajada (alluvial plain) below the Mule Mountains, and limestone slopes in the Huachuca Mountains. Cover data for all vascular species were obtained from 238 0.1 ha (20×50 m) sample quadrats located over ranges of elevation and topographic position within the study areas. Overall, 69 C_4 species representing 6 angiosperm families were encountered. C_4 species accounted for 13.5% to 22.3% of vascular species within the study areas. C_4 species frequency in quadrats (on the basis of all species or of grasses only) increased from mesic to xeric community types in all study areas except the calcareous bajada. Similar, but less consistent, trends were evident in the relative cover contributed by C_4 species. In two of the study areas (granite slopes in the Mule Mountains, limestone slopes in the Huachuca Mountains) regression analyses revealed statistically significant trends of C_4 species frequency and relative cover along environmental (elevation/solar-irradiation scalar) and compositional (reciprocal averaging ordination) gradients. A lack of consistent trends on limestone slopes in the Mule Mountains may be the result of grazing and/or recent invasion of low-elevation limestone areas by a Chihuahuan Desert flora dominated by C_3 dicot shrubs. The calcareous bajada below the Mule Mountains was studied less intensively, but its flora was found to contain the highest frequency of C_4 species of the 4 study areas. In contrast, C_4 cover on the bajada was low, presumably as a consequence of heavy grazing pressure on the grasses. The results of the present investigation support the prediction that C_4 species should be proportionally more successful in habitats characterized by high temperatures, high irradiance and low moisture.

* Nomenclature for most species follows Kearney & Peebles (1969). Gould (1951) and Benson (1969) were used for identification and nomenclature of Gramineae and Cactaceae, respectively. Gould (1968) was used for taxonomy of Gramineae at the tribe and subfamily levels. Voucher specimens are on deposit in the L. H. Bailey Hortorium of Cornell University.
** I wish to express my appreciation to J. A. Teeri for suggesting that I investigate distributions of C_4 plants in southeastern Arizona, and to J. D. Elson for valuable assistance with preliminary data analyses. The data discussed in this paper were collected as part of a Ph.D. dissertation completed at Cornell University under the direction of R. H. Whittaker. U. Blum, S. C. Huber, R. K. Peet and J. F. Reynolds provided helpful reviews of the manuscript. I gratefully acknowledge financial support provided by the North Carolina Agricultural Research Service, the National Science Foundation, Cornell University, and E. I. DuPont de Nemours and Company.
*** Paper No. 8258 of the Journal Series of the North Carolina Agricultural Research Service, Raleigh, NC 27650 USA.

Introduction

The C_4 photosynthetic pathway is now known to occur in at least 15 angiosperm families (Björkmann, 1976). C_4 photosynthesis is particularly common in the *Gramineae*, occurring in ca one-half of all species (Smith & Brown, 1973). Relative to C_3 plants, most C_4 plants share a set of distinctive characteristics including well-developed leaf vein bundle sheaths, low CO_2 compensation points, little or no photorespiration, high overall photosynthetic efficiencies, high maximum rates of photosynthesis, high photosynthetic temperature optima and high light intensities for photosynthetic saturation (Black, 1971, 1973; Bidwell, 1979). Several investigators have also reported higher water use efficiency for C_4 plants (e.g., Slatyer, 1970; Downes, 1969). The unique physiological, biochemical and structural features of C_4 plants have prompted numerous investigators to consider possible ecological consequences (Black, 1971; Tieszen *et al.*, 1979). Björkmann (1971, 1976) and coworkers (Björkmann *et al.*, 1975) have proposed that the C_4 pathway is an adaptation to, and provides advantages in, environments characterized by high temperatures, high irradiance, and conditions (e.g., water stress) promoting high stomatal resistance to gas exchange. C_4 plants also make more efficient use of nitrogen than do C_3 plants (more biomass is produced per unit plant nitrogen), a trait which may adapt them to environments with scarce nitrogen (Brown, 1978). Caswell *et al.* (1973) have hypothesized that C_4 plants are inferior to C_3 plants as herbivore food sources and will therefore be favored under conditions of selective herbivory.

C_3 plants, in contrast, should be at a net photosynthetic advantage relative to C_4 plants at low temperatures (given adequate moisture) because of their greater potential quantum yield (Black, 1973) and reduced photorespiration at low temperatures (Ehleringer & Björkman, 1977; Ehleringer, 1978). This advantage is eliminated above temperatures of ca 30 °C because of the temperature dependence of photorespiration in C_3 species (Ehleringer & Björkmann, 1977; Ehleringer, 1978).

Numerous investigations (e.g. Hofstra *et al.*, 1972; Syvertsen *et al.*, 1976; Teeri & Stowe, 1976; Mulroy & Rundel, 1977; Chazdon, 1978; Eickmeier, 1978; Stowe & Teeri, 1978; Doliner & Joliffe, 1979; Tieszen *et al.*, 1979; Boutton *et al.*, 1980; Rundel, 1980) have provided information about the distributions of C_4 species along environmental gradients. Most of these investigations have considered environmental gradients extending over relatively large distances of a few hundred to several thousand km (exceptions: Syvertsen *et al.*, 1976; Rundel, 1980), and many have dealt primarily with floristic data (numbers of frequencies of C_4 species in samples). With the exception of Syvertsen *et al.* (1976) and Eickmeier (1978), all distributional studies cited have supported predictions that C_4 species should be more abundant in habitats having higher temperatures, higher irradiance and lower moisture. Eickmeier's (1978) work is of particular

interest, as he showed peak cover of C_4 species (primarily grasses) at intermediate moisture and temperature conditions on an elevational transect in Big Bend National Park, Texas.

The present study was conducted with the objective of providing further insights regarding the ecology of C_4 plants. Data available from vegetation studies in two small mountain ranges in se Arizona (Wentworth, 1976, 1981, and unpublished) provided the basis for investigating aspects of C_4 plant distributions on a local scale. Several attributes made these data particularly useful for such an investigation: (1) the regional flora is rich in C_4 species, (2) complete floristic lists and plant cover data had been obtained from numerous sample quadrats, (3) quantitative indices of both environmental and compositional gradients were available, and (4) comparisons could be made between mountain ranges, between rock types within a mountain range, and between mountain slopes and the bajada (plain) below.

Methods

Study areas

Vegetation sampling was conducted in and around the Mule and Huachuca Mountains of se Arizona (Cochise County), at 31° 30′ N latitude. These small, isolated ranges are located in the Mexican Highland section of the Basin and Range Province (Fenneman, 1931), 5–15 km e and w of the San Pedro River, respectively. Elevations extend from 1 400–1 500 m in both ranges to 2 246 m in the Mules and to 2 885 m in the more massive Huachucas. Steep, rugged topography in these mountains contrasts sharply with the gentle slope of the alluvial plains, or bajadas, below. There is a diversity of rock formations exposed in the mountains of the study areas. Of particular interest are limestones, which occupy substantial areas of both the Mule and Huachuca Mountains. Soils derived from limestone parent materials support vegetation strikingly different in several respects from that found on soils derived from the other rock types (granite, sandstone, siltstone, conglomerate, gneiss, quartzite, schist) of the region (Wentworth, 1976, 1981).

Se Arizona has a mild, semiarid climate with two distinct precipitation seasons (July–September,

December–February). At Bisbee, at 1 658 m in the Mule Mountains, mean annual precipitation is 468 mm, 65% of which falls during April through September. Mean annual temperature at Bisbee is 16.3 °C, with average temperatures of 7.7 °C and 24.7 °C in January and July, respectively (Green & Sellers, 1964). Pronounced regional temperature and precipitation gradients are related to elevation, as indicated by lapse rates of –0.0073 °C/m for mean July temperature and 0.347 mm/m for mean annual precipitation (Wentworth, 1976).

Vegetation patterns in portions of the Mule and Huachuca Mountains were determined from gradient analyses of the same data sets used in the present paper (Wentworth, 1976, 1981, and unpublished). On open granite slopes between 1 500 and 1 900 m in the Mule Mountains, major community types grade from pygmy conifer-oak scrub (*Pinus cembroides, Juniperus deppeana, Quercus arizonica, Q. hypoleucoides* and *Q. emoryi* codominants) in mesic situations through open oak woodland (scattered *Quercus emoryi* and *Q. oblongifolia* in a grass-shrub matrix) to desert grassland (*Elyonurus barbiculmis, Bouteloua radicosa, Heteropogon contortus* and several other perennial bunchgrasses codominant) in xeric situations. The corresponding sequence on limestone is *Cercocarpus breviflorus* scrub, mesic phase Chihuahuan desertscrub (*Mortonia scrabella* and *Quercus pungens* codominants) and xeric phase Chihuahuan desertscrub (*Larrea tridentata, Acacia vernicosa, Flourensia cernua* and *Parthenium incanum* codominants). The calcareous bajada to the sw of the Mule Mountains supports desert scrub similar in many respects to that of the lower limestone slopes. At higher elevations (1 650–2 050 m) in the Huachuca Mountains, open limestone slopes support vegetation which grades from pine-oak woodland (*Pinus cembroides* and *Quercus arizonica* codominants) in mesic situations through mesic phase *Cercocarpus breviflorus* scrub (with *Pinus cembroides* and *Rhus choriophylla*) to xeric phase *Cercocarpus breviflorus* scrub (with *Rhus choriophylla* and *Bouteloua curtipendula*) in xeric situations.

Sampling

Sampling sites were chosen to represent the ranges of topographic variation (slope, aspect, position, exposure) and elevation in each of the 4 study areas: granite slopes, Mule Mountains (65 sites, mean elevation 1 723 m); limestone slopes, Mule Mountains (72 sites, mean elevation 1 571 m); calcareous bajada, Mule Mountains (33 sites, mean elevation 1 388 m); and limestone slopes, Huachuca Mountains (68 sites, mean elevation 1 852 m). At each site, I located a 0.1 ha sample quadrat (20 × 50 m) extending 10 m to either side of a 50 m tape. Stem counts and basal area estimates within each quadrat provided data for analysis of woody vegetation (Wentworth, 1981 and unpublished). Data discussed in the present paper are visual cover estimates for all vascular plant species in 25 meter-square subquadrats arranged at 1 m intervals on alternate sides of the 50 m tape. In a few woodland samples, it was necessary to estimate cover of canopy species by line-interception along a 50 m transect. Cover over the 25 subquadrats was summed and expressed as a percentage (of total potential cover of 25 m²) for each species. These steps provided an absolute measure of species performance, referred to as cover throughout the paper. Each 0.1 ha quadrat was also searched for species not occurring in the subquadrats.

All species were classified as C_4 or non-C_4 following lists kindly provided by L. G. Stowe and J. A. Teeri of the University of Chicago. The non-C_4 group (henceforth referred to as C_3) consisted predominantly of C_3 angiosperms but included also some CAM angiosperms as well as the gymnosperms and cryptograms (ferns). The Stowe and Teeri classification is based upon published results of photosynthetic pathway determinations (using criteria of leaf anatomy, CO_2 compensation point, carbon isotope ratio etc.), and is described in Teeri & Stowe (1976) for *Gramineae,* and in Stowe & Teeri (1978) for *Dicotyledonae.* Since many species have not yet been evaluated for pathway type, Stowe & Teeri classified these on the basis of taxonomic affinity. In the *Gramineae,* for example, all members of the subfamily *Festucoideae* are C_3, whereas all members of the subfamily *Eragrostoideae* are C_4 (Teeri & Stowe, 1976). The only grass subfamily known to contain both C_3 and C_4 species is the *Panicoideae,* and within this subfamily only the genus *Panicum* has both C_3 and C_4 species (Smith & Brown, 1973); however, the two *Panicum* species encountered in the present study are known to be C_4 (Smith & Brown, 1973; Downton, 1975). Besides the *Gramineae,* the only monocotyledonous family with C_4 representatives is the

Cyperaceae (Downton, 1975). No species of this family encountered in the present study are listed by Downton (1975) as having the C_4 pathway.

Among the *Dicotyledonae,* Stowe & Teeri (1978) found that ca 75% of the North American species in families known to have C_4 representatives had been evaluated for pathway type; pathways of the remaining species were inferred from taxonomic affinities. All of the C_4 dicotyledonous species listed in the present study (Table 1) have been evaluated for pathway type (Downton, 1975; Krenzer *et al.,* 1975).

Table 1. Occurrences of C_4 plant species in Granite Mule Mts(1); Limestone Mule Mts (2); Bajada Mule Mts (3) and Limestone Huachuca Mts (4). A '+' indicates that a species occurred in at least one 0.1 ha sample quadrat.

	1	2	3	4
A. Gramineae				
1. subfamily Panicoideae				
a. tribe Paniceae				
Digitaria sanguinalis				+
Trichachne californica	+			
Leptoloma cognatum	+			+
Panicum bulbosum	+			+
P. hallii		+	+	+
Setaria macrostachya		+	+	
b. tribe Andropogoneae				
Sorghum halepense		+		
Andropogon barbinodis	+	+		+
A. hirtiflorus	+			
A. cirratus	+			
A. scoparius	+			
Trachypogon montufari	+			+
Elyonorus barbiculmis	+	+		+
Heteropon contortus	+	+	+	+
Tripsacum lanceolatum	+			
2. subfamily Eragrostoideae				
a. tribe Eragrosteae				
Eragrostis lehmanniana	+		+	
E. intermedia	+	+		+
Tridens pulchellus		+	+	
T. grandiflorus	+	+		+
T. muticus	+	+	+	+
Scleropogon brevifolius			+	
Lycurus phleoides	+	+		+
Muhlenbergia sinuosa	+			
M. minutissima	+			
M. arizonica	+	+		
M. porteri		+	+	
M. arenicola		+		
M. glauca	+	+		+
M. polycaulis	+			
M. montana	+			
M. pauciflora	+	+		+
M. monticola	+	+		+
M. rigida		+	+	+
M. rigens	+			+
M. longiligula				+
M. emersleyi	+	+		+
Sporobolus contractus			+	
S. airoides			+	
Blepharoneuron tricholepsis	+			
b. tribe Chlorideae				
Leptochloa dubia	+	+		+
Cynodon dactylon	+			
Microchloa kunthii	+			
Bouteloua curtipendula	+	+	+	+
B. radicosa	+	+		
B. hirsuta	+	+	+	+
B. gracilis	+			
B. eriopoda		+	+	
B. rothrockii			+	
Aegopogon tenellus	+			
Hilaria mutica			+	
c. tribe Aristideae				
Aristida barbata	+	+	+	
A. divaricata		+		
A. orcuttiana	+	+		+
A. ternipes	+	+		
A. hamulosa		+		
A. pansa		+	+	
A. adscensionis	+			
A. arizonica				+
A. glauca		+	+	+
A. longiseta	+		+	
B. Chenopodiaceae				
Atriplex canescens			+	
Salsola kali			+	
C. Amaranthaceae				
Brayulinea densa			+	
Froelichia arizonica	+	+		
D. Nyctaginaceae				
Allionia incarnata			+	+
Boerhaavia coccinea	+	+		
E. Portulacaceae				
Portulaca suffrutescens	+			
F. Euphorbiaceae				
Euphorbia vermiculata			+	
E. albomarginata			+	+

Analysis

Sample quadrats in each of the 3 montane sample sets were first ordinated by direct and indirect methods of gradient analysis (Whittaker, 1967). Direct ordination was accomplished by construction of an environmental scalar incorporating measures of elevation, potential direct-beam solar-irradiation, exposure and position into a single index (Wentworth, 1976, 1981). Relationships among the various factors in the scalar were determined on the basis of field experience with vegetation and a criterion that the scalar provide an arrangement of species' population density data with minimum scatter about fitted moving average and Gaussian curves (Wentworth, 1976).

Indirect ordination was accomplished by means of reciprocal averaging (RA; Hill, 1973, 1974) using log-transformed and relativized cover data for those species having ⩾10% constancy in a sample set. Sample quadrats were classified by subdivision of their first-axis arrangement into community types corresponding to those in common usage for the region. Several quadrats in each sample set were dropped in the course of classification because these were found to have low similarity to most other quadrats. Data from the calcareous bajada of the Mule Mountains were not subjected to gradient analysis, but were treated as representative of a single community type.

The following measures of C_4 species performance were calculated for vascular plants in each of the sample quadrats: total number of species (0.1 ha basis), number of C_4 species, % C_4 species of the total (referred to hereafter as C_4 species frequency), total cover (25 m² basis), cover of C_4 species, and % C_4 cover of the total (referred to hereafter as C_4 relative cover). The calculations were performed twice, once for all species in a quadrat and again for grasses only. Means of these measures were also determined over all classified quadrats in a sample

set and over all quadrats in each community type within a set.

I used regression analysis to determine the relationships of the various indices of C_4 performance to both the direct and indirect quadrat ordinations, which were placed on a 0–100 scale with higher values representing increasingly warmer and drier environmental conditions. Simple linear as well as quadratic models were investigated, with ordination scores as the independent variables. For the sake of simplicity, only linear models are presented in this paper. The arcsin transformation was used to stabilize variance when a proportion (derived from C_4 species frequency or from C_4 relative cover) was used as the dependent variable (Snedecor & Cochran, 1967). Regressions on proportional data for grasses were calculated for only those quadrats having both C_3 and C_4 grasses. The latter decision was made because many sample quadrats had only C_4 grasses, a situation which occurred with increasing frequency toward warmer and drier sites. The C_4-grass quadrats formed a distinctive group, and it was judged inappropriate to combine these with quadrats having both C_3 and C_4 grasses for regression analyses.

Results

Sixty-nine C_4 species representing 6 angiosperm families were encountered in the 238 0.1 ha quad-

Table 2. Total species, C_4 species and percentage C_4 species (%) in various taxa in total study and 4 study areas.

	Total study (N = 238)			Granite, Mule Mts. (N = 65)			Limestone, Mule Mts. (N = 72)			Bajada, Mule Mts. (N = 33)			Limestone, Huachuca Mts. (N = 68)		
	Total	C_4		Total	C_4		Total	C_4		Total	C_4		Total	C_4	
	no.	no.	%	no.	no.	%	no.	no.	%	no.	no.	%	no.	no.	%
vascular plants	443	69	15.6	258	43	16.7	202	34	16.8	121	27	22.3	185	25	13.5
spermatophytes	427	69	16.2	246	43	17.5	199	34	17.1	121	27	22.3	177	25	14.1
angiosperms	419	69	16.5	244	43	17.6	197	34	17.3	120	27	22.5	170	25	14.7
monocots	98	60	61.2	65	40	61.5	39	29	74.4	28	22	78.6	49	25	51.0
Gramineae	73	60	82.2	46	40	87.0	31	29	93.6	23	22	95.7	36	25	69.4
dicots	321	9	2.8	179	3	1.7	158	5	3.2	92	5	5.4	121	0	0.0
Chenopodiaceae	3	2	66.7	1	0	0.0	0	–	–	2	2	100.0	0	–	–
Amaranthaceae	2	2	100.0	1	1	100.0	1	1	100.0	1	1	100.0	0	–	–
Nyctaginaceae	5	2	40.0	4	1	25.0	5	2	40.0	1	1	100.0	1	0	0.0
Portulacaceae	4	1	25.0	3	1	33.3	1	0	0.0	0	–	–	0	–	–
Euphorbiaceae	12	2	16.7	3	0	0.0	9	2	22.2	4	1	25.0	5	0	0.0

rats of this study (Table 1). The C_4 species accounted for 15.6% of the 443 vascular species found in these quadrats (Table 2). Five dicot families: *Chenopodiaceae, Amaranthaceae, Nyctaginaceae, Portulacaceae* and *Euphorbiaceae*, were represented by a total of 9 C_4 species. The remaining C_4 species were grasses in the tribes *Paniceae, Andropogoneae, Eragrosteae, Chlorideae* and *Aristideae* of Gould (1968).

The distributions of C_4 species in various taxa are summarized in Table 2 of for each of the 4 study areas. The percentage of C_4 species in the vascular flora ranged from a low of 13.5 on limestone slopes in the Huachuca Mountains to a high of 22.3 on calcareous bajada below the Mule Mountains. C_4 species were relatively uncommon among dicots, ranging from none on limestone slopes in the Huachuca Mountains to 5.4% (5 species) on calcareous bajada of the Mule Mountains. C_4 species had considerably better representation among monocots (51.0–78.6%) because of the abundance of C_4 grasses in the region. The percentage of grasses that were C_4

Table 3a. Means for total species, C_4 species, C_4 species frequency (%), total cover, C_4 cover and C_4 relative cover (%) over all classified sample quadrats and over quadrats within community types in 4 sample sets. Based on data for all vascular species.

| | Mean no. of species per 0.1 ha | | | Mean cover per 25 m² | | |
| | Total no. | C_4 | | Total cover | C_4 | |
		no.	%		cover	%
Granite, Mule Mts. (N = 59)	41.5	8.5	20.3	47.5	12.5	30.3
pygmy conifer-oak scrub (N = 24)	34.9	5.9	16.7	55.3	4.7	9.7
open oak woodland (N = 30)	47.3	9.9	21.2	43.0	17.1	41.9
desert grassland (N = 5)	38.2	12.4	32.6	37.0	22.4	60.3
Limestone, Mule Mts. (N = 64)	32.3	6.9	21.0	50.9	10.7	20.9
Cercocarpus scrub (N = 10)	31.3	6.3	19.9	56.3	7.6	14.6
Chihuahuan desertscrub, mesic phase (N = 29)	34.6	7.4	21.2	49.0	11.7	23.5
Chihuahuan desertscrub, xeric phase (N = 25)	30.0	6.5	21.3	50.8	10.8	20.4
Bajada, Mule Mts. (N = 33)	22.6	4.6	20.6	49.8	3.9	8.1
Limestone, Huachuca Mts. (N = 60)	24.9	4.7	19.7	64.5	9.2	16.7
Pine-oak woodland (N = 15)	29.7	3.2	11.4	80.4	8.4	10.2
Cercocarpus scrub, mesic phase (N = 23)	22.1	4.3	19.7	71.7	6.3	9.5
Cercocarpus scrub, xeric phase (N = 22)	24.5	6.1	25.5	46.2	12.9	28.7

Table 3b. Means for total species, C_4 species, C_4 species frequency (%), total cover, C_4 cover and C_4 relative cover (%) over all classified sample quadrats and over quadrats within community types in 4 sample sets. Based on data for grasses only.

| | Mean no. of species per 0.1 ha | | | Mean cover per 25 m² | | |
| | Total no. | C_4 | | Total cover | C_4 | |
		no.	%		cover	%
Granite, Mule Mts. (N = 59)	8.7	8.2	93.4	12.9	12.5	94.6
pygmy conifer-oak scrub (N = 24)	6.5	5.8	89.3	5.5	4.7	87.1
open oak woodland (N = 30)	9.9	9.4	95.6	17.2	17.1	99.5
desert grassland (N = 5)	12.0	12.0	100.0	22.4	22.4	100.0
Limestone, Mule Mts. (N = 64)	7.1	6.5	93.7	11.3	10.7	94.0
Cercocarpus scrub (N = 10)	6.9	6.3	91.3	8.2	7.6	90.1
Chihuahuan desertscrub, mesic phase (N = 29)	7.7	7.1	92.7	12.2	11.6	94.5
Chihuahuan desertscrub, xeric phase (N = 25)	6.4	6.0	95.9	11.5	10.8	95.1
Bajada, Mule Mts. (N = 33)	4.3	4.3	99.3	3.8	3.8	98.9
Limestone, Huachuca Mts. (N = 60)	6.2	4.7	76.5	11.6	9.2	80.1
Pine-oak woodland (N = 15)	5.9	3.2	54.0	11.2	8.4	69.0
Cercocarpus scrub, mesic phase (N = 23)	5.7	4.3	78.0	10.0	6.3	72.8
Cercocarpus scrub, xeric phase (N = 22)	6.9	6.1	90.3	13.5	12.9	95.3

ranged from 69.4 on limestone slopes in the Huachuca Mountains to 95.7 on calcareous bajada of the Mule Mountains.

Means for the four measures of C_4 performance (number of species, species frequency, cover and relative cover) are summarized in Table 3a over all classified quadrats in each sample set and over the quadrats in the various community types; means for total species and total cover are also provided. Table 3b repeats this summary for grasses. Quadrats on granite in the Mule Mountains averaged greater total species (41.5 per 0.1 ha) and greater C_4 species (8.5 per 0.1 ha) than did those in the other sample sets (Table 3a). However, C_4 species frequency varied little from one sample set to the next. Total cover was substantially greater on limestone in the Huachuca Mountains (64.5) than in the other sample sets, but C_4 cover (12.5) and relative cover (30.3%) were highest on granite in the Mule Mountains (Table 3a). Trends of C_4 performance over the community types within a study area were most clearly evident on granite in the Mule Mountains.

Species number, species frequency, cover and relative cover increased from pygmy-conifer oak scrub to desert grassland (Table 3a). Similar trends were not evident on limestone in the Mule Mountains, but number and frequency of C_4 species both increased from pine-oak woodland to xeric phase *Cercocarpus* scrub on limestone in the Huachuca Mountains (Table 3a).

Patterns in the performance of grasses (Table 3b) showed some differences from those discussed for all species (Table 3a). One difference was that the sample quadrats on limestone in the Huachucas averaged a lower frequency of C_4 grasses (76.5%) than did those in the other sample sets. Total cover of grasses was similar in the three montane sets (11.3–12.9) but substantially lower (3.8) on the calcareous bajada of the Mule Mountains. All sample sets averaged high relative cover of C_4 grasses, although limestone slopes in the Huachucas were somewhat lower in this respect. Trends in number and cover of C_4 grasses of community types within a sample set (Table 3b) were essentially the same as

Table 4a. Slope parameters (b) and R^2 values for linear regression equations based on data for all vascular plant species in 3 samples sets. Dependent variables are total species, C_4 species, arcsin transform of proportion C_4 species, total cover, C_4 cover and arcsin transform of proportion C_4 cover. Significance levels of t-tests on slope parameters (null hypothesis: $\beta = 0$) are indicated by asterisks (*, significant at 0.05 level; **, significant at 0.01 level; NS, not significant).

	Number of species per 0.1 ha			Cover per 25 m²		
		C_4			C_4	
	Total no.	no.	arcsin (proportion)	Total cover	cover	arcsin (proportion)
Granite, Mule Mts. (N = 65)						
Environmental Scalar						
b	NS	0.0920**	0.140**	-0.245*	0.234**	0.475**
R^2		0.38	0.38	0.07	0.33	0.37
RA Ordination						
b	0.166**	0.0883**	0.113**	-0.339**	0.251**	0.508**
R^2	0.25	0.61	0.43	0.22	0.66	0.73
Limestone, Mule Mts. (N = 72)						
Environmental Scalar						
b	NS	NS	0.0403*	NS	NS	NS
R^2			0.06			
RA Ordination						
b	-0.107**	NS	NS	NS	NS	NS
R^2	0.16					
Limestone, Huachuca Mts. (N = 68)						
Environmental Scalar						
b	-0.0884*	0.0590**	0.208**	-0.612**	NS	0.220**
R^2	0.09	0.44	0.56	0.44		0.21
RA Ordination						
b	0.126**	0.0589**	0.222**	-0.634**	NS	0.216**
R^2	0.16	0.41	0.59	0.43		0.19

Table 4b. Slope parameters (b) and R² values for linear regression equations based on data for all grass species in 3 sample sets. Dependent variables are total species, C₄ species, arcsin transform of proportion C₄ species, total cover, C₄ cover and arcsin transform of proportion C₄ cover. Significance levels of t-tests on slope parameters (null hypothesis: $\beta = 0$) are indicated by asterisks (*, significant at 0.05 level; **, significant at 0.01 level; NS, not significant).

| | Number of species per 0.1 ha | | | Cover per 25 m² | | |
| | | C₄ | | | C₄ | |
	Total no.	no.	arcsin (proportion)	Total cover	cover	arcsin (proportion)
Granite, Mule Mts. (N = 65)						
Environmental Scalar						
b	0.0663**	0.0848**	0.129*	0.212**	0.234**	0.786**
R²	0.22	0.37	0.16	0.28	0.33	0.34
RA Ordination						
b	0.0735**	0.0821**	0.0991*	0.237**	0.251**	0.434*
R²	0.48	0.60	0.21	0.61	0.66	0.32
Limestone, Mule Mts. (N = 72)						
Environmental Scalar						
b	NS	NS	0.0678*	NS	NS	NS
R²			0.19			
RA Ordination						
b	-0.0273*	-0.0204*	0.0709*	NS	NS	NS
R²	0.09	0.06	0.18			
Limestone, Huachuca Mts. (N = 68)						
Environmental Scalar						
b	0.0271*	0.0590**	0.324**	NS	NS	0.264*
R²	0.09	0.44	0.60			0.12
RA Ordination						
b	NS	0.0589**	0.315**	NS	NS	0.257*
R²		0.41	0.58			0.11

those for all C₄ species (Table 3a). However, an interesting trend emerged in the proportional measures (species frequency and relative cover) for C₄ grasses: in every case there was an increase in these measures from the most mesic community type to the most xeric (Table 3b).

The direct and indirect ordinations were highly correlated within a given sample set; Spearman rank-correlation coefficients between the environmental scalar and RA first-axis ordinations were 0.79 for granite in the Mule Mountains, 0.83 for limestone in the Mule Mountains and 0.86 for limestone in the Huachuca Mountains (all p < 0.0001). The close agreement between the scalar and RA ordinations suggested that the dominant compositional trends observed reflect direct or indirect responses of many species to the environmental factors incorporated into the scalar. Statistics for linear regressions for the 4 measures of C₄ species performance, as well as total species number and total cover, against the ordinations are presented in

Tables 4a and 4b. In most cases significant regressions were found, with R² values ranging from 0.06 to 0.73.

The closest relationships between measures of C₄ performance and ordination scores were found in the Mule Mountains granite data. All such regressions were significant, but R² values were generally (although not always) greater when the RA ordination was used as the independent variable. All slope parameters related to C₄ performance were positive. Figure 1 illustrates the relationships between the proportional measures of C₄ performance and the first reciprocal averaging axis scores for granite slopes in the Mule Mountains.

Relationships between all dependent variables and the ordinations were poor in the Mule Mountains limestone data. Only one-quarter of the regressions were statistically significant, and R² values were low (maximum of 0.19). Relationships were stronger for data from limestone slopes in the Huachuca Mountains. Most regressions were statistically significant, with moderate R^2 values ranging as high as 0.60. All slope parameters related to C_4 performance were positive.

Comparisons of the Arizona C_4 distributional patterns with those of other studies required that the untransformed proportional measures of C_4 performance be plotted against elevation. This has been done in Figures 2A and 2C for granite slopes in the Mule Mountains and in Figures 2B and 2D for limestone slopes in the Huachuca Mountains.

To simplify data presentation and to effect a moderate smoothing, sample quadrats were grouped into 50-m elevation classes and means of the proportional measures were taken within these classes.

Discussion

Because sampling was conducted during the summer and fall months, the results of this investi-

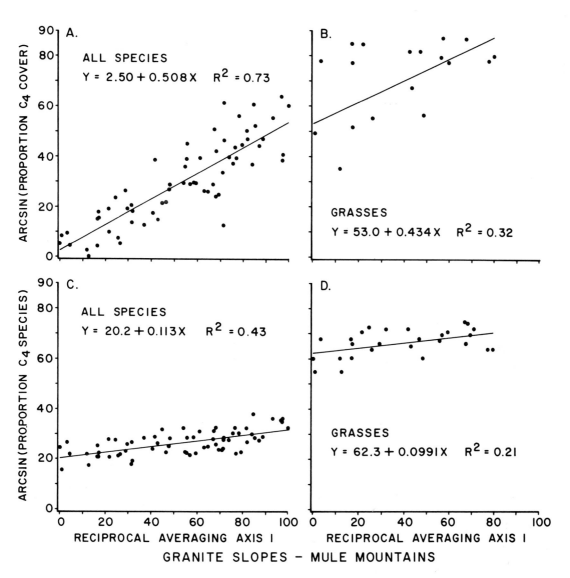

Fig. 1. Arcsine transform of proportional measures of C_4 species performance vs. the first reciprocal averaging axis for quadrat samples from granite slopes in the Mule Mountains: (A.) arcsin (proportion C_4 cover) for all species, (B.) arcsin (proportion C_4 cover) for grasses only (C.) arcsin (proportion C_4 species) for all species and (D.) arcsin (proportion C_4 species) for grasses only.

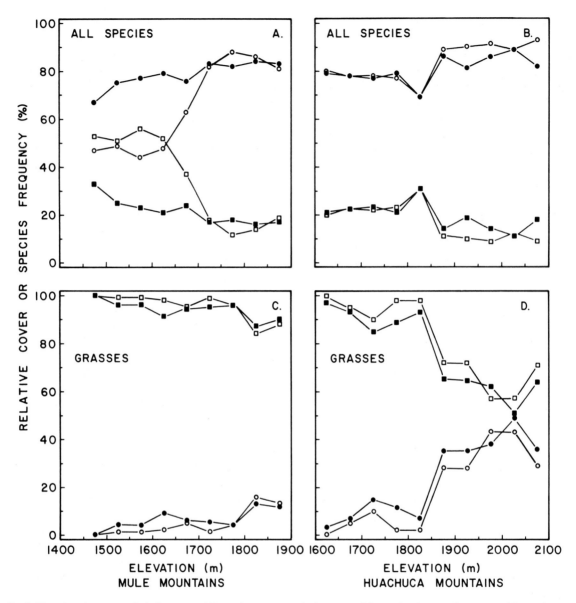

Fig. 2. Elevational patterns of relative cover of C_4 species (□——□), relative cover of C_3 species (○——○), frequency of C_4 species (■——■) and frequency of C_3 species (●——●). Points represent means for quadrat samples in 50-m elevation classes for: (A.) all species on granite slopes, Mule Mountains, (B.) all species on limestone slopes, Huachuca Mountains, (C.) grasses on granite slopes, Mule Mountains and (D.) grasses on limestone slopes, Huachuca Mountains.

gation are based upon data for species active or at least visible during the precipitation maximum (July–September) of se Arizona.

All but one of the study areas (limestone slopes in the Huachuca Mountains) exceeded the highest values reported by Teeri & Stowe (1976) for the percentage of C_4 species among grasses in a North American region north of Mexico (Table 2). The

latter result was perhaps to be expected because the data were collected during the warm season, when C_4 species are generally most active (see Ode *et al.,* 1980). The data of Teeri & Stowe (1976), however, are from published floras and should have no seasonal bias in favor of C_4 species.

Because Bisbee, located at 1658 m, is only 10 m lower than the mean elevation of the 238 sample quadrats (1668 m), its climatic data may be used in a regression equation developed by Teeri & Stowe (1976) for predicting the percentage of C_4 species among grasses in regions of North America. The relevant data, from Green & Sellers (1964), are normal July minimum temperature (64.2 °F or 17.9 °C), mean annual degree days (2581, base 65 °F or 18.3 °C) and log of the mean length in days of the annual freeze-free period ($\log_{10} 247 = 2.39$). The predicted percentage is 81.0, which agrees almost exactly with the observed percentage, 82.2 (Table 2). The latter percentage is higher than the value of 57% reported by Teeri & Stowe (1976) for the Arizona flora, but identical to their value of 82% for the flora of the nearby Sonoran Desert. The study areas thus appear reasonably consistent with other North American regions in their percentage of C_4 species among grasses.

The percentage of C_4 dicot species in the study areas varied somewhat (0.0% to 4.1% of the total spermatophyte flora), but, except for limestone slopes in the Huachuca Mountains, fell within the range of values listed by Stowe & Teeri (1978) for floras of several western states.

The prediction that $\bar{C}_4$ species should increase in importance relative to other species toward warmer, drier conditions may be evaluated from several perspectives. A strong source of support comes from results presented in Tables 3a and 3b. In three study areas representing slopes in two mountain ranges on two different rock types, C_4 species generally increased in frequency and relative cover from the most mesic to most xeric community-types. This was true for all vascular species (Table 3a) as well as for grasses (Table 3b). Further support for the prediction may be seen in regressions based upon data from granite slopes in the Mule Mountains and from limestone slopes in the Huachuca Mountains, presented in Tables 4a and 4b. In these regressions, C_4 species showed statistically significant increasing trends of both absolute and proportional measures of floristic richness and cover against both environmental scalar and RA ordination axes. In addition, these results were obtained for all species (Table 4a) as well as for grasses (Table 4b). The exception to these generalizations involved data for cover on limestone in the Huachuca Mountains. The nature of these relationships can be more readily appreciated for the selected data presented in Figure 1.

The demonstration that performances of C_4 species varied in a predictable manner along compositional and environmental gradients constitutes an important finding of the present study, albeit one consistent with previous work. If this finding held true only on the basis of the total flora, the role of the C_4 pathway and associated adaptations in determining such results would be in some doubt. Because nearly all the C_4 species encountered were grasses, a reasonable alternative interpretation would be that the observed patterns were the result of certain unique characteristics of the grasses, other than their predominantly C_4 photosynthetic pathway. However, the fact that distributional patterns of C_4 grasses were evident within the *Gramineae* (Tables 3b, 4b; Fig. 1) suggests that the observed responses of the C_4 species were a consequence of their photosynthetic pathway and not of some other characteristic(s) peculiar to grasses.

Another significant finding of the present investigation is that distributional patterns of C_4 species were evident in both presence-absence and cover data. It thus appears that conditions of high temperature and moisture stress favor not only a greater proportion of C_4 species, but also a greater contribution of such species to community biomass and productivity. Comparison of the Arizona distributional patterns with those of other studies is possible, using the data as presented in Figures 2A–2D. Trends of decreasing contribution of C_4 species to species richness and cover with increasing elevation are evident, and these patterns may be seen for all species (Figures 2A and 2B) as well as for grasses (Figures 2C and 2D). An interesting feature of these figures is that corresponding curves appear roughly similar in the zone of elevational overlap for transects from the two mountain ranges; this similarity suggests that curves from each range are small segments of larger, consistent trends that could be mapped over a greater elevational range in the region.

Rundel (1980) and Tieszen *et al.* (1979) have presented data for the species frequency of C_4 grasses along elevational transects in Hawaii and Kenya, respectively. The comparable curves from the Arizona data (Figures 2C and 2D) lie between the Hawaiian and Kenyan curves, with a floristic crossover point (50% each of C_3 and C_4 species) of 2 100 m, estimated from slight extrapolation of a regression line for the Huachuca Mountains data.

Table 5. Comparison of floristic crossover points for C_3 and C_4 grasses in three regions.

Location	Approximate latitude	Source of data	Floristic crossover elevation	Mean minimum temperature of warmest month	Mean maximum temperature of warmest month
Kenya	0°	Tieszen *et al.* (1979), cited from Rundel (1980)	2 300 m	8 °C	22 °C
Hawaii Volcanoes National Park	20 °N	Rundel (1980)	1 400 m	9 °C	21 °C
Arizona, Huachuca Mountains	32 °N	Present study	2 100 m	13 °C	26 °C

Crossover points for the Hawaiian and Kenyan curves are at approximately 1 400 m and 2 300 m, respectively (Rundel, 1980). Teeri & Stowe (1976) have found that mean minimum temperature of the warmest month is the single best predictor of species frequency of C_4 grasses in North America, and their data show that a minimum July temperature of ca 18 °C corresponds to the floristic crossover point of C_3 and C_4 grasses (Rundel, 1980). Extrapolation of a regression equation for 15 se Arizona stations (elevations 1 090–1 658 m, data of Green & Sellers, 1964) indicates an approximate mean minimum July temperature of 13 °C at the 2 100 m floristic crossover point in Arizona, midway between the 18 °C value of Teeri and Stowe, and the 9 °C and 8 °C mean minimum temperatures in the warmest month for the Hawaiian and Kenyan floristic crossover points, respectively (Table 5, data from Rundel, 1980). Rundel (1980) has suggested that C_4 grasses achieve floristic dominance at lower environmental temperatures in the tropics than in temperate regions, possibly because of more uniform temperature conditions throughout the year. The Arizona data appear to support this suggestion.

Rundel (1980) also presents Hawaiian elevational transect data for relative cover of C_4 grasses, which may be compared with the corresponding Arizona data of Figures 2C and 2D. The Arizona relative cover curves for C_4 grasses closely paralleled the species frequency curves already discussed, with an identical cover crossover point of 2 100 m in the Huachuca Mountains. Rundel's Hawaiian cover crossover point was near 1 200 m, and his mean maximum temperature for the warmest month of 22 °C at 1 211 m is somewhat lower than the corresponding value of 26 °C for the Arizona cover crossover point. This difference is consistent with

the temperate-tropical pattern of floristic crossover temperatures discussed above.

An apparent inconsistency in the results was the failure of data from limestone slopes in the Mule Mountains to yield strong, statistically significant ecological trends of C_4 species performance (Tables 4a, 4b). A possible source of this inconsistency may be grazing disturbance, as both the calcareous bajada and limestone slopes of the Mule Mountains were open to cattle, while both other study areas were more protected. Grazing was most severe on the calcaeous bajada, decreasing in intensity toward the steeper and generally less accessible higher elevations. Moreover, grazing in the study areas was a patchy disturbance, varying at any elevation with local accessibility and management practices of the landowners. In se Arizona, cattle selectively remove grasses, the predominant C_4 species, thereby promoting increase (and sometimes invasion) of the predominantly C_3 shrubby species (Humphrey, 1958; Hastings & Turner, 1965). This problem is complicated by the fact that Chihuahuan Desert species, most notably shrubs, have recently invaded lower elevation calcareous sites that supported grassland less than a century ago. The causes of shrub invasion are poorly understood, but subtle climatic shifts, as well as overgrazing, have been implicated (Humphrey, 1958; Hastings & Turner, 1965). Clearly the results on the two study areas (calcareous bajada and lower elevation limestone slopes of the Mule Mountains) most severely affected by such events must be considered in light of potential consequences for C_4 grasses. The failure of data from limestone slopes in the Mule Mountains to yield clear trends of C_4 species performance was probably not a consequence of the limestone substratum. Data from limestone slopes in the Huachuca Mountains, where grazing was generally light or nonexistent, showed trends similar to those demonstrated for granite slope in the Mule Mountains (Tables 4a, 4b). One indication that limestone slopes and calcareous bajada of the Mule Mountains have at least a potential for C_4 success is the fact that both had consistently high values for C_4 species frequency in their floras (Tables 2, 3a, 3b). This was particularly true for the calcareous bajada which, despite the fact that it had by far the lowest mean relative cover of C_4 species (8.1%) of the 4 areas investigated (Table 3a) had the highest percentage C_4 species in its flora (Table 2). It is noteworthy that Syvertsen et al. (1976) reported low importance of C_4 species (as determined by biomass) in several grazed lowland Chihuahuan Desert sites, while Eickmeier (1978) showed declining C_4 cover (with replacement by CAM species) toward low-elevation desert areas of Big Bend National Park, that part of the park most severely altered by overgrazing (Whitson, 1974).

The limestone slopes in the Huachuca Mountains differed from either limestone or granite slopes in the Mule Mountains in having lower percentages of C_4 species in various taxa; C_4 dicotyledonous species were altogether lacking (Table 2). Although the 3 montane study areas had similar C_4 species frequencies among all vascular species (Table 3a), the Huachucas were lower in C_4 relative cover among all vascular species (Table 3a) and in both C_4 species frequency and C_4 relative cover among grasses (Table 3b). These differences between the Huachuca and the Mule Mountains were probably a consequence of cooler and moister environments in the Huachuca Mountains study area, resulting from slightly higher elevation of the study area and reinforced by generally cooler and moister conditions found in the higher, more massive Huachucas. Not only would such environments be expected to favor C_3 relative to C_4 species, but greater stature and canopy closure of vegetation in the Huachucas would reduce irradiance in the understory (where all C_4 species occurred), further favoring the C_3 species (see Ehleringer, 1978).

The findings of the present study are consistent with those reported by most other investigators: indices of C_4 species performance show a general pattern of increase toward warmer and drier environments, an apparent distributional reflection of the physiological characteristics of the C_4 photosynthetic pathway. The demonstration of clear patterns of C_4 performance from relatively short

(500 m) elevational transects over distances of 10–20 km complements the results of other studies conducted, for the most part, on considerably larger geographic scales. The results point to some consistency between these patterns across rock types and between mountain ranges within the same region. Contrasts in pattern between these different situations are also evident resulting apparently from factors such as disturbance and mountain mass, which are well-known for their effects on vegetation pattern in general. Finally, the results of this study provide validation of the Teeri and Stowe (1976) climatic predictive equation for species frequency of C_4 grasses and provide some support for Rundel's (1980) recognition of differences between tropical and temperate distributions of C_4 grasses.

References

Benson, L., 1969. The cacti of Arizona. Third edition. 218 pp. University of Arizona Press, Tucson, AZ, USA.

Bidwell, R. G. S., 1979. Plant physiology. Second edition. 726 pp. MacMillan, New York, NY, USA.

Björkmann, O., 1971. Comparative photosynthetic CO_2 exchange in higher plants. In: Hatch, M. D., Osmond, C. B. & Slatyer, R. O. (eds.), Photosynthesis and photorespiration. pp. 18–32. Wiley-Interscience, New York, NY, USA.

Björkmann, O., 1976. Adaptive and genetic aspects of C_4 photosynthesis. In: Burris, R. H. & Black, C. C. (eds.), Proceedings of the Fifth Annual Harry Steenbock Symposium, Madison, Wisconsin, June 9–11, 1975. pp. 287–309. University Park Press, Baltimore, MD, USA.

Björkmann, O., Mooney, H. A. & Ehleringer, J., 1975. Comparison of photosynthetic characteristics of intact plants. Carnegie Institution of Washington Year Book 74: 743–751.

Black, C. C., 1971. Ecological implications of dividing plants into groups with distinct photosynthetic production capacities. Adv. Ecol. Res. 7: 87–113.

Black, C. C, 1973. Photosynthetic carbon fixation in relation to net CO_2 uptake. Ann. Rev. Pl. Physiol. 24: 253–286.

Boutton, T. W., Harrison, A. T. & Smith, B. N., 1980. Distribution of biomass of species differing in photosynthetic pathway along an altitudinal transect in southeastern Wyoming grassland. Oecologia 45: 287–298.

Brown, R. H., 1978. A difference in N use efficiency in C_3 and C_4 plants and its implications in adaptation and evolution. Crop. Sci. 18: 93–98.

Caswell, H., Reed, F., Stephenson, S. N. & Werner, P. A., 1973. Photosynthetic pathways and selective herbivory: a hypothesis. Amer. Nat. 107: 465–480.

Chazdon, R. L., 1978. Ecological aspects of the distribution of C_4 grasses in selected habitats of Costa Rica. Biotropica 10: 265–269.

Doliner, L. H. & Jolliffe, P. A., 1979. Ecological evidence concerning the adaptive significance of the C_4 dicarboxylic acid pathway of photosynthesis. Oecologia 38: 23–34.

124

Downes, R. W., 1969. Differences in transpiration rates between tropical and temperate grasses under controlled conditions. Planta 88: 261–273.

Downton, W. J. S., 1975. The occurrence of C_4 photosynthesis among plants. Photosynthetica 9, 96–105.

Ehleringer, J. R., 1978. Implications of quantum yield differences on the distributions of C_3 and C_4 grasses. Oecologia 31: 255–267.

Ehleringer, J. & Björkmann, O., 1977. Quantum yields for CO_2 uptake in C_3 and C_4 plants. Pl. Physiol. 59: 86–90.

Eickmeier, W. G., 1978. Photosynthetic pathway distributions along an aridity gradient in Big Bend National Park, and implications for enhanced resource partitioning. Photosynthetica 12: 290–297.

Fenneman, N. M., 1931. Physiography of western United States. 534 pp. McGraw-Hill, New York, NY, USA.

Gould, F. W., 1951. Grasses of the southwestern United States. Biological Science Bulletin 7. 352 pp. Univ. of Arizona, Tucson, AZ, USA.

Gould, F. W., 1968. Grass systematics. 382 pp. McGraw-Hill, New York, NY, USA.

Green, C. R. & Sellers, W. D., 1964. Arizona climate. 503 pp. Univ. of Arizona Press, Tucson, AZ. USA.

Hastings, J. R. & Turner, R. M., 1965. The changing mile: an ecological study of vegetation change with time in the lower mile of an arid and semiarid region. 317 pp. Univ. of Arizona Press, Tucson, AZ, USA.

Hill, M. O., 1973. Reciprocal averaging: an eigenvector method of ordination. J. Ecol. 61: 237–249.

Hill, M. O., 1974. Correspondence analysis: a neglected multivariate method. J. R. Statist. Soc., Series C 23: 340–354.

Hofstra, J. J., Aksornkoae, S., Atmowidjojo, S., Banaag, J. F., Santosa, R., Sastrohoetomo, A. & Thu, L. T. N., 1972. A study on the occurrence of plants with a low CO_2 compensation point in different habitats in the tropics. Ann. Bogor. 5: 143–157.

Humphrey, R. R., 1958. The desert grassland. A history of vegetational change and an analysis of causes. Bot. Rev. 24: 193–252.

Kearney, T. H. & Peebles, R. H., 1969. Arizona flora. Second edition with supplement by J. T. Howell, E. McClintock, et al. 1085 pp. Univ. of California Press, Berkeley and Los Angeles, CA, USA.

Krenzer, E. G., Jr., Moss, D. N. & Crookston, R. K., 1975. Carbon dioxide compensation points of flowering plants. Pl. Physiol. 56: 194–206.

Mulroy, T. W. & Rundel, P. W., 1977. Annual plants: adaptations to desert environments. Bioscience 27: 109–114.

Ode, D. J., Tieszen, L. L. & Lerman, J. C., 1980. The seasonal contribution of C_3 and C_4 plant species to primary production in a mixed prairie. Ecology 61: 1304–1311.

Rundel, P. W., 1980. The ecological distribution of C_4 and C_3 grasses in the Hawaiian Islands. Oecologia 45: 354–359.

Slatyer, R. O., 1970. Comparative photosynthesis, growth and transpiration of two species of Atriplex. Planta 93: 175–189.

Smith, B. N. & Brown, W. V., 1973. The Kranz syndrome in the Gramineae as indicated by carbon isotopic ratios. Amer. J. Bot. 60. 505–513.

Snedecor, G. W. & Cochran, W. G., 1967. Statistical methods. 593 pp. Iowa State Univ. Press, Ames, IA, USA.

Stowe, L. G. & Teeri, J. A., 1978. The geographic distribution of C_4 species of the dicotyledonae in relation to climate. Am. Nat. 112: 609–623.

Syvertsen, J. P., Nickell, G. L., Spellenberg, R. W. & Cunningham, G. L., 1976. Carbon reduction pathways and standing crop in three Chihuahuan Desert plant communities. SW. Nat. 21: 311–320.

Teeri, J. A. & Stowe, L. G., 1976. Climatic patterns and the distribution of C_4 grasses in North America. Oecologia 23: 1–12.

Tieszen, L. L., Senyimba, M. M., Imbamba, S. K. & Troughton, J. H., 1979. The distribution of C_3 and C_4 grasses and carbon isotope discrimination along an altitudinal and moisture gradient in Kenya. Oecologia 37: 337–350.

Wentworth, T. R., 1976. The vegetation of limestone and granite soils in the mountains of southeastern Arizona. Doctoral dissertation, Cornell Univ., Ithaca, NY, USA.

Wentworth, T. R., 1981. Vegetation on limestone and granite in the Mule Mountains, Arizona. Ecology 62: 469–482.

Whitson, P. D., 1974. The impact of human use upon the Chisos Basin and adjacent lands. U.S. Nat. Pk. Serv. Monogr. 4. 92 pp.

Whittaker, R. H., 1967. Gradient analysis of vegetation. Biol. Rev. 42: 207–264.

Accepted 12.7. 1982.

Gradient analysis of the vegetation of the Byron-Bergen swamp, a rich fen in western New York*

John M. Bernard[1], Franz K. Seischab[2] & Hugh G. Gauch[3], Jr.**
[1] *Department of Biology, Ithaca College, Ithaca, NY 14850, U.S.A.*
[2] *Department of Biology, Rochester Institute of Technology, Rochester, NY 14623, U.S.A.*
[3] *Department of Ecology and Systematics, Cornell University, Ithaca, NY 14853, U.S.A.*

Keywords: Detrended correspondence analysis, Gradient analysis, Marl wetland, Mire, Rich fen, Vegetation gradient

Abstract

Transects of contiguous one square meter quadrats were sampled across the marl and peat mosaic of the Byron-Bergen swamp, a rich fen in western New York. The data were analyzed by detrended correspondence analysis (DCA).

Ordination of species and samples produced arrangements reflecting a complex environmental gradient of hydrology, soil organic matter and soil carbonate-carbon concentration. They successfully separated fens underlain with peat from those underlain with marl and showed a physiognomic gradient associated with hummock development.

Introduction

The majority of vegetation studies analyze differences between communities, whereas the multiple species analysis of within community vegetation patterns is relatively neglected. Whittaker found multivariate methods particularly useful for describing within-community patterns (Shmida & Whittaker, 1981; Whittaker, Gilbert & Connell, 1979; Whittaker & Naveh, 1979; Whittaker, Niering &

Crisp, 1979; also see Gauch 1982: 90–92). Although terrestrial vegetation pattern is a two-dimensional phenomenon, to simplify sampling requirements Whittaker emphasized almost one-dimensional transects composed of 100 to 1000 contiguous small quadrats. For each sample quadrat the percentage cover of all vascular plant species was estimated visually. Consequently, when using this sampling method, some means of synthesizing the species data into vegetation types (classes) or a small number of continuous gradients (e.g. ordination axes) must be employed. This conversion is typically accomplished by multivariate analysis, particularly ordination.

The multivariate method employed by Whittaker was reciprocal averaging (RA) ordination. Subsequently an improved algorithm, detrended correspondence analysis (DCA), has been developed to alleviate RA's main shortcomings (Hill & Gauch, 1980). In addition, all of Whittaker's studies involved dry area vegetation with two prominent vegetation phases, shrub clumps occurring in an herbaceous matrix. The purpose of this study is to expand on Whittaker's work by testing the applica-

* Vascular plant nomenclature follows Fernald (1950). Bryophyte nomenclature follows Crum & Anderson (1981).

** Two decades of research in ordination and classification by Robert Whittaker was seen by him as a prelude to the study of pattern within communities. We acknowledge Whittaker's leadership in this field, and his contribution to the research presented here.

This research was supported in part by National Science Foundation Grant DEB-7809340 to R. H. Whittaker, especially an amendment to support John M. Bernard under NSF Support for Small College Faculty through Grants at Large Institutions program.

The Bergen Swamp Preservation Society, Inc. allowed us to sample on their property.

bility of DCA in identifying pattern in more complex vegetation than Whittaker studied. Specifically we examined the vegetation pattern of the 750 ha Byron-Bergen swamp in W New York (43°6'N, 78°1'W) which is characterized by multiple vegetation types reflecting soil chemistry and microtopography (Seischab, 1977).

There are marl fen areas in the center of this complex which are surrounded by a coniferous mire forest of *Thuja occidentalis, Pinus strobus* and *Acer rubrum*. Geologically it is located in the Salina depression which is underlain by Camillus shale carrying a calcareous overburden of glacial till (Fairchild, 1928). Small streams and seepage water from the Niagara escarpment of Lockport dolomite to the north and the Onondaga escarpment of Onondaga limestone to the south drain highly calcareous water into this depression.

Water flows at or near the surface as it passes through this ecosystem in a NE direction. Marl is actively being deposited from these calcareous waters near the center of the mire with depositions of up to two meters being accumulated. In some of the fens calcite marl chips, well mixed with peat, have been deposited to a depth of 4–5 m.

Most of the area of these fens, on either marl or peat substrate, is dominated by sedges and other herbaceous plants, but there are also shallow pools dominated by the alga *Chara vulgaris*. Scattered across the fens are small islands of vegetation built on hummocks of peat. These islands are of various height (10 to 50 cm) with the highest dominated by *Sphagnum* spp., *Thuja occidentalis, Pinus strobus,* and *Larix laricina* rather than by sedges.

Muenscher (1946, 1948) and his collaborators (Brown, 1948; Hotchkiss, 1950; Rogerson & Muenscher, 1950; and Winne, 1950) have provided extensive species lists for the entire area. Stewart & Merrell (1937) organized the vegetation into associations and zones which were not 'distinct and easily separable units' and produced a successional sequence proceeding from open marl areas to *Sphagnum* associations, then to pine-hemlock forest and finally to beech-maple climax. Walker (1974) also studied areas of the fens and collected environmental data. Recently, Seischab (1977) examined the same fens investigated in this study. He ordinated the vegetation samples using both Bray-Curtis and Gaussian ordination techniques. These ordinations were correlated with a number of environmental factors measured along the transects including percent organic matter and percent carbonate-carbon in the soils, soil pH, depth to and fluctuation in the water table, and height of vegetation mounds. A successional sequence was inferred for each of the fen areas based on environmental parameters, vegetational growth and the ordination of the vegetation.

Methods

Field transects of contiguous, one square meter quadrats were established across four different areas of fen. Two were on a marl substrate and two were on a substrate of mixed peat and marl. The two marl areas had transects of 148 and 132 quadrats and the two peat areas 150 and 75 quadrats, for a total of 505 samples. The transects were placed so that samples were taken from pool, level fen, and hummock mounds in each area. The percentage cover of each plant species in each quadrat was recorded between July 17–24, 1980, and the location of each hummock was noted. The maximum height of the mounds in each quadrat was measured along one peat transect, noting the dominant species on each mound.

Detrended correspondence analysis (DCA), was used for ordination (Hill, 1979; Hill & Gauch, 1980). Unlike earlier methods, DCA represents samples in an ordination space in which distances have a consistent meaning in terms of amount of change in species composition; in addition spurious axes are minimized. As gradients of community change become long (more than 4 or 5 half changes), these advantages of DCA become important. Within-community studies already available indicate that within-community gradients are frequently long, so the development of DCA is of technical importance for such studies. DCA first-axis sample scores summarize data on numerous species (here 54) into a single number for each sample, effectively converting species data into a vegetation variable.

One reason for selecting the Byron-Bergen site was that environmental data were available for these fens from a previous study (Seischab, 1977). In the previous study weekly data on water table depth were gathered from three wells in peat fens, four wells in marl fens and four wells on shrub and tree sites. These data were gathered during the

1972–73 hydrologic year. Soil data were gathered during the same time period to a depth of 60 cm on both peat and marl sites. The 60 cm approximates the rooting depth of trees in the area.

Every 30 m along each transect a substrate core, 7.5 cm in diameter and 60 cm in depth, was removed. Cores were dried to a constant weight at 105 °C. Each core was subdivided according to morphological differences in core strata. Each stratum was examined for pH, specific conductance, organic matter content, carbonate-carbon content, cation-exchange capacity and bulk density.

The pH was determined with a Fisher Model 120 meter following the procedure of Peech (1965). The same samples were examined for specific conductance using YSI conductivity meter. Percent carbonate-carbon and percent organic matter were determined by wet oxidation according to Broadbent (1965) and Allison (1965). A sodium saturation technique of Chapman (1965) was used to determine CEC. For further details of soils procedures see Seischab (1977).

Results

The DCA of species from two marl and two peat areas gave clear results. The alga Chara and bare ground, both characteristic of shallow pools, occur on one side of the ordination. From this, the successions on marl and peat differ. Important marl site early invaders are *Rhynchospora capillacea*, *Eleocharis rostellata*, *Cladium mariscoides* and *Scirpus acutus*; on peat substrates *E. rostellata* is most important. All these species form a matrix within which the other species occur (Fig. 1).

Mounds form in each area, marl sites having *Potentilla* as the dominant mound former, while *Scirpus cespitosis* is more characteristic of the smaller mounds of peat sites. Mounds on marl are typically 8 cm to 100 cm in diameter, whereas on peat 5 cm to 50 cm. Both areas change gradually to shrub-tree sites as organic matter increases and hummocks become larger. *Thuja occidentalis* and *Acer rubrum* are most important on mounds over marl, while more typical poor fen species such as *Sphagnum fuscum*, *S. girgensohnii*, *S. warnstorfii*, *Gaylussacia baccata*, *Larix laricina*, and *Pinus strobus* are more characteristic on peat site mounds. As stated previously *Thuja occidentalis*, *Pinus strobus* and *Acer rubrum* make up the surrounding forest community.

The DCA ordination was successful in separating marl fen from peat fen. Previous ordinations using Bray-Curtis techniques as well as reciprocal averaging failed to make this separation in axes 1 and 2. In addition the RA ordination produced an arch in the display. Arching and other distortions in ordination displays has been particularly troublesome in data sets exhibiting high beta diversity. Such high diversity is the case here, the peat fen having beta diversity of 4.67 Sd, the marl fen 8.13 and the total data set a beta diversity of 6.99.

Table 1 shows that there are major environmental differences between the marl and peat fen areas. The marl areas tend to be lower in soil organic matter, having smaller and lower vegetation mounds. Marl sites also have expectedly higher carbonate-carbon levels and higher soil specific conductance but lower cation-exchange capacities. Also seen in Table 1 is a substantial difference in the water table fluctuation between treed and shrubbed areas and each of the fens.

Table 1. Water table fluctuation, vegetation mound height, percent organic matter and carbonate-carbon, cation-exchange capacity, specific conductance and bulk density in three areas of the Byron-Bergen swamp. Data are from Seischab (1977). One standard deviation is shown.

	Shrubs & Trees	Sedge fen	Marl pools
Annual water table fluctuation (cm)	34.3 ± 0.6	18.8 ± 10.8	14.9 ± 1.6
Height of vegetation mounds (cm)	26.7 ± 7.3	8.1 ± 2.6	0.06 ± 0.04
Soil organic matter (%)	21.2 ± 1.2	21.4 ± 3.1	6.4 ± 0.8
Soil carbonate-carbon (%)	4.7 ± 1.5	6.1 ± 1.3	8.2 ± 1.2
Cation-exchange capacity (me/100 g)	113.8 ± 21.7	77.2 ± 13.0	13.8 ± 0.1
Soil specific conductance (μmhos/cm)	202.0 ± 31.0	216.0 ± 28.0	318.0 ± 67.0
pH	7.4 ± 0.0	7.3 ± 0.1	7.5 ± 0.1
Bulk density (g/cm)	0.2 ± 0.0	0.2 ± 0.1	0.6 ± 0.0

The ordination is suggestive of two gradients. Along axis 1 is a hydrologic as well as physiognomic gradient. Axis 2 is suggestive of a soil organic matter, carbonate-carbon, nutrient gradient.

For a more intensive analysis of within- and between-community pattern, we concentrated on the peat site for which elevation of each quadrat was measured. Figure 2 shows the DCA ordination for the 150 samples. The points on the left of the figure (fen) represent those samples which had small mounds of less than ca 15 cm in height. The points on the right have mound heights greater than 15 cm, and those in the hummock section represent sample quadrats of hummocks with heights to ca 45 cm and surrounded by fen. The data points in the forest section are samples where the mounds average 15–30

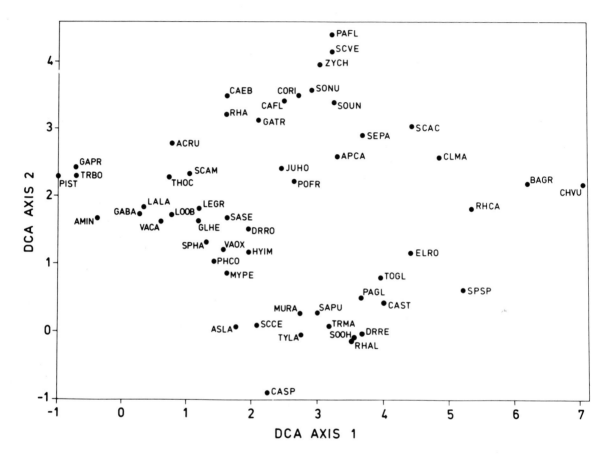

Fig. 1. Scatter figure for 54 species on peat and marl sites as ordinated by decorana. The first axis proceeds from shrub an tree sites, through sedge fen sites, to marl pools on the right. The second axis separates peat (below) and marl (above) samples in the mid (sedge) area of the first axis. The species names are: ACRU = *Acer rubrum,* AMIN = *Amelanchire intermedia,* APCA = *Apocynum cannabinum,* ASLA = *Aster lateriflorus,* BAGR = Bare ground, CAEB = *Carex eburnea,* CAFL = *Carex flava,* CASP = *Carex spicata,* CAST = *Campylium stellatum,* CHVU = *Chara vulgaris,* CLMA = *Cladium mariscoides,* CORI = *Comandra richardsiana,* DRRE = *Drepanocladus revolvens,* DRRO = *Drosera rotundifolia,* ELRO = *Eleocharis rostellata,* GABA = *Gaylussacia baccata,* GAPR = *Gaultheria procumbens,* GATR = *Galium triflorum,* GLHE = *Glechoma hederacea,* HYIM = *Hypnum imponens,* JUHO = *Juniperus horizontalis,* LALA = *Larix laricini,* LEGR = *Ledum groenlandicum,* LOOB = *Lonicera oblongifolia,* MURA = *Muhlenbergia racemosa,* MYPE = *Myrica pensylvanica,* PAFL = *Panicum flexile,* PAGL = *Parnassia glauca,* PHCO = *Phragmites communis,* PIST = *Pinus strobus,* POFR = *Potentilla fruticosa,* RHAL = *Rhynchospora alba,* RHA = *Rhamnus anifolia,* RHCA = *Rhynchospora capillacea,* SAPU = *Sarracenia purpurea,* SASE = *Salix serissima,* SCAC = *Scirpus acutus,* SCAM = *Scirpus americanus,* SCCE = *Scirpus cespitosus,* SCVE = *Scleria verticillata,* SEPA = *Senecio pauperculus,* SONU = *Sorghastrum nutans,* SOOH = *Solidago ohioensis,* SOUN = *Solidago uniligulata,* SPHA = *Sphagnum* sp., SPSP = *Spiranthes* sp., THOC = *Thuja occidentalis,* TOGL = *Tofieldia glutinosa,* TRBO = *Trientalis borealis,* TRMA = *Triglochin maritima,* TYLA = *Typha latifolia,* VACA = *Vaccinium canadensis,* VAOX = *Vaccinium oxycoccus,* and ZYCH = *Zygadenus chloranthus.*

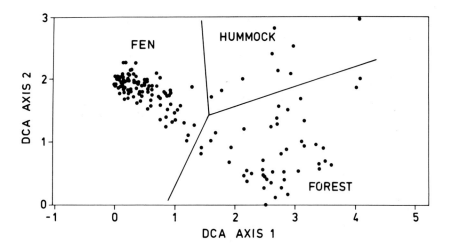

Fig. 2. Scatter figure from DCA ordination of 150 samples from one peat fen area. The fen section represents samples dominated by herbaceous species, the hummock section represents samples on hummocks in the herbaceous fen area, and the forest section represents samples on hummocks at the edge of the encroaching forest.

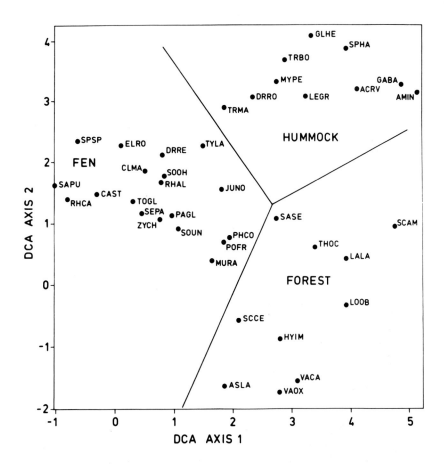

Fig. 3. Scatter figure from DCA ordination of 39 species from one peat fen area. The sections are the same as in Fig. 2 and species codes the same as in Fig. 1.

130

cm in height and are within the area of encroaching forest.

The beginnings of mound formation occur in a similar manner both at the edge of the encroaching forest and in the pattern of randomly arranged hummocks within the fen. These mounds are initiated autogenically by the mosses *Drepanocladus revolvens* and *Campylium stellatum*, the sedge *Scirpus cespitosus*, and the shrub *Potentilla fruticosa* (Seischab, 1977). Additional species invade the mounds as they grow higher. Typical mound invaders in the fen are shown in the hummock section of Figure 3, and include *Myrica pensylvanica, Ledum groenlandicum, Acer rubrum, Sphagnum* spp. and *Gaylussacia baccata. Thuja occidentalis* and *Larix laricina* are typical invaders of mounds within the area of the encroaching forest. Mound formation in either fen or encroaching

forest generates another environmental gradient from wet fen surface to promontory top. Since the promontories are then invaded by shrubs and trees (Fig. 3) these microtopographic alterations facilitate further succession.

Since one of the DCA axes was suggestive of the accumulation of organic matter and the development of mounds we used a direct measure of hummock height to further clarify the pattern. The relationship between the first axis DCA scores, mound height, and dominant mound species is shown in Figure 4. The tallest mounds are on the right side of the figure and tend to be dominated by *Sphagnum* spp. and *Gaylussacia baccata*. The fen to encroaching forest gradient occurs on small promontories (left side of Figure 4). The lowest portions of the fen are dominated by *Eleocharis rostellata*, but as these form small mounds *Campylium stellatum* also en-

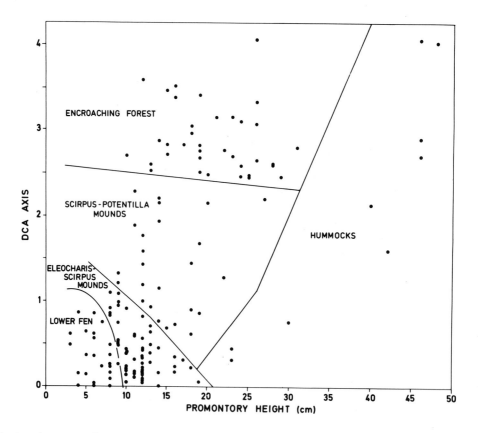

Fig. 4. Important species on mounds as plotted against DCA sample score and promontory height in five areas of peat fen. Lower fen dominants are *Eleocharis rostellata* and *Campylium stellatum. Eleocharis–Scirpus* mounds are dominated by *E. rostellata* and *S. cespitosus* and *Scirpus-Potentilla* mounds are dominated by *S. cespitosus* and *Potentilla fruticosa*. Encroaching forest areas are dominated by *Thuja occidentalis, Larix laricina*, and *Phragmites communis* on the forest edge. Hummocks are dominated by *Sphagum fuscum* and *Gaylussacia baccata*.

ters the community. Together these two species form the low (< 10 cm mounds) matrix of the fen. *Scirpus cespitosus* and *Potentilla fruticosa* are larger mound-forming species that enter the community when the mound is approximately 10 cm in height. These mounds continue growth and are invaded by shrubs and shrubby trees when they reach a height of ca 15–20 cm.

Discussion and conclusions

One objective of ordination is to describe major vegetation gradients. Figures 1–3 and Table 1 indicate complex environmental gradients of hydrology, nutrients, organic matter and carbonate concentration as well as a corresponding physiognomic gradient. These are gradients important in wetlands as shown by Heinselman (1970).

Our ordination successfully separated sedge fen from marl fen vegetation in spite of the high beta diversity (6.99) and further separated hummock from encroaching forest sites. Neither of these separations was accomplished by alternative ordination techniques.

We conclude that DCA offers a significant advantage over RA and earlier ordination techniques for the study of within- and between-community pattern because of its superior performance with long vegetation gradients (more than 4 or 5 half-changes, as observed in this study and indicated by the high beta diversity shown). DCA serves to convert species composition data into a vegetational variable whose patterns can then be compared with environmental patterns in order to generate hypotheses about the causes of within-community vegetation patterns.

References

Allison, L. E., 1965. Organic carbon. In: Black et al. (eds.). Methods of Soil Analysis. Amer. Soc. of Agronomy, Inc., Madison, Wisc.

Broadbent, F. E., 1965. Organic matter. In: Black et al. (eds.). Methods in Soil Analysis. Amer. Soc. of Agronomy, Inc., Madison, Wisc.

Brown, B. I., 1948. The vegetation of Bergen Swamp II. The epiphytic plants. Proc. Roch. Acad. Sci. 9: 119–130.

Chapman, H. D., 1965. Cation-exchange capacity. In: Black et al. (eds.). Methods in Soil Analysis. Amer. Soc. of Agronomy, Inc., Madison, Wisc.

Crum, H. & Anderson, L. E., 1981. Mosses of Eastern Norh America. Columbia Univ. Press, N. Y. 1328 pp.

Fairchild, H. L., 1928. Geologic Story of the Genesee Valley and Western New York. Rochester, N. Y. 125 pp.

Fernald, M. L., 1950. Gray's manual of Botany, 8th ed. American Bk Co, N. Y. 1632 pp.

Gauch, H. G., 1982. Multivariate Analysis in Community Ecology. Cambridge University Press, Cambridge.

Heinselman, M. L., 1970. Landscape evolution, peatland types and the environment in the Lake Agassiz Peatlands Natural Area. Ecol. Monogr. 40: 235–261.

Hill, M. O., 1979. DECORANA – A FORTRAN program for detrended correspondence analysis and reciprocal averaging. Cornell Ecology Programs, Ecology and Systematics, Cornell University, Ithaca, N. Y. 52 pp.

Hill, M. O. & Gauch, H. G., 1980. Detrended correspondence analysis, an improved ordination technique. Vegetatio 42: 47–58.

Hotchkiss, A. T., 1950. The vegetation of Bergen Swamp IV. The algae. Proc. Roch. Acad. Sci. 9: 237–266.

Muenscher, W. C., 1946. The vegetation of Bergen Swamp I. The vascular plants. Proc. Roch. Acad. Sci. 9: 64–117.

Muenscher, W. C., 1948. The vegetation of Bergen Swamp III. The myxomycetes. Proc. Roch. Acad. Sci. 9: 131–137.

Peech, M., 1965. Hydrogen ion activity. In: Black et al. (eds.). Methods in Soil Analysis. Amer. Soc. of Agronomy, Inc., Madison, Wisc.

Rogerson, C. T. & Muenscher, W. C., 1950. The vegetation of Bergen Swamp. VI. The fungi. Proc. Roch. Acad. Sci. 9: 277–314.

Seischab, F. K., 1977. Plant community development in the Byron-Bergen Swamp: A rheotrophic mire in Genesee County, N.Y. Ph.D. Thesis, State Univ. Col. of Environmental Science & Forestry, Syracuse, N.Y. Univ. Microfilms Int. Publ. No.: 78: 6183.

Shmida, A. & Whittaker, R. H., 1981. Pattern and biological microsite effects in two shrub communities, southern California. Ecology 62: 234–251.

Stewart, P. A. & Merrell, W. D., 1937. The Bergen Swamp: An ecological study. Proc. Roch. Acad. Sci. 7: 209–262.

Walker, R. S., 1974. The vascular plants and ecological factors along a transect in the Bergen-Byron Swamp. Proc. Roch. Acad. Sci. 12: 241–270.

Whittaker, R. H., Gilbert, L. E. & Connell, H. H., 1979. Analysis of two-phase pattern in a mesquite grassland, Texas. J. Ecol. 67: 935–952.

Whittaker, R. H. & Naveh, Z., 1979. Analysis of two-phase patterns. In: Patil, G. P. & Rosenzweig, M. (eds.). Contemporary Quantitative Ecology and Related Ecometrics, pp. 157–165. Statistical Ecology, Vol. 12. Internat. Coop. Publ. House, Fairland, Maryland.

Whittaker, R. H., Niering, W. A. & Crisp, M. D., 1979. Structure, pattern and diversity of a mallee community in New South Wales. Vegetatio 39: 65–76.

Winne, W. T., 1950. The vegetation of Bergen Swamp VII. The Bryophytes. Proc. Roch. Acad. Sci. 9: 315–326.

Accepted 15. 3. 1983.

Vegetation patterns related to environmental factors in a Negev Desert watershed*

L. Olsvig-Whittaker[1], M. Shachak[2] & A. Yair[3]**
[1] *Section of Ecology and Systematics, Cornell University, Ithaca, NY 14853, U.S.A.*
[2] *Desert Research Institute, Ben Gurion University of the Negev, Sede Boquer, Israel*
[3] *Department of Geography, Hebrew University, Jerusalem, Israel*

Keywords: Desert, Detrended correspondence analysis, Ecosystem, Microsites, Negev Desert, Ordination, Pattern analysis, Spatial heterogeneity, Species diversity, Vascular plants, Watershed

Abstract

Three strip transects, each ca 100 contiguous 0.5×1 m^2 quadrats, were sampled during the spring bloom of March 1981 across four surface structural units of a Negev Desert research watershed at Sede Boqer, Israel. Presence of all vascular plants was recorded. Data were subjected to detrended correspondence analysis (DCA ordination), and resulting spatial patterns of species distribution and abundance were compared. Large-scale gradients of vegetation were related to differences in soil moisture availability among the four structural units. Where micro-scale vegetation patterns were important, these correlated with rock and crevice microtopography. Species richness was influenced by high numbers of therophytes on the dry upper slope of the watershed and their reduced importance on the lower three units. Relationships between vegetational patterns and known ecosystem properties of the watershed are discussed.

Introduction

In arid land ecosystems such as the Negev Desert (Israel), variation in soil moisture availability may be one of the most important causes of spatial heterogeneity in plant communities (Evenari *et al.*, 1971). Previous studies conducted in a Negev Desert watershed at Sede Boqer have demonstrated that surface properties of rock and soil (e.g. relative cover percentages, rock surface structure, soil surface permeability) are the main factors controlling the spatial distribution of soil moisture at various scales (Yair, Sharon & Lavee, 1978, 1980). From this work a basic question has emerged: do the spatial patterns of distribution and abundance of the biotic communities (both plants and animals) in this watershed correspond to patterns of surface properties? This paper is one in a series of Sede Boqer studies on this question (see Yair & Danin, 1980; Yair & Shachak, 1982). The assumption that spatial patterns in the vascular plant community correspond to patterns in soil surface properties of the watershed is examined in the present study using ordination-based pattern analysis developed by R. H. Whittaker (Whittaker, Gilbert & Connell, 1979; Whittaker & Naveh, 1979; Whittaker, Niering & Crisp, 1979; Shmida & Whittaker, 1981).

* Nomenclature follows Zohary (1962).
** We thank the staff of the Desert Research Institute at Sede Boqer for facilities and field assistance in the execution of the transect work for this study. In particular, we wish to thank Dr Yitzchak Gutterman, Ms Jael Bar, and Mr Bert Boeken for their help in the field and identification of species. We also thank Dr Avinoam Danin for help with species identification. Computer analyses were conducted by the senior author at Cornell University, where we thank Mr Hugh Gauch, Jr, for advice and assistance with the computer programs, and his comments on interpretation of the data. We thank Dr Danin and Prof. Zev Naveh for advice and comment.

This paper is dedicated to the memories of the senior author's husband, Robert H. Whittaker, and father, Edward E. Olsvig, who anticipated and encouraged the early stages of this research.

Site description

The Sede Boqer research watershed is located in the northern Negev Desert of Israel, about 40 km south of Beersheva, at an altitude of 510 m (Fig. 1, inset). Average annual rainfall is 92 mm, with extreme recorded values of 34 and 167 mm. Rainfall is limited to winter (October–April) with the number of rain days varying between 15 and 42. Mean monthly temperatures vary from 9 °C in January to 25 °C in August (climatological data from Evenari *et al.*, 1971).

The study site (Fig. 1) covers an area of 1.1 ha on the north-facing hillside of a first order drainage basin. Local stratigraphy is Upper Cretaceous (Turonian of Arkin & Braun, 1965), represented by three limestone formations called (from top of slope to bottom): Netser, Shivta, and Drorim. Although similar in composition, these formations differ greatly in structural properties, which create four different meso-scale physical environments for the plant communities: (A) The upper part of the Netser formation, at the top of the watershed, is thinly bedded, densely jointed chalky limestone with very shallow, patchy soil. (B) The lower portion of the Netser formation is more massive, resembling the strongly step-like, crystalline limestone Shivta formation which lies underneath. Lower

Netser and Shivta should be considered one unit. The Drorim formation is subdivided into two structural units: (C) an upper unit of massive limestone with extensive bedrock outcrops, and (D) a lower unit which is densely jointed and covered with an extensive colluvial mantle (Yair & Danin, 1980; Yair & Shachak, 1982). These four structural units, in the sequence just given, will henceforth be referred to as Units A, B, C, and D.

Throughout the site, soil material is mainly loess (Yaalon & Dan, 1974) which is high in silt and sand (85–95%). The clay fraction forms a higher percentage (14.5%) of soil in joints and crevices than in the soil covering flatter bedrock surfaces (7–10%; Yair & de Ploey, 1980).

The vegetation of this region is considered to be a transition between the Irano-Turanian plant geographical region and the Saharo-Arabian region, with some Mediterranean components in the most mesic sites (Danin *et al.*, 1975; Yair & Danin, 1980). The watershed slope has a range of communities from semi-desert (10–30% perennial vegetation cover) on the rocky upper slopes, to some patches of true desert (less than 10% perennial cover) on the lower colluvium. Perennials include the shrubs and semishrubs *Artemisia herba-alba, Gymnocarpos decander, Hammada scoparia, Noaea mucronata, Reamuria negevensis,* and *Zygophyllum dumosum*

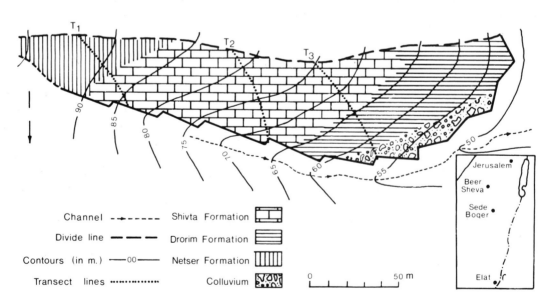

Fig. 1. Sede Boqer research watershed, showing the four major structural units: Netser (Unit A), Shivta (Unit B), Drorim (Unit C), and Colluvium (Unit D). Inset: Map of southern Israel, showing location of Sede Boqer. T1, T2, T3 are transects 1, 2, and 3, respectively.

as dominant species. In addition, during the rainy winter season, there is an assortment of annuals, geophytes, and hemicryptophytes (Table 1). Units A and D are both rich in annuals. Geophytes such as *Iris, Gagea, Ornithogalum, Scilla,* and *Tulipa* are a striking feature of Unit B and are important in the vertebrate ecology, nutrient cycling, water dynamics, and soil movement on the watershed slope (Yair & Shachak, 1982).

Methods

Pattern analysis, general methods

Most traditional measures of within-community patterns test contagion and species association on only one or two species at a time (e.g. Kershaw, 1957; Greig-Smith, 1964; Pielou, 1974). Since our present research is concerned with more general ecosystem patterns, we will confine our analyses to those parameters which integrate information from the whole flora thereby revealing major patterns of differentiation in the plant community. These measurements include species diversity statistics, for example, as well as the results of community ordination and classification.

The system of pattern analysis used in this study was developed over the past decade by R. H. Whittaker and other colleagues in our research group (see Whittaker, Gilbert & Connell, 1979; Whittaker & Naveh, 1979; Whittaker, Niering & Crisp, 1979; Olsvig(-Whittaker), 1980; Shmida & Whittaker, 1981). Previous studies have mostly concentrated on biologically induced patterns in shrubland mosaic vegetation, although such methods are equally appropriate to examining the kind of physically induced patterns which conern us here.

Field data for this type of pattern analysis is taken from a strip transect, i.e. a series of contiguous small plots. One hundred plots seem in practice to be the usual minimum sample size required for useful results. In each plot, presence or coverage of species is recorded, along with pertinent additional data.

Earlier studies by our group used reciprocal averaging (RA; Hill, 1973, 1974; Gauch *et al.*, 1977) as the main ordination method for pattern analysis. More recently we have used detrended correspondence analysis (DCA; Hill, 1979; Hill & Gauch,

1980), an improved version of RA. Both algorithms ordinate samples and species simultaneously, so that the final output matrix places samples with similar composition, and species with similar distribution in the samples, adjacent to each other (for details, see Gauch, 1982). However, with DCA the axes have been rescaled so that the difference in value between two sample scores reflects their relative difference in species composition. Because typically only the first and second DCA axes are ecological interpretable, we use only the first two axes scores for pattern analysis. Generally, vegetation differences evident in these scores are examined in one of two ways: 1) by a scatter plot of sample or species scores on the first two DCA axes (e.g. Fig. 2) or 2) by a plot of sample scores from a single DCA axis against sample position on the strip transect (e.g. Fig. 3). Such a plot of ordination position to real sample position has been termed an ordination 'trace' (Whittaker, Gilbert & Connell, 1979; Whittaker & Naveh, 1979; Whittaker, Niering & Crisp, 1979).

Sample to sample variations along a DCA ordination trace are a reflection of micro-scale variation in species composition. The meso-scale changes in vegetation, which may also be evident in the DCA trace, are more difficult to spot. If there is a major change in vegetation along the transect, it may be continuous. In this case, it will show on the first axis DCA trace as a relatively smooth trend of change in sample scores from one end of the transect to the other (Fig. 3B). In these circumstances, a regression line fitted to the trace will illustrate the meso-scale change quite well. The slope of the line, like the maximum difference in scores, reflects the degree of change along the derived coenocline. The regression coefficient, r, indicates the significance of the observed trend.

If the meso-scale change is an abrupt one, with two or more relatively homogeneous sets of samples (i.e. vegetation type patches) which are quite different from each other, the DCA trace will show clusters of sample scores with sharp score changes between sets. An extreme of this situation occurs when the between-patch variation forms a strong, repeating 'wave' in the DCA trace (e.g. Fig. 3C). This situation has been termed 'two-phase vegetation' (Whittaker & Naveh, 1979). In Figure 3C, clusters of samples in one microsite type (rock) have similar scores, which differ greatly from scores for the other microsite type (soil pockets). In such a

Table 1. Average frequency of occurrence of species.

Species / Transect:	1			2				3	
Units:	A	INT	B	B_1	B_2	C	D	B_1	B_2
Therophytes									
Aizoon hispanicum	9.3								
Anagallis arvensis		16.6	5.1						
Anthemis pseudocotula		33.3		3.8		12.5			
Biscutella didyma	2.3								
borage sp.	2.3								
Bromus sp. A	9.3	16.6	10.3		3.8			2.0	24.0
Bromus sp. B			12.8						
Calendula arvensis								2.0	2.0
Carrichtera annua	7.0				11.5				4.0
composite sp.									
Cuscuta sp.	7.0	83.3	33.3	26.9	26.9	37.5	41.7		
Cutandia memphatica	34.9								
Erodium desertorum	18.6								
Filago contracta				11.5					
F. desertorum	34.9	33.3	2.6		19.2	29.2	20.8		
Gymnarrhena micrantha	14.0							4.0	
Helianthemum ledifolium									4.0
Lappula sessiliflora	2.3								
Lathyrus pseudocicera (?)	4.7								
Linaria haelava	2.3								
Lophochloa pumila	2.3			11.5	15.4	8.3	4.2		
Matthiola livida	39.5								
Minuartia picta	55.8								
Ononis sicula				3.8			4.2	2.0	
Picris damasena				3.8	3.8	20.8	62.5	28.0	28.6
Plantago coronopus								4.0	24.0
P. ovata	18.6				19.2	4.2			
Pteranthus dichotomus	7.0								
Reboudia pinnata	2.3		2.6						
Scabiosa porphyromeura				7.7					6.0
Schismus arabicus							4.2		2.0
Senecio desfontanii (?)	7.0		2.6			4.2	4.2	2.0	
Stipa capensis				3.8	19.2	16.7	12.5	8.0	26.0
Trigonella arabica	18.6					4.2			
T. stellata				7.7					
Total number of species	21	5	7	9	8	9	8	8	9
Average frequency (%)	14.3	36.6	9.9	8.9	14.9	15.6	19.3	6.5	13.4
Geophytes									
Allium sp.			2.6	11.5					
Bellevalia desertorum			2.6					2.0	
Colchicum tuviae			5.1						
Crocus sp. (?)	2.3		2.6					4.0	26.0
Gagea chlorantha				15.4					
G. reticulata	16.3		7.7	7.7	3.8	8.3			
Iris edumea				3.8					
I. sisyrinchium				3.8					
I. regni-uzyae								6.0	
Ornithogalum narbonense									2.0
Scilla hanburyi			2.6	3.8	11.5	8.3			6.0
Tulipa montana			10.3	11.5		25.0	41.7	2.0	4.0
Total number of species	2		7	7	2	3	1	4	4
Average frequency (%)	9.3		4.8	8.2	7.7	13.9	41.7	3.5	9.5

Table 1. (Continued).

Species / Transect:	1			2				3	
Units:	A	INT	B	B₁	B₂	C	D	B₁	B₂
Hemicryptophytes									
Anemone coronaria					3.8			2.0	
Carex pachystylis			2.6						
Centaurea aegyptiaca							4.2		
C. eryngioides	4.7	33.3	71.8	26.9	23.1			10.0	2.0
Diplotaxus harra	32.6		7.7	7.7	19.2	16.7	16.7	10.0	6.0
Erodium hirtum	76.7	100	64.1	65.4	69.2	45.8	25.0	38.0	28.0
Haplophyllum tuberculatum	2.3								
Helianthemum sp.	2.3		5.1		11.5	16.7	12.5		
Launea nudicaulus	60.5	33.3	30.8	7.7					
Piptatherum mileacea				3.8				4.0	
Ranunculus asiaticus						8.3			
Scorzonera judaica	44.2	16.6	10.3				29.2		
Total number of species	7	4	7	4	5	4	5	6	3
Average frequency (%)	31.9	45.8	27.5	26.9	25.4	21.9	17.5	12.8	12.0
Chamaephytes									
Artemisia herba-alba	51.5	50.0	61.5	61.5	69.2	75.0	83.3	26.0	40.0
Asparagus stipularis								2.0	2.0
Astragalus sanctus	20.9		23.2	3.8	3.8	8.3	8.3	4.0	10.0
Echinops polyceras				3.8				2.0	4.0
Ephedra aphylla								8.0	
Gymnocarpos decandrum	4.7		46.2	34.6	30.8	29.2		18.0	22.0
Hammada scoparia	7.0								
Helianthemum kahiricum	34.9	100	64.1	15.4	3.8	20.8	8.3	4.0	
H. ventosum		16.6	23.1	3.8		16.7	45.8	8.0	18.0
H. vesicarium	9.3		41.0	11.5	11.5	58.3	75.0	16.6	6.0
Limonium pruinosum			2.6						
Noaea mucronata			23.1	19.2	46.2	33.3	66.7	32.0	28.0
Paronychia syriaca								4.0	
Pituranthos tortuosus								2.0	
Reamuria negevensis	23.3	16.6	12.8	23.1				6.0	2.0
Salvia lanigera							8.3		
Stachys aegyptiaca				7.7				2.0	6.0
Varthemia iphionoides						4.2			2.0
Zygophyllum dumosum	23.3								2.0
Total number of species	8	4	9	10	6	8	7	14	12
Average frequency (%)	21.9	45.8	33.1	18.4	27.6	30.7	42.2	9.6	11.8
Total species, all life forms	39	13	30	30	21	24	21	31	27

situation, the trend of change in vegetation from one end of the transect to the other is relatively insignificant compared to the microsite differences.

Pattern analysis, Sede Boqer

For this study, three separate transects of contiguous (0.5 × 1 m²) plots were laid out on the north-facing slope of the watershed, during March 1981. Although our major interest was in the nature of the transition between geological/structural units, the character of the watershed made it impossible to include all four structural units in a single transect without changing other important factors, such as slope or aspect. Therefore, two parallel transects, each intersecting two or more structural boundaries, were used. Transect 1 extended for 88 quadrats within watershed study plot 10 (Fig. 1); there were in this area only 44 m between crest and wadi, of which roughly half were in Unit A and half in Unit

B. Transect 2 included 100 quadrats, beginning in Unit B and ending in Unit D, within study plot 5. The additional transect 3 was entirely within Unit B, study plot 7, and was included to compare patterns entirely within one structural unit against patterns which occur when two or more units are encountered. Each transect began on the upper slope of the watershed, and extended to the wadi. In each quadrat, the presence of all vascular plant species was recorded. In addition, field notes on topography and visual and probe estimates of soil depth were recorded for each quadrat. DCA ordination was done for each transect separately. Ordinations combining all three transects were also conducted, but added no new information.

Species diversity

Most diversity indices, such as the Shannon-Wiener, combine different diversity parameters, leading to ambigous results if the component parts are not responding in the same way to environmental patterns. For this reason, it is useful to examine separately the basic components of diversity.

Within a site, diversity parameters include 1) species richness on the micro-scale (point or quadrat species richness); 2) species richness on the meso-scale (site floristic richness or alpha diversity *sensu* Whittaker, 1972, 1977); 3) the degree of change in species composition along a coenocline or environmental gradient within the site; and 4) total site heterogeneity. Both terms 3) and 4) have been called *beta diversity* (Whittaker, 1975) when applied to standard coenocline data, and *pattern diversity* when applied to within-site transect data (Whittaker & Naveh, 1979). However, change in species composition along a single ordination-based coenocline is a derivation, i.e. an abstraction which does not usually reflect the full compositional variation of the data set. Therefore term 4) will be distinguished in this paper as *site heterogeneity*. Only 3), the turnover of species along the derived coenocline, will be termed *beta* or *pattern diversity*.

The following methods were used to obtain the four diversity parameters discussed above:

1. Point diversity, $\bar{S}$, was determined as the average number of species in a 0.5 m^2 quadrat, averaged over 10 quadrats per structural unit, per transect.

2. Alpha diversity, S_c, was determined as the total number of species in each structural unit of each transect, or the combined species richness. Since the value of S_c is clearly dependent on sample size, and comparisons are valid only between sample sets of equal size, an additional statistic was used. S_c10 was the combined species richness of 10 quadrats per structual unit. It was used as a check on trends observed in S_c.

3. Pattern diversity, or species turnover on a coenocline, can be determined using a number of different indices (Gauch, 1982; Wilson & Mohler, 1982). Turnover of species along a coenocline can be quantified in half-changes (*HC*), or 50% changes in species composition (Whittaker, 1960; Gauch & Whittaker, 1972). Wilson & Mohler (1982, 1983) propose the Gleason (*G*) as the compositional turnover occurring if all change were concentrated in a single species whose abundance changed 100%. 'For example, a 1 *G* coenocline might consist of three species, two of which change from 50% to 0% relative abundance as the third species changes from 0% to 100% relative abundance.' DCA ordination provides a third index, the number of SD units, or average standard deviations of species distributions, which is used to scale the ordination axis length and correlates with species turnover (Hill, 1979; Hill & Gauch, 1980; Gauch, 1982). In the present study, we used three calculations of *HC* values: 1) Whittaker's graphical estimate, 2) Wilson & Mohler's refinement of that estimate (the GRADBETA program; Wilson & Mohler, 1982, 1983), and 3) Hill's formula:

$$HC = 2.568 \sqrt{EV/(1 - EV)}$$

where *HC* is the number of half-changes, and *EV* is the eigenvalue of the DCA ordination axis selected for the coenocline (M. O. Hill, personal communication).

Gleason values were obtained using Wilson & Mohler's GRADBETA program, and were analyzed on the basis of 10 composite samples from each transect. The composite samples were formed by dividing the DCA ordination coenocline in 10 equal units, and pooling all samples in each unit.

SD units were obtained directly from the DECORANA ordination output of each transect.

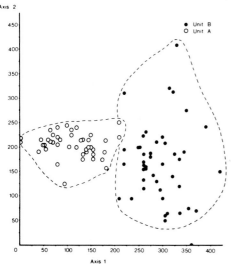

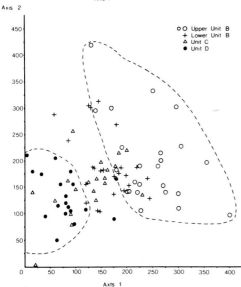

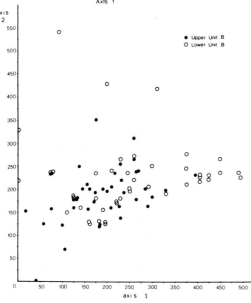

4. Site heterogeneity was calculated by $S_c/\bar{S}$ for each structural unit. The same caveat about sample size discussed for S_c applied here, and therefore $S_c 10/\bar{S}$ was also calculated. An alternative, visual demonstration of site heterogeneity, was the ordination trace (Fig. 3), which provided information on the range of sample scores as well as their pattern of distribution.

Results

DCA ordination: scatter plots

Scatter plots of sample scores from DCA axes 1 and 2 (Fig. 2) showed definite separation of samples from the four structural units. However, the discreteness of this separation varied, depending on the structural units involved. There was complete separation between samples from Units A and B (Transect 1, Fig. 2A), but overlap between Units B, C, and D (Transect 2, Fig. 2B). When only one structural unit was involved (Unit B, Transect 3, Fig. 2C), there was little meso-scale differentiation between top and bottom of the slope. Thus, at least in Unit B, slope position did not seem important in within-unit differentiation of vegetation.

DCA ordination: sample score traces

The DCA traces for the three transects (Fig. 3) provide information about pattern on both the micro-scale and meso-scale. Only Transects 1 and 2 showed a major change in vegetation (as indicated in the first axis of DCA ordination) from one end of the transect to the other. The slope of the regression line through sample points was steepest (indicating greatest meso-scaling change in vegetation) on Transect 1, i.e. the transition from Unit A to B. Note that this transect also showed the clearest separation of samples on the scatter plots of Figure 2. In contrast, Transect 3 (Fig. 3) showed no significant meso-scale change in vegetation.

Fig. 2. Scatter plots for first two axes of DCA sample ordinations. (A) Transect 1: Unit A (Upper Netser) to Unit B (Lower Netser and Shivta). (B) Transect 2: Unit B (Shivta) to Units C (Upper Drorim) and D (Lower Drorim). (C) Transect 3: Unit B. On Transect 3, samples were equally divided between Upper and Lower Shivta. Axis lengths are scaled in SD units of DCA ordination.

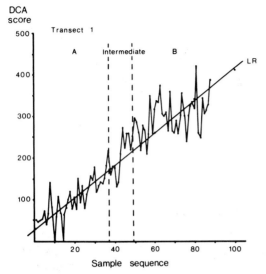

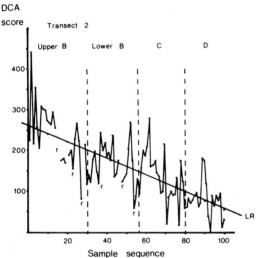

Transect 3, confined to structural unit B, was meant to test the relative importance of meso-scale patterns (especially position on the watershed slope) versus micro-scale patterns. As observed above, meso-scale differentiation did not seem important within the unit. Instead, the greatest contrast in sample scores reflected micro-scale conditions: the lowest value generally corresponded to rock crevice samples, and the highest to vegetation on thin surface soil (Fig. 3C). Thus, within Unit B, the most important differentiation of vegetation was caused by the physical micro-scale structure of the substrate. It should be noted that this is also physically much more heterogeneous than the other three units, with sharp contrast between the properties of rock outcrop and crevice (Fig. 4). Hence relatively greater microsite differentiation of vegetation is not surprising.

Point species diversity

The species richness obtained as an average of 10 quadrats within each transect segment (structural unit) is presented in Table 2. Unit A has by far the greatest total species richness, with an average of 8.4 species per 0.5 m^2. Separation of growth forms shows that this is mainly due to the large number of annual therophytes and hemicryptophytes in Unit A.

Alpha diversity

The total number of species in each structural unit (Tables 1 and 2) is affected by the number of samples from each. Since there is some variation in sample size among the four units, these data must be treated with caution. At every scale, however, the species richness of Unit A is much greater than that of other units (Table 2). This is mainly due to the large number of annuals (therophytes) in Unit A. These samples were taken in a high rainfall year, however, when the annual flow was rich. A low rainfall year might not produce the same results.

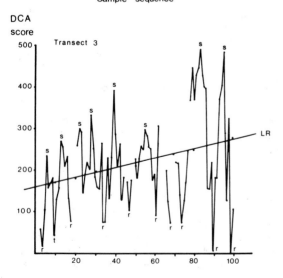

Fig. 3. DCA traces of first axis sample ordination scores. (A) Transect 1: Units A to B. (B) Transect 2: Units B, C, and D. (C). Transect 3: Unit B only. Abbreviations: LR = linear regression line through DCA trace; r = rock outcrop, s = soil pocket or crevice. Axis 1 is the sample sequence on the transect; Axis 2 is the sample score on the first DCA ordination axis.

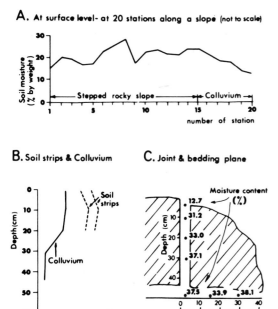

A. At surface level- at 20 stations along a slope (not to scale)

B. Soil strips & Colluvium

C. Joint & bedding plane

Fig. 4. Soil moisture on the Sede Boqer watershed slope. (A) Percentage surface soil moisture from Unit B (Shivta) at left, to Unit D (Colluvium) at right. Note high variation in Unit B soil moisture, and the steady decline in soil moisture through the colluvium. (B) Contrast in soil moisture between Colluvium and soil strips (crevices) of Unit B. Note the increase in soil moisture with depth in Unit B, with a maximum at 10 cm depth. (C) Pattern of soil moisture within crevices of Units B and C. Note that the soil accumulated under rock in the horizontal bedding plane has the highest soil moisture (37.5–38.1%). Data from A. Yair & M. Shachak.

Spatial heterogeneity

$S_c/\bar{S}$ is dependent on the total number of samples used to calculate S_c. Therefore both S_c of whole structural units and S_c10, the combined richness over 10 samples per unit, are presented in Table 2A. In both cases, the highest values for spatial heterogeneity occurred in Unit B, and the lowest in Unit D. This trend seems due to the relatively low point diversity of Unit B. However, it is not simply related to the reduction in annual species which occurs in Unit B, since the same trends are observed when $S_c/\bar{S}$ and $S_c10/\bar{S}$ are calculated without annuals (Table 2B).

Pattern diversity

Five indices of pattern diversity are presented for the three transects (Table 2B). Three of these indi-

ces, the Whittaker measure of half-change (HC), Wilson & Mohler's refinement of Whittaker's index, and Wilson & Mohler's Gleason index (G), show the same trend: highest values for Transect 2 and lowest values for Transect 3. Apparently pattern diversity is directly correlated to the number of structural units represented in the coenocline.

Hill's eigenvector estimate of half-change HC-(EV), shows the reverse trend. Since the first three indices require compositing of samples, we checked the effect of making composite samples on $HC(EV)$ as well. The same trend was observed as before. DCA ordination SD lengths correlated with HC (EV) on composite samples, but not on original samples. We must conclude that both $HC(EV)$ and the SD units are unreliable estimates of pattern diversity defined as species turnover along a gradient.

Patterns of soil moisture

The information currently available about soil moisture and related parameters, based on published and ongoing studies of the Sede Boqer watershed, is presented in Table 2 and Figure 4. Soil moisture following rainfall is highest in Unit B, and equally low in Units A and D; electrical conductivity, primarily reflecting salinity, shows an inverse relationship (Table 3). Spatial heterogeneity, as previously discussed, seems related to relatively high moisture conditions, although species richness (including annuals) is not.

Figure 4A clearly shows the high variability of soil moisture within the rocky slope area (Units B and C) compared to a steady decline in soil moisture within the colluvium (Unit D). Where soil has accumulated within crevices of Unit B, there is a marked increase in soil moisture with depth (Fig. 4C) providing an improved moisture resource for those plants which can reach it. Even at the surface, the soil of Unit B and C contains more moisture than the colluvium of Unit D (Fig. 4B).

Discussion

Correlations

At the Sede Boqer watershed, geological surface structure is correlated with most aspects of vegetation, including patterns of species richness and

Table 2. Species and pattern diversity.

A. Species diversity and spatial heterogeneity

Transect:	1		2				3	
Units:	A	B	B$_1$	B$_2$	C	D	B$_1$	B$_2$
1. Point diversity, $\bar{S}$ (.05 m^2)								
with therophytes	8.4	6.5	4.4	4.2	5.2	4.9	2.2	4.1
without therophytes	5.1	5.8	3.4	3.5	4.2	3.8	2.1	2.1
2. Alpha diversity, 10 quadrats combined, S_c10 (5 m^2)								
with theorphytes	32	21	16	14	17	15	12	20
without therophytes	12	19	12	10	12	10	12	11
All quadrats in unit, S_c								
with therophytes	39	30	30	21	24	21	31	27
without therophytes	17	23	21	13	15	13	24	19
In entire transect								
with therophytes	54		47				40	
without therophytes	30		31				28	
3. Spatial heterogeneity, $S_c/\bar{S}$								
with therophytes	4.6	4.6	6.8	5.0	4.6	4.3	14.1	6.6
without therophytes	3.3	4.0	6.2	3.7	3.6	3.4	11.4	9.0
S_c10/$\bar{S}$								
with therophytes	3.8	3.2	3.6	3.3	3.3	3.1	5.4	4.9
without therophytes	2.4	3.3	3.5	2.9	2.9	2.6	5.7	5.2

B. Pattern diversity

Transect:	1	2	3
Whittaker's *HC*	2.7	4.4	2.2
Wilson & Mohler's *HC*	3.0	4.8	1.9
Wilson & Mohler's Gleasons	3.1	6.1	2.9
Hill's *HC(EV)*			
on composite samples	3.4	3.7	4.2
on original samples	2.9	2.5	3.8
DCA ordination SD units			
on composite samples	420	443	495
on original samples	517	471	503

composition. A general summary is presented in Table 3. The reasons for these correlations are complex, but seem founded in a basic hydrogeological feature of the Negev, and probably many other deserts as well: the pattern of runoff generation. If the environment for the desert community were uniform, flat, and comprised only of even-texture soil (a loessal plain comes close to this ideal), then the spatial distribution of soil moisture would correlate with the spatial distribution of rainfall. But in a desert on rocky hillsides, with patches of rock alternating with patches of soil, the situation is quite different. The rocky patches generate runoff, while the soil patches absorb the water. Therefore, the distribution of soil moisture will be dependent on the distribution and ratio of rock surface to soil volume. High soil moisture can be predicted where the rock/soil ratio is high and surface properties can be considered the controlling factors determining the spatial distribution of soil moisture. This relationship may be expressed by a formula:

$$\text{relative soil moisture} = f\left(\frac{SA_{\text{rock}}}{SA_{\text{soil}}} \times \frac{1}{\text{soil depth}}\right)$$

where SA is surface area.

Relatively aridity may be generated by a low $SA_{\text{rock}}/SA_{\text{soil}}$ ratio, as in Unit A of this watershed, or it may be generated by great soil depth, as in the colluvium of Unit D. In contrast, where $SA_{\text{rock}}/SA_{\text{soil}}$ ratio is high and soil depth is generally low,

Table 3. Summarization of results.

Structural unit	A	B	C	D
	Upper Netser	Lower Netser-Shivta	Upper Drorim	Lower Drorim
Structure	thin soil cover, fine-grained	massive, blocky, step-like	rock outcrop	colluvial accumulation
Transition to lower unit	sharp	gradual	gradual	--
Internal (micro-scale) heterogeneity	weak	strong	intermediate	weak
% soil moisture (Yair & Danin, 1980) after rainfall event	low, 15%	high; 25-30% in soil strips, to 38% in crevices	intermediate, 20-25%	low, 10-15%
EC (salinity) mMHO/cm (Yair, unpublished data)	high, 3.9	low; 0.4-0.7 lower Netser, 0.4-0.5 Shivta	low, 0.4	high, 5
Species richness				
Annuals	high	low	low	low
Hemicryptophytes	high	high	low	low
Geophytes	low	high	high	low
Shrubs	low	high	high	high
Spatial heterogeneity (pattern diversity)	low	high	intermediate	intermediate
Animal activity	low	intermediate-high; concentrated in Shivta	intermediate	low

as in the step-like Unit B, a relatively mesic environment is created (Table 3 and Fig. 4). This relationship seems important at all scales of environment. On the micro-scale, variation in this ratio determines the high degree of heterogeneity in Unit B (Fig. 4A). Soil moisture, salinity (Yair & Danin, 1980), and texture differ greatly between the soil in crevices and the soil strips on flat bedrock in this unit. On the macro-scale, the landscape, this same relationship may be as important as rainfall in determining aridity. For example, studies by Yair in Ramat Hovav, which is north of Sede Boqer and receives more rain, indicate that the soil is actually more arid than at the Sede Boqer watershed. This situation is apparently caused by the absence of the runoff-generating Unit B-type rock (Shivta formation).

Life forms

The results suggest that the strength of apparent vegetation response to water regime is strongly affected by the life forms of the plant species involved. Species in this site may be conveniently divided into four major groups: therophytes, hemicryptophytes, geophytes, and chamaephytes, each distinctive in its distribution pattern.

Desert therophytes in the Negev Highland exploit the top centimeters of the soil, and are abundant only in favorable years (Loria & Noy-Meir, 1979) when surface soil moisture is relatively high during the growing season. Moisture conditions of the surface soil vary less spatially than they do temporally (A. Carnelli, unpublished data). Hence spatial heterogeneity of therophytes as a response to the surface structure of their environment may be relatively low, although countered by biologically caused patterning (Ellner & Shmida, 1982).

Hemicryptophytes are larger than therophytes, on the average, and, with their deeper rooting systems, will probably be more affected by spatial heterogeneity of soil moisture below the top centimeters. These species also show more patterning (for example, are more frequent in the Unit B crevices, averaging 0.4 individuals per plot, than in the soil strips, where they average 0.1 individual per plot). This is probably related to the greater heterogeneity of the total soil moisture which they can exploit.

From our data, both therophytes and hemicryptophytes seem at a disadvantage in this site when interacting with chamaephytes and geophytes. The richness and frequency of therophytes in particular seems inversely related to that of the latter three groups (Table 1). Previous studies (Friedman *et al.*, 1977) have indicated suppression of certain annual species by *Artemisa herba-alba* in Negev vegetation, which could explain the reduced importance of therophytes in Units B, C and D.

Chamaephytes can exploit deeper soil for moisture than the other species reach. This group predominates in the upper colluvium (Unit C), where moisture resources are more limiting. Likewise, chamaephyte frequency (Table 1) is high in Unit B. Here, clay accumulation in the bedding plains beneath rocks (Yair & de Ploey, 1979) makes the available moisture harder to extract, but there is less evaporation loss. Because of larger root volume and reduced osmotic potential, the perennial groups may extract this moisture more successfully than therophytes. Thus, through competition for water and through physical limitation, the therophytes may be limited to the shallow soil strips in this unit.

Geophytes generally are the most mesic group of the four discussed here, belonging primarily to the Mediterranean component of the flora (Zohary, 1962). Most desert geophyte species are restricted to relatively mesic locations, such as Units B and C as seen in Table 1. This particular aspect of vegetation pattern also affects other ecosystem functions. In this area, porcupines (*Hystrix indica*) are a major herbivore preditor of geophytes (Yair & Lavee, 1981; Yair & Shachak, 1982). Their activity is therefore concentrated in those relatively mesic areas of Units B and C where geophytes are found. Porcupine digging loosens soil, which increases erosion from these areas but also enhances water infiltration. This in turn maintains the relatively high moisture and low salinity which the geophytes require. Hence a self-reinforcing mechanism is generated, which helps to maintain the existing vegetation pattern (Yair & Lavee, 1981; Yair & Shachak, 1982).

Transition zones

Our original hypothesis about relationships between the plant community and its environment were only partly substantiated. Surface structure, the ratio of rock to soil in particular, seems the most important factor determining the spatial pattern of the vegetation. However, the correlation is far from perfect.

The transition in vegetation from the topmost Unit A to the next structural unit downslope, B, is quite sharp (Fig. 2A). (Note that there is no overlap between samples from Units A and B in the two dimensional plot.) However, the change in vegetation composition from Unit B through Units C and D is continuous, as demonstrated in Figure 2B.

These patterns may result from the amount and importance of runoff input from the upslope unit to adjoining downslope areas. Unit A, with a slow SA_{rock}/SA_{soil} ratio, is a net 'holder', i.e. generates little runoff, while Unit B is a net 'mover', with high runoff generation (Pomeroy, 1974; Shachak, 1983). Input of the small amount of runoff from Unit A to Unit B probably has little impact on growth conditions, since Unit B is the most mesic unit in any case. The runoff input from Unit B to the xeric Units C and D, on the other hand, is high (Yair *et al.*, 1978, 1980) and probably has a strong influence on the vegetation. In other words, runoff from a xeric, therefore saline unit (holder) to a more mesic, less saline unit (mover) has little influence and hence the transition between the two units will be sharp and the units functionally independent. Runoff input from a more mesic unit to a relatively xeric unit, in contrast, raises the moisture availability in the lower unit to a degree proportionate to the amount of runoff received. This runoff will be distributed in the lower unit in the form of a moisture gradient (Fig. 4A) along which vegetation change will be gradual.

In conclusion, pattern analysis has effectively demonstrated relationships between spatial patterns of vegetation, surface structure, and movements of soil and water in this Negev Desert watershed. The pattern analysis approach developed by Whittaker to describe vegetation has proven effective in describing these relationships, and has been useful as a heuristic method for the detection of correlations upon which future research in this watershed project can be based. Conversely, the necessity of obtaining environmental information for the interpretation of vegetational pattern analysis is equally clear.

References

Arkin, Y. & Braun, M., 1965. Type sections of Upper Cretaceous formations in the northern Negev. Geol. Surv., Stratigraphic Section 2A. Jerusalem.

Danin, A., Orshan, G. & Zohary, M., 1975. The vegetation of the northern Negev and the Judean Desert of Israel. Isr. J. Bot. 24: 118–172.

Ellner, S. & Shmida, A., 1982. Why are adaptations for long range seed dispersal rare in desert plants? Oecologia 51: 133–134.

Evenari, M., Shanan, L. & Tadmor, N., 1971. The Negev: the challenge of a desert. Harvard University Press, Cambridge, Massachusetts. 343 pp.

Friedman, J., Orshan, G. & Ziger-Cfir, Y., 1977. Suppression of annuals by Artemisia herba-alba in the Negev Desert of Israel. J. Ecol. 65: 413–426.

Gauch, H. G., 1982. Multivariate Analysis in Community Ecology. Cambridge Univ. Press, Cambridge. 298 pp.

Gauch, H. G. & Whittaker, R. H., 1972. Coenocline simulation. Ecology 53: 446–451.

Gauch, H. G., Whittaker, R. H. & Wentworth, T. R., 1977. A comparative study of reciprocal averaging and other ordination techniques. J. Ecol. 65: 157–174.

Greig-Smith, P., 1964. Quantitative Plant Ecology, 2nd ed. Butterworths, London. 256 pp.

Hill, M. O., 1973. Reciprocal averaging: an eigenvector method of ordination. J. Ecol. 61: 237–249.

Hill, M. O., 1974. Correspondence analysis: a neglected multivariate method. J. Roy. Stat. Soc. Ser. C 23: 240–254.

Hill, M. O., 1979. DECORANA – A FORTRAN program for detrended correspondence analysis and reciprocal averaging. Section of Ecology and Systematics, Cornell University, Ithaca, New York 14853.

Hill, M. O. & Gauch, H. G., 1980. Detrended correspondence analysis, an improved ordination technique. Vegetatio 42: 47–58.

Kershaw, K. A., 1957. The use of cover and frequency in the detection of pattern in plant communities. Ecology 38: 291–299.

Loria, M. & Noy-Meir, I., 1979. Dynamics of some annual populations in a desert loess plain. Isr. J. Bot. 28: 211–225.

Olsvig(-Whittaker), L., 1979. Pattern and diversity analysis of the irradiated oak-pine forest, Brookhaven, New York. Vegetatio 40: 67–78.

Pielou, E. C., 1974. Population and Community Ecology: principles and methods. Gordon & Breach, New York. 424 pp.

Pomeroy, L. R., 1974. Cyclings of Essential Elements. Dowden, Hutchinson & Ross, Stroudsburg, Pennsylvania. 373 pp.

Shachak, M., 1983. Some aspects of the relationship between experimental scale and ecosystem properties. Submitted to Oecologia.

Shmida, A. & Whittaker, R. H., 1981. Pattern and biological microsite effects in two shrub communities, southern California. Ecology 62: 234–251.

Whittaker, R. H., 1960. Vegetation of the Siskiyou Mountains, Oregon and California. Ecol. Monogr. 30: 279–338.

Whittaker, R. H., 1972. Evolution and measurement of species diversity. Taxon 21: 213–251.

Whittaker, R. H., 1975. Communities and Ecosystems, 2nd ed. Macmillan, New York. 385 pp.

Whittaker, R. H., 1977. Evolution of species diversity in land communities. Evol. Biol. 10: 1–67.

Whittaker, R. H., Gilbert, L. E. & Connell, J. H., 1979. Analysis of two-phase pattern in a mesquite grassland, Texas. J. Ecol. 67: 935–952.

Whittaker, R. H. & Naveh, Z., 1979. Analysis of two phase patterns. In: Patil, G. P. & Rosenzweig, M. L. (eds.), Contemporary Quantitative Ecology and Related Ecometrics, pp. 157–165. Stat. Ecol. Ser. 12. International Cooperative Publishing House, Burtonsville, Maryland.

Whittaker, R. H., Niering, W. A. & Crisp, M. D., 1979. Structure, pattern, and diversity of a mallee community in New South Wales. Vegetatio 39: 65–76.

Wilson, M. V. & Mohler, C. L., 1982. GRADBETA – A FORTRAN program for measuring compositional change along gradients. Section of Ecology and Systematics, Cornell University, Ithaca, New York 14853.

Wilson, M. W. & Mohler, C. L., 1983. Measuring compositional change along gradients. Vegetatio (in press).

Yaalon, D. H. & Dan, J., 1974. Accumulation of loess derived deposits in the semidesert and desert fringe areas of Israel. Zeitschr. für Geom. N. P. 20: 91–105.

Yair, A. & Danin, A., 1980. Spatial variation in vegetation as related to the soil moisture regime over an arid limestone hillside, northern Negev, Israel. Oecologia 47: 83–88.

Yair, A. & de Ploey, J., 1979. Field observations and laboratory experiments concerning the creep process of rock blocks in an arid environment. Catena 6: 245–258.

Yair, A. & Lavee, H., 1981. An investigation of source areas of sediment and sediment transport by overland flow along arid hillslopes. In: Proc. Florence Symp. Erosion and Sediment Transport Measurement, pp. 433–446. IAHS Publ. 133.

Yair, A. & Shachak, M., 1982. A case study of energy, water, and soil flow chains in an arid ecosystem. Oecologia (in press).

Yair, A., Sharon, D. & Lavee, H., 1978. An intrumented watershed for the study of partial area contribution of runoff in an arid area. Zeitschr. für Geom. Suppl. B 29: 71–82.

Yair, A., Sharon, D. & Lavee, H., 1980. Trends in runoff and erosion processes over an arid limestone hillside, northern Negev, Israel. Hydrol. Sci. Bull. 25: 243–255.

Zohary, M., 1962. Plant Life of Palestine. Ronald Press, New York. 262 pp.

Accepted 30. 5. 1983.

Disturbance and vegetation response in relation to environmental gradients in the Great Smoky Mountains*

M. E. Harmon[1], S. P. Bratton[2] & P. S. White[3]
[1] Department of Botany and Plant Pathology, Oregon State University, Corvallis, OR 97331, U.S.A.
[2] U.S. National Park Service, Cooperative Unit, Institute of Ecology, University of Georgia, Athens, GA 30602, U.S.A.
[3] Uplands Field Research Lab, Great Smoky Mountains National Park, Gatlinburg, TN 37738, U.S.A.

Keywords: Community structure, Disturbance, Gradient analysis, Successional strategies, Temperate forest

Abstract

In constructing models of species and community distributions along environmental gradients in the Great Smoky Mountains, R. H. Whittaker (1956) focused on old-aged, apparently stable, natural communities. More recent studies indicate that disturbance gradients potentially influence and are influenced by the complex environmental gradients of Whittaker's original models. Using primarily fire and exotic species invasion as examples, this paper shows: 1) disturbance parameters vary along the topographic, elevation and moisture gradients in the Great Smoky Mountains in much the same way as temperature, moisture and solar radiation change; 2) species composition at different locations along the major environmental gradients is partially determined by the disturbance parameters; 3) species characteristics such as mode of reproduction are often correlated with specific disturbance parameters; 4) functional aspects of ecosystem response to disturbance vary along environmental gradients; and 5) man-caused disturbance may vary along environmental or biotic gradients. Since disturbance gradients may parallel physical environmental gradients, the two may be difficult to distinguish. Modification of disturbance frequencies along major environmental gradients may result in slow shifts in the distribution of both individual species and whole communities.

Introduction

R. H. Whittaker's classic study of vegetation of the Great Smoky Mountains (Whittaker, 1956) documented population and community distributions along major environmental gradients. In this work, Whittaker presented an enduring model of the vegetation–landscape relations of this diverse region and provided central paradigms for the field of plant ecology. As a simplification, Whittaker focused his work on old-aged, apparently stable, natural communities and avoided sites obviously influenced by recent disturbances such as fire, logging, and windfall. In the last decade, however, plant ecologists have intensified work on the role of natural disturbance in community structure and

dynamics (see e.g. White, 1979; Pickett & Thompson, 1978). Biologists in such preserves as Great Smoky Mountains National Park have also become concerned with the influence of direct and indirect human disturbance (White & Bratton, 1980).

Whittaker's original study sites are protected within the second largest eastern North American National Park (Great Smoky Mountains National Park; hereafter, GRSM). However, as is the case for most preserves worldwide, preservation does not connote stasis (Bratton & White, 1980). Natural and anthropogenic disturbances overlay Whittaker's original model of the Great Smokies coenoplane. Our purpose in this paper is to show how these disturbance gradients potentially influence and are influenced by the complex environmental gradients of Whittaker's original model. We devel-

* Botanical nomenclature follows Radford et al. (1968).

op this theme using several of the better documented disturbance regimes in this landscape.

Characteristics of disturbance gradients

In this paper we present five characteristics of disturbance–vegetation interaction along gradients.

First, disturbance parameters vary along the topographic, elevation and moisture gradients in GRSM in much the same way as temperature, moisture and solar radiation change (White, 1979; Bratton *et al.*, 1981). These disturbance parameters include

a) *Frequency* – average number of disturbance events per time period within a given area (most commonly used as point frequency, the number of events per time period at one locus in the landscape).

b) *Predictability* – inversely related to variation around mean frequency.

c) *Area* – average area disturbed per event.

d) *Cycle* – time interval required to disturb an area equivalent to the arbitrarily defined area of interest (i.e., the study area, see Heinselman, 1973). This does not assume each site was disturbed once during the cycle; some sites may not have been disturbed while others may have been disturbed several times.

e) *Severity and intensity* – these two are distinguished as follows: severity represents the impact of the disturbance on the community, thus, severity is defined in terms of change in community properties such as basal area, density, species composition and biomass; intensity refers to properties of the disturbance itself. In the case of fire, intensity equals heat released per length of fireline per time, whereas, in the case of rainstorms, intensity equals volume or depth of precipitation per time period.

Second, the species composition at different locations along the major environmental gradients is partially determined by the disturbance parameters. In other words, to fully understand the distribution of species along environmental gradients one must often consider disturbance effects. For example, with widespread burning, *Pinus* covers a broad portion of the moisture gradient, but with little fire it occurs only on the most xeric sites (Barden & Woods, 1976; Barden, 1977). Disturbances such as severe windstorms and fires, may influence both local (within stand) species composition and the position of large areas of successional vegetation along topographic, moisture and elevation gradients.

Third, species characteristics such as mode of reproduction, are often correlated with specific disturbance parameters. The distribution of such characteristics will likely be the result of a combined response to environmental factors such as moisture availability and disturbance events such as fires.

Fourth, functional aspects of ecosystem response to disturbance vary along environmental gradients. This is particularly true of soil recovery and decay of dead organic matter.

Fifth, man-caused disturbances, including those which have appeared since the establishment of the National Park may vary along environmental or biotic gradients, and gradient analyses, such as those of Whittaker, may be used to predict the frequency, intensity, and severity, of these unnatural disturbances on a landscape basis. The differentiation of man-caused and natural disturbance is fundamental in U.S. National Park Service management philosophy in that the former are, at least in a theoretical sense, to be excluded, while the latter are, ideally, to be allowed. In reality, however, some man-caused disturbances have permeated National Park borders (e.g., chestnut blight), while some natural disturbances (e.g., lightning fire) have often been suppressed. Some man-caused disturbances closely resemble natural disturbances (e.g., fire in GRSM, at least in some kinds of pine stands), whereas other man-caused disturbances have no close analog among natural disturbances (e.g., rarely are natural pathogens as dramatic in effect as chestnut blight).

Natural disturbances occur as a result of an interaction of climate, topography, geology and biota. In GRSM, windstorms, icestorms, floods, droughts, late frosts, debris slides, erosional processes, karst processes and the activities of both vertebrates and invertebrates have caused, at various times, major disruptions of community structure or ecosystem function. We analyze some of the better documented disturbances, including both man-caused and lightning-caused fire and exotic species invasion to illustrate the five above-mentioned points.

Study area

The 208 000 ha GRSM National Park, located at N 35°37', W 83°31' in the southern Appalachian Mountains, has had only minor boundary revisions since R. H. Whittaker first studied the area in 1947. Elevations are among the highest in the southeastern United States, ranging from 256 to 2 025 m. However, the mountains are not high enough for a climatic tree line. The climate is continental warm temperate. Rainfall is high, ranging from 1 500 mm at 445 m to 2 500 mm at 1 520 m (Stephens, 1969). The bedrock is largely metamorphosed sandstones and shales of Precambrian age. Ordovician limestone is exposed in 'windows' through the Precambrian overthrust in the western part of the park (King *et al.*, 1968). Whittaker (1956) did not investigate stands below about 500 m elevation and did not consider the diverse, and largely successional communities on limestone.

Whittaker (1956) listed 14 major mature vegetation types (Figs. 1a and 1b).

As indicated by Whittaker's community type names, chestnut *(Castanea dentata)* was formerly a co-dominant in four communities. It has been eliminated as a canopy tree due to chestnut blight (Woods & Shanks, 1959).

Whittaker's basic gradient model of vegetation types is shown in Figures 1a and 1b. In this paper, we assume that Whittaker's model is largely correct, but the five points we have proposed concerning disturbance are not dependent on Whittaker's data *per se*. Recent work by Baron & Matthews (1977) showed that Whittaker's scheme applied to the generally more xeric North Carolina side of the park (where Whittaker did little sampling), but that communities are displaced somewhat in regard to topographic position. For example, mesic communities such as hemlock forests tended to be limited to stream valleys whereas oak forests extend onto the valley floors. Data of Eager (1978) and Johnson (1977) indicate a higher elevation range for red spruce than that reported by Whittaker

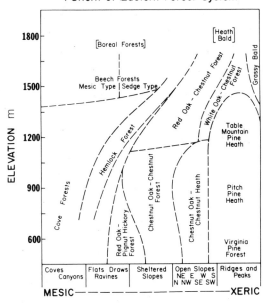

Fig. *1a.* The original gradient analytic model of Whittaker (1956). Major vegetation types are shown relative to their topographic and elevation positions. Vegetation in the eastern and western halves of the Great Smoky Mountains is dissimilar in that spruce-fir forest dominates the high ridges in the east but hardwood types dominate in the west. On the abscissa, the diagram shows the shift from sheltered valley (coves) with low evapotranspiration to drier exposed ridges. The ordinate shows changes with increasing elevation.

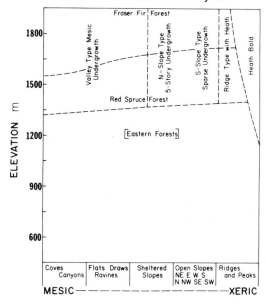

Fig. *1b.* The original gradient analytic model of Whittaker (1956). Major vegetation types are shown relative to their topographic and elevation positions. The diagram is separated from Figure 1a, because hardwood types dominate between 4 500 and 5 800 feet in the western half of the park.

(1956). The grassy balds appear to be anthropogenic communities, and should be removed from the topographic analysis (Lindsay & Bratton, 1979a; 1979b; 1980). Stratton & White (1982) indicate that heath balds occur on less than 50% of the seemingly appropriate sites as defined by elevation and topography. Disturbance is probably a key element in their occurrence although soil depth and moisture availability are doubtless also critical. Our observations indicate there are flood plain communities at low elevations, which were not described by Whittaker, and therefore his data do not cover a full range of edaphic conditions in the National Park.

Methods

The results presented in this study have been derived from several sources. First, the fire history statistics and the distribution of fires along environmental gradients are derived from U.S. National Park Service records collected from 1940–1979 and analyzed by Harmon (1981). These records include the date, the probable point of ignition, the probable source of ignition and the maximum extent of an individual fire. Topographic and forest type information in the park records was verified by plotting on topographic maps, and in some cases, by visiting the site of the fire. The data were supplemented by fire frequency estimates obtained by removing wedges of wood from the trunks of fire-scarred trees and reading tree rings to date fire scars going back to about 1850 (Harmon, 1982a). The diagrams of species response to disturbance were constructed using Whittaker's (1956) models of individual tree species distribution. Species were classified as sprouters or non-sprouters and then densities for individual species were summed to give totals for the response classes.

The discussion of recovery after fire adapts Olson's (1963) model for prediction of forest floor recovery rate which is based on relationships between production, decay rate and the time required to reach a steady-state litter biomass. In the simplest case, Olson assumed that litter production was not affected by the presence of fire and that 100% of the original litter layer was removed. In this case the time required to reach 95% of the steady-state biomass $t_{0.95} = \dfrac{3}{k}$ where k is the exponential decay rate.

The amount of biomass accumulated at steady-state is equal to $\dfrac{I}{k}$ where I is the input or production rate. Although few fires would meet the assumptions of this simple model, these relationships represent a relative index by which ecosystems can be compared. Figure 7 was prepared by estimating the $t_{0.95}$ for the species shown in Whittaker's (1956) species nomograms. Data from Shanks & Olson (1961) and Singh & Gupta (1977) were used to estimate $t_{0.95}$ for specific species. Since our knowledge of species decay rates is very rudimentary we grouped $t_{0.95}$ into 5-year intervals. The mosaic charts were then used to calculate a weighted average $t_{0.95}$. That is, each species contributed litter relative to its density. We have made two simplifications in presenting this model: we have assumed 1) that biomass is proportional to stem density, and 2) that decay rates can be represented as a single value per species (more realistically they vary with environment).

The diagrams of the spread of the exotic chestnut blight and balsam woolly aphid impact were obtained directly from Whittaker's (1956) data on the host species distributions. The pattern of wild boar *(Sus scrofa)* impact was derived from an indirect gradient analysis using data from 60 50 m × 20 m vegetation plots placed at representative positions across the major elevational and topographic gradients. The ordination was completed using the Cornell Ecology Programs package (Gauch, 1973) for principal components analysis of the total basal area, by species, for all woody stems 1 cm in diameter or greater, 1.4 m above the ground. The percentage of hog rooting was estimated for the entire surface area of the plots in 1977 and 1978 (see Bratton *et al.*, 1982 for details).

Results and discussion

Fire as a cause of disturbance

Our first two characteristics of vegetation-disturbance gradients (that disturbance parameters vary along gradients and species composition along gradients is partially determined by disturbance) are well illustrated in GRSM by fire pattern and history. Fire in GRSM has been both natural and anthropogenic. The pattern and impact of fires has changed dramatically over the history of GRSM and much of this change has been associated with the management practices of man. Recent records of fire occurrence, size and cause are very complete when compared to disturbances such as wind or frost damage. However, records predating the establishment of GRSM are incomplete and mostly qualitative in nature. Man has inhabited the southern Appalachians for at least 12 000 years (Dickens, 1976) and has probably been a major cause of fires during this time. When Euro-American man first contacted the Cherokee tribes, man-set fire was probably an important landscape influence (Goodwin, 1977). Euro-American settlers inhabited GRSM between 1790 and 1930 and frequently burned submesic and xeric forests below 1 500 m elevation (Ayres & Ashe, 1905; Lambert, 1958). Harmon (1982a) found that *Pinus*-dominated forests in the western portion of GRSM were burned with a mean frequency of once every 12 year during the period ca 1850–1930. The fire interval frequency for Harmon's entire study area, which included many mesic forest types, was estimated to be 25 yr. Many of the fires set by man during this period were light surface fires (Ayres & Ashe, 1905). This contrasts strongly with the period of logging fires between 1910 and 1930 (Lambert, 1958) when large quantities of logging slash combined with dry weather conditions led to severe and widespread fires. Since 1930 much of GRSM has been protected from fire. The characteristics of all GRSM fires between 1940 and 1979 were studied by Harmon (1981). Man caused the majority of fires during this period both in terms of number (87%) and area burned (97%). This pattern reflects the overall trend in the southern Appalachians as a whole (Barden & Woods, 1974). At present the GRSM fire cycle is very long: 840 yr for man-caused and over 30 000 yr for lightning-caused fires. Clearly, fire suppression activities of park management have all but eliminated fire as a disturbance.

The distribution of fires over the landscape is not even and varies with the cause (Figs. 2 and 3). For both natural and man-caused fires, occurrence decreases as elevation increases, but most lightning fires occur on xeric sites while most man-caused fires start on mesic sites. The decrease in fire frequency with elevation is probably associated with many factors including decreasing temperatures, decreasing evapotranspiration, increasing precipi-

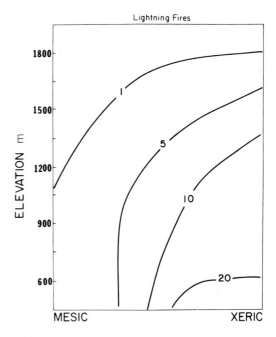

Fig. 2. Distribution of lightning fire ignitions based on data from Harmon (1981). The isolines represent the relative abundance of total lightening-caused fires observed between 1940 and 1979.

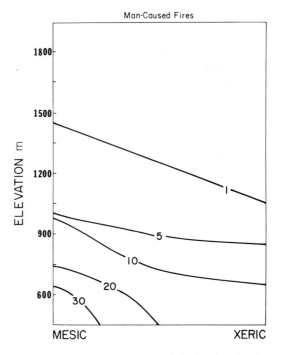

Fig. 3. The distribution of man-caused fire ignitions based on data from Harmon (1981). The isolines represent the relative abundance of total man-caused fires observed between 1940 and 1979.

tation, and increasing period of snow cover. Lightning tends to strike topographic high points which also are drier and thus more flammable whereas many man-caused fires start along roads which tend to run through valleys and not along ridges.

Calculated fire return cycles for the sites of highest fire occurrence are 180 and 9 400 yr respectively for man-caused and lightning fires. Before the establishment of the park many forests were exposed to more frequent fires and these fires influenced community composition and tree population age structure. Lightning fires would have to increase in size 1 000 times to become equivalent to the pre-park frequency of largely man-set fires. Since an increase of this magnitude is unlikely, simply allowing lightning fires to burn will still lead to major changes in composition as the forest equilibrates to reflect the new frequency. Moreover, because the severity of lightning fires tends to be lower than man-caused fires (Barden & Woods, 1974), the importance of shade intolerant species will probably decrease even on sites burned by lightning fires.

We infer from this discussion that the species composition of pine and oak forests as sampled by Whittaker (1956), was influenced by disturbances prior to park establishment in 1934. Without fire, pine forests are likely to be replaced by oaks and other hardwoods on all but the most xeric and exposed sites (Barden, 1974). Vegetation types will shift through time relative to their positions in Figure 1. It is also worth noting that Figure 3 represents the pattern of man-caused fires during the park era only. In earlier, post-colonial times (1790–1930), pine and oak forests were purposefully burned to encourage fruit crops, such as blueberries, and to improve forage for cattle (Lindsay, 1976). Prior to the establishment of the park man-caused ignitions were probably: 1) more frequent, and 2) extended to higher and more xeric slopes. Thus, the gradient positions for pine and oak forest reported by Whittaker (1956) may not represent 'natural' in the sense of pre-European settlement conditions. A change in park policy to allow lightning fires to burn would be unlikely to be sufficient to maintain the vegetation patterns present in the 1940's.

Species responses to disturbances

Turning to the third disturbance-vegetation characteristic, that reproductive patterns are related to the major environmental and disturbance gradients, we consider sprouting ability in the GRSM landscape. In GRSM, such disturbances as fire, insect outbreaks, wind, frost, logging, or pathogen epidemics often cause top-kill or defoliation. Given disturbances of equal intensity, a community rich in sprouting species would be expected to respond differently than one with few sprouters. For example, in GRSM, mortality of trees is lower in communities rich in sprouters. Vigorous sprouting also can inhibit seed reproduction in the post-disturbance stand.

The ability to sprout or form new foliage is a complex function that varies with species and individual stem age and size (Kramer & Kozlowski, 1979). For the purposes of this paper we will assume sprouting is an all-or-nothing response. We will also assume that conifers cannot sprout although some *Pinus* (e.g. *Pinus rigida* and *P. echinata* in GRSM) can form sprouts (Stone & Stone, 1954). In our experience epicormic sprouting in *Pinus rigida* and *P. echinata* may result in refoliation after crown scorch but large individuals of these two species do not recover after more severe disturbance such as stem girdling. Most important hardwoods found in GRSM have the ability to sprout. Such sprouting can produce dense patches of hardwood saplings two to four years after moderate disturbance.

Using stem density data presented in Whittaker's (1956) analysis of GRSM vegetation, we have plotted the percentage of stems capable of sprouting as a function of position within the Whittaker mosaic chart (Figs. 4 and 5). Underlying these figures is the fact that stands with high percentages of conifers do not respond to moderate fire disturbance in the same way that hardwood stands respond. Because dense sprout saplings outcompete seedling reproduction (which often depends on exposure of mineral soil as well), the two recovery modes (sprouts and via seedlings) are non-randomly distributed within the mosaic diagram. In eastern GRSM the proportion of stems in a stand capable of sprouting peaks in submesic, oak-dominated stands and decreases to xeric stands. An exception is the relatively low importance of sprouting in mesic stands at mid elevations, corresponding to peak *Tsuga* density. Potential sprouting also decreases from low to high elevations. Sprouting patterns are the same in

Percentage of Stems
Capable of Sprouting
Eastern Great Smoky Mountains

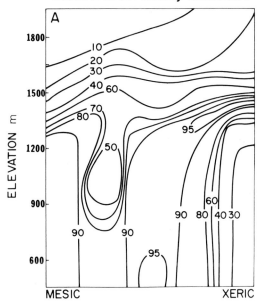

Fig. 4. The percentage of stems for all species capable of sprouting after top-kill or total defoliation for GRSM east of Double Springs Gap. Species incapable of sprouting are *Abies fraseri, Picea rubens, Pinus pungens, P. rigida, P. virginiana* and *Tsuga canadensis.* In the eastern Great Smokies, conifers dominate the high elevations.

Percentage of Stems
Capable of Sprouting
Western Great Smoky Mountains

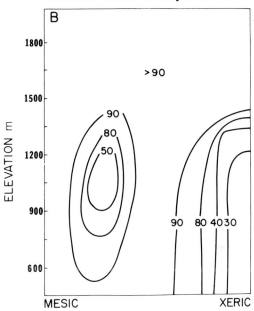

Fig. 5. The percentage of stems capable of sprouting after top-kill or total defoliation in GRSM west of Double Springs Gap. The high elevations in this section of the park are dominated by hardwoods.

the western GRSM except that there is no decrease with elevation since upper elevations are oak and beech dominated.

A comparison of the sprouting gradients to the occurrence of fires reveals some interesting relationships. Lightning fires are frequent in the chestnut oak–heath and the pine communities. The former community has many sprouting species while the latter has few. Present-day man-caused fires tend to occur in communities (mesic and submesic hardwoods) where the proportion of sprouting species is high.

The sprouting response gradients also have some important implications for man-caused disturbances such as logging and agricultural clearing. Logging in areas low in sprouters, such as the hemlock and spruce–fir communities, should lead to an abundance of seed reproduction by shade-intolerant species. In contrast, logging in the chestnut oak–chestnut community should lead to heavy

dominance by sprout reproduction. Agricultural clearing, which removes sprouters as well as non-sprouters, should lead to an increase in reproduction of intolerants from seed.

Comparison of the pattern of sprouting to the distribution of vagile, early successional tree species that colonize by seedling establishment (some of which also reproduce by sprouts), reveals that four major groups are collectively able to occupy all major positions on the environmental gradients: 1) *Prunus pensylvanica,* 2) *Liriodendron tulipifera,* 3) *Betula* spp. and 4) *Pinus* spp. (Fig. 6). *Pinus* predominates on lower, more xeric slopes, *Liriodendron* is dominant on the lowest stream flats and in mesic positions in logged over sites, and is replaced by *Betula lutea* at the middle elevations. *Prunus pensylvanica* and *Betula lutea* dominate cut over sites well into the high elevation spruce–fir zone.

Forest floor recovery rate

Our fourth point, that functional aspects of eco-

154

Gradient Distribution of Light Seeded Trees

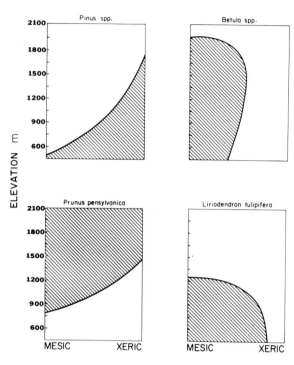

Fig. 6. The distributions of important, light seeded tree species in GRSM.

Number of Years for Litter Recovery After Fire

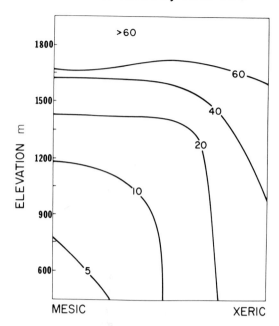

Fig. 7. Number of years estimated for the forest floor to recover from litter removal by ground fire which did not kill the canopy. The number of years to reach the steady-state represents a weighted average for the species in a portion of the mosaic chart.

system resources are related to environmental gradients, is illustrated by fire effects. Fire differs from many other disturbances in that it removes dead organic matter (fuel) when it kills live portions of the canopy (windfall by contrast increases dead organic matter).

Litter accumulation response times range from 5 yr on mesic low elevation sites to over 60 yr on high elevation sites in eastern GRSM (Fig. 7). The time required to reach steady-state also increases from 5–30 yr as one progresses along the moisture gradient from mesic to xeric at low elevations. Man-caused fires occur on sites which recover rapidly from fuel removal, whereas lightning fires occur on sites with recovery times ranging from 20–40 yr. Man-caused fires occur on sites with low total fuel accumulations, lightning fires burn areas with moderate fuel accumulations, and the zones with the highest biomass accumulations rarely burn. Although considerably less is known about decay rates of wood, a similar diagram could be prepared for the time required for woody stems

killed by disturbance to disappear. The overall patterns would probably be similar to the litter accumulation gradients with increasing times as one proceeds from mesic to xeric sites and from low to high elevations. For example, in one study (Harmon, 1982b), logs were predicted to take less than 20 yr to disappear on low elevation sites whereas at a high elevation spruce–fir site was observed to have equivalent diameter logs remaining from a fire in 1910.

Post-park disturbances

Finally, to approach our fifth point, that recent, anthropogenic, disturbances vary along environmental and biotic gradients, we consider exotic species invasions. In the case of species-specific pathogens and parasites, the range of the new disturbances will be determined by existing biotic gradients. The chestnut blight, *Endotheca parasitica,* for example, was originally introduced to the United States in 1919 and by 1930 it had killed most

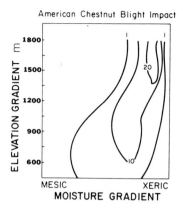

Fig. 8. The impact of the chestnut blight on the forest canopy, in percentage of stems removed from stand.

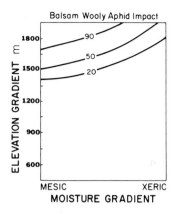

Fig. 9. Abies fraseri as a percentage of the forest canopy stems (after Whittaker, 1956). If balsam woolly aphid kills all the Abies fraseri in the park, these percentages should also represent the mortality for this tree species.

of the mature individuals of *Castanea dentata* in eastern North America including the study area. At that time there were about 75 000 ha of oak–chestnut forest in GRSM. Since mortality of mature trees was total, the pattern of canopy opening followed Whittaker's gradient diagram. The disease is apparently somewhat cold intolerant, so there was also a gradient response in the pattern of disease spread. The trees in the upper elevations were the last to die – some still had live branches when Whittaker sampled in the late 1940's (Fig. 8).

The recovery response of the forest community was also related to environmental gradients, since the nature of *Castanea* replacement was a function of its original density (Woods & Shanks, 1959). Where density was high, both replacement by established saplings of shade-tolerant species and seed reproduction by shade-intolerant species were important. In sites with lower densities of *Castanea*, canopy expansion by adjacent trees was the dominant process. This prevented the establishment of shade-intolerant species and the release of advanced reproduction. In general, *Acer rubrum, Quercus prinus* and *Quercus rubra* increased in importance as *Castanea* declined. From Whittaker's work, one could predict that most of the intolerant replacement occurred in high elevation and low elevation subxeric forests.

A second example of a host-specific introduced parasite is the balsam woolly aphid *(Adelges piceae),* the most damaging exotic insect now in the park. First found in GRSM in 1963, the aphid threatens to eliminate mature *Abies fraseri* in its

6 000 ha range (Fig. 9). Initial infestations occurred at the spruce–fir/hardwood ecotone (Eager, 1978) but have subsequently spread to include stands all along the elevation gradient. Trees may take years to succumb, but death sometimes occurs in as few as three growing seasons (Johnson, 1977). Because *Abies fraseri* increases in importance with elevation, the pattern of invasion is the reverse of the pattern of potential impact.

Attack by balsam woolly aphid causes an increase in light and alters microclimate in dense fir stands. Understory shrubs, herbs, and reproduction respond quickly to this disturbance by growing more vigorously (Eagar, 1978). Lichens in dead fir canopies also dramatically increase after needle drop. An increase of seed reproduction has not yet been documented, but an increase in *Betula lutea, Sambucus canadensis, Rubus canadensis, Sorbus americana,* and *Amelanchier laevis* should be expected. *Abies fraseri* may reach a mature size and reproduce before aphid attacks. If this occurs, then one can envision a mosaic of *Abies fraseri* populations: some mature with aphids and others immature and relatively free of aphids. However, Hay *et al.* (1978) believe it is only a matter of time before *Abies fraseri* is entirely eliminated, removing an endemic species from the already species-poor high elevation flora. *Abies fraseri* tends to support more epiphytic mosses and liverworts than red spruce; hence, the loss of fir may also lead to local extinctions in the cryptogamic flora (Fig. 9).

A more complicated case is the effect of the exotic European wildboar *(Sus scrofa)*, a disturbance which is not related to any single plant species or single environmental gradient. The movements and habitat preferences of wild boar, are a function of the animal's direct response to variables such as temperature (the wild boar move to lower elevations and bed on warmer southwest facing slopes in winter), and the availability of favored food items (Bratton, 1975). Wild boar rooting reaches maximum disturbance levels in grey beech forest from 1 450–1 800 m (Fig. 10). Bratton *et al.* (1982) related boar rooting to understory composition. They found some understories to have almost no hog activity but others to have high levels of rooting. This distribution was negatively correlated with density of *Rhododendron maximum* and other ericads, and positively correlated to occurrence of mesic herbs and more open understories. As a complicating factor, recently burned stands have more hog rooting, presumably due to increased availability of herbs, insects or both (Bratton *et al.*, 1982).

Conclusions

We have used Whittaker's original (1956) analysis to show that superimposed on the gradient pattern of forest composition in GRSM are similar disturbance patterns which also can be treated as gradients. The vegetation pattern reported by Whittaker is not independent of these disturbance factors, nor are these factors independent of the vegetation. Modification of disturbance frequencies and intensities by management activities such as fire suppression may result in slow shifts in the distributions of both individual species and of whole communities. Species responses to disturbance are probably the result of both physical environmental considerations (e.g. conifers are tolerant of xeric conditions), and of disturbance patterns, (e.g. pines recolonize bare soil quickly after severe fire). Certain aspects of recovery, such as accumulation of leaf litter or humus are correlated with physical environmental gradients and may in turn influence disturbance frequencies and intensities. Recent anthropogenic disturbances also show a relation to landscape gradients.

Ecosystem responses to disturbance on a landscape basis are complex and integrate multiple factors. Since disturbance gradients may parallel physical environmental gradients, the two may be difficult to distinguish. Adding disturbance parameters to Whittaker's original study of the Great Smoky Mountains can provide us with an increased ability to understand species composition and vegetation change.

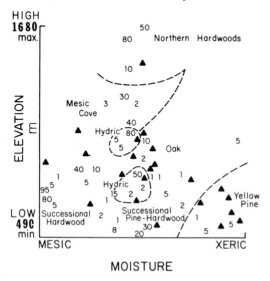

Percent of Soil Surface Disturbed by Wild Boar Rooting In Western Great Smoky Mountains

Fig. 10. Percent of soil surface disturbed by wild boar rooting. Triangles indicate no hog activity. Note the greater intensity of rooting in mesic sites (after Bratton *et al.*, 1982).

References

Ayres, H. B. & Ashe, W. W., 1905. The southern Appalachian forests. USDI, USGS Prof. Pap. 37.

Barden, L. S., 1974. Lightning fires in Southern Appalachian forests. Doctoral Dissertation, Univ. of Tennessee, Knoxville.

Barden, L. S., 1977. Self-maintaining populations of *Pinus pungens* in the southern Appalachian mountains. Castanea 42: 316–23.

Barden, L. S. & Woods, F. W., 1974. Characteristics of lightning fires in southern Appalachian forests. Proc. Tall Timbers Fire Ecol. Conf. 13: 345–361.

Barden, L. S. & Woods, F. W., 1976. Effects of fire on pine and pine-hardwood forests in the Southern Appalachians. Forest Sci. 22: 399–403.

Baron, J. S. & Matthews, R. C., 1977. Enviromental analysis of the proposed foothill parkway. U.S. Dept. Interior, National Park Service, SE Region, Uplands Field Research Lab., GSMNP, Management Report No. 19. 59 pp.

Bratton, S. P., 1975. The effect of the European wild boar *(Sus scrofa)* on gray beech forest in the Great Smoky Mountains. Ecology 55: 1356–1366.

Bratton, S. P., Harmon, M. E. & White, P. S., 1982. Rooting impacts of the European wild boar on the vegetation of Great Smoky Mountains National Park during a year of mast failure. Castanea 47: 230–242.

Bratton, S. P. & White, P. S., 1980. Rare plant management – After preservation what? Rhodora 82: 49–75.

Bratton, S. P., White, P. S. & Harmon, M. E., 1981. In: J. F. Franklin & M. A. Hemstrom (eds.). Disturbance and recovery of plant communities in Great Smoky Mountains National Park: successional dynamics and concept of naturalness, pp. 42–79. Proc. 2nd U.S.-U.S.S.R. Symposium in Biosphere Reserves.

Dickens, R. S., 1976. Cherokee prehistory: the Pisgah Phase in the Appalachian Summit Region. Univ. Tennessee Press, Knoxville.

Eager, C. C., 1978. Distribution and characteristics of balsam woolly aphid infestations in the Great Smoky Mountains. Master's Thesis, Univ. Tennessee, Knoxville.

Gauch, H. G., Jr., 1973. The Cornell Ecology Program Series. Bull. Ecol. Soc. Am. 54: 10–11.

Goodwin, G. C., 1977. Cherokees in transition: A study of changing culture and environment prior to 1775. Res. Pap. 181. Geography Dept., Univ. Chicago.

Harmon, M. E., 1981. Fire history of Great Smoky Mountains National Park. 1940–1979. Uplands Field Research Lab. Research/Resource Manag. Rep. No. 46.

Harmon, M. E., 1982a. Fire history of the westermost portion of Great Smoky Mountains National Park. Bull. Torrey Bot. Club 109: 74–79.

Harmon, M. E., 1982b. Decomposition of standing dead trees in the Southern Appalachian Mountains. Oecologia 52: 214–215.

Hay, R. L., Eager, C. C. & Johnson, K. D., 1978. Fraser fir in the Great Smoky Mountains National Park; its demise by the balsam woolly aphid. Rep. to the National Park Service.

Heinselman, M. L., 1973. Fire in the virgin forests of the Boundary Waters Canoe Area, Minnesota. Quat. Res. 3: 329–382.

Johnson, K. D., 1977. Balsam woolly aphid infestations of Fraser fir in the Great Smoky Mountains. Master's Thesis, Univ. Tennessee, Knoxville.

King, P. B., Newman, R. B. & Hadley, J. B., 1968. Geology of the Great Smoky Mountains National Park, Tennessee and North Carolina. USDI USGS Prof. Pap. 587.

Kramer, P. J. & Kozlowski, T. T., 1979. Physiology of woody plants. Academic Press, New York. 811 pp.

Lambert, R. S., 1958. Logging in the Great Smoky Mountains National Park. Rep. to the superintendent, Great Smoky Mountains National Park. Gatlinburg, Tenn.

Lindsay, M. M., 1976. History of the grassy balds in the Great Smoky Mountains National Park. USDI NPS. Uplands Field Research Lab. Management Rep. 4.

Lindsay, M. M. & Bratton, S. P., 1979a. Grassy balds of the Great Smoky Moutains: their history and flora in relation to potential management. Environ. Manage. 3: 417–430.

Lindsay, M. M. & Bratton, S. P., 1979b. The vegetation of grassy balds and other high elevation disturbed areas in the Great Smoky Mountains National Park. Bull. Torrey Bot. Club 106: 264–275.

Lindsay, M. M. & Bratton, S. P., 1980. The rate of woody plant invasion on two grassy balds. Castanea 45: 75–87.

Olson, J. S., 1963. Energy storage and the balance of producers and decomposers in ecological systems. Ecology 44: 322–331.

Pickett, S. T. A. & Thompson, J. N., 1978. Patch dynamics and the design of nature preserves. Biol. Conserv. 13: 27–37.

Radford, A. E., Ahles, H. E. & Bell, C. R., 1968. Manual of the Vascular Flora of the Carolinas. The University of North Carolina Press, Chapel Hill. 1183 pp.

Shanks, R. E. & Olson, J. S., 1961. First-year breakdown of leaf litter in southern Appalachian forests. Science 134: 194–195.

Singh, J. S. & Gupta, S. R., 1977. Plant decomposition and soil respiration in terrestrial ecosystems. Bot. Rev. 43: 449–528.

Stephens, L. A., 1969. A comparison of climatic elements at four elevations in the Great Smoky Mountains, 1969. Master's Thesis. Univ. Tennessee, Knoxville.

Stone, E. L. & Stone, M. H., 1954. Root collar sprouts in pine. J. Forestry 52: 487–491.

Stratton, D. A. & White, P. S., 1982. Heath balds of the Great Smoky Mountains: flora, distribution, and ecology. National Park Service, Southeast Regional Office Res./Resour. Manage. Rep. 58. 33 pp.

White, P. S., 1979. Pattern, process and natural disturbance in vegetation. Bot. Rev. 45: 229–299.

White, P. S. & Bratton, S. P., 1980. After preservation: Philosophical and practical problems of change. Biol. Conserv. 18: 241–255.

Whittaker, R. H., 1956. Vegetation of the Great Smoky Mountains. Ecol. Monogr. 26: 1–80.

Woods, F. W. & Shanks, R. E., 1959. Natural replacement of chestnut by other species in the Great Smoky Mountains National Park. Ecology 40: 349–361.

Accepted 2.7.1983.

Vegetation of the Santa Catalina Mountains: community types and dynamics*

William A. Niering[1] & Charles H. Lowe[2]
[1] *Department of Botany, Connecticut College, New London, CT 06320, U.S.A.;*
[2] *Department of Ecology and Evolutionary Biology, University of Arizona, Tucson, AZ 85721, U.S.A.*

Keywords: Fire, Montane forests, Pine–oak woodlands, Plant communities, Santa Catalina Mts. (Arizona), Sonoran desert, Succession

Abstract

This paper interprets plant community dynamics within three major vegetation regions – the Sonoran desert, the Encinal and coniferous forest – which extend from 700 m to the summit (2 766 m) on the south slope of the Santa Catalina Mts., Arizona. On the bajada, *Larrea tridentata* and disturbed desert scrub communities have been degraded by overgrazing and *Carnegiea gigantea* reproduction is failing on many sites. In the Spinose, suffrutescent desert scrub on protected lower mountain slopes *Carnegiea* is reproducing but mature populations are periodically decimated by freezing temperatures. In the desert grassland, graminoids or *Agave schottii* dominate, the former favored by fire, the latter by fire protection. In the Encinal above 1 220 m an open oak woodland dominated by *Quercus oblongifolia* is transitional to the pine, oak woodland where fire and drought result in several community segregates. A relict *Cupressus arizonica* forest is restricted to certain canyons. Above the Encinal (2 100 m) a *Pinus ponderosa-Q. hypoleucoides* forest is replaced by a less xeric *P. ponderosa* forest. In the latter a dense pine understory develops with fire protection and savanna-like pine stands are favored by fire. At higher elevations a mature *Pseudotsuga menziesii, Abies concolor* forest dominates the north-facing slopes where fire plays a significant factor in its perpetuation. An even-aged subalpine *Abies lasiocarpa* stand on the north slope below the summit suggests post-fire origin. In the Pinaleno Mts. to the northeast, mixed conifer and spruce, fir forests complete the vegetation gradient typical of these southwest mountain ranges. Here fire and windthrows interact in maintaining a mosaic of pure or mixed even or uneven-aged stands.

Although Shreve's (1915) description of this mountain vegetation is still valid after more than half a century, the role of grazing and freezing at lower elevations and fire at higher elevations adds a new dimension to those factors operative in this series of relatively stable yet highly dynamic communities.

Introduction

The Santa Catalina Mountains are representative of relatively undisturbed mountain ranges of the southwestern United States. The vegetation pattern, from Sonoran desert scrub through evergreen woodland to montane forest, was first described in a classic study by Shreve (1915). More recent papers have treated community and floristic

composition, biogeography, saguaro population dynamics, vegetation gradients, soils and community productivity (Marshall, 1957; Lowe, 1961, 1964; Niering *et al.,* 1963; Niering & Whittaker, 1963, 1965; Hastings & Turner, 1965; Whittaker & Niering, 1964, 1965, 1968a, b; Patton *et al.,* 1966; Whittaker *et al.,* 1968; Steenbergh & Lowe, 1969, 1976, 1977, 1982). Shreve's (1915) treatment of the vegetation of the Catalina Mountains remains primary, a fact that is all the more remarkable when it is realized that he did it all on horseback and on

* Botanical nomenclature follows Kearney & Peebles (1973).

160

foot more than a quarter of a century before our access road, the present Mt. Lemmon or Catalina Highway was built. The composite set of photographs is included in memory of Forrest Shreve to give the reader a true feeling of the 'constantly changing panorama of vegetation'.

This paper complements and completes an extensive body of work carried out by Whittaker and the first author in the Catalinas during the 1960's. Bob always visualized a community dynamics paper in which such factors as overgrazing and fire would be considered in relation to other factors as they influence the vegetation gradient. This paper is written in memory of Robert H. Whittaker with whom we had the pleasure of working for nearly a decade.

Physical features

The Santa Catalina Mts. and neighboring ranges are a prominent physiographic feature in the Basin and Range Province of southeastern Arizona. The Santa Catalinas, located northeast of Tucson, are roughly triangular in outline; the east–west base of the triangle is about 32 km and the apex is about 32 km north from the base. Elevations range from 850–980 m at the southwestern base of the range near Tucson to 2 766 m at the summit (Mt. Lemmon). Over most of the south slope the parent material is Catalina gneiss, but near the summit it grades into granite. Sediments of complex history comprise the bajada, which slopes at ca. 3° from the mountain rock base at 915 m to Tucson at 735 m in the Santa Cruz Valley.

Climatically the Basin and Range Province in southern Arizona is arid to semi-arid with two rainy seasons – one from December to March, and the other from July to September (Smith, 1956; Sellers, 1960). Mean annual precipitation at Tucson on the southwest, and Oracle (1 370 m) on the northeast side of the range, is 27.8 cm and 49.2 cm with about half occurring in the summer. Shreve (1915) reported a precipitation increase with elevation: 19 cm at 915 m to 42–47 cm at 2 340 and 2 440 m. Shreve also found an average lapse rate of 7.5 °C per 1 000 m. Year-round soil temperatures at 20 cm, taken across the range from Tucson to Oracle, indicate a soil temperature gradient with a mean decrease of 8.9 °C/1 000 m (Whittaker et al., 1968).

The bajada soils show gradation from relatively impermeable soils underlain by caliche on the lower bajada to coarse porous soils on the upper bajada. Soils of the mountain slopes are mostly shallow lithosols, and as one proceeds upward litter cover, organic matter, nitrogen content and carbon/nitrogen ratios increase while pH and soil Ca, Mg and K decrease (Whittaker et al., 1968).

Land use history

Spanish control ceased after 1820 and, following the arrival of the railroad and the suppression of the Apaches in 1870, the area was opened to more intensive settlement. By 1880 ranchers were grazing extensive herds of cattle on the desert and desert grasslands (Wagoner, 1952). Cattle numbers increased until the severe drought of 1892–93 when cattle losses occurred estimated at 50% or up to 75% on some ranches (Haskett, 1935). About this time floods of severe intensity eroded deep channels or arroyos, thus lowering the water table (Rich, 1911; Thornthwaite et al., 1942). Although climatic change has been cited as a possible causal factor (cited by Niering et al., 1963), overgrazing is also considered to have contributed significantly to this erosion (Griffiths, 1910; Thornber, 1910; Haskell, 1945).

Grazing has dramatically affected our study area from the bajada near Tucson to the base of the Catalinas. Urban development has also continued to expand up the bajada. Most of the mountain range has been a forest reserve since 1902 and part of the Coronado National Forest since 1908. The Sabino Canyon drainage and Mt. Lemmon highway area, including our study transect on the south slope, have been closed to grazing since 1947. No lumbering is permitted in the National Forest (although tree salvage occurs at higher elevations). Fires of lightning origin are frequent: four-fifths of all fires prior to 1960 were of natural origin (personal communication, J. W. Waters). Prior to grazing, grass cover was undoubtedly greater, thus allowing fire to spread more rapidly and possibly to occur at a higher frequency.

Vegetation setting

The vegetation of the Santa Catalinas brings together four regional vegetation types: Sonoran desert scrub related to the vegetation of the south and west, desert grasslands with affinities to the southeast, Mexican oak and pine, oak woodlands communities with southern affinities, and montane forests with northern affinities (Lowe, 1961). The vegetation of the Catalinas bears a marked resemblance to that found in other southwest ranges' the Rincons (Blumer, 1910), Pinalenos (Shreve, 1919; Martin & Fletcher, 1943), Chiricahuas (Blumer, 1910; Reeves, 1977) and Huachucas (Wallmo, 1955) in southern Arizona and mountains of southern New Mexico (Moir, 1963; Gehlbach, 1967).

This similarity to these other ranges is particularly evident on the northeast slope of the Catalinas where the pattern is from desert grassland through woodland to forest. On the southwest slope the communities are more diverse, ranging from Creosotebush (*Larrea tridentata*) and paloverde, bursage desert scrub (*Cercidium, Franseria* desert scrub) on the bajada, through spinose, suffrutescent desert scrub on the lower mountain slope, desert grasslands, open oak and pine, oak woodlands, and montane pine and fir forests to a limited subalpine fir (*Abies lasiocarpa arizonica*) forest

near the summit (Figs. 1, 2). These are also among the natural vegetation types recognized by Brown & Lowe (1974a, b).

The plant communities recognized in this paper are basically similar to those recognized earlier by Whittaker & Niering (1965) in their mosaic chart (Fig. 2). The Sonoran desert scrub includes the spinose, suffrutescent desert scrub community; pine, oak forest includes *Pinus ponderosa, Quercus hypoleucoides* forest; pine forest includes both *Pinus ponderosa* and *Pinus ponderosa, Pinus strobiformis* forests; montane fir forest includes *Pseudotsuga, Abies* and *Abies concolor* ravine forests; subalpine forest includes *Abies lasiocarpa arizonica,* mixed conifer, and *Picea engelmannii, Abies lasiocarpa arizonica* forests. The several bajada communities included here were not recognized on the mountain slope mosaic. In this paper the communities are grouped under three physiognomic types as recognized by Shreve (1915): Sonoran desert, Encinal and forest regions.

Due to steepness of gradients and the marked correlation between moisture and temperature with elevation, the vertical zonation of the vegetation is

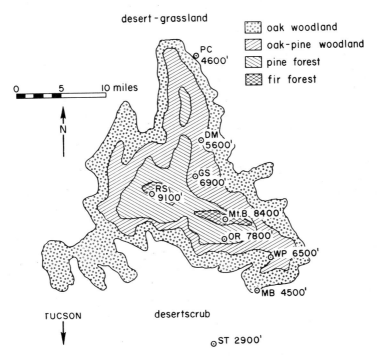

Fig. 1. Vegetation of the Santa Catalina Mountains, Arizona (in Patton, Heed & Lowe, 1966). Symbol locations: Soldiers Trail (ST), Molino Basin (MB), Windy Point (WP), Organization Ridge (OR), Radar Station (RS), Mt. Bigelow (Mt B), Green Spring (GS), Daily Mine (DM), Peppersauce Canyon (PC).

162

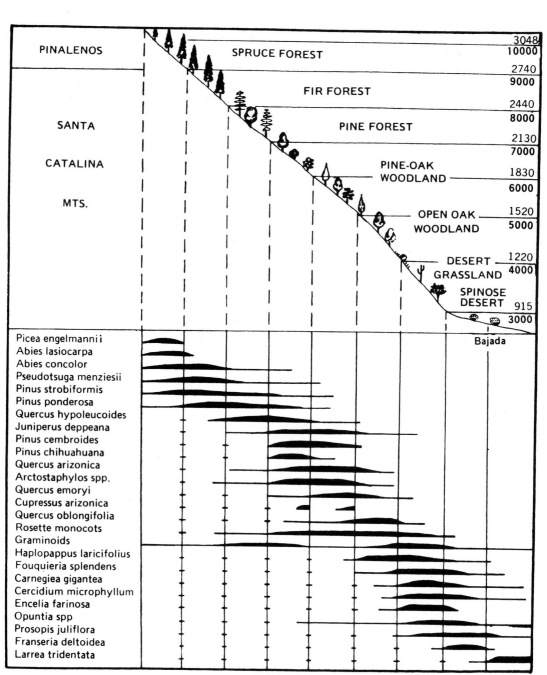

Fig. 2. Diagrammatic bisect of south slope of Santa Catalina Mts. and higher elevations of Pinalenos showing distribution of primary taxa. Width of solid lines indicates relative importance of taxa at various elevations. Elevations given in meters above and feet below the lines demarcating vegetation zone.

especially dramatic in these south-west ranges. Striking differences in north and south slope vegetation at a given elevation are describable as a shifting of the elevational position of the vegetational zones. A given vegetation type found on a south-facing slope at one elevation may be on the north-facing slope a thousand or more feet lower. Among the climatic factors conditioning the vegetation, Shreve (1915) concluded that duration of low temperatures and aridity were especially significant at lower elevations and increasing precipitation accompanied by decreasing evaporation at higher elevations. He reported the irregularity of the vegetation gradient due to slope exposure, along with the relatively high moisture relations in arroyos and canyons.

Methods

The initial field data were collected by Whittaker & Niering. Subsequent work by Lowe and associates of the University of Arizona at Tucson provided supplementary data. Basic methods employed in vegetation sampling are described more extensively in Whittaker & Niering (1965). Some 350 samples were taken in relatively undisturbed stands on the south slope of the range from the lower bajada to the mountain summit. Fifty samples were also taken above 2 750 m in the Pinaleno Mts. 95 km northeast of the Catalinas in order to complete the spectrum of vegetation types found in these southwest ranges. On the mountain slopes, within 300 m elevation belts, samples were taken from various slope exposures including ravines. Data were collected using 0.1 ha quadrats. Trees exceeding 1 cm diameter at breast height were tallied by species and diameter. Tree seedlings and shrubs were counted for the quadrat area. Herbs, shrubs and tree seedlings were recorded in 25 – 1 m^2 quadrats along the center tape. Increment borings were taken in representative stands.

In compiling the data for this paper 10 stands throughout the elevation range of each community type were used to characterize the major plant communities (Tables 1–3). These are diagrammatically depicted in Figures 2, 3, 4 and 5 which were prepared with the constructive assistance of Robert Whittaker. Additional vegetation data by elevational belts are available as Document No. 8392 from American Documentation Institute, Auxiliary Publ. Project, Photoduplication Service, Library of Congress, Washington, D.C. 20540.

Although some nomenclatural changes have occurred since our earlier papers, we have used the original binomials for consistency. The following are among the several taxonomic changes that have occurred: *Franseria* = *Ambrosia*; *Haplopappus* = *Aplopappus*; *Opuntia engelmannii* = *O. phaecantha discata*. The following community terms are equivalents: semi-desert grassland = desert grassland; juniper, pinyon woodland = pinyon, juniper woodland; subalpine forest = spruce–fir forest. In this paper desert = desert scrub (Figs. 3 and 4).

Table 1. Sonoran desert region, Santa Catalina Mts. and contiguous bajada. Shrub density and tree density by size classes per hectare. Data for *Carnegiea* by height, classes are as follows: (1) up to 1 ft (0.3 m); (2) 2–6 ft (0.6–1.8 m); (3) 7–12 ft (2.1–3.6 m); (4) 13–18 ft (3.9–5.5 m); (5) 19–24 ft (5.7–7.3 m); (6) 24–36 ft (7.3–11.0 m).

Species[a]		Community types					
		Bajada			Lower mt. slopes		
		Creosote-bush desert scrub	Paloverde bursage desert scrub	Disturbed desert scrub	Arroyo margin woodland	Spinose suffrutescent desert scrub	Semi-desert grassland
	feet	2500–2700	2800–3000	2800–3000	2500–3000	3000–4000	4000–5500
	meters	760–825	855–915	855–915	760–915	915–1220	1220–1700
Shrubs							
Agave schottii							23 541
Baccharis sarothroides				6	60		3
Calliandra eriophylla		162	11			1101	732
Carlowrightia arizonica		88	10			4	
Cassia covesii		2		14		35	3
Celtis pallida		8	92		2	46	10
Condalia lycioides		4	20		14	1	
Dasylirion wheeleri							32
Echinocereus fendleri		222	11			20	9
Encelia farinosa		420	1472			1909	16
Eriogonum wrightii						159	131

Table 1. Continued.

Species[a]	dbh cm	Bajada			Lower mt. slopes		
		Creosote-bush desert scrub	Paloverde bursage desert scrub	Disturbed desert scrub	Arroyo margin woodland	Spinose suffrutescent desert scrub	Semi-desert grassland
feet		2500–2700	2800–3000	2800–3000	2500–3000	3000–4000	4000–5500
meters		760–825	855–915	855–915	760–915	915–1220	1220–1700
Ferocactus wislizeni		1	42	14		82	33
Franseria ambrosioides				76	6	1	24
Franseria deltoidea			9764	1			
Gutierrezia californica			6	109			
Haplopappus laricifolius				9		25	673
Haplopappus tenuisectus				2020			
Hymenoclea monogyra					180		
Janusia gracilis			88	30		897	11
Jatropha cardiophylla			92	16		184	148
Larrea tridentata		1602	84	1	2		
Lycium berlandieri			20	9	8	122	
Mammillaria microcarpa			1178	6		38	27
Opuntia bigelovii				1		8	
Opuntia engelmannii				41		137	18
Opuntia fulgida		9	94	102		29	
Opuntia phaeacantha			318	25	4		
Opuntia spinosior		2		12		105	58
Opuntia versicolor			72	24		24	
Psilostrophe cooperi		173	2	103			
Simmondsia chinensis			90				
Zinnia pumila		49	52	9			
Trees							
Acacia greggii	<2.5	4	4	68	12	118	
	2.5–15		8	28	2	84	
Carnegiea gigantea	1		2	1		28	4
	2		8			64	10
	3		6			29	2
	4			1		22	2
	5		2	7		11	
	6		2	9		2	
Cercidium microphyllum	<2.5		38	16		290	
	2.5–15		418	211		287	
	15–30			2			
Fouquieria splendens	<2.5		8			87	226
	2.5–15		830	17		828	458
Prosopis juliflora	<2.5	2	2	31	22	36	2
	2.5–15	25	10	191	142	83	14
	15–30	1		17	48	2	2
	30–60				22		
	60–90				2		

[a] Additional species by community: Arroyo margin woodland, *Celtis reticulata* <2.5 cm = 90, 2.5–15 = 80, 15–30 = 58, 30–60 = 12, 60–90 = 2; *Fraxinus pennsylvanica velutina* <2.5 = 2, 2.5–15 = 22, 15–30 = 28, 30–60 = 5; *Platanus wrightii* <2.5 = 2, 2.5–15 = 4; 15–30 = 10, 30–60 = 1; *Populus fremontii* <2.5 = 2, 94–122 = 2. Semi-desert grassland, *Quercus oblongifolia* <2.5 = 2, 2.5–15 = 4, 15–30 = 5, 30–60 = 2.

Table 2. Pine–oak woodland region, Santa Catalina Mts. Tree density by size classes and shrub density per hectare for principal species.

Species[a]	dbh cm	Open oak woodland 4500–5500 1400–1700	Pygmy conifer oak 5000–6000 1520–1830	Pinus cembroides, oak woodland 6000–7000 1830–2130	Pinus chihuahuana, oak woodland 6000–7000 1830–2130	Pygmy conifer, oak scrub 6000–7000 1830–2130	Cypress canyon forest 4000–6000 1220–1830
Trees							
Cupressus arizonica	<2.5						28
	2.5–15						88
	15–30						38
	30–60						58
	60–90						26
	90–120						2
Juniperus deppeana	<2.5	2	5	8	18	5	
	2.5–15	3	50	26	32	28	
	15–30	6	19	22	8	29	
	30–60	4	14	36	4	10	
	60–90		1	4	1		
	90–120			2			
Pinus cembroides	<2.5		53	110	30	78	74
	2.5–15		109	232	14	111	24
	15–30		29	74	2	23	
	30–60		7	8		2	
Pinus chihuahuana	<2.5				26		
	2.5–15				88		
	15–30				44		
	30–60				6		
Quercus arizonica	<2.5	27	76	18	42	72	
	2.5–15	206	182	72	218	60	
	15–30	30	30	2	12	12	
	30–60	2	2	2	11		
Quercus emoryi	<2.5	15	19	90		1	15
	2.5–15	28	66	56	10	18	28
	15–30	9	8	4		2	9
	30–60	4					4
Quercus hypoleucoides	<2.5			168	100	225	80
	2.5–15			146	1048	214	86
	15–30				44	3	10
	30–60						2
Quercus oblongifolia	<2.5						4
	2.5–15						26
	15–30						43
	30–60						21
	60–90						1
	90–120						
Quercus rugosa	<2.5			2	8	26	94
	2.5–15			42	24	19	118
	15–30						8
	30–60						2
Vauquelinia californica	<2.5	9	1				
	2.5–15	20	2				
	15–30	7					

Table 2. Continued.

Species[a]		Community types					
		Open oak woodland	Pygmy conifer oak	Pinus cembroides, oak woodland	Pinus chihuahuana, oak woodland	Pygmy conifer, oak scrub	Cypress canyon forest
	feet	4500–5500	5000–6000	6000–7000	6000–7000	6000–7000	4000–6000
	meters	1400–1700	1520–1830	1830–2130	1830–2130	1830–2130	1220–1830
Shrubs							
Agave palmeri		44	72	462	160	102	76
Agave schottii		7777	1700				
Arctostaphylos pringlei	<2.5	8	69	16		409	
	2.5–15	11	7	36		312	
	15–30					3	
Arctostaphylos pungens	<2.5	85	154	918	372[2]	925	
	2.5–15		91	22	168	47	
Berberis wilcoxii							1094
Dasylirion wheeleri		107	15	60	2	39	
Echinocereus triglochidiatus				196	48	5	
Garrya wrightii	<2.5	106	293	154	4	153	24
	2.5–15	106	115	114	10	40	
Haplopappus laricifolius		1796	1				
Nolina microcarpa		922	3036	248	390	179	8
Yucca schottii		96	308	78	144	67	92

[a] Additional species by community type: Arizona cypress forest, *Alnus oblongifolia*, <2.5 cm = 294, 2.5–15 = 70, 15–30 = 56, 30–60 46, 60–90 = 4; *Fraxinus pennsylvanica velutina* <2.5 = 36, 2.5–15 = 76, 15–30 = 18, 30–60 = 8; *Juglans major* <2.5 = 54, 2.5–15 = 12; *Platanus wrightii* <2.5 = 4, 2.5–15 = 6, 15–30 = 16, 30–60 = 28.

Community types and dynamics

Sonoran desert region

The Sonoran desert is recognized as one of the most structurally diverse and floristically rich deserts in the world. Although centered in Sonora, Mexico, it extends northward into southern Arizona where it forms the Arizona upland desert (Shreve, 1951). Annual precipitation for this Upland desert may range from 33–38 cm.

The Santa Catalina and Rincon Mountains, near the eastern limit of the Sonoran desert (at this latitude) exhibit a rich spectrum of desert communities on their lower rocky slopes and surrounding bajadas. Here the saguaro (*Carnegiea gigantea*) is near

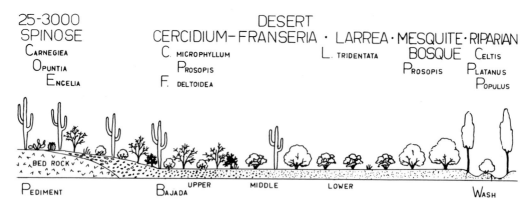

Fig. 2. Bisect of Sonoran desert vegetation types on bajada and pediment 760–915 m (2500–3000 ft) on south slope of Santa Catalina Mts. See Figure 5 for key to symbols.

Table 3. Forest region tree density by size classes per hectare for principal species and dominant shrubs and saplings.

Species[a]		Community types							
		Santa Catalina Mts.						Pinaleno Mts.	
		Ponderosa pine, oak forest	Ponderosa pine forest	Ponderosa pine, white pine forest	Douglas fir white fir forest	White fir ravine forest	Corkbark fir forest	Mixed conifer forest	Englemann spruce, fir forest
	feet	7000–8000	8000–9000	8500–9000	8000–9000	7500–8500	8500–9000	9000–9600	9700–10 700
	meters	2130–2440	2440–2740	2590–2740	2440–2740	2290–2590	2590–2740	2740–2920	2950– 3 260
Trees	dbh cm								
Abies concolor	<2.5	1	10		2	252	0	840	1
	2.5–15		1		76	83	8	207	
	15–30				22	32	16	64	
	30–60				33	47	14	43	
	60–90				7	24	12	24	
	90–120				4	10		4	
Abies lasio-carpa	<2.5							270	751
	2.5–15						112	66	403
	15–30						118	35	144
	30–60						122	21	28
	60–90					1		1	1
Acer glabrum	<2.5				8	4			
	2.5–15				3	18	4	36	4
	15–30					8	22	6	22
Acer grandi-dentatum	<2.5					40			
	2.5–15				10	334			
	15–30				6	38			
	30–60					6			
Jamesia americana	<2.5		2		13	17			7
	2.5–15				335	47	274		
Picea engel-manni	<2.5							199	434
	2.5–15							94	527
	15–30							38	181
	30–60							46	142
	60–90							12	24
	90–120							1	3
Pinus ponderosa	<2.5	513	304	50		4		4	
	2.5–15	271	489	504		2		3	
	15–30	49	107	110					
	30–60	60	114	102	1	2		2	
	60–90	8	11	10				2	
Pinus strobi-formis	<2.5	51	97	24	10	66		43	4
	2.5–15	25	164	800	22	43		27	16
	15–30	3	23	86	3	13		4	6
	30–60	1	11	48	5	11	1	13	8
	60–90		2		1	2		8	1
	90–120				1	3			
Populus tremuloides	<2.5					2			2
	2.5–15							2	14
	15–30						6	10	40
	30–60				2		14	17	4
	60–90							1	
Pseudotsuga menziesii	<2.5	5	11	2	12	201	6	94	3
	2.5–15	9	29	80	48	36	2	36	9
	15–30		2	2	28	4	14	12	15
	30–60	1			61	9	12	39	9

Table 3. (Continued).

Species[a]		Community types								
		Santa Catalina Mts.						Pinaleno Mts.	
		Ponderosa pine, oak forest	Ponderosa pine forest	Ponderosa pine, white pine forest	Douglas fir white fir forest	White fir ravine forest	Corkbark fir forest	Mixed conifer forest	Englemann spruce, fir forest
	feet	7000–8000	8000–9000	8500–9000	8000–9000	7500–8500	8500–9000	9000–9600	9700–10 700
	meters	2130–2440	2440–2740	2590–2740	2440–2740	2290–2590	2590–2740	2740–2920	2950– 3 260
	60–90				95	6	16	37	1
	90–120				14	5	10	11	2
	120–150				11	6		3	
	>150				4				
Quercus arizonica	<2.5	35							
	2.5–15	51	30						
	15–30	12							
	30–60	2							
Quercus gambelii	<2.5		71			7			
	2.5–15		76			24			
	15–30		9			3			
	30–60		1						
Quercus hypoleucoides	<2.5	352	22						
	2.5–15	890	92	32					
	15–30	58							
	30–60	31							
Quercus rugosa	<2.5	17	24	28					
	2.5–15	39	30	44					
Robinia neomexicana	<2.5		93		44	48			
	2.5–15		4		6	4		1	2
	15–30					2			
Shrub stratum									
Abies concolor		1			2	309		891	1
Abies lasiocarpa								326	750
Acer grandidentatum						36			
Holodiscus dumosus			23	26					
Jamesia americana			2		13	17			
Picea engelmannii								228	398
Pinus ponderosa		418	348	50				4	
Pinus strobiformis		51	98	24	10	54		61	4
Pseudotsuga menziesii		7	11	2	18	232		124	3
Quercus arizonica		35							
Quercus gambelii			71			7			
Quercus hypoleucoides		289	22	36					
Quercus rugosa		18	24	48					
Robinia neomexicana					44	25			

Additional species by community: pine-oak, *Arbutus arizonica,* white fir ravine forest, <2.5 cm = 4, 2.5–15 = 15, 15–30 = 10; *Alnus oblongifolia,* <2.5 cm = 8, 2.5–15 = 6, 15–30 = 10, 30–60 = 2. Mixed conifer forest, *Alnus tenuifolia* <2.5 = 2, 2.5–15 = 63, 15–30 = 6.

ts northeast limit and it, along with foothill palo-verde (*Cercidium microphyllum*) and associates comprise the distinctive vegetation on the upper bajada and lower mountain slopes. On the bajada, Creosotebush (*Larrea tridentata*) desert scrub is dominant and on the upper bajada it is replaced by either paloverde, bursage (*Cercidium, Franseria*) or disturbed desert scrub communities (referring to a deteriorated stand that has been adversely affected by land abuse as in overgrazing ('disclimax')). A spinose, suffrutescent desert scrub with saguaro and paloverde is characteristic on the lower mountain slopes and it is replaced by semi-desert grass-land (Fig. 4) with increasing elevation.

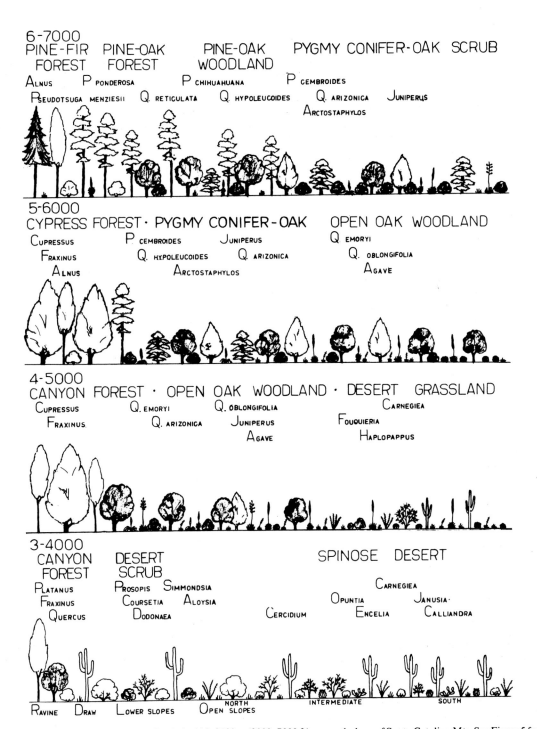

Fig. 4. Bisect of vegetation types by elevation belts 915–2130 m (3000–7000 ft) on south slope of Santa Catalina Mts. See Figure 5 for key to symbols.

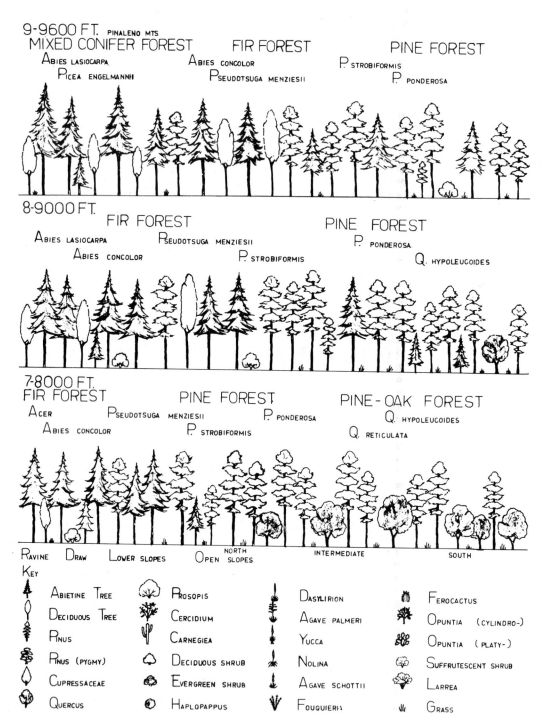

9-9600 FT. PINALENO MTS.
MIXED CONIFER FOREST   **FIR FOREST**   **PINE FOREST**
Abies lasiocarpa   Abies concolor   P. strobiformis
Picea engelmannii   Pseudotsuga menziesii   P. ponderosa

8-9000 FT.
FIR FOREST   **PINE FOREST**
Abies lasiocarpa   Pseudotsuga menziesii   P. ponderosa
Abies concolor   P. strobiformis   Q. hypoleucoides

7-8000 FT.
FIR FOREST   **PINE FOREST**   **PINE-OAK FOREST**
Acer   Pseudotsuga menziesii   P. ponderosa   Q. hypoleucoides
Abies concolor   P. strobiformis   Q. reticulata

Ravine   Draw   Lower slopes   Open   NORTH
                                                    slopes   INTERMEDIATE   SOUTH

Key

Abietine Tree	Prosopis	Dasylirion	Ferocactus
Deciduous Tree	Cercidium	Agave palmeri	Opuntia (cylindro-)
Pinus	Carnegiea	Yucca	Opuntia (platy-)
Pinus (pygmy)	Deciduous shrub	Nolina	Suffrutescent shrub
Cupressaceae	Evergreen shrub	Agave schottii	Larrea
Quercus	Haplopappus	Fouquieria	Grass

Fig. 5. Bisect of vegetation types by elevation belts 2130–2740 m (7000–9000 ft) on south slope of Santa Catalina Mts. and 2740–2920 m (9000–9600 ft) from Pinaleno Mts.

Larrea desert scrub

Northeast of Tucson from 760–825 m Creosotebush desert scrub covers many square miles of the gently sloping lower bajada (Table 1, column 1). Scattered *Larrea* 1–2 m in height form a relatively homogeneous community with the shrubs covering about 20–30% of the ground (Figs. 3, 6–7). Widely scattered large saguaro occur within this community.

During the dry season the ground is essentially devoid of herbaceous cover except for scattered tufts of fluff grass (*Tridens pulchellus*) and the sparse spreading stems of the purple-flowering *Allionia incarnata*. Perennial grasses such as bush muhly (*Muhlenbergia porteri*) and three-awn (*Aristida ternipes*) are rare. However, with the arrival of the summer rains, there quickly appear two annual grasses, *Bouteloua aristidoides* and the less common *B. barbata*, that create a striking green tinge over the otherwise bare soil. Following the winter rains, the introduced annual filaree (*Erodium cicutarium*) may form a conspicuous open ground cover.

On calcareous valley-fill soil in the nearby Reddington Road area, *Larrea* desertscrub occurs on south-facing slopes. Sandpaperbush (*Mortonia scabrella*), typical of the Chihuahuan desert (Benson & Darrow, 1944; Wentworth, 1981), is dominant on northerly exposures, but these stands are geographically marginal and contain few Chihuahuan species.

Larrea desert scrub in Arizona has a marked affinity for caliche-underlain soils (Youngs, 1931; Shreve & Mallery, 1933), although not restricted to such sites. Along the upper bajada transect there is occasionally a sharp line of demarcation between *Larrea* and the degraded desert on sandy soils where one can step from a *Larrea* stand directly into disturbed desert scrub stands of mesquite (*Prosopis juliflora* and burroweed (*Haplopappus tenuisectus*). On Tumamoc Hill in the nearby Tucson Mountains of volcanic origin, creosotebush is often associated with the underlying caliche exposed by erosion. On the granitic slopes of the Catalinas above the bajada, *Larrea* is essentially absent. Elsewhere throughout its range the species exhibits a broad edaphic tolerance that may in part be related to ecotypic variation within the species. There is no evidence that these stands represent large ancient clones as described for the Mohave Desert where they are estimated to be over 6 000 years old (Vasek, 1980).

Geographically, *Larrea* represents one of the most widespread desert shrubs in the New World deserts (Barbour & Diaz, 1973; Barbour *et al.*, 1974, 1977; MacMahon, 1979). In the Avra Valley west of Tucson on sandy soils and the finer silty clays below 760 m it is co-dominant with white bursage (*Franseria dumosa*). On these soils capillary water and soluble salts are high while water availability to plants is low (Yang & Lowe, 1956; Yang, 1957).

Past land use, especially grazing, has modified this low bajada community. In the mid-19th century, Emory (1858) described scanty growths of *Muhlenbergia porteri* under the ever-abundant creosotebush on the desert plain of the Santa Cruz. Prior to 1890 at Fort Lowell, near our study site, Mexicans and Indians gathered bunches of this grass with hoes from under the shrubs to feed government livestock. Poorly adapted to cattle grazing, this grass is now infrequently seen. However, with grazing protection, it, along with other graminoids, dramatically increases in coverage (Gardner, 1950; Gibble, 1950; Blydenstein *et al.*, 1957). The absence of young saguaros to replace the widely scattered dying adults also appears indicative of changed land use patterns. Because the lower bajada is especially prone to subfreezing winter temperatures, the effects of periodic freezes on saguaro seedlings may have been accentuated by the loss of protective ground cover due to grazing. Except for the absence of saguaro seedlings, this lower bajada *Larrea* community on the outskirts of Tucson appears relatively stable. The greatest threat to its survival is urban development.

Cercidium, Franseria desert scrub

The paloverde, bursage community is the more typical of two community types found on the upper bajada on the south slope (Table 1, column 2). It is usually found on gravelly soils and is part of the more widespread paloverde, saguaro Sonoran desert vegetation which extends onto the lower rocky mountain slopes. This community is especially well developed in the Tucson Mountains, and north of Tucson where samples were taken from 825–915 m near the base of the Catalinas (Fig. 3). In contrast to the fine textured soil supporting the *Larrea* desert scrub on the lower bajada, here the soil surface is covered by a coarse gravel underlain by finer clay-silts and gravel. The community is characterized by three strata: a low shrubby cover of triangle bursage (*Franseria deltoidea*) (Table 1), scattered paloverde 3–6 m in height and occasional large 8–9 m saguaros extending above the paloverdes. In the early 1900's Shreve (1915) describing the general aspect of this community observed that 'the openness of the stand is such that it is possible in all places to ride a horse through the vegetation and to take whatever course the rider may wish with occasional digression . . . from the general direction of travel'. Today the aspect is very similar to that which Forrest Shreve saw from horseback more than 60 years ago.

Total plant cover is 20–40%. Important associated species include ocotillo (*Fouquieria splendens*) and sizable sprawling patches of prickly pear (*Opuntia phaecantha*). Within a single hectare 35–45 perennial species are frequently recorded. There is a marked tendency for certain species such as brittlebush (*Encelia farinosa*), *Janusia gracilis*, *Mammillaria microcarpa* and saguaro seedlings, to be localized under the paloverdes and triangle bursage. It is not uncommon to observe a dozen or more pin cushion cacti (*Mammillaria microcarpa*) under a single *Cercidium*. These miniature cacti reach their maximum abundance in this community, locally over 1 000 per ha. This tendency for species to be localized under 'nurse' plants is related to concealment from predators, increased organic accumulation, more favorable moisture and microclimatic amelioration, all interacting to favor seedling establishment.

Due to the gravelly nature of the soils, rodent activity, especially burrowing evidence, is rare compared to that found on the sandy soils of the disturbed desert scrub community discussed next. Saguaro establishment does occur here, although not as successfully as on the adjacent mountain slopes (Table 1). Since bursage is unpalatable to livestock, this vegetation may not have been severely affected by past grazing.

This plant community with its abundance of bursage and relatively low saguaro density is distinctively different from the more diverse spinose, suffrutescent desert on the adjacent rocky mountain slopes. In its present relatively undisturbed state it represents a stable self-perpetuating foothills community. In the Catalinas, the greatest threat to this integrity is urban development. Fortunately, it is such an esthetically attractive plant community that it is usually preserved as a form of naturalistic landscaping.

Disturbed desert scrub

A second plant community has also developed on the upper bajada between 825–915 m, but on more sandy soils than the *Cercidium, Franseria* community that is restricted to gravelly soils. This upper bajada area at the base of Mt. Lemmon highway is somewhat atypical in its proximity to Soldiers Canyon which may account for the sandy nature of the soil and the less xerophytic aspect compared to the adjacent stony upper bajada (Table 1, column 3). The vegetation is also less xerophytic than on the stony upper bajada and is obviously transitional between desert scrub and semi-desert grassland. It has been modified by decades of grazing and such evidence is still present today. Comparable examples of this vegetation also occur at the base of the adjacent Rincon Mountains (Niering *et al.,* 1963).

In this upper bajada community, paloverde and saguaro may reach their maximum size (Fig. 6–8). Paloverde attains 6–9 m in height and its multistems reach 15–20 cm in diameter. Occasional large barrel cacti may reach 1.2 m in height. Although the larger saguaros reach 9–12 m in height, a third of the population sampled along the highway below Soldiers Canyon was dead from freezing and reproduction was wanting. Burroweed (*Haplopappus tenuisectus*), cholla (*Opuntia spinosior*) and prickly pear (*Opuntia engelmannii*) are especially abundant (Table 1) and have been favored by past grazing (Humphrey, 1937; Benson & Darrow, 1944; Mehrhoff, 1955). Brittlebush (*Encelia farinosa*) is also frequent and recognized as a pioneer shrub on disturbed sites (Vasek, 1979/1980). Perennial herbs, especially grasses, are rare; however, following the rainy seasons a carpet of annual grasses may cover more than 50% of the ground. Rodent activity is widespread; woodrat dens are frequent and ground squirrel burrowing activities are common.

The decline of the saguaro population in this community and its lack of reproduction on many similar bajada sites have been attributed to various factors: bacterial necrosis, overgrazing, high rodent populations and climatic change (Alcorn & May, 1962; Niering *et al.,* 1963; Turner *et al.,* 1969). Recently Steenbergh & Lowe (1977) pinpointed the significant role of periodic freezes as a critical factor in the decline of both mature and seedling populations. They have also discounted the role of 'bacterial necrosis disease' in saguaro population mortality, indicating that the so-called 'disease' follows freezing as a natural decomposition process. Grasses and other plant cover appear to play a significant role in ameliorating the effects of low temperatures to which the bajada is periodically subjected. As we report-

ed in the early 1960's, saguaro reproduction was also wanting on comparable severely degraded sites in the Saguaro National Monument (Niering *et al.,* 1963). However, a resurvey in the Monument a decade later revealed saguaro reproduction in areas where it appeared absent in 1962. A subsequent freeze in the Tucson area in 1971 apparently did not affect those newly established plants under paloverde. Grazing was removed from this part of the Monument in 1958, only four years before our initial studies in the early 1960's. By the 1970's grass cover, primarily *Muhlenbergia porteri,* had increased dramatically, especially under the paloverde where saguaro reproduction is now present. Cessation of grazing followed by an increase in ground cover may have aided in modifying microclimatic conditions to protect saguaro seedlings from freezing.

Young saguaro are also killed by certain rodents, especially wood rats and ground squirrels, which are frequently abundant in degraded or highly disturbed stands (Turner *et al.,* 1969; Steenbergh & Lowe, 1977). Wood rats are favored by increasing populations of prickly pear and cholla cacti which provide food and cover. Ground squirrels as burrowing animals are also favored by sandy bajada sites. Some mature plants become riddled with spiral runways made by wood rats and eventually die. Past grazing drastically reduced ground cover, and rabbit browse of saguaro nurse plants is locally severe. Predator control may also be a factor since coyote control has been widespread. Overgrazing here, as well as in Saguaro National Monument and elsewhere has had a destructive effect on the structure of natural desert communities. It would appear that grazing, rodents, and freezing temperatures are intricately interrelated in the decline of saguaro in such sites.

Spinose, suffrutescent desert scrub

A spinose, suffrutescent desert scrub community (also referred to as paloverde, saguaro desert scrub) dominated by paloverde and saguaro, characterizes the lower mountain slopes within the Sonoran desert (Table 1, column 5). In the Catalinas it extends from the rocky pediment at the base of the mountain near 915 m up the lower rocky slopes to about 1 220 m where it is replaced by semi-desert grassland. The community is best developed on the warmer slopes, especially on south-facing exposures where the saguaro reaches maximum density (Niering *et al.,* 1963) (Fig. 6–3). *Franseria deltoides,* typical in the paloverde, bursage community below, is here replaced by the suffrutescent shrub brittlebush (*Encelia farinosa*) (Table 1, column 5). Other

---

←

Fig. 6. Sonoran desert region, Santa Catalina Mountains. 1. Desert grassland, SE 25 ° slope, 1220 m below Molino Basin. Stand 222. *Fouquieria splendens* and *Haplopappus laricifolius* principal shrubs. *Agave palmeri* flowering. 1962. 2. Desert grassland SSE 22 °, 1220 m below Molino Basin. Unburned grassland (foreground). *Agave schottii* principal species. Recently burned grassland (background), *Bouteloua curtipendula* dominant. Dark low shrub is *Haplopappus laricifolius.* 1963. 3. Spinose-suffrutescent desert taken from 915 m on lower mountain slopes of Catalinas along Mt. Lemmon highway. *Carnegiea gigantea* and *Cercidium microphyllum* dominant species. 1963. 4. Spinose-suffrutescent desert, E slope, 990 m along Mt. Lemmon highway. *Carnegiea gigantea* and *Cercidium microphyllum* principal species. Large dead saguaro, left. In Nov. 1962 mortality of larger plants as a result of Jan. 1962 freeze was 25%. Note good ground cover on rocky slopes protected from grazing. Nov. 1962. 5. Riparian forest, 838 m in lower Sabino Canyon. *Populus fremonti* and *Platanus wrightii* dominant trees. Saguaro evident on adjacent rocky slopes. Dec. 1962. 6. Arroyo margin woodland, 609 m on upper bajada along Soldiers Trail. Stand 351. *Platanus wrightii* left, *Celtis reticulata* right. 1962. 7. *Larrea tridentata* desert scrub lower bajada. *Prosopis juliflora* extends above creosotebush left center. Dried desert annuals on soil surfaces. July 1962. 8. Degraded desert, 855 m on upper bajada. Large *Carnegiea gigantea;* cholla left foreground. Low shrubs primarily *Haplopappus tenuisectus; Cercidium microphyllum* and *Prosopis juliflora* beyond. Santa Catalina Mts. in background. 1963.

important suffrutescents include *Calliandra eriophylla, Janusia gracilis* and *Jatropha cardiophylla*. The spinose deciduous *Fouquieria splendens* is an important and distinctive woody species along with the thorny *Celtis pallida* and *Lycium berlandieri parviflorum*. Numerous succulents (*Opuntia phaecantha, O. fulgida, O. spinosior, O. biglovii* and *Ferocactus wislizeni*) add a further spiny aspect to this community. Over 80 species of vascular plants, excluding annuals, have been recorded in this community type and it is not uncommon to record 30 or more in a 0.1 ha plot.

On north-facing slopes a distinctively less xeric shrub phase occurs with scattered saguaro, paloverde, ocotillo and various shrubs contributing most of the community coverage. *Coursetia microphylla* and *Dodonaea viscosa* are especially distinctive on the north-facing slopes of Sabino Canyon where shrub coverage can reach 50% or more. Here saguaros are infrequent which further reflects their sensitivity to winter low temperatures.

Although most of the species populations in this community appear to be in equilibrium with their environment, on the southerly slopes a considerable portion of the saguaro population studied in the summer of 1962 had been killed by a prolonged freeze in January 1962 (Fig. 6–4). Six months later, mortality of the plants over 2 m in height averaged 30%, and reached over 50% in some stands. Individuals less than 2 m high were more resistant, although some frost damage was evident on many of them. Under the paloverde where most of the seedling saguaros survived the 1962 freeze, air temperatures at ground level were at least 2 °C warmer at night in winter than in open exposed sites. In a permanently established 0.1-ha plot on an open south slope at 1050 m the fate of these seedlings, 4–30 cm in height, was followed by Niering for nearly a decade. Of the 50 seedlings initially recorded in 1963, 50% were dead, uprooted, or not found in 1972. However, 10 new seedlings occurred that were initially either not present or not recorded. In 1972 the seedlings were vigorous and most had doubled or quadrupled their height growth over this 9-year period. Although this population was also exposed to a second freeze in 1971, there were no marked effects evident here in the 1972 census (for additional results for the 1971 freeze see Steenbergh & Lowe, 1976). Although several severe freezes have been recorded in the Tucson area before and since, the 1962 event was truly catastrophic for many of the stands on the south slope of the Catalinas. Many of the larger saguaros killed were in excess of 100 years of age. Such periodic events not only tend to restrict saguaros to the warmer slopes, but also limit their upward extension to a maximum of 1585 m on rocky southern exposures on the southwest corner of the range.

It should also be noted that these slopes are within the Coronado National Forest and have been protected from grazing or other disturbance since 1947. An excellent ground cover has developed on these protected slopes compared to grazed slopes in the nearby Rincon Mountains where saguaro reproduction appeared to be poor during the 1950's and 1960's. Protection from grazing adds to the potential survival of the saguaro and other cacti in terms of protective ground cover and consequent microclimatic protection. All evidence indicates that the saguaro populations and other associated species on these undisturbed rocky slopes represent a relatively stable community in whose evolution droughts and periodic freezing temperatures have played an intricate part. Although catastrophic freezes dramati-

cally reduce the more mature saguaro population, there is ample reproduction on undisturbed slopes to maintain this species.

Arroyo and canyon vegetation

Along the margins of larger arroyos or washes, which may reach a hundred or more feet in width, is found a deciduous woodland composed of hackberry (*Celtis reticulata*), ash (*Fraxinus pennsylvanica velutina*), sycamore (*Platanus wrightii*) and cottonwood (*Populus fremontii*) (Table 1, column 4; Figs. 3, 6–6). These riparian woodlands may reach 11–14 m in height with tree trunks in excess of 2 m diameter. We recall Whittaker's pleasure in measuring a 2.5 m dbh cottonwood which was standing in association with hackberry and mesquite in a sandy wash near Tucson. On the sandy islands within the braided stream bed, two tall shrubs, *Baccharis sarothroides* and *Hymenoclea monogyra*, are locally abundant as pioneer species. Herb cover varies from relatively sparse in the wash to seasonally dense on the banks.

This magnificent woodland is comprised of large trees and is a remarkably stable entity as a linear community periodically swept by torrents of water during flash floods. A mesquite woodland on the floodplains usually borders this arroyo vegetation (see Fig. 3), and may expand into mesquite bosques that reach several acres in extent along the major washes draining the Catalinas. Original stands had large trees 9–12 m in height. Only a few remnants of these once great mesquite bosques still persist (Johnson & Carothers, 1982).

Above the arroyos in the adjacent mountain canyons between 840–1000 m, where surface or subsurface water is found throughout the year, an impressive riparian forest has developed. Cottonwood and sycamore 30–60 cm or more in diameter line the bouldery stream beds. Two evergreen oaks (*Quercus emoryi* and *Q. oblongifolia*) from the encinal woodland reach their lower limit here. *Baccharis sarothroides* and *Bidens aurea* are among the more frequently found shrubs. Herbaceous cover is sparse. In the fall this riparian forest lends a golden-yellow autumnal aspect to these desert canyons which is reminiscent of parts of the eastern deciduous forest (Fig. 7–5).

Semi-desert grassland

The desert or semi-desert grassland typically occurs between the spinose desert and open oak woodland. In Arizona these grasslands are usually found between 1200–1600 m in the higher valleys and occasionally on the mountain slopes. The annual rainfall ranges from 23–46 cm. Many of these grasslands have been severely damaged by overgrazing (Wooton, 1915; Sampson & Weyl, 1918; Taylor *et al.*, 1935).

On the slopes of the Catalinas a narrow belt of desert grassland occurs between 1200–1400 m but extends upward to 1700 m on the most xeric slopes. The overall aspect is an open grassland with scattered woody species. With overgrazing and in the absence of fire, a local dominance of amole (*Agave schottii*) develops on many of these granitic sites (Table 1, column 6; Fig. 6–2) and in the nearby Rincons. In such stands amole can form a relatively continuous ground cover 15–30 cm in height. Scattered grasses are associated with these *Agave*-dominated stands, and graminoids are predominant in areas recently burned since *Agave* is fire sensitive (Fig. 6–2). Among the principal grasses are *Bouteloua curtipendula, B. hirsuta, Aristida ternipes, Heteropogon contortus, Andropogon barbinodis* and *Muhlenbergia emersleyi.* Following fire grass cover can reach 50% or more and species richness is increased.

In addition to *A. schottii* other distinctive rosette monocot shrubs, sotol (*Dasylirion wheeleri*) and bear grass (*Nolina microcarpa*), appear here. A distinctive low tree, rosewood (*Vauquelinia californica*), occurs on more mesic sites. Scattered cholla and barrel cacti extend up into this grassland community and locally ocotillo dominates on more xeric grassland slopes (Fig. 6–1). The evergreen, needle-leaved turpentine bush (*Haplopappus laricifolius*) is frequent, especially where fire has not occurred for some time.

Various phases of semi-desert grassland are also well developed on the north side of the range. On non-calcareous slopes between 1460–1600 m, *Bouteloua curtipendula*, *Artemisia ludoviciana albula* and *Eriogonum wrightii* are the principal species. On calcareous slopes between 1600–1700 m, *B. curtipendula* is also dominant along with *Calliandra eriophylla* and *Dasylirion wheeleri*. These species are also present on the south slope.

In the San Pedro Valley near Oracle, on relatively level noncalcareous deep soil sites, one encounters a *Yucca elata* grassland at 1160 m. Here there is a sharp contrast between severely grazed and contiguous protected sites. In the heavily grazed areas two typical grazing increasers, snakeweed (*Gutierrezia sarothrae*) and burroweed (*Haplopappus tenuisectus*), are dominant. The former reaches densities of over 20000 and the latter 3000 plants per hectare. *Opuntia fulgida* and *O. engelmannii* are also frequent along with scattered mesquite and *Yucca elata*. Grass cover is usually much less than 5%. This disturbed grassland resembles, in some respects, the previously discussed desert scrub community on the upper bajada where *H. tenuisectus* and *Opuntia* are abundant. Contiguous, but separated by only a barbed wire fence, is the Page Ranch where the owner controlled overgrazing for several decades with the aim of rehabilitating a degraded landscape (Haskell, 1945). By the early 1960's the area was grass-dominated with mean grass coverage 47%, and locally reaching 50–60%. In addition to *Bouteloua curtipendula,* both *Andropogon barbinodis* and *Aristida ternipes* were frequent. Although typical grazing increasers are present in this protected grassland, their density is low. Total woody cover is less than 10% compared to 20–40% in the disturbed community. These contrasting communities dramatically illustrate how protection from grazing, or low grazing intensity, even in this extreme environment can result in rapid recovery.

Encinal region

The Encinal is a region characterized by various evergreen trees, primarily oaks, junipers, and pinyons. These evergreen woodlands reach maximum development in the Sierra Madre Occidental of northwestern Mexico and continue northward into mountains of southern Arizona. It is a region that experiences moderate minimum winter temperatures and a bi-seasonal pattern of rainfall that totals 30–60 cm.

Shreve (1915) described the lower more open, and upper more closed, Encinal of the Catalinas.

The region extends from above the desert grassland at 1400 m to the pine forest at about 2130 m. Nowhere in the Santa Catalina Mountains is a vegetation continuum more in evidence, and community classification more arbitrary than in the Encinal. As explained previously '. . . these species are in a woodland and forest continuum which begins below as an 'open encinal' or oak–grass landscape between 4000 and 4500 feet, and increases upslope in density and in cover on the moisture–temperature gradient to an oak–juniper–grass landscape between 4500 and 5500 feet, and with increased density, coverage, and floral components to an oak–juniper–pine landscape (oak–pine woodland) that merges with the ponderosa forest above' (Lowe, 1964).

The lower Encinal is an open oak woodland with alligator juniper (*Juniperus deppeana*) (Table 2, 1400–1700 m). The upper Encinal is an oak, juniper, pinyon (pygmy conifer, oak) woodland (Table 2, 1520–1830 m). At 1830–2130 m, the still denser *Pinus cembroides,* oak woodland (Table 2, 1830–2130 m) is transitional to a limited *Pinus chihuahuana,* oak woodland on the south slope of the Santa Catalina Mountains (Table 2, 1830–2130 m). In the canyons throughout this elevation range the riparian vegetation is composed of a mixture of evergreen oaks and broad-leaved deciduous species. Locally, forests of Arizona cypress and ponderosa pine occur in the deeper canyons (Lowe & Brown, 1973; Parker, 1980).

Open oak woodland

The open oak woodland occurs above the desert grassland from 1300 m on the more mesic slopes up to 1700 m on the more xeric slopes (Fig. 7–8). These orchard-like woodlands are dominated by scattered Mexican blue oak (*Quercus oblongifolia*), Emory oak (*Q. emoryi*) and alligator juniper (*Juniperus deppeana*) (Fig. 4, Table 2; Fig. 7–9). The low rounded oaks, 3.5–5.5 m in height and 15–60 cm in diameter, seldom contribute more than 10–20% canopy coverage.

Although present in the desert grassland, *Vauquelinia californica* is of greater importance in the open oak woodland (Fig. 7–6). Two broad-leaved evergreen shrubs, *Garrya wrightii* and manzanita (*Arctostaphylos pungens*) along with four rosette monocot shrubs, give this woodland a distinctive aspect. *Agave schottii* extends up from the desert grassland, clumps of beargrass (*Nolina microcarpa*) are frequent, *Agave palmeri* is occasional, and sotol (*Dasylirion wheeleri*) is conspicuous on rocky sites. The low needle-like evergreen shrub *Haplopappus laricifolius,* present in the desert grassland, reaches its maximum development on the more southerly and rocky exposures (Fig. 7–6).

Perennial grasses, especially *Bouteloua curtipendula* which

176

may contribute 1–2% or more coverage, extend up from the desert grassland. Several xerophytic evergreen ferns, *Notholaena sinuata, Cheilanthes fendleri* and *C. lindheimeri,* are typical of the Encinal and are frequent in rock crevices. Mats of *Selaginella rupincola* form distinctive terraces on the more mesic slopes and play a significant role in stabilizing the granitic parent material (Fig. 7–7).

Like the contiguous semi-desert grassland, the open oak woodland is quite capable of carrying fire. Data collected in 1963 following a fire above Molino Basin revealed that ocotillo, *Haplopappus, Dasylirion, Arctostaphylos, Vauquelinia, Nolina* and *A. schottii* decreased following the fire. Blue oak resprouted abundantly after being stem-killed. A number of *Ferocactus wislizeni,* which here reach their upper limit, were killed whereas others, although charred, had survived and were flowering. McLaughlin & Bowers (1982) also found a reduction in *Ferocactus* following an unusual wildfire in *Cercidium, Franseria* desert scrub south of Florence, Arizona. As in the contiguous grassland, the graminoids such as *Bouteloua* and *Andropogon* increased markedly following the fire and shrub cover declined. In these woodlands fire and drought are the salient factors which account for shifts in species dominance.

On the north slope of the range, open oak woodland is well developed between 1520–1830 m on granitic and other acid parent materials. Here blue oak is much less important, whereas Emory and Arizona oak are the principal species. *Vauquelinia* and *Haplopappus laricifolius* are also much less common and *Agave schottii* and *Selaginella* are apparently absent. On limestone sites the oaks are replaced by a distinctive mountain mahogany (*Cercocarpus breviflorus*) scrub community (Whittaker & Niering, 1968b).

Pygmy conifer, oak woodland

The pygmy conifer, oak woodland occurs as a transitional community between the open oak woodland below and the more closed pine, oak woodland above. It is best developed between 1520–1830 m on intermediate and north-facing slopes. *Quercus oblongifolia* is replaced by *Q. arizonica* and *Q. emoryi* (Table 2). Mexican pinyon pine (*Pinus cembroides*) appears as a second coniferous associate with *Juniperus deppeana.*

Two of the rosette monocot shrubs, *Nolina microcarpa* and *Yucca schottii,* reach their greatest abundance in this transition zone. *Arctostaphylos* increases but *Haplopappus,* frequent in the open oak woodland below, is uncommon here. Tree and shrub cover may reach 50% or more in some stands with *Arctostaphylos* contributing 20–30%. Although herb cover is sparse, scattered grasses, *Muhlenbergia emersleyi* and *Aristida orcuttiana,* are present in the openings. Xerophytic ferns are frequent and *Selaginella rupincola* continues to mat on mini-terraces, reaching its best development on lower slopes. There is considerable mortality within the oak populations, especially *Q. emoryi.* This may be related to prolonged drought. Otherwise, this community appears to represent a relatively stable transitional woodland.

Pine, oak woodland

Above 1830 m as one enters the pine, oak woodland along the Mt. Lemmon highway, the slopes become increasingly steep, rocky and more exposed. On the more mesic slopes there occur localized stands of *Pinus chihuahuana,* oak woodland (Table 2, column 4), which grade into *Pinus cembroides,* oak woodland on intermediate, drier sites (Table 2, column 3). On the steeper rocky slopes and crests one encounters a distinctively scrubby pygmy conifer, oak vegetation that represents the most xeric phase within this elevational zone (Table 2, column 5).

The *Pinus chihuahuana,* oak community, best developed on lower slopes, is characterized by *P. chihuahuana* up to 12 m in height and 30 cm or more in diameter (Fig. 7–1), in association with *P. cembroides, Juniperus deppeana* and *Quercus hypoleucoides.* The last is the dominant understory tree in association with *Quercus arizonica* and madrone (*Arbutus arizonica*). The scattered Chihuahua pines extend above an oak understory 3–7.5 m in height. In some stands, pine contributes 15–20% and oak 50% or more coverage. Although this community is restricted to the more xeric, south slope of the range, it is well developed between 1830–2130 m on acidic sites on the more mesic north-facing slopes, as reported by Shreve (1915) and by Whittaker & Niering (1968b).

On intermediate slopes the *Pinus cembroides,* oak woodland of lower stature is characterized by *P. cembroides, Juniperus, Quercus emoryi, Q. arizonica* and *Q. hypoleucoides;* in addition there are many thickets of manzanita (Fig. 7–2). Although tree growth may reach 5.5–7.5 m in height, the overall aspect is that of a lower statured woodland type than the *Pinus chihuahuana,* oak community. Even though the conifers are low in stature, some pinyons reach 60 cm in diameter and junipers are frequently larger (Table 2). Manzanita (0.9–1.8 m high) contributes to the woody cover along with several monocot shrubs; *Agave palmeri*

Fig. 7. Encinal region of Santa Catalina Mountains. 1. *Pinus chihuahuana,* oak woodland, S 18° slope, 1920 m above General Hitchcock Picnic Area. Stand 364. *Quercus hypoleucoides* is principal understory tree. 1962. 2. *Pinus cembroides,* oak woodland, lower S slope, 1700 m along Mt. Lemmon highway. *Juniperus deppeana* left; *Pinus cembroides* center; *Quercus arizonica* right. Rock spires in pine, oak woodland zone in background. 1963. 3. Pygmy conifer, oak scrub, W slope 13°, 2012 m near Hitchcock Monument. Open stratum of *Pinus cembroides, Juniperus deppeana* and *Quercus hypoleucoides.* Sclerophyllous shrubs, *Arctostaphylos pringlei, A. pungens* and *Garrya wrightii.* Rosette monocot shrub in foreground. Productivity stand (see Whittaker & Niering 1975). 1964. 4. *Cupressus arizonica* forest, 1722 m in Bear Canyon along Mt. Lemmon highway. Smaller cypress evident up north slope in background. 1962. 5. Pygmy conifer, oak scrub 2073 m on rocky crest above Hitchcock Monument. Stand 108. This *Arctostaphylos pringlei* phase has reached maturity with natural fire protection afforded by surrounding rock ledges. Larger stems reach 20 cm dbh. Note large dead stems in foreground. 1962. 6. Open oak woodland, rocky E upper slope, 1700 m. *Dasylirion wheeleri* rosette monocot with flowering spike, *Haplopappus laricifolius,* low shrubs among rocks and *Vauquelinia californica* upper right. Stand 174. 1962. 7. *Selaginella rupincola* terracing NNW upper open slope. 1493 m. Sycamore Canyon. Upper limit of open oak woodland. 1962. 8. Open oak woodland, lower encinal, on distant slopes. Taken from 1700 m. 1962. 9. Open oak woodland, SE 16° slope, 1463 m. Stand 165. *Quercus oblongifolia* left; *Quercus emoryi,* darker, right; *Agave schottii* foreground. 1962.

is most common, *Nolina microcarpa* and *Dasylirion wheeleri* are frequent to occasional. Increment borings indicate that *P. chihuahuana* 20 cm dbh are at least 120 years old and that *Juniperus* 46 cm dbh are over 150 years old. This differential growth rate may account for the greater number of larger *Juniperus*. The presence of fire signs indicates that fire plays a role in these woodland types.

Pygmy conifer, oak scrub

The distinctively scrubby pygmy conifer, oak scrub stands which develop in response to fire and drought on steeper, open rocky slopes and crests in the woodland and forest represent the most xeric community in this woodland zone (Table 2, column 5). It is recognized by its low scrubby character, the abundance of *Arctostaphylos* thickets, and a very open tree stratum of low-growing *Pinus cembroides*, *Juniperus* and *Quercus*. The trees are mostly less than 4.5 m in height (Fig. 7–3). In some stands manzanita and shrubby oaks contribute 60–70% coverage with little tree growth present. The presence of burned pine snags and other dead pine without fire scars suggests that drought and fire interact in controlling the structure and composition of this community. A quarter of the larger pines are dead, indicating that oak and manzanita are favored over pine under the current climatic and fire regime.

Another fascinating aspect of this scrub community occurs locally on the rocky crests around 2100 m. Here pure stands of large manzanita (*Arctostaphylos pringlei*) form 'mini-woodlands' up to 3.5 m in height. Shrub stems 15–18 cm dbh are common; the largest specimen recorded was 28 cm in diameter 30 cm above the ground. This tall shrub cover may reach 70–80% with scattered pines up to 28 cm dbh contributing 20% coverage. Based on ring counts, these shrubs may be 75 or more years of age. The presence of large dead stems and the absence of fire scars suggest that these stands have developed with protection from fire afforded by the extreme rockiness. Periodic severe droughts may well account for the natural demise of such stands (Fig. 7–5).

Similar pygmy conifer, oak scrub communities occur in other southwest ranges and in California chaparral, in the Mediterranean and elsewhere on serpentine (Whittaker, 1954). The absence of such a community from the north slope of the Catalinas suggests that its presence on the south slope may be related to the extreme xerophytism associated with the steeper rocky slopes and exposure to the hot dry winds of the Sonoran desert during periods of moisture stress.

In comparing the north and south slope upper Encinal one finds a near absence of this pygmy conifer, oak scrub along with two major species, *Pinus cembroides* and *Arctostaphylos pringlei*. Equivalent positions on the north slope are occupied by less xeric pine, oak woodlands.

Cypress canyon forest

In the canyons of this Encinal region there occurs a varied riparian vegetation. Between 1520–1830 m evergreen oaks and broad-leaved trees such as *Juglans*, *Platanus* and *Alnus* frequently intermingle. In the deeper canyons between 1700–2130 m, ponderosa pine occurs a thousand feet or more below its normal range. Favored by cold air drainage, this occurs in Bear Canyon along the Mt. Lemmon highway where *Pinus ponderosa* and occasional *Pseudotsuga menziesii* form an impressive riparian needle-leaved forest. One of the most interesting deep canyon communities is dominated by Arizona cypress (*Cupressus arizonica*) and scattered deciduous trees (Table 2, column 6; Fig. 7–4). *Cupressus* 33–65 cm in diameter ranged from 200–250 years of age. Associated shrubs include *Berberis wilcoxii*, *Rhus radicans*, *Rhamnus californica ursina* and a vine, *Vitis arizonica*. Along the stream bed one frequently encounters sedges (*Carex* spp.) and forest herbs such as *Aquilegia chrysantha*, *Oxalis metcalfei* and *Galium asperrimum* which are normally found at higher elevations.

Arizona cypress is a relict that has neared stand-extinction on the few local sites remaining at near the northern winterfreeze limit of its range, from southeastern Arizona to the Big Bend region of Texas-Chihuahua. The center of the species distribution is well south in the Sierra Madre of northern Mexico. Today the few remaining relict stands in southeastern Arizona live under strong edaphic moisture–temperature constraints in topographic settings as described above. In essence, this highly restricted community represents a relatively stable vegetation type dependent upon adequate moisture and fire protection afforded by certain canyon environments.

The forest region

In the mountain ranges of southern Arizona, montane forests characterized by ponderosa pine (*Pinus ponderosa*) and Douglas fir (*Pseudotsuga menziesii*) occur between 1800–2900 m but may extend down to 1700 m in canyons and up to 3050 m on ridges and south-facing slopes. Here this forest zone represents one of the driest coniferous forests in North America with an annual precipitation of only 50–71 cm. On the higher mountains like the Pinalenos, one encounters a more mesic subalpine spruce–fir forest where rainfall reaches 102 cm (Lowe & Brown, 1973).

In the Catalinas, Shreve (1915) noted a striking change in the vegetation pattern as he moved from the pine oak woodlands into the pine forests (Fig. 5). Along the Mt. Lemmon highway this transition occurs around 2130 m. Here the dry ponderosa pine, oak forests are characterized by an understory of evergreen oak. This drier pine community is replaced at higher elevations by relatively pure ponderosa pine forests. On northerly slopes above 2300 m Douglas fir dominates. Near the summit there is also a highly restricted *Abies lasiocarpa arizonica* community on north-facing slopes, the only subalpine element represented in the Catalinas. In the Pinaleno Mts. to the east we complete our altitudinal vegetation pattern with a mixed conifer forest above 2740 m and spruce-fir forest extending above 2920 m to the summit (Moir & Ludwig, 1979).

Ponderosa pine, oak forest

The Ponderosa pine, oak forest, best developed on southerly slopes between 2100–2450 m, represents a low elevation xeric phase of pine forest (Table 3, column 1). It is characterized by widely scattered ponderosa pines and an understory of evergreen oaks, primarily silverleaf oak (*Quercus hypoleucoides*). The overstory trees are primarily 30–60 cm dbh and reach 15–21 m in height. The understory oak are 15–30 cm in diameter and 6–9 m in height. Tree coverage is highly variable. The pine stratum may contribute 20–40% or more and the understory oak 40–70% coverage. Scattered *Quercus arizonica, Arbutus arizonica* and *Quercus rugosa* may be present in the undergrowth. Certain dry-tropic elements, *Agave palmeri, Yucca schottii* and *Echinocereus triglochidiatus,* also extend upward into this dry forest community. Herb cover is sparse, usually less than 5% but locally, grassy glades of *Muhlenbergia emersleyi* occur. In a typical stand selected for productivity studies, the pine ranged from 120–170 years of age and the oak 30–50 years (Whittaker & Niering, 1975). The largest pine recorded on a southeast slope at 2320 m was 183 cm in diameter. In these pine, oak stands one encounters for the first time a relatively continuous layer of leaf litter (Whittaker *et al.,* 1968).

The presence of fire scars indicates that this forest has been subjected to periodic ground fire. The occurrence of evergreen oak woodlands with scattered burned pine snags suggests that a woodland phase can result from localized severe burns. In this community, fire and drought appear to interact to create a highly variable vegetation pattern in terms of pine and oak density and associated ground cover.

Ponderosa pine forest

With increasing elevation, *Quercus hypoleucoides* decreases in importance and ponderosa pine becomes the principal forest type. Pine-dominated stands can occur between 2100–2440 m (Table 3, column 2; Fig. 8–8) but are especially well developed from 2440–2740 m on southerly slopes. In stands unburned for several decades, a distinctive understory of pine transgressives occurs. The overall aspect of such stands is large scattered overstory pine, 60–90 cm dbh and 21–28 m in height, with a pine understory which may contribute 50% or more coverage. With increasing elevation, Mexican white pine (*Pinus strobiformis*) becomes a more important associate with the ponderosa pine in the understory. Scattered *Q. hypoleucoides* still persist along with *Quercus gambelii* and *Q, rugosa. Muhlenbergia virescens* and *Pteridium aquilinum* are dominant species of the sparse ground cover. Tree ring counts indicate that the older pine stands are 200–300 years of age.

Frequent fire scars indicate that fire has played an important role in determining the aspect of these pine forests, and their understory has, as previously mentioned, developed largely in response to fire protection. In northern Arizona, Cooper (1960) found that an open ponderosa pine savanna was the more typical forest type prior to fire protection and that, in the absence of fire, a pine understory developed.

Observations on burns in these pine forests provide additional insights. The Red Ridge and Box Camp Canyon burns were lightning-caused fires and some 15 years after these burns the aspect was open and park-like, very different from most of the fire-protected stands sampled (Fig. 8–5, 8–6). In these burned stands there occurred small patches of pine transgressives

2–2.5 m high and 30–60 cm dbh, but most of the area was open and dominated by grasses. With many decades of fire protection a severe potential fire hazard has developed in many parts of this pine forest region. Elsewhere in the southwest, prescribed burning has been successfully employed to reduce this type of hazard (Biswell *et al.,* 1973).

Oak, fire scrub

Within this pine belt one occasionally encounters an oak, fire scrub community that has developed primarily as a result of fire and/or drought, especially on steep south-facing upper slope sites. Shreve (1915) also recognized this community type (Plate 26). In such sites any associated pine would tend to have been eliminated and the sprouting oaks favored. Thickets of *Quercus hypoleucoides* and *Q. rugosa* 1.2–1.8 m in height characterize these scrub communities where severe crown fires in the past have destroyed the pine overstory.

Based on the paucity of pine reproduction within this oak scrub, these thickets appear to represent relatively stable entities that may persist for a long time. This emphasizes the possibility that two different relatively stable communities such as a pine forest or oak scrub can exist on the same site but at different times in the dynamics of an ecosystem (Niering & Goodwin, 1974). It is probable that some of the oak thickets or woodland may have been induced by locally severe fires of natural origin. Of the forest burns investigated, those of lightning origin tended not to crown and favored the development of a pine–savanna vegetation. Man-caused fires which frequently occur during the dry season when forest conditions may be especially conducive to crown fires are generally more severe and frequently produce a fire scrub like that previously described. Only one known man-caused fire resulted in a pine–savanna. Prior to the advent of fire protection, a large part of this pine forest was undoubtedly open and park-like, as observed by Shreve, since the frequency of lightning-induced fires is high in this mountain range.

Ponderosa pine, white pine forest

This localized Ponderosa pine, white pine (*Pinus strobiformis*) community occurs between 2650–2750 m near the summit on southerly exposures underlain by quartzite. Stands are frequently two-layered: the larger usually more abundant ponderosa extend above the smaller white pine (Table 3, column 3; Fig. 8–9). Transgressives of the latter are especially abundant and frequently contribute 50–60% coverage. Woody undergrowth is sparse but scattered *Quercus hypoleucoides, Q. rugosa* and *Holodiscus dumosus* may be present. Herb cover is scattered but locally may reach 5–10% with the same species as are present in the ponderosa pine forest. In this high elevation pine forest, white pine locally exceeds ponderosa on some exposures and, with continued fire protection, white pine may increase in importance in these stands.

Douglas fir, white fir forest

The Douglas fir, white fir (*Pseudotsuga menziesii, Abies concolor*) forest is restricted to north-facing sites above 2450 m, although elements of this community extend down into the canyons of the pine forest region. Douglas fir reaches its best development between 2620–2775 m on the north slopes of Mt. Lemmon (Shreve, 1915). In upper Sabino Canyon above Summerhaven, one encounters impressive stands of Douglas fir

120–150 cm in diameter and towering 30–36 m in height (Table 3, column 4; Fig. 8–3). Large white firs (*Abies concolor*) 60–90 cm dbh make impressive associates. The relatively dense overstory greatly limits undergrowth, although *Jamesia americana,* the principal shrub may form thickets in the openings. Herb cover is sparse; *Bromus richardsonii* and *Pteridium aquilinum* are the principal species.

The forest floor is frequently littered with large windthrows typical of those conditions expected in a mature forest. From ring counts it would appear that some of the largest Douglas fir are over 400 years old. Lightning strikes and localized fires often result in small aspen groves where the trend is toward conifer forest development. The presence of large fire-scarred firs plus charcoal in the well-developed soil of the fir stands suggests that fire, along with windthrows, has been among the important factors favoring the perpetuation of this forest type. Preliminary tree ring data suggest that many stands are even-aged and thus are likely to be of post-fire origin. Currently, there appears to be adequate reproduction to maintain the present composition of this mature forest ecosystem.

White fir ravine forest

In the ravines where white fir reaches its best development, trees are up to a meter or more in diameter and 38 m in height (Table 3, column 5). The largest fir recorded was 142 cm in diameter. Associated with the white fir are large Douglas fir and smaller Mexican white pine. In these riparian sites several deciduous trees, *Acer grandidentatum, A. glabrum neomexicanum* and *Robinia neomexicana,* are also found along with *Quercus hypoleucoides* and *Q. gambelii.* The increased mesophytism also favors a rich assemblage of forest herbs, especially along the streams. In addition to those listed above, others frequently found include *Aquilegia chrysantha, Galium asperrimum, Geranium richardsonii, Glyceria elata, Oxalis metcalfei, Senecio wootonii, Thalictrum fendleri* and *Viola canadensis.* In most stands, fire scars are also evident indicating that fires have swept through these sites in the past. Some windthrows occur, but they are less frequent than on the slopes. Here fir reproduction increases markedly compared to the surrounding slopes and all trends suggest that white fir will maintain its dominance in these more favorable sites.

Corkbark fir forest

In the Catalinas a highly restricted corkbark fir (*Abies lasiocarpa arizonica*) community first recorded by Shreve (1915),

occurs on the north-facing slope just below the summit (Table 3, column 6; Fig. 8–1). Corkbark fir, the dominant tree, occurs in association with large Douglas firs, white firs and quaking aspen (*Populus tremuloides*). These firs are primarily 60–90 cm dbh and contribute 60–70% of the relatively continuous tree canopy. The undergrowth is typically sparse; *Jamesia americana* is the principal shrub and *Pyrola virens* and *P. secunda* the most common herbs.

There is considerable mortality in the smaller size classes of the corkbark fir; nearly 50% of the stems in the 2.5–15 cm size class and one third of those in the 18–30 cm size class were dead. In the 1950's and 1960's sapling reproduction was absent and seedlings rare, a condition continuing to the present. The presence of scattered mature Douglas fir in this stand suggests that this fir community may have become established following a fire in which the large Douglas fir survived. As usual, aspen may have been the initial invader since some of these larger trees in the 25–50 cm size class are 100–130 years of age. Probably a disturbance over a century ago favored the development of this community. Since corkbark fir must compete with two other aggressive conifers, periodic disturbance may be critical to the perpetuation of this subalpine species.

High elevation forests of the Pinaleno Mountains

Mixed conifer forest

In the Pinaleno Mountains from 2740–2920 m there occurs a mixed conifer forest on the lower and north-facing slopes (Table 3; Fig. 5). Here five conifers, corkbark fir, Douglas fir, white fir, Mexican white pine and Engelmann spruce (*Picea engelmannii*), may occur together. The trees range from 60–90 cm in diameter and usually form a relatively continuous canopy except for occasional openings created by windthrows. As in the Douglas fir, white fir forest of the Catalinas, the principal herbs are *Bromus richardsonii* and *Pteridium aquilinum.* On intermediate slopes, Douglas fir increases, while spruce and corkbark fir decrease in importance. As one moves on to more xeric southerly slopes, this mixed conifer forest is replaced by ponderosa and Mexican white pine with Douglas fir as an associate. *Robinia neomexicana* and *Salix scouleriana* are the principal deciduous trees. Diversity and coverage of herbaceous species also increase in these higher elevation ponderosa pine forests.

Spruce, fir forest

In the Pinalenos from 2920 m to the summit near 3260 m one

←
Fig. 8. Forest region of Santa Catalina and Pinaleno Mountains. 1. *Abies lasiocarpa* forest with snow cover, N slope 2713 m just below summit along Mt. Lemmon highway. Santa Catalina Mts. Jan. 1963. 2. *Picea engelmannii* forest, SSW slope 3094 m on High Peak Road. Stand 438. Open grown stand in former meadow, *Picea* saplings evident. Pinaleno Mts. 1962. 3. *Pseudotsuga menziesii* forest, NNW 20° slope 2591 m in upper Sabino Canyon. Left to right: *Abies concolor* with person standing 141 cm dbh, two darker *Pseudotsuga menziesii* 158 cm and 132 cm dbh; clump of *Acer glabrum neomexicanum* foreground. Douglas fir log on forest floor. Santa Catalina Mts. 1962. 4. *Quercus rugosa* fire scrub, SW 2576 m Carter Canyon. This 1948 manignited fire resulted in the loss of ponderosa pine forest (note charred snags) and favored oak shrub development. Santa Catalina Mts. 1964. 5. *Pinus ponderosa* forest, S slope 2440 m along Mt. Lemmon highway. Dense pine understory has developed with fire protection. Santa Catalina Mts. 1963. 6. *Pinus ponderosa* WNW slope 2424 m in Box Canyon. Fire of lightning origin occurred in 1946. Note savanna-like aspect created with *Muhlenbergia virescens* as principal grass. Larger trees are 75 cm dbh. Santa Catalina Mts. 1964. 7. *Pinus ponderosa, Quercus hypoleucoides* forest, SW slope 2164 m above Rose Canyon. Stand 375. Silverleaf oak dominant in understory. Santa Catalina Mts. 1962. 8. Mature *Pinus ponderosa* forest, NE slope 2225 m. Butterfly Burn Trail. Trees 45–90 cm dbh. Santa Catalina Mts. 1964. 9. *Pinus strobiformis, Pinus ponderosa* forest with snow cover, SE slope 2713 m near summit of Mt. Lemmon. A Mexican white pine evident on right with smooth bark. Santa Catalina Mts. Jan. 1963.

encounters a spruce, fir (*Picea engelmannii, Abies lasiocarpa arizonica*) forest at its southern distribution in the southwest (Dye & Moir, 1977). Pure or mixed stands of Engelmann spruce and/or corkbark fir occur, primarily on northerly exposures with the fir tending to dominate the more mesic and the spruce the more xeric sites. Trees reach 60–90 cm dbh and 24 m or more in height on the better sites (Table 3). Below 3050 m Douglas fir and Mexican white pine may be present. Locally where fires or blowdowns have occurred, aspen is the principal tree. Although the undergrowth is sparse, *Vaccinium oreophilum* is locally important. Herbs contribute less than 2% coverage; *Bromus richardsonii* and *Carex* spp. are most frequent. Windthrows litter the forest floor and past fire evidence is present in most stands. As in the Douglas fir forest of the Catalinas, wind and fire interact to favor the development of a mosaic of even- or uneven-aged, pure or mixed stands. Crown fires tend to result in even-aged stands, whereas periodic low intensity fires and wind-throws accompanied by gap phase reproduction give rise to uneven-aged stands. Some spruce stands appear to have become established in former open meadows. This too may represent post-fire development (Fig. 8–2). Of the subalpine forest elements represented here, only the corkbark fir is locally present in the Catalinas. Although some cutting has occurred in these high elevation forests, most stands were relatively undisturbed. These mixed conifer and spruce, fir forests in the Pinalenos represent a dynamic set of self-perpetuating communities in which species shifts may occur, but in which the overall pattern should persist for many decades, and, of course, indefinitely under the present climate and continued restriction of human influence.

In conclusion

Shreve's treatment of this mountain vegetation emphasized the role of climatic factors. To his principle features of altitudinal climatic change – shortening of the frost-free season, critical season of aridity and increasing precipitation – must be added such anthropogenic influences as overgrazing on the bajada, and fire protection on the mountain slopes. Although he recognized the deleterious role of grazing on saguaro reproduction on the bajada and rocky slopes at Tumamoc Hill where his desert laboratory was located (Shreve, 1910), there is no mention of its effects in his Catalinas work. One can only surmise that continued grazing over the decades thereafter resulted in the development of the disturbed desert scrub community which is now widespread on the bajada. Although he recognized prolonged freezing temperatures as a critical factor in establishing the upper limit of Sonoran desert species, including saguaro, the effects of low temperatures can also result in a dramatic change in stand structure, as we observed following the 1962 freeze on the lower slopes of the Catalinas. This

highlights the often overlooked role of catastrophies in the dynamics of plant communities (cf. Egler, 1977).

In the Encinal and forest regions fires of lightning origin have been modifying the vegetation pattern for centuries. Shreve observed 'clear park-like stretches of Pine Forest' which one can only relate to re-occurring natural fires. Following several decades of fire protection there are now two-layered pine stands in addition to the pine savanna observed by Shreve. The oak scrub on upper south-facing slopes observed by Shreve may well have resulted from the interaction of drought and severe burns which eliminated the pine. In the Encinal and desert grassland, fire, both natural and man-caused, interacts with drought in modifying vegetation structure and composition.

In the mature fir forest near the summit the authors and Shreve probably observed a very similar forest community where fires, both localized and more extensive, have played a major role over the centuries. Although this mountain vegetation, in the words of R. H. Whittaker, 'may be described in terms of community gradients or coenoclines along topographic moisture gradients', discrete communities can also be recognized that are highly dynamic yet relatively stable.

In this paper we have attempted to update Shreve's classic work on the vegetation of the Santa Catalina Mountains with an emphasis on community dynamics. This also represents an additional contribution in an impressive aggregate developed under the leadership and guidance of Robert H. Whittaker. He contributed significantly to our understanding of the floristic aspects, the vegetation gradients of both the north and south slopes, soil relationships, primary productivity and diversity patterns. Collectively his work in the Catalinas represents one of the most comprehensive sets of studies conducted in such a mountain range and should serve as an inspiration to future vegetation scientists.

References

Alcorn, S. M. & May, C., 1962. Attrition of a saguaro forest. Plant Disease Reporter. 46: 156–158.

Barbour, M. G. & Diaz, D. V., 1973. Larrea plant communities on bajada and moisture gradients in the United States and Argentina. Vegetatio 28: 335–352.

Barbour, M. G., Diaz, D. V. & Breidenbach, R. W., 1974. Contributions to the biology of Larrea species. Ecology 55: 1199–1215.

Barbour, M. G., MacMahon, J. A., Bamberg, S. A. & Ludwig, J. A., 1977. The structure and distribution of Larrea communities. In: Mabry, T. J., J. H. Hunziker & D. R. DiFeo, Jr. (eds.), Creosote Bush: biology and chemistry of Larrea in New World deserts, pp. 225–251. Dowden, Hutchinson & Rose, Inc., Stroudsburg, Pa.

Benson, L. & Darrow, R. A., 1944. A manual of southwestern desert trees and shrubs. Univ. of Ariz. Bull., Biol. Sci. Bull. 6, vol. 15, 2. 411 pp.

Biswell, H. H., Kallender, H. R., Komarek, R., Vogl, R. J. & Weaver, H., 1973. Ponderosa fire management: a task force evaluation of controlled burning in ponderosa pine forests of central Arizona. Tall Timbers Res. Sta. Misc. Pub. 2. Tallahassee, Fla. 49 pp.

Blumer, J. C., 1910. Comparison between two mountain sides. Plant World 13: 134–140.

Blydenstein, J., Hungerford, C. R., Day, G. I. & Humphrey, R. R., 1957. Effect of domestic livestock exclusion on vegetation in the Sonoran Desert. Ecology 38: 522–526.

Brown, D. E. & Lowe, C. H., 1974a. A digitized computer-compatible classification for natural and potential vegetation in the Southwest with particular reference to Arizona. J. Ariz. Acad. Sci. 9, Suppl. 2: 1–11.

Brown, D. E. & Lowe, C. H., 1974b. The Arizona system for natural and potential vegetation – illustrated summary through the fifth digit for the North American Southwest. J. Ariz. Acad. Sci. 9, Suppl. 3: 1–56.

Cooper, C. F., 1960. Changes in vegetation, structure and growth of southwestern pine forests since white settlement. Ecol. Monogr. 30: 129–164.

Dye, A. J. & Moir, W. H., 1977. Spruce-fir forest at its southern distribution in the Rocky Mountains, New Mexico. Am. Mid. Nat. 97: 133–146.

Egler, F. E., 1977. The nature of vegetation: its management and mismanagement – an introduction to vegetation science. Connecticut Conservation Assoc., Bridgewater, Ct.

Emory, W. H., 1858. Report on the United States and Mexican boundary survey. Vol. 2, 34th Congress, 1st Session. Ex doc. no. 135.

Gardner, J. L., 1950. The effects of thirty years of protection from grazing desert grassland. Ecology 31: 44–50.

Gehlbach, F. R., 1967. Vegetation of the Guadalupe Escarpment, New Mexico-Texas. Ecology 48: 404–419.

Gibble, W. P., 1950. Nineteen years of vegetational change in a desert habitat. Masters Thesis, Univ. of Arizona, Tucson.

Griffiths, D., 1910. A protected stock range in Arizona, U.S.D.A., Bureau Plant Industry. Bull. 177.

Haskell, H. S., 1945. Effects of conservative grazing on a desert grassland range as shown by vegetational analysis. Masters Thesis, Univ. of Arizona, Tucson.

Haskett, B., 1935. Early history of cattle industry in Arizona. Ariz. Hist. Rev. 6: 3–42.

Hastings, J. R. & Turner, R. M., 1965. The changing mile. Univ. of Ariz. Press, Tucson. 317 pp.

Humphrey, R. R., 1937. Ecology of the burroweed. Ecology 18: 1–9.

Johnson, R. R. & Carothers, S. W., 1982. Southwestern riparian habitats and recreation: interrelationships and impacts in the Rocky Mountain region. U.S.D.A. Forest Service. Rocky Mtn. For. Rge. Expt. Sta., Eisenhower Consortium Bull. 12 (in press).

Kearney, R. H. & Peebles, R. H., 1973. Arizona flora. 2nd. ed. Univ. of California Press, Berkeley. 1085 pp.

Lowe, C. H., 1961. Biotic communities in the Sub-Mogollon region of the inland southwest, J. Ariz. Acad. Sci. 2: 40–49.

Lowe, C. H., 1964. The vertebrates of Arizona: annotated check lists of the vertebrates of the State: the species and where they live. Univ. of Arizona Press, Tucson. 132 pp.

Lowe, C. H. & Brown, D. E., 1973. The natural vegetation of Arizona, Ariz. Res. Inf. System. Coop. Pub. 2. 53 pp.

McLaughlin, S. P. & Bowers, J. E., 1982. Effects of wildfire on Sonoran Desert plant community. Ecology 63: 246–248.

MacMahon, J. A., 1979. North American deserts: their floral and faunal components. In: Goodall, D. W. & R. A. Perry (eds.) Arid-land ecosystems: their structure, functioning and management, pp. 21–82. Cambridge Univ. Press, Cambridge.

Marshall, J. T., 1957. Birds of pine-oak woodland in southern Arizona and adjacent Mexico. Cooper Ornithological Soc., Berkeley. Pacific Coast Avifauna 32.

Martin, W. P. & Fletcher, J. E., 1943. Vertical zonation of great soil groups on Mt. Graham, Arizona as correlated with climate, vegetation and profile characteristics. Ariz. Ag. Expt. Sta. Tech. Bull. 99.

Mehrhoff, L. A., 1955. Vegetation changes on a southern Arizona grassland range – an analysis of courses. Masters Thesis, Univ. of Arizona, Tucson.

Moir, W. H., 1963. Vegetational analyses of three southern New Mexico mountain ranges. M. S. Thesis, New Mexico State Univ., Las Cruces.

Moir, W. H. & Ludwig, M. A., 1979. A classification of spruce-fir and mixed conifer habitat types of Arizona and New Mexico. Rocky Mt. For. Expt. Sta., U.S. Forest Service Res. Paper RM-207.

Niering, W. A. & Goodwin, R. H., 1974. Creation of relatively stable shrublands with herbicides: arresting 'succession' on rights-of-way and pastureland. Ecology 55: 784–795.

Niering, W. A. & Whittaker, R. H., 1963. Vegetation of the Santa Catalina Mountains. Progressive Agriculture in Arizona 15: 4–6.

Niering, W. A. & Whittaker, R. H., 1965. The saguaro problem and grazing in southwestern National Monuments. National Parks Mag. 39: 4–9.

Niering, W. A., Whittaker, R. H. & Lowe, C. H., 1963. The saguaro: a population in relation to environment. Science 142: 15–23.

Parker, A. J., 1980. Site preference and community characteristics of Cupressus arizonica Greene. (Cupressaceae) in southeastern Arizona. Southwest Nat. 25: 9–22.

Patton, J. L., Heed, W. B. & Lowe, C. H., 1966. Inversion frequency analysis of Drosophila pseudoobscura in the Santa Catalina Mountains, Arizona, J. Ariz. Acad. Sci. 4: 105–117.

Reeves, T., 1977. Vegetation and flora of Chiricahua National Monument, Cochise County, Arizona. M. S. Thesis, Arizona State Univ., Tempe.

Rich, J. L., 1911. Recent stream trenching in the semi arid region of southwestern, New Mexico, a result of removal of vegetative cover. Am. Jour. Sci. 32: 237–245.

Sampson, A. W. & Weyl, L. H., 1918. Range preservation and its relation to erosion control on western grazing lands. U.S.D.A. Bull. 675.

Sellers, W. D., 1960. Arizona climate. Univ. Arizona Press, Tucson.

Shreve, F., 1910. The rate of establishment of the giant cactus. Plant World 13: 235–240.

Shreve, F., 1915. Vegetation of a desert mountain range as conditioned by climatic factors. Carnegie Inst. Wash., Pub. 217.

Shreve, F., 1919. A comparison of the vegetational features of two desert mountain ranges. Plant World 22: 291–307.

Shreve, F., 1951. Vegetation of the Sonoran Desert. Carnegie Inst. Wash., Pub. 591: 1–129.

Shreve, F. & Mallery, T. D., 1933. The relation of caliche to desert plants. Soil Sci. 35: 99–112.

Smith, H. V., 1956. The climate of Arizona. Univ. of Ariz., Ag. Expt. Sta. Bull. 279.

Steenbergh, W. F. & Lowe, C. H., 1969. Critical factors during the first years of life of the saguaro (Cereus giganteus) at Saguaro National Monument, Arizona, Ecology 50: 825–834.

Steenbergh, W. F. & Lowe, C. H., 1976. Ecology of the saguaro. I. The role of freezing weather on a warm-desert plant population. In: Research in the Parks, National Park Service Symposium Ser. 1: 49–92.

Steenbergh, W. F. & Lowe, C. H., 1977. Ecology of the saguaro. II. Reproduction, germination, establishment, growth and survival of the young plant. Nat. Park Serv., Scientific Monogr. Ser. 8. 242 pp.

Steenbergh, W. F. & Lowe, C. H., 1982. Ecology of the saguaro. III. Growth and demography. U.S. National Park Service Sci. Monog. Ser. (in press).

Taylor, W. P., Vorhies, C. T. & Lister, P. B., 1935. The relation of jack rabbits to grazing in southern Arizona. J. Forestry 33: 490–498.

Thornber, J. J., 1910. The grazing ranges of Arizona. Univ. of Ariz., Ag. Expt. Sta., Bull. 65: 245–357.

Thornthwaite, C. W., Sharpe, C. F. S. & Dasch, E. F., 1942. Climate and accelerated erosion in the arid and semi-arid southwest, with special reference to the Polaca Wash drainage basin, Ariz. U.S. Dept. Ag. Tech. Bull. 808: 1–134.

Turner, R. M., Alcorn, S. M. & Olin, G., 1969. Mortality of transplanted saguaro seedlings. Ecology 50: 835–844.

Vasek, F. C., 1979/80. Early successional stages in Mojave Desert scrub vegetation. Israel J. Bot. 28: 33–148.

Vasek, F. C., 1980. Creosote bush: long-lived clones in the Mojave Desert. Am. J. Bot. 67: 246–255.

Wagoner, J. J., 1952. History of the cattle industry in southern Arizona, 1540–1940. Univ. of Ariz. Social Sci. Bull. 20.

Wallmo, O. C., 1955. Vegetation of the Huachuca Mts., Arizona. Am. Mid. Nat. 54: 466–480.

Wentworth, T. R., 1981. Vegetation on limestone and granite in the Mule Mountains, Arizona. Ecology 62: 469–482.

Whittaker, R. H., 1954. The ecology of serpentine soils. IV. The vegetational response to serpentine soils. Ecology 35: 275–288.

Whittaker, R. H., Buol, S. W., Niering, W. A. & Havens, Y. H., 1968. A soil and vegetation pattern in the Santa Catalina Mountains, Arizona. Soil Sci. 105: 440–450.

Whittaker, R. H. & Niering, W. A., 1964. Vegetation of the Santa Catalina Mountains, Arizona. I. Ecological classification and distribution of species. J. Ariz. Acad. Sci. 3: 9–34.

Whittaker, R. H. & Niering, W. A., 1965. Vegetation of the Santa Catalina Mountains, Arizona. II. A gradient analysis of the south slope. Ecology 46: 429–452.

Whittaker, R. H. & Niering, W. A., 1968a. Vegetation of the Santa Catalina Mountains, Arizona. III. Species distribution and floristic relations on the north slope. J. Ariz. Acad. Sci. 5: 3–21.

Whittaker, R. H. & Niering, W. A., 1968b. Vegetation of the Santa Catalina Mountains, Arizona. IV. Limestone and acid soils. J. Ecol. 56: 523–544.

Whittaker, R. H. & Niering, W. A., 1975. Vegetation of the Santa Catalina Mountains, Arizona. V. Biomass, production and diversity along the elevation gradient. Ecology 56: 771–790.

Wooton, E. O., 1915. Factors affecting range management in New Mexico, U.S.D.A. Bull. 211.

Yang, T. W. & Lowe, C. H., 1956. Correlation of major vegetation climaxes with soil characteristics in the Sonoran Desert. Science 123: 542.

Yang, T. W., 1957. Vegetational, edaphic and faunal correlations on the western slope of the Tucson Mountains and the adjoining Avra Valley. Ph. D. thesis Univ. of Ariz.

Youngs, F. O., 1931. Soil survey of the Tucson area. Univ. of Ariz. Ag. Expt. Sta. Pub. 19: 1–60.

Accepted 6.2.1984.

Patterns of tree replacement: canopy effects on understory pattern in hemlock – northern hardwood forests*

Kerry D. Woods**
Environmental Studies, University of California, Santa Barbara, CA 93196, U.S.A.

Keywords: *Acer saccharum,* Beech–maple forest region, *Betula lutea,* Canopy–understory interaction, Climax, Community stability, *Fagus grandifolia,* Gap phase, Hemlock–northern hardwood forest region, *Tilia americana,* Tree replacement pattern, *Tsuga canadensis*

Abstract

The effect of canopy trees on understory seedling and sapling distribution is examined in near-climax hemlock–northern hardwood forests in order to predict tree replacement patterns and assess compositional stability.

Canopy trees and saplings were mapped in 65 0.1-ha plots in 16 tracts of old-growth forests dominated by *Tsuga canadensis, Acer saccharum, Fagus grandifolia, Tilia americana,* and *Betula lutea* in the northeastern United States. Seedlings were tallied in sub-plots. Canopy influence on individual saplings and sub-plots was calculated, using several indices for canopy species individually and in total. For each species sapling and seedling distributions were compared to those distributions expected if saplings were located independently of canopy influence. Non-random distributions indicated that sapling and seedling establishment or mortality were related to the species of nearby canopy trees. Hemlock canopy trees discriminate against beech and maple saplings while sugar maple canopy favors beech saplings relative to other species. Basswood canopy discourages growth of saplings of other species, but produces basal sprouts. Yellow birch saplings were rarely seen beneath intact canopy.

Since trees in these forests are usually replaced by suppressed seedlings or saplings, canopy–understory interactions should influence replacement probabilities and, ultimately, stand composition. I suggest that hemlock and basswood tend to be self-replacing, maple and beech tend to replace each other, and birch survives as a fugitive by occupying occasional suitable gaps. This suggests that these species may co-exist within stands for long periods with little likelihood of successional elimination of any species. There is some suggestion of geographical variation in these patterns.

Introduction

Species co-existence and biological interactions

Studies of species co-existence may be approached on various spatio-temporal scales. Explanations for the maintenance of floristic richness over whole landscapes rely on analyses of species distribution on coarse-scale successional or environmental gradients. In contrast, understanding of the factors controlling species co-occurrence within stands may require consideration of fine-scale bio-

* Nomenclature follows Fernald (1950).
** Acknowledgements: There are many whose help has contributed to this paper but Robert Whittaker played the primary advisory role throughout. The research described sprang from his ideas and his help was crucial in the performance of the research. I am also grateful to those who helped in the field work and the agencies and individuals who allowed access to field sites. I thank Peter Marks and Margaret Davis for their assistance and suggestions and Robert Peet and Steward Pickett for thorough and helpful review of the manuscript. The research was financed – in part – through a National Science Fellowship and by a McIntyre-Stennis grant.

logical interactions. Although intra-stand variability in the physical environment, such as substrate and microtopography, certainly facilitates niche differentiation and species co-existence (Bratton, 1976a, 1976b; Hicks, 1981), at this scale it is likely that interactions between plants and populations will be relatively more important in controlling community dynamics and composition. It may also be hypothesized that control of species co-existence within stands will pass from external physical factors to internal biological interactions as disturbances shift from severe, where large numbers of dominant individuals are killed, to relatively mild, where death of individual dominants occurs largely independently of the death of others. Biological interactions should also be more strongly asserted as the interval between disturbances lengthens.

The distinction between external, abiotic controlling factors and biological interactions is certainly less than absolute. Broad patterns of climate, substrate, and intra-stand patterns of biological influences are parts of the same continuous spectrum of environmental influences to which all individuals are subjected and which determine vegetation patterns at all scales. The ultimate controlling factors are, in any case, the same; a plant can become established only in microsites which propagules reach and which have adequate light, water, and nutrients.

This paper presents results of a study aimed at discovering biological interactions which could cause co-existence of several species in stands subject, primarily, to small, infrequent disturbances. Observations from a number of stands are used to suggest a pattern of population interactions in a particular range of forest communities.

Canopy–understory interactions in forests

Jones (1945) proposed that forests dominated by shade-tolerant species are generally subject to only mild disturbance in the form of single or few tree gaps since dominance by shade tolerants is likely to be achieved only after long periods without massive disturbance. Jones further argued that these small gaps are usually filled by suppressed saplings of the dominant species. Such replacement may, in the absence of large disturbances over a period of two or more generations of canopy dominants, lead to increased stability in stand species composition.

Two further propositions, added to that of Jones, suggest a way in which co-existence may arise from interactions between individuals. The first proposition is that important niche differences between co-existing plants may be manifested only at certain stages of life-histories, particularly the early stages (Grubb, 1977). In the case of forest trees niche differences allowing co-existence may apply primarily to seedlings and saplings. Changes in niche relationships with age (or size) may involve ontogenetic changes or may simply be due to seedlings and saplings experiencing the environment at a finer scale than mature trees and so experiencing greater variability in microsite conditions. Several workers have discussed the significance of early parts of the life-cycle in understanding the distribution of plants in niche space and in real space and time. The 'microsite mosaic' of Whittaker (1975) and Whittaker & Levin (1977), the 'safe site' of Harper (1977), and the 'regeneration niche' constructed by Grubb (1977) are all similar in their emphasis on the importance of environmental conditions affecting germination, establishment, and early growth.

The second proposition invokes, as a source of variability among microsites, the influence of nearby canopy trees. Canopy trees of different species might differently affect subjacent microsites such as through differences in shade or litter quality, nutrient or water use, or root competition. In this manner canopy individuals could influence seedling and sapling success which, in turn, could lead to co-existence of species through self-replacement, cyclic succession on microsite, or a combination of these patterns.

Effects of canopy trees on understory environment have frequently been implicated as factors controlling successional processes and even the dynamics of relatively stable communities. Whittaker (1975) proposed a microsite mosaic model wherein community stability could be maintained by modification of microsites by canopy trees. Various mechanisms by which such modification can be effected have also been proposed. Connell (1970) and Janzen (1970) have suggested that species-specific herbivores on canopy trees might inhibit growth of conspecific seedlings on subjacent microsites. Alexandre (1977) has shown evidence of such a pattern involving a nematode parasitic on a rain forest tree, but such mechanisms have not yet been documented in temperate forests. Several attempts have been made to deduce the existence of canopy-understory interactions in temperate forests in the northeastern United States through analysis of distributions of canopy trees, saplings, and seedlings (Forcier, 1975; Fox, 1977; Horn, 1971, 1975). Others (Runkle, 1981; Barden, 1979; Brewer & Merritt, 1978) have attempted to predict replacement patterns by associating sapling species in gaps with tree species which died to form the gaps.

Hemlock–northern hardwood forests

Prior to European settlement mesic uplands in a broad belt from the western Great Lakes region to New England were occupied by forests dominated by shade-tolerant, long-lived trees. The primary dominants in undisturbed stands of these forests are the shade-tolerant sugar maple (*Acer saccharum*), American beech (*Fagus grandifolia*), and eastern hemlock (*Tsuga canadensis*), plus the less tolerant yellow birch (*Betula lutea*) and basswood (*Tilia americana*). The range of forests dominated primarily by these species corresponds roughly to the combined area of the maple–basswood, beech–maple, and hemlock–white pine–northern hardwoods types of Braun (1950) shown in Figure 1. Composition varies over this range and hemlock–northern hardwood forests grade continuously into others. Beech is most important in the east and south, its range barely reaching the Upper Great Lakes. Hemlock reaches neither the southern nor western limits of the geographical area defined. Yellow birch is most important in the northern and basswood in the western extent of the region. Only sugar maple is an important dominant throughout.

The predominance of shade-tolerant species suggests that these forests may fit the model of a community in which biological interactions are important in controlling community dynamics. Large-scale disturbances in mesic uplands appear to be quite infrequent. None of the five species discussed show resistance or reproductive adaptations to fire. Their total dominance in a stand is evidence that no significant fire has occurred in several centuries. Extensive wind-throw may be the most frequent major disturbance in mesic upland stands of the region, particularly in coastal areas subject to hurricanes. Inland forests are also apparently subject to destruction by strong down-burst winds associated with thunderstorms and by torna-

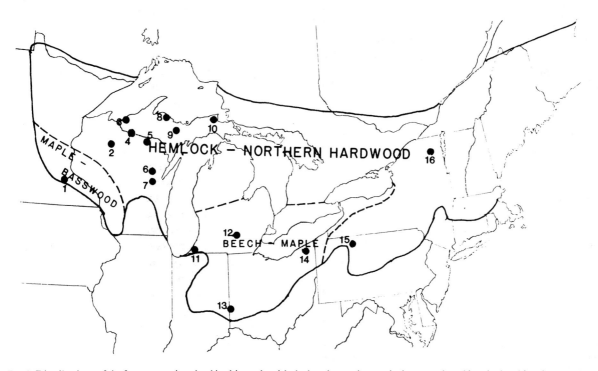

Fig. 1. Distributions of the forest types involved in this study with the beech–maple, maple–basswood, and hemlock–white pine–northern hardwood regions of Braun (1950) delineated. Numbered dots correspond to stands listed in Table 1.

does. However, Canham & Loucks (1983) have determined that, in northern Wisconsin, such disturbances have a return time in excess of 1 000 years – three to five times the period generally spent in the canopy by individuals of the species under study. Small canopy gaps associated with deaths of individual trees are apparently the chief mode of canopy disruption for relatively long periods. Runkle (1981), Barden (1979), Fox (1977), and Brewer & Merritt (1978) have shown that these gaps are usually filled by individuals of the dominant species as predicted by Jones. Forest on less mesic sites or in large areas without natural firebreaks such as the western Great Lakes region are subject to more frequent fire and their composition reflects this.

Methods

Field methods

Stands to be sampled were required to be 1) strongly dominated by some combination of the five major species listed, 2) free of evidence of disturbance other than gaps caused by natural death of one or a few trees, and 3) fairly homogeneous in substrate and topography. These criteria should best permit isolation of effects of canopy trees on the distribution of seedlings and saplings. No contention is made that such stands were generally abundant in primeval forests, but these effects should be present in all forests, though they may be strongly confounded in more disturbed stands. Nine hemlock–hardwood stands were sampled in northern Michigan and Wisconsin, and two in Pennsylvania and New York. Four beech–maple stands were sampled in southern Michigan and Ohio, and one maple–basswood forest in southern Minnesota (Fig. 1, Table 1). Sites are concentrated in northern Wisconsin and Michigan due to difficulties in satisfying the second and third criteria for stand selection in areas of greater topographic diversity and more pronounced anthropogenic disturbance.

Two to eight 0.1-ha (20 × 50 m) quadrats were placed in each stand. This usually involved pacing a predetermined distance along a transect or trail, then moving a predetermined distance in a predetermined direction from the point reached. In each quadrat saplings (individuals >2 m tall and <15 cm

Table 1. Names and locations of sample sites, with sample numbers.

Site number	Name	Location	Sample numbers
1	Nerstrand Woods State Park	Rice Co., Minn. T110N, R19W	52–54
2	Flambeau 'Big Block' Scientific Area	Flambeau St. For. Sawyer Co., Wisc. T37N, R3W	12–16
3	Porcupine Mts. State Park	Ontonagon Co., Mich. T50N, R44W	7–11
4	Sylvania Tract	Ottawa Nat. For. Gogebic Co., Mich. T44N, R40W	1–6 42–43
5	Bose Lake Scientific Area	Nicolet Nat. For. Forest Co., Wisc. T40N, R12E	17–20, 44
6	Jung Beech–Hemlock Scientific Area	Shawano Co., Wisc. T27N, R14E	37–41
7	Tellock's Hill Scientific Area	Waupaca Co., Wisc. T24N, R13E	35–36
8	Huron Mountain Club	Marquette Co., Mich. T52N, R28W	45–49
9	Dukes Experimental Forest	Hiawatha Nat. For. Marquette Co., Mich. T46N, R23W	26–29
10	Tahquamenon Falls State Park	Luce Co., Mich. T49N, R8W	21–25
11	Warren Dunes State Park	Berrien Co., Mich. T7S, R20W	55–60
12	Irwin's Woods	Jackson Co., Mich. T2S, R2E	63
13	Hueston Woods State Park	Preble Co., Ohio	61–62
14	Hiram College Biological Station	Portage Co., Ohio	64–65
15	Tionesta Tract	Allegheny Nat. For. McKean Co., Pa.	30–34
16	Saranac Lake	Adirondack State Park, N.Y.	50–51

diameter breast height) and canopy trees (>15 cm dbh) were recorded by species and diameter and mapped (canopy trees outside quadrats were included in the sample if they were within 8 m of saplings within the quadrat). Sapling and canopy stems were mapped using tapes and an optical range-finder to within 1 m of their true location (accuracy was generally greater). In each 0.1-ha quadrat 25 or 50 1-square meter sub-plots were placed along the long axis (in later quadrats only every other sub-plot was sampled). In each of these sub-plots, seedlings of all woody species were counted and divided into 3 height classes: less than 20 cm, between 20 cm and 1 m, and between 1 m and 2 m.

Analytic methods

Previous attempts to describe the effects of canopy trees on understory patterns have used more or less subjective, qualitative evaluations of canopy influence. Most have involved identification of the single canopy species most dominant at a point on the forest floor. In practice this usually has meant the species of the canopy individual nearest a seedling or sapling or the species of the tree which died to form a gap. An objective in this study was to quantify canopy influence by incorporating the size and distance from an understory point of all canopy trees within a set distance. Canopy influence, whether through shading, allelopathy, competition in the rhizosphere, or any other mechanism, should be in some way additive and dependent on tree size and spacing.

Attempts have been made to quantify competitive interactions among trees. Previously developed indices (e.g. Bella, 1971; Alemdag, 1978; Lorimer, 1981) were not used here because they either do not quantify influence in the way specified above, are not appropriate for dealing with individual microsites, deal with species not involved in this study, or are based on models of early successional communities. Three indices were devised for this study each based on a different model of the decline in influence of a particular canopy tree with increasing distance (Fig. 2). I_L assumes influence to decrease linearly with distance, I_N is represented by a bell-shaped curve and assumes a plateau of high influence extending some distance from the tree, and I_R assumes influence to vary with the reciprocal of distance (the 'tails' on I_R and I_N which are lacking on I_L proved to affect results little). These curves were subjectively drawn to imitate a range of biological models. Influence of individual trees as expressed by these indices may be calculated and added for any number of trees.

Values were calculated by each of the above indices for influence of individual canopy species and for total canopy influence on each sapling within the 0.1-ha quadrats and on each square-meter seedling quadrat. Only trees within 8 m of a sapling or sub-plot were used in calculating indices; this distance nearly always allowed inclusion of any tree whose canopy extended above the sapling and usually included at least three or four canopy trees. Similar values were obtained using a more conservative, non-distance-weighted index which was simply the sum of basal areas of canopy trees within the 8 m limit.

An expectation for these indices under the null hypothesis that seedlings or saplings are distributed independently of canopy tree influence may be calculated if density and diameter distributions are known for each canopy species. The formula for this expectation as derived in Woods & Whittaker (1981) is

$$E(I) = \sum_{i=1}^{n} p_i \, 2\pi \int_o^x Rf(R) dR$$

where the summation is over size classes i to n, p_i is density in size class i, x is the maximum distance for inclusion of trees (here 8 m), and $f(R)$ is the index of canopy influence as a function of distance (with diameter held constant).

Patterns of relationships between canopy and sapling distributions were then examined by comparing observed values of indices to corresponding calculated expectations (Woods & Whittaker, 1981). Comparisons were made of influence of individual canopy species and of total canopy influence.

In another analysis (see Woods, 1979) pairwise comparisons of sapling species responses to indices were formed by taking the difference between indices of the influence of two canopy species.

$$I_n = I_A - I_B$$

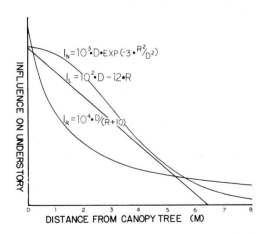

Fig. 2. Graphical representation with formulae for models of variation of canopy influence with distance from canopy tree. All curves are calculated for a canopy tree of 40 cm diameter. In formulae D is diameter of canopy tree, and R is distance from tree.

190

I_A and I_B are influence of species A and species B. I_n, or net influence will be zero if influence of the two species is equal, positive if species A is more important ($I_A > I_B$), and negative if species B is more important ($I_B > I_A$). Pairs of sapling species were compared in their response to net canopy influence of pairs of canopy species by ranking all saplings of both species by values of a particular I_n and comparing rankings of the two sapling species with a Wilcoxon rank-sum test. Significant results allow rejection of the null hypothesis that the two sapling species are distributed similarly with respect to the joint influence (I_n) of the two canopy species involved. Rejection means that one or both sapling species is relatively more abundant near trees of one

of the two canopy species. Similar tests for all pairs of sapling species for all possible I_n's permitted comparison of the relative responses of particular sapling species to particular canopy species.

Indices of canopy influence on seedling sub-plots were tested for correlation with number of seedlings in each species and height class.

For all of these treatments the 0.1-ha quadrats were grouped into twelve composite samples. For results to be. meaningful, the analyses described need to be performed only on groups of samples of similar composition. Nine such composite samples were formed by pooling samples closely associated in a detrended correspondence analysis (DECO-RANA; Hill & Gauch, 1980) ordination (Fig. 3).

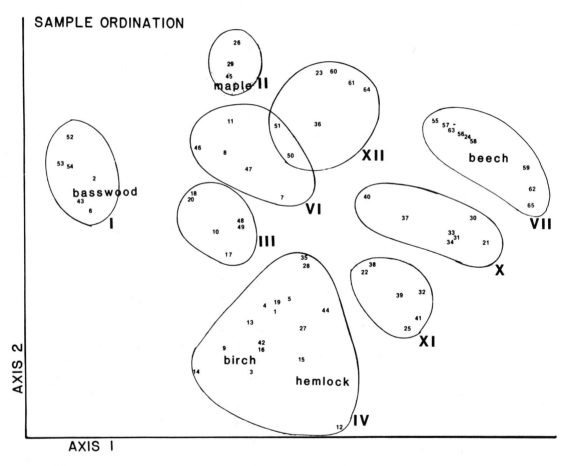

Fig. 3. Distribution of all samples from stands shown in Figure 1 on the first two axes of DECORANA ordination. Groups are composite samples. Areas of maximum abundance are indicated for each species (beech at the right of the ordination, basswood at the left, and hemlock and yellow birch at lower center). Maple is significant throughout the ordination but shows maximum abundance at upper center. Composites 5, 8, and 9 are not illustrated, but are composed of quadrats from the Huron Mountains (45–49), Jung (37–41), and Tionesta (30–34) stands respectively.

These groups were formed by pooling nearest quadrats or groups with nearness determined by distance between the nearest pair of quadrats in the two groups (single linkage classification, sensu Pielou, 1977). Composites 6 and 12 overlap because quadrat 50 is equidistant from 36 and 47. Three other composite samples (5, 8, and 9) were made by pooling quadrats from relatively uniform stands. It was hoped that these would reveal any geographical changes in patterns of interaction that would be obscured by forming composites from quadrats from different stands. Composite samples will hereafter be referred to as samples. The samples are not all independent as some quadrats are placed in more than one sample (Fig. 3). Other groupings into composite samples were tried and yielded similar results (results in Woods & Whittaker, 1981, are from a different grouping).

Results of these tests were compared with those of a simpler set of tests. For each sapling species tests for correlation of sapling abundance with basal area of the various canopy species were performed using Spearman's rank-correlation coefficient. Coefficients were calculated using whole 0.1-ha quadrats as samples and on sample sets obtained by dividing large quadrats into two 20 × 25 m plots and into four 10 × 25 m plots. Each correlation of sapling abundance and canopy basal area was calculated using only samples from stands within the geographic range of both species involved in the test.

All tests of association between canopy trees and sapling distributions were done with and without inclusion of beech root sprouts.

Results

Basal areas for the composite samples are listed by species in Table 2. Total basal area ranged from 35 to 56 m²/ha, densities of canopy trees from 200–500 stems per ha and sapling densities from 100–4 000 stems per ha.

Saplings and seedlings of the dominant species showed extreme variation in abundance. At least 500 saplings of each of the shade-tolerant species were involved in the study. Maple saplings were three times as abundant as beech and hemlock saplings were somewhat rarer (beech root suckers are not included in this total, but represented only

Table 2. Distribution of basal area by species in composite samples.

Sample number	Species basal area (m²/ha)						
	Acer	*Betula*	*Fagus*	*Tilia*	*Tsuga*	Other[a]	Sum
1	14.4	2.4	0.0	20.7	0.5	6.9	44.9
2	36.3	3.0	0.0	3.1	3.8	0.2	46.4
3	18.8	7.9	0.0	8.3	18.2	0.4	53.6
4	7.2	10.9	0.5	2.3	30.3	5.1	56.3
5	25.0	3.2	0.0	6.0	12.0	0.4	46.6
6	29.8	3.6	0.3	2.9	11.9	0.4	48.9
7	8.2	0.0	25.7	0.0	0.0	3.0	36.9
8	8.8	1.4	16.6	0.0	25.8	0.0	52.7
9	2.2	1.1	19.2	0.0	12.0	0.9	35.6
10	5.6	1.0	18.9	0.0	10.8	0.7	37.0
11	3.7	2.5	13.1	0.0	24.5	1.8	45.5
12	24.2	0.5	7.1	0.2	4.4	1.3	37.7

[a] Other species include *Abies balsamea, Acer rubrum, Acer pensylvanicum, Betula lenta, Carpinus caroliniana, Carya cordiformis, Fraxinus americana, Fraxinus nigra, Juglans cinerea, Liriodendron tulipifera, Ostrya virginiana, Picea rubens, Pinus strobus, Prunus serotina, Quercus borealis, Thuja occidentalis, Ulmus americana,* and *Ulmus rubra.*

about 10% of beech saplings). Only 40 yellow birch saplings were found, mostly in small canopy gaps. Basswood basal sprouts were abundant but only 16 independent saplings were found in the samples. Seedling abundances showed similar variation among species.

Comparison of observed and expected canopy influence

Comparisons of observed and expected canopy influence gave similar results for each of the indices including the non-distance-weighted basal area measure. There were slight differences due to peculiarities of the models (Fig. 2) but such differences did not significantly affect the results. The results displayed below were obtained using I_R with exclusion of beech root suckers. Differences observed when suckers were included were small and are discussed separately.

Comparisons of observed and expected hemlock canopy influence for all sapling species are displayed in Figure 4. This figure and Figures 5 and 6 are similar to those presented in Woods & Whittaker (1981) but illustrate results obtained from an improved grouping into composite samples. In this and similar figures each row of symbols represents a

% EXPECTED HEMLOCK CANOPY
INFLUENCE ON SAPLINGS

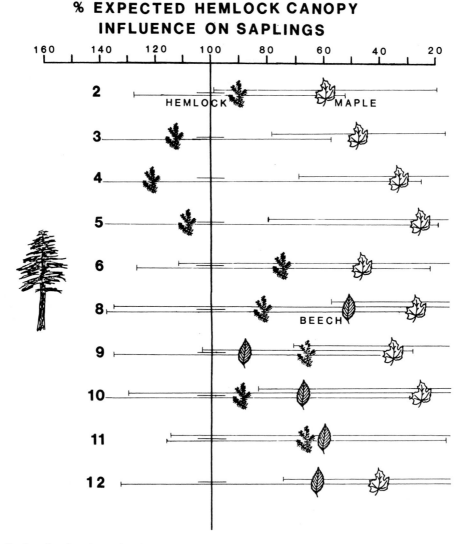

Fig. 4. Distribution of saplings by species with respect to hemlock canopy influence. Leaf symbols represent sapling species, rows single composite samples (sample numbers given at left). The horizontal axis (top) is hemlock canopy influence as percent of its calculated expectation and is thus independently scaled for each sample. Leaf symbols are located at mean values of relative hemlock canopy influence observed for all saplings of that species in that sample. Sapling species for which symbols are located further to the right in the diagram are found, on the average, at lower levels of hemlock canopy influence. Error bars associated with each sapling symbol indicate one standard deviation (error bars are not shown in succeeding figures, but those shown here are representative). For each sample, the bottom error bar corresponds to the sapling species at highest relative canopy influence. Only samples with significant representation of hemlock in the canopy are shown.

single composite sample. Only samples in which the canopy dominant (hemlock in this case) constitutes >10% total basal area are shown.

The null hypothesis level of hemlock canopy influence was exceeded on the average only by hemlock saplings, and hemlock saplings were, in eight of nine instances, found at higher hemlock canopy influence than any other species (in the exception, Tionesta, beech was at higher hemlock influence). This consistency of ranking permits rejection at $p < 0.05$ of the null hypothesis of similar sapling response, even if 'overlapping' samples 5, 8, and 9 are not considered. In nine of nine samples maple saplings were found at the lowest average hemlock

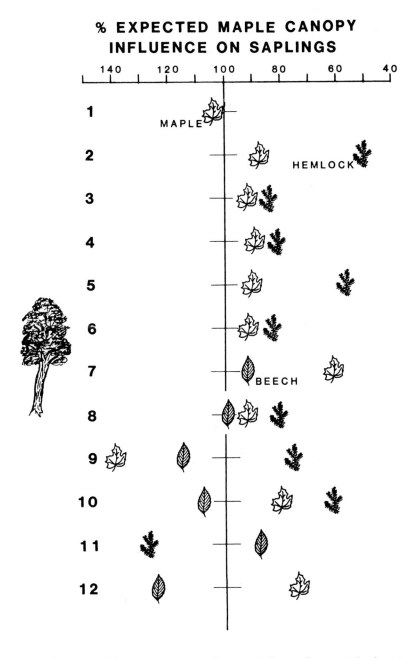

% EXPECTED MAPLE CANOPY INFLUENCE ON SAPLINGS

Fig. 5. Distribution of saplings by species with respect to sugar maple canopy influence. Representation is as described for Figure 4.

canopy influence of all species (this consistency allows rejection of the same null hypothesis, with the same considerations, at $p < 0.05$). In only one case did the mean for maple saplings fall within one standard deviation of the expected value. Finally, beech saplings were found, on average, at higher hemlock influence than were maple saplings, though

not as high as would be expected by chance. Avoidance of hemlock canopy was shown by maple saplings and, to a lesser degree, beech saplings, while hemlock saplings showed no clear trend.

The organizing effect of maple canopy influence (Fig. 5) on the understory was apparently not as great as that of hemlock. Maple saplings tended to

194

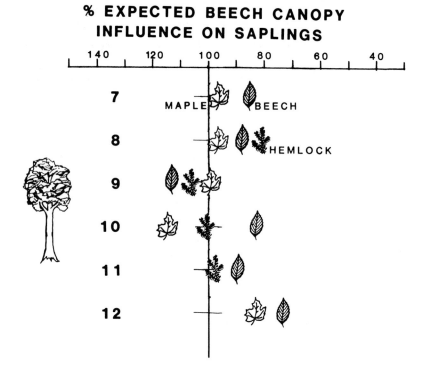

Fig. 6. Distribution of saplings by species with respect to beech canopy influence. Representation as in previous figures.

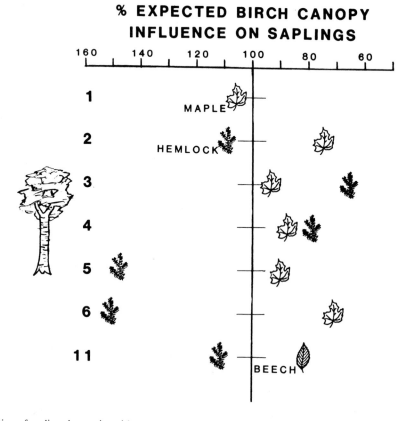

Fig. 7. Distribution of saplings by species with respect to yellow birch canopy influence. Representation as in previous figures.

be found near but slightly below expected levels of maple influence. In four of five cases, beech saplings were at higher maple influence than were maple saplings (consistency of ranking significant at $p = 0.05$). The most notable response was by hemlock saplings where in eight of nine cases they were found below expectation and at lower maple influence than any other sapling species (the exception is again Tionesta).

Beech canopy (Fig. 6) and yellow birch canopy (Fig. 7) elicited little clear response in sapling distribution, mean influence values for all sapling species being scattered about the expected values. Possible tendencies for beech saplings to be found at lower beech canopy influence than maple saplings (Tionesta again an exception) and hemlock saplings to be found at high birch canopy influence are consistent with patterns discussed below.

Saplings of maple and hemlock were apparently inhibited by basswood canopy influence (Fig. 8). However, the basswood saplings were almost exclusively basal sprouts of canopy trees and so were found at extremely high levels of basswood canopy influence.

The plot of sapling distributions with respect to total canopy influence (Fig. 9) suggests that maple saplings are found at lower canopy influence than saplings of the other shade-tolerant dominants. Beech and hemlock saplings also generally occur at lower than expected values of canopy influence, though average values for beech saplings tend to be higher than those for hemlock. All sapling species were found at relatively low canopy influence, a not surprising confirmation that saplings are most abundant in gaps or beneath light canopy.

Figure 10 shows average canopy influence by various species on beech saplings. The ranking of shade-tolerant canopy trees by their mean influence on beech saplings is highly consistent; the ratio of observed to expected influence of maple canopy on beech saplings was higher than that of beech canopy in all but one sample (sample 9). In short, beech saplings show less avoidance of maple canopy than beech canopy. Similar analyses were conducted for maple and hemlock saplings with results consistent with those shown in Figures 4–9; hemlock saplings were found at high relative values for conspecific influence and low values of maple influence while

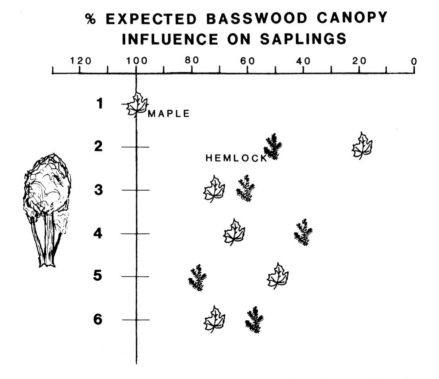

% EXPECTED BASSWOOD CANOPY INFLUENCE ON SAPLINGS

Fig. 8. Distribution of saplings by species with respect to basswood canopy influence. Representation as in previous figures.

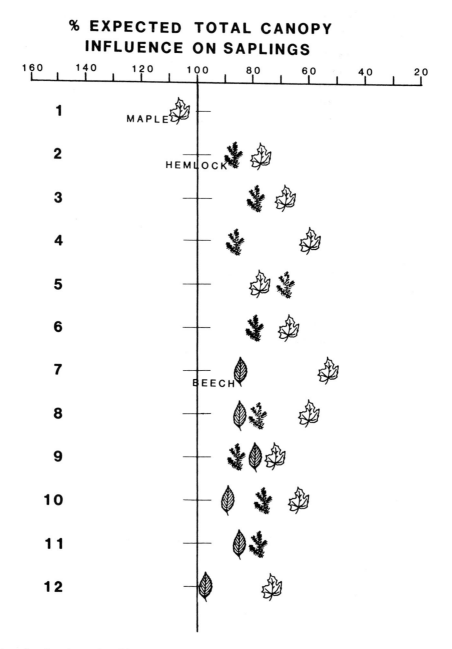

Fig. 9. Distribution of saplings by species with respect to total canopy influence of all species. Representation as in previous figures.

maple saplings were always at much lower values of hemlock canopy influence than would be expected by chance.

Similar analyses of dead sapling distributions were compared with these for live saplings to identify differential mortality caused by differences in canopy influence. Dead maple saplings showed distributions similar to those of live saplings. Both beech and hemlock saplings showed increased relative abundance of living saplings at high maple canopy influence. Living yellow birch saplings tended to be found at low canopy influence of all species, while higher canopy influence was associated with greater relative abundance of dead yellow birch saplings.

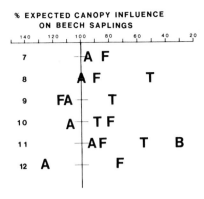

% EXPECTED CANOPY INFLUENCE ON BEECH SAPLINGS

Fig. 10. Distribution of canopy species with respect to their average influence on beech saplings. Representation is similar to that in previous figures with horizontal lines representing canopy influence as percent of expectation, vertical line representing expectation, and rows of symbols corresponding to samples. In each sample symbols representing canopy species are located according to average influence, relative to expectation, of that canopy species on beech saplings. Symbols are: A = sugar maple, F = beech, T = hemlock, and B = yellow birch. Thus, canopy symbols at right of diagram indicate low relative influence on beech saplings. Only samples with more than 10 beech saplings are shown.

Rank-sum comparisons of net influence

Results of the rank-sum tests (Fig. 11) were consistent with the canopy influence comparisons discussed above. In all tests involving hemlock canopy influence (5, 6, 8, and 9 in Fig. 11), hemlock saplings occurred at greater hemlock influence than saplings of the other species. This strengthens the suggestion of relative avoidance of hemlock canopy by deciduous species. In the absence of hemlock, beech and maple saplings were found at relatively lower influence of conspecific canopy trees than of the other canopy species in the comparison (significant in all but one case). This is most apparent in the maple–beech comparison (1) where this mutual self-avoidance was shown to depart with high significance from random expectation.

Correlations of sapling abundance with canopy basal area

Correlations of sapling abundance with canopy basal area (Fig. 12) further establish hemlock's suppression of saplings and the reciprocal relationship between beech and maple. The null hypothesis

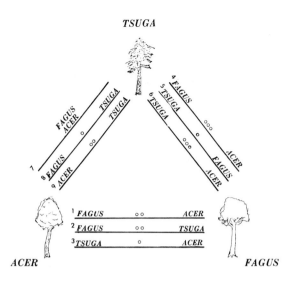

Fig. 11. Pairwise comparisons of response of sapling species to composite indices of canopy influence. Tree symbols at corners of diagram represent canopy influence of indicated species. Each line between tree species represents tests done on one of the possible pairwise comparisons of sapling species response to net influence of the two canopy species connected by the line. All lines represent several comparisons in various samples. Where rank-sum tests indicated significant differences in response of saplings species, saplings species names are shown nearest the canopy species which they, relatively, favored. Levels of significance at which the null hypothesis of similar distribution of sapling species were rejected are indicated by the number of dots above each line: one dot indicated rejection in at least 3 of 4 tests at alpha = 0.05, two dots rejection in all tests at alpha between 0.005 and 0.05, three dots rejection in all tests with alpha less than 0.005.

is that canopy trees do not effect distribution of understory individuals. Expected correlations between sapling abundance and conspecific canopy basal area are positive for species with limited dispersal and patchy seed sources, and zero for species with uniformly dispersed seeds.

For all plot sizes hemlock saplings showed positive correlations with conspecific basal area and negative correlations with basal area of both maple and basswood. These become significant at small plot size. Maple saplings were positively correlated with beech canopy and negatively associated with total basal area at all plot sizes. Negative correlations of maple sapling abundance with hemlock basal area were consistent and approached significance at smaller plot sizes. Beech saplings had no significant correlations though they did show positive association with maple basal area and negative

198

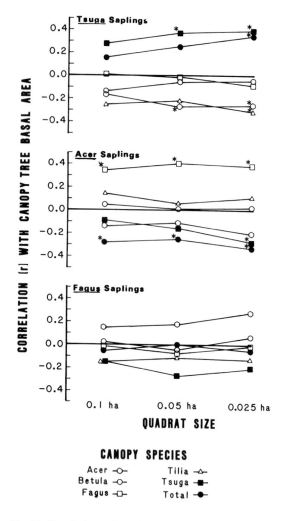

CANOPY SPECIES

Acer –○–   Tilia –△–
Betula –○–   Tsuga –■–
Fagus –□–   Total –●–

Fig. 12. Correlations of sapling abundance with canopy tree basal area by species of canopy tree and sapling. Each graph shows results involving one of the shade-tolerant sapling species for three plot sizes. Lines connect graphed coefficients for same canopy species. Stars indicate significance of correlation coefficient at alpha = 0.05.

association with hemlock and basswood (the last based on a very small number of co-occurrences).

In nearly all cases particular correlations became stronger with smaller plot size. This was especially notable for hemlock basal area.

Correlations of seedling abundance with canopy influence

Maple seedlings showed strong positive correlation with conspecific influence in the smallest size

classes, but this was lost in the larger classes (Table 3). Correlations with total canopy influence were initially positive but became strongly negative. Maple seedlings were, throughout, negatively associated with hemlock canopy. Similar trends were seen for beech seedlings in correlations with conspecific and total canopy influence, and beech seedlings also showed increasingly negative correlation with hemlock canopy influence in larger seedling size classes. Hemlock seedlings were comparatively rare in the study but may have been positively correlated with conspecific canopy.

Discussion

Canopy influence on understory pattern and tree replacement

The distribution of seedlings and saplings in forest understories may be controlled by fine-scale physical variation, by seed dispersal, or by other canopy tree influences. Samples in this study were chosen to minimize microsite variation. The effect of seed dispersal alone should range from distinct 'seed shadows' for poorly dispersed species to little at all in species with well-dispersed seeds. Consequently, negative association of seedlings and saplings with conspecific canopy trees can be interpreted as a result of canopy influence on reproductive success.

Results show that saplings and seedlings are not distributed through the stand independently of the influence of canopy trees. In some cases association with conspecific canopy is shown, as would be expected if distributions resulted from seed shadows. In other cases saplings were relatively under-represented near conspecific canopy trees. These distributions of seedlings and saplings, with the assumption that saplings are likely successors of current canopy trees, suggest that the probabilities of various species replacing a canopy tree are affected by the species of that tree. Observed distributions may be used to suggest a complex of likely canopy transitions with varying probabilities of occurrence.

Other workers have suggested that canopy influence plays an important role in determining patterns in beech–maple forests. Horn (1975), Whittaker & Levin (1977), and Fox (1977) have described beech–maple forests in which saplings have greater

Table 3. Correlations of seedling abundance with canopy influence. Number of tests is number of samples with more than 25 seedlings of appropriate species and size. Positive correlations are indicated by * ($p = 0.01$) or + ($p = 0.05$), negative correlations by = ($p = 0.01$) or – ($p = 0.05$). Only significant correlations are indicated. Size classes are: 1, first year seedlings with cotyledons; 2, seedlings less than 20 cm tall; 3, seedlings between 20 cm and 1 m tall; and 4, seedlings between 1 m and 2 m tall.

No. of tests	Seedling species	Size class	Canopy influence					
			Acer	Betula	Fagus	Tilia	Tsuga	Total
9	Acer	1	****	*		*++		*
			+++			=	===	++
9		2	****	**		***	+ –	*
			==	=		=	====--	==
7		3	**	*		**		+
			+	=		=	==	===-
3		4		+		*		
							=-	=-
3	Betula	1						
			=					=
3	Fagus	1	*	*	*+		*	*
			–					–
2		2		*+			**+	*
			=					–
4		3			*			
			–				–	=-
2		4					=	=
1	Tsuga	2					*	
			–					

relative abundance beneath canopy of the other species. In a previous study I have shown (Woods, 1979) sapling frequency distributions of beech and maple to be relatively skewed toward more frequent occurrence beneath the canopy of the other species. Figures 5, 6, 10, and 11 suggest that the interactions between beech and maple described for the special case of strong two-species dominance have some generality in more complex stands. Dead beech saplings showed particular association with high beech canopy influence. In 75% of the samples dead beech saplings were found at higher average beech canopy influence than were living beech saplings. This could be due to greater frequency of establishment beneath parent trees and thus more saplings available to die, but it is also consistent with the hypothesis of high differential mortality of beech beneath conspecific canopy. The relative abundance of small dead saplings beneath conspecific canopy trees suggests that differential mortality may occur just before or after entry into the sapling size class. Gradual attrition of maple seedlings at high maple canopy influence may be indicated, especially for larger seedling size classes, by data in Table 3.

The causes of the differential success of saplings and seedlings are unknown. Root competition might be especially significant for a seedling or sapling beneath a conspecific canopy tree. The 'natural enemies' hypothesis of Janzen (1970) and Connell (1970) could cause similar patterns. Horn (1971) proposed that differences in shade tolerance and canopy density could permit co-existence. If maple is more shade-tolerant in the understory and casts relatively light shade in the canopy, while beech canopy casts denser shade and grows more rapidly as saplings in higher light, each may prosper beneath the canopy of the other. Poulson (personal communication) has used detailed work on extension growth of seedlings and saplings in a Michigan beech–maple forest to suggest a mechanism essentially the reverse of Horn's; beech is more shade-tolerant (grows more rapidly beneath intact canopy) while maple, although barely able to survive as suppressed seedlings under full canopy, can grow much more rapidly than beech at high light levels. Poulson claims that maple canopy inhibits both species but beech may survive as saplings while the less tolerant maple does not survive beyond seedling size, with the result that maple trees may be more frequently replaced by beech. Beech canopy, however, inhibits growth of both species in the understory thus confining most regeneration to canopy gaps where more rapidly growing maple is usually more successful. Thus distributions of beech and maple saplings suggest replacement of each species in the canopy by the other, a phenomenon termed reciprocal replacement by Woods (1979). This pattern is supported by studies of forest gaps by Runkle (1981). In order to calculate transition probabilities for species pairs Runkle recorded dominant saplings in small gaps in old-growth forests, assuming these would be most likely to fill the gaps. Transition probabilities calculated by Runkle for Hueston Woods, a beech–maple stand also sampled in this study, indicate a strong tendency toward reciprocal replacement.

Conclusions regarding beech replacement rest, in part, on the exclusion of beech root sprouts from the analysis. When sprouts were included in the rank-sum tests the null hypothesis that maple and beech saplings were distributed similarly with respect to the index of joint canopy influence was still rejected, but at $p = 0.05$ as compared to $p = 0.001$ when sprouts were excluded. Other test results were unchanged. Root sprouts are generally very near the parent tree and so strongly influence the distance-weighted indices. The contribution of root sprouts to the establishment of new canopy beech trees remains unclear and probably varies geographically (Ward, 1961). There is a long-standing debate concerning the ability of root sprouts to develop

into canopy trees (Fowells, 1965). Claims have been made that beech sprouts, particularly those near the parent tree, rarely survive beyond the sapling stage. This may be due to transmission of pathogens from the parent tree (Campbell, 1938) or lack of development of an independent root system (Poulson, personal communication).

Beech and sugar maple were the only shade-tolerant species in the stands studied to exhibit nearly complete dominance, so interactions of other species must be studied in a more complex context.

Hemlock canopy seems to exert the strongest influence on the seedling and sapling strata, apparently inhibiting beech and, especially, maple saplings relative to hemlock saplings. This is unlikely to result from poor seed dispersal on the part of beech and maple since neither species was found at particularly high abundance beneath conspecific canopy compared to other species than hemlock. Beech seedlings were positively associated with hemlock canopy (Table 3) and maple seeds are so widely dispersed that dispersal limitation seems improbable. The most likely explanation is higher mortality of beech and maple seedlings beneath hemlock canopy. Beech seedlings may not experience this mortality in their smallest sizes while maple seedlings are apparently affected very soon after establishment. Dead saplings of beech and maple were not particularly common beneath hemlock suggesting earlier mortality. Dead hemlock saplings, on the other hand, showed increased frequency at high hemlock canopy influence suggesting a gradual attrition rather than a strong limitation at early stages.

The abundance of hemlock saplings beneath conspecific canopy is only relative; they were relatively few in number. Once established, however, hemlocks typically show greater shade tolerance. Increment cores of saplings showed as many as four distinct episodes of suppression. Dead hemlock saplings were comparatively rare. The association between hemlock saplings and canopy becomes more pronounced at finer scales (Fig. 12). Hemlock canopy trees were shown by Peterson's (1976) index of segregation to exhibit strong positive contagion which should intensify the canopy–understory interaction.

Hemlock canopy has been suggested to have a negative influence on understory species in Indiana (Daubenmire, 1936), central New York (Lewin,

1974), hemlock's northern range (Rogers, 1980), and in the Great Smoky Mountains (Hicks, 1981). In this study, total sapling density was significantly lower in quadrats with high hemlock dominance. Suggested mechanisms for this suppressing effect of hemlock include shade effects of its dense evergreen canopy and allelochemical or mechanical inhibition by hemlock litter. Taken together, these patterns suggest that hemlock is most frequently self-replacing.

Maple canopy has effects on the understory beyond the relative success of beech over maple saplings. Hemlock saplings in nearly all cases were disfavored by maple canopy. Again, this pattern could be a result of poor dispersal of seed from the locality of hemlock parent trees, but this is not supported by the distribution of hemlock saplings with respect to other canopy species. Mechanisms which might cause enhanced mortality of hemlock beneath maple canopy are not obvious. Tubbs (1973, 1976) did find allelopathic suppression of several species of conifers by root exudates of maple seedlings, but hemlock was not one of the species tested. Maple seedlings in areas with maple dominance usually form a dense understory stratum magnifying the potential for allelopathic effects.

Yellow birch is a 'gap-phase' species, not tolerant of heavy shade (Fowells, 1965) but capable of rapid growth and exploitation of relatively small canopy gaps where light levels are moderate to high (Runkle, 1981; Barden, 1979; Tubbs, 1969; Trimble, 1970; Forcier, 1975). In this study, since large gaps were intentionally unsampled, saplings of birch were relatively rare and could usually be associated with some recent canopy opening. The few birch saplings were found at much lower leves of total canopy influence than expected. If gaps are of great enough frequency, yellow birch should be able to persist as an important canopy species, though its importance may, to some extent, reflect previous severe disturbances. Hemlock and yellow birch were strongly associated in the canopy and showed similar distributions in ordinations while all other species pairs were negatively correlated. The association of these two species has been noted by Brown & Curtis (1952), Whittaker (1956), McIntosh (1972), and Rogers (1978). Since hemlock canopy suppresses other species in the understory, causing low densities of subjacent saplings, gaps

formed in areas of high hemlock dominance should be more open to successful colonization by birch. Barden (1979) reported birch as the most frequent replacement of hemlock canopy trees in the Great Smoky Mountains (22 of 73 hemlock gaps observed).

Basswood is also shade-intolerant and independent saplings or seedlings were quite rare. This rarity was also noted by Daubenmire (1930) and Bray (1956) in old-growth stands of maple–basswood forests. Pigott & Huntley (1978) make the same observation regarding *Tilia cordata* in Great Britain. Basswood trees do produce abundant and vigorous basal sprouts which are densely clustered around parent trees and which appear to have a high likelihood of replacing the parent trees (Fowells, 1965). Most of the basswood canopy trees observed in this study were obviously sprouts originating from older stems. Sometimes several trees originated from a single parent stem. Cores of basswood sprouts showed growth rates several times those seen in other sapling species beneath closed canopy suggesting subsidization by the parent tree. Runkle (1979) found that basswood trees generally break off rather than uproot, a tendency which might be important in reducing mortality of sprouts upon the death of the parent tree. This literal self-replacement by basswood permits an established individual to have an extremely great life expectancy and the capability of occupying a single site for several generations of stems. Basswood canopy also appears to have some inhibitory effect on hemlock and sugar maple saplings (Fig. 8) which would encourage its self-replacement. In the plots studied beech and basswood were very rarely co-dominant at the scale sampled so no judgement of the significance of their interactions is possible.

Suggested species replacement patterns together with estimation of relative probabilities are shown in Figure 13. Additional transitions certainly occur, but those shown appear to be the most frequent. No strong directional or successional trend is suggested, though relative species importances should be expected to fluctuate over long periods of time. The probabilities suggested are likely altered somewhat by a 'mass action' effect of relative abundance of reproduction of different species. Figure 13, for instance, may be interpreted to suggest increasing importance of hemlock, but this tendency may be countered by the comparatively low abundance of

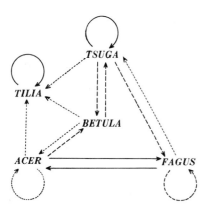

Fig. 13. Diagram showing patterns of canopy tree replacement suggested by distributions of saplings and seedlings with respect to influence of canopy species. Arrows are directed toward the species being replaced. Solid lines indicate transitions which occur most frequently, dashed lines less frequent but common transitions, and dotted lines those transitions which probably occur regularly but only occasionally. Other possible transitions almost certainly occur but more rarely.

hemlock reproduction in most forests. Maple and birch, both prolific seed producers, are often abundant as small seedlings and may be more successful than otherwise indicated. Even though hemlock saplings are usually near hemlock canopy trees, not all hemlock canopy trees have subjacent hemlock saplings and these may be replaced by other species.

The most notable and persistent deviations from the above patterns are seen in sample 9, composed of quadrats from the Tionesta stand. Here the relationship between beech and maple appears to be reversed with saplings of each species relatively more abundant near conspecific canopy trees. The relationship between beech and hemlock is also altered; saplings near canopy of both species tend to be of the opposite species. Sample 9 is also the only instance in which hemlock saplings are found at higher beech canopy influence than are beech saplings. In Tionesta maple and beech seem to be self-replacing while hemlock may replace beech with some frequency. Tionesta is the easternmost stand sampled and is widely separated from other hemlock stands studied. Geographical variation in the interactions among species is certainly possible.

The replacement patterns reported are based on present distribution patterns. If environmental changes affecting these patterns occur (and do not fluctuate around a mean with a period shorter than tree life-spans), then replacement patterns will be altered and directional change in species composition may occur. One phenomenon which may have a strong effect on these patterns is deer browsing. Deer show strong browsing preferences among the species studied here (Hough, 1965; Moore & Johnson, 1967; Anderson & Loucks, 1979; Graham, 1954, 1958; Webb, King & Patric, 1956; Marquis, 1975). In particular, hemlock seedlings and basswood sprouts are so preferred as to be virtually eliminated in areas of high deer population. Workers differ in evaluation of effects of deer on beech, sugar maple, and yellow birch. When deer populations are extremely high these species are also browsed, but at lower deer population levels they may increase in relative importance. Recent decades have seen an increase in deer populations in most areas studied here and deer browse was noted on hemlock and basswood in most stands sampled. In a few cases reproduction by these species was virtually eliminated. If deer populations decline significantly in the near future, the decline of these tree species could be quickly reversed while extended high deer populations could have lasting impact on canopy composition.

Scale of perception and stability

Assertions of stability or instability are frequently made but are not meaningful or comparable unless accompanied by considerations of a spatial and temporal scale. Temporal scale may be defined so as to either verify or refute stability. No community is stable if the amount of time considered is great enough, while over periods much briefer than the life-spans of individual trees no significant change in community composition occurs. As the forest stands studied were judged to have been free of massive disturbance for at least several centuries, patterns observed should be largely determined by biological interactions rather than by residual effects of large disturbances.

Spatial scale is equally important. Forests exist as mosaics of patches in various stages of regeneration or change. No forest community could reasonably be called stable if samples were of such a size as to include only one or two trees as disturbance at this scale is a necessary consequence of individual mortality. Coarser scales of sampling integrate the states of many microsites occupied by individual trees. Hence, perception of stability depends on the relative size of patches in the mosaic and sample size. At a sample size large enough to include patches of all ages, any forest type should satisfy criteria for stability. The spatial units of perception in this study are stands of a few to a few tens of hectares. At this scale population interactions should determine in the absence of massive disturbances the stability of the stand.

Explanation of patterns perceived at one scale may require consideration of phenomena occurring at other scales. Stability of forests at the scale of the stand may depend on instability at the scale of the microsite. Properties such as stability may appear at particular scales but become illusory when other scales are considered. Conflicts regarding description of community stability may often be resolved by standardization of scale.

The stability suggested for these forests is not a monolithic one involving a set of tightly co-evolved species. The modern ranges of the species studied do not coincide precisely at any scale and current species assemblages have existed only since the arrival of various species following their independent northward migration during the Holocene (Davis, 1976). Climatic change over the few millennia during which current species combinations have persisted has probably repeatedly altered community dynamics. However, whenever and wherever environmental conditions remain fairly stable for more than a generation or two, groups of tree species whose pre-existing life histories allow co-existence should become prevalent. The approximate stability suggested by the replacement patterns described must be regarded as fragile; the strongest statement that can be made is that a pattern is formed by which stable co-existence might be maintained at the defined spatial scale if the environment which produced the pattern does not and has not recently changed. Other workers (Runkle, 1981; Fox, 1977) have proposed patterns of saplings distributions and tree replacement suggesting some degree of stability in similar forests.

Forcier (1975), examining the distributions of seedlings and saplings with respect to canopy trees in a New Hampshire forest suggested a directional change involving gradual replacement of yellow birch and maple by beech. Horn (1975), in a Markov analysis of a New Jersey forest also suggested successional trends toward beech dominance. These last two results are not necessarily in conflict with the current study, since both studies were done in stands which might be considered successional, and involved sites in the northeastern United States which was poorly sampled in this study. Results from the single eastern stand studied here (Tionesta) are potentially consistent with Forcier's study.

Conclusions

In all of the above studies canopy-understory interactions have been identified as primary forces determining both successional trends and the patterns of species co-existence in stable forests, thus emphasizing the importance of two general concerns in the study of forest communities. First, an understanding of all life cycle stages is crucial for interpretation of community structure and dynamics. The specific patterns and interactions suggested for the populations and communities studied here could not have been suggested by observations of canopy or understory patterns alone. The study of the behavior of tree species in the understory and their requirements in early life-history stages has often been neglected, with the result that understanding of niche differentiation among them is poorly developed. Further understanding of community dynamics and properties will depend on study of complete life-histories of individual species.

The second point concerns scale of perception. Interactions between life-history strategies and environmental fluctuation control community dynamics; these driving forces are universal (Denslow, 1980). Our perception of differences among communities is often due not to intrinsic differences in community mechanisms, but to variation in the spatial and temporal scales at which they function. The driving forces of succession are rooted in the modification of environment by existing species and in chance factors such as seed source and disturbance history (Drury & Nisbet, 1973; Connell & Slatyer, 1977). This study and others show that the dynamics of species interactions and patterns of tree replacement are controlled by essentially similar factors in mature, stable forests. The forces and mechanisms driving successional change and maintaining stability are not qualitatively different. Rather, similar forces acting on different parts of a continuum of spatial and temporal scales lead to communities which differ in perceived stability.

References

Alemdag, I. S., 1978. Evaluation of some competition indexes for the prediction of diameter increment in planted white spruce. Can. For. Serv. For. Manage. Inst., Inf. Rep. FMR-X-108.

Alexandre, D. Y., 1977. Regeneration naturelle d'un arbre caracteristique de la forêt equatoriale de Côte-d'Ivoire: Turraeanthus africana Pellegr. Oecol. Plant. 12: 241–262.

Anderson, R. C. & Loucks, O. L., 1979. White-tail deer (Odocoileus virginianus) influence on structure and composition of Tsuga canadensis forests. J. Applied Ecol. 16: 855–861.

Barden, L. S., 1979. Tree replacement in small canopy gaps of a Tsuga canadensis forest in the southern Appalachians, Tennessee. Oecologia 44: 141–142.

Bella, I. E., 1971. A new competition model for individual tree. Forest Sci. 17: 364–372.

Bratton, S. P., 1976a. Resource distribution in an understory herb community: responses to temporal and microtopographic gradients. Am. Nat. 110: 679–693.

Bratton, S. P., 1976b. The response of understory herbs to soil depth gradients in high and low diversity communities. Bull. Torrey Bot. Club. 103: 165–172.

Braun, E. L., 1950. Deciduous Forests of Eastern North America. Hafner, New York.

Bray, J. R., 1956. Gap-phase replacement in a maple–basswood forest. Ecology 37: 598–600.

Brewer, R. & Merritt, P. G., 1978. Wind throw and tree replacement in a climax beech–maple forest. Oikos 30: 149–152.

Brown, R. T. & Curtis, J. T., 1952. The upland conifer–hardwood forests of northern Wisconsin. Ecology 33: 217–234.

Campbell, W. A., 1938. Preliminary report on decay in sprout northern hardwoods in relation to timber stand improvement. U.S. Dep. Agric. For. Serv. Occasional Paper 7. Northeastern Forest Experiment Station, Upper Darby.

Canham, C. D. & Loucks, O. L., 1983. Catastrophic windthrow in the presettlement forests of Wisconsin. Ecology (in press).

Connell, J. H., 1970. On the role of natural enemies in preventing competitive exclusion in some marine animals and in rain forest trees. In: P. J. den Boer & G. R. Gradwell (eds.), Proc. Adv. Study Inst. Dynamics of Populations, pp. 298–312. Oosterbeek.

Connell, J. H. & Slatyer, R. O., 1977. Mechanisms of succession in natural communities and their role in community stability and organization. Am. Nat. 111: 1119–1144.

Daubenmire, R. F., 1930. The relation of certain ecological factors to the inhibition of forest floor herbs under hemlock. Butler Univ. Bot. Studies 1: 61–76.

Daubenmire, R. F., 1936. The 'Big Woods' of Minnesota: its structure, and relation to climate, fire, and soils. Ecol. Monogr. 6: 233–268.

Davis, M. B., 1976. Pleistocene biogeography of temperate deciduous forests. Geoscience and Man 13: 13–26.

Denslow, J. S., 1980. Patterns of plant species diversity during succession under different disturbance regimes. Oecologia 46: 18–21.

Drury, W. H. & Nisbet, I. C. T., 1973. Succession. J. Arnold Arboretum 54: 331–368.

Fernald, M. I., 1950. Gray's Manual of Botany, 11th ed. American Book Co., New York.

Forcier, L. K., 1975. Reproductive strategies and co-occurrence of climax tree species. Science 189: 808–810.

Fowells, H. A., 1965. Silvics of Forest Trees of the United States. Agric. Handbook 271. U.S. Gov. Printing Office.

Fox, J. F., 1977. Alternation and coexistence of tree species. Am. Nat. 111: 69–89.

Graham, S. A., 1954. Changes in northern Michigan forests from browsing by deer. Trans. No. Am. Wildl. Confer. 19: 526–531.

Graham, S. A., 1958. Results of deer exclosure experiments in the Ottawa National Forest. Trans. No. Am. Wildl. Confer. 23: 478–490.

Grubb, P. J., 1977. The maintenance of species richness in plant communities. The importance of the regeneration niche. Biol. Rev. 52: 107–145.

Harper, J. L., 1977. Population Biology of Plants. Academic Press, London.

Hicks, D. J., 1981. Intrastand distribution patterns of southern Appalachian cove forest herbaceous species. Am. Midland Nat. 104: 209–233.

Hill, M. O. & Gauch, H. G., 1980. Detrended correspondence analysis: An improved ordination technique. Vegetatio 42: 47–58.

Horn, H. S., 1971. The Adaptive Geometry of Trees. Princeton University Press, Princeton.

Horn, H. S., 1975. Markovian properties of forest succession. In: M. L. Cody & J. Diamond (eds.), Ecology of Species and Communities, pp. 196–211. Harvard University Press, Cambridge.

Hough, A. F., 1965. A twenty-year record of understory vegetational change in a virgin Pennsylvania forest. Ecology 46: 370–373.

Janzen, D. H., 1970. Herbivores and the number of tree species in tropical forests. Am. Nat. 104: 501–527.

Jones, E. W., 1945. The structure and reproduction of the virgin forest of the north temperate zone. New Phytol. 44: 130–148.

Lewin, D. C., 1974. The vegetation of the ravines of the southern Finger Lakes, New York region. Am. Midland. Nat. 91: 315–342.

Lorimer, C. G., 1981. Survival and growth of understory trees in oak forests of the Hudson Highlands, New York. Can. J. For. Res. 11: 689–695.

Marquis, D. A., 1975. The impact of deer browsing on Allegheny hardwood regeneration. U.S. Dep. Agric. For. Serv. Res. Pap. NE-308.

McIntosh, R. P., 1972. Forests of the Catskill Mountains, New York. Ecol. Monogr. 42: 143–161.

Moore, W. H. & Johnson, F. M., 1967. Nature of deer browsing on hardwood seedlings and sprouts. J. Wildl. Manage. 31: 351–353.

Peterson, C. H., 1976. Measurement of community pattern by indices of local segregation and species diversity. J. Ecol. 64: 157–170.

Pielou, E. C., 1977. Mathematical Ecology. Wiley, New York.

Pigott, C. D. & Huntley, J. P., 1978. Factors controlling the distribution of Tilia cordata at the northern limits of its geographical range. I. Distribution in northwest England. New Phytol. 81: 429–441.

Rogers, R. S., 1978. Forests dominated by hemlock (Tsuga canadensis): distribution as related to site and postsettlement history. Can. J. Bot. 56: 843–854.

Rogers, R. S., 1980. Hemlock stands from Wisconsin to Nova Scotia: transitions in understory composition along a floristic gradient. Ecology 61: 178–193.

Runkle, J. R., 1979. Gap phase dynamics in climax mesic forests. Ph.D. Thesis, Cornell University, Ithaca, New York.

Runkle, J. R., 1981. Gap regeneration in some old-growth forests of the eastern U.S. Ecology 62: 1041–1051.

Trimble, G. R., 1970. Twenty years of intensive uneven-aged management: effect on growth, yield, and species of composition in two hardwood stands in West Virginia. U.S. Dep. Agric. For. Serv. Res. Pap. NE-154.

Tubbs, C. H., 1967. Natural regeneration of yellow birch in the Lake States. In: Birch Symp. Proc. Northeast For. Exp. Sta., Durham, New Hampshire, pp. 74–78.

Tubbs, C. H., 1973. Allelopathic relationship between yellow birch and sugar maple seedlings. Forest Sci. 19: 139–145.

Tubbs, C. H., 1976. Effect of sugar maple root exudate on seedlings of northern conifer species. U.S. Dep. Agric. For. Serv. Res. Note NC-213.

Ward, R. T., 1961. Some aspects of the regeneration habits of the American beech. Ecology 42: 828–832.

Webb, W. L., King, R. T. & Patric, E. F., 1956. Effect of white-tailed deer on a mature northern hardwood forest. J. Forestry 54: 391–398.

Whittaker, R. H., 1956. Vegetation of the Great Smoky Mountains. Ecol. Monogr. 26: 1–80.

Whittaker, R. H., 1975. The design and stability of plant communities. In: W. H. van Dobben & R. H. Lowe-McConnell (eds.), Unifying Concepts in Ecology, pp. 169–181. Junk, The Hague.

Whittaker, R. H. & Levin, S. A., 1977. The role of mosaic phenomena in natural communities. Theor. Pop. Biol. 12: 117–139.

Woods, K. D., 1979. Reciprocal replacement and the maintenance of codominance in a beech–maple forest. Oikos 33: 31–39.

Woods, K. D. & Whittaker, R. H., 1981. Canopy-understory interaction and the internal dynamics of mature hardwood and hemlock–hardwood forests. In: D. C. West, H. H. Shugart & D. B. Botkin (eds.), Forest Succession: Concepts and Application, pp. 305–323. Springer-Verlag, New York.

Accepted 18.11.1983.

Measuring compositional change along gradients

Mark V. Wilson[1] & C. L. Mohler[2],*
[1] Environmental Studies Program, University of California, Santa Barbara, CA 93106, U.S.A.
[2] Ecology and Systematics, Cornell University, Ithaca, NY 14853, U.S.A.

Keywords: Beta diversity, Gradient analysis, Multivariate methods, Niche, Ordination, Siskiyou Mts., Species distributions, Species turnover, Succession

Abstract

A new procedure for measuring compositional change along gradients is proposed. Given a matrix of species-by-samples and an initial ordering of samples on an axis, the 'gradient rescaling' method calculates 1) gradient length (beta diversity), 2) rates of species turnover as a function of position on the gradient, and 3) an ecologically meaningful spacing of samples along the gradient. A new unit of beta diversity, the gleason, is proposed. Gradient rescaling is evaluated with both simulated and field data and is shown to perform well under many ecological conditions. Applications to the study of succession, phenology, and niche relations are briefly discussed.

Introduction

The observation that community composition varies along environmental gradients is basic to much work in community ecology. Gradients not only form the context for many field studies (see Whittaker 1967, 1972) but figure heavily in much theoretical work on niche relations (Hutchinson, 1958; McNaughton & Wolf, 1970; Whittaker et al., 1973). Despite the large role of compositional gradients, or coenoclines as they are sometimes called (Whittaker, 1967), statistical methods for the analysis of continuous compositional variation remain rather primitive.

We address three related problems. First, one may want to know the biological length of some coenocline, which is to say, the total amount of compositional change associated with the gradient. Such a measure is particularly useful when comparing two or more gradients (Whittaker, 1960; Peet, 1978). For example, if Janzen (1976) is correct in believing that mountain passes of given elevation are biologically 'higher' in the tropics, then the amount of compositional change from sea level to the pass should be greatest at tropical latitudes and decline toward the poles. Whittaker (1960, 1965, 1972) refers to compositional differentiation along gradients as beta diversity, in contrast to the alpha diversity or richness of a single sample. He and subsequent authors have measured beta diversity in units of half-changes (HC), analogous to the half-life of radioactive elements. Roughly speaking, a coenocline of 1 HC has endpoints which are 50% similar, a coenocline of 2 HC's has endpoints which are each 50% similar to the midpoint, etc. However, measurement of coenocline length in half-changes is complicated by apparently random compositional variation among replicate samples, henceforth referred to as 'noise', by the non-linear relation

* The authors thank E. W. Beals, R. Furnas, P. L. Marks, R. K. Peet, O. D. Sholes and the late R. H. Whittaker for helpful comments. This work was supported in part by a National Science Foundation grant to Robert H. Whittaker of the Section of Ecology and Systematics at Cornell University, and in part by McIntire-Stennis Grant No. 183–7551 and a grant from the National Park Service, both to Peter L. Marks of the Section of Ecology and Systematics at Cornell University.

between similarity measures and sample separation along the gradient (Gauch, 1973), and by the presence of other, unmeasured gradients. To our knowledge, previous workers have not attempted to test the accuracy of Whittaker's procedure for determining beta diversity in half-changes, and alternative procedures are few.

Second, one may want to compare rates of species turnover at various points on a gradient, for example, to define objectively ecotones between community types (Whittaker, 1960; Beals, 1969) or to locate periods of rapid change during a successional or phenological sequence. Turnover rate may be expressed as beta diversity per unit gradient as measured over an infinitesimal length of gradient. (Conversely, beta diversity is turnover rate integrated over the gradient (Bratton, 1975).) Whittaker, Beals & Bratton all estimated compositional turnover using the measures percentage similarity (PS) or percentage difference (PD = 100 – PS) (Goodall, 1978). This approach may lead to problems because: 1) random variation in composition contributes to PD; 2) non-monotonic change in species abundance may cause PD to underestimate compositional change; and 3) spacing of samples along the gradient reflects the investigator's perception of the environment rather than that of the species which compose the coenocline.

Third, one may want a natural, species-defined spacing of samples. In general, any ordination produces a species-based but undefined spacing; in this paper we consider in particular the spacing defined explicitly by a constant species turnover rate along the gradient. When turnover rate is constant the separation between samples is an expression of sample distinctiveness, that can be used in weighting niche metrics (Colwell & Futuyma, 1971; Pielou, 1972). Furthermore, if habitat axes are rescaled in beta diversity units, breadth and overlap measures for different systems can be compared and multi-dimensional niche metrics can be computed directly rather than by the addition or multiplication of several one-dimensional measures (cf. Levins, 1968; Pianka, 1973; Cody, 1974).

We believe the technique presented here, gradient rescaling, resolves the problems of measuring compositional change, rates of species turnover, and axis scale. The gradient rescaling approach is treated in the following sequence. First, we give the rationale of using species turnover as a basis for gradient analysis and develop the computational procedures for measuring beta diversity in half-changes. Second, we describe our evaluation of the technique using simulated data sets. Third, we illustrate the use of the technique with field data. Finally, we discuss and compare methods and mention some applications of our approach to the study of compositional change along gradients.

Gradient rescaling

Rationale

The relationship between beta diversity, rate of species turnover, and gradient scale may be expressed as:

$$\beta = \int_a^b R(x)\mathrm{d}x \qquad (1)$$

where β is the beta diversity of the compositional gradient between points a and b, R is the species turnover rate and x are values along X, the initial scaling of the gradient. Field measurements of environmental factors along the gradient X are, of course, in a scale chosen by the investigator (e.g., elevation in meters or acidity in pH units). The choice of scale must be arbitrary, reflecting the investigator's notion of what is easy, traditional or interesting to record. Since the investigator's scale of measurement does not necessarily reflect how organisms respond to changes in environmental conditions, several authors have proposed rescaling gradient axes using species distributions as an 'ecoassay' (e.g. Colwell & Futuyma, 1971) of the environmental gradient. Gauch et al. (1974) and Ihm & Groenewoud (1975) proposed scaling axes such that species distributions most closely approximate Gaussian curves. Austin (1976) discounted the generality of Gaussian curves in nature and used the more flexible beta distribution of statistics in an algorithm similar to that of Gauch et al. (1974). In contrast to these curve-fitting approaches, Hill & Gauch (1980) scaled reciprocal averaging ordination axes so that, on the average, the species in each sample have distribution curves of unit standard deviation.

Our approach is based on the view that compositional turnover is the essence of community gradients, and that environmental change is ecologically significant primarily to the extent that it influences the relative abundances of species. Accordingly, we scale community gradients so that the rate of compositional change is constant throughout. In this scaling, each portion of the gradient is ecologically equivalent to all other portions. This scaling allows the biological length of the compositional gradient and the rate of species turnover along physically-defined gradients (e.g., elevation), to be accurately calculated. Our approach starts with the relationship between gradient scale, rate of turnover and beta diversity shown in equation (1). There are three phases. We first find a new scaling, X', such that $R(x') = k$, for all x', where k is a constant. That is, we find a new set of relative sample positions such that the rate of species turnover along the gradient is constant. Second, we calculate the compositional length or beta diversity of the gradient as:

$$\beta = k \cdot (b - a)$$

where $b - a$ is the distance between endpoints in terms of the scaling X'. The value of β is independent of the initial scaling. Finally, $R^*(x)$, the rate of compositional change along the original gradient, is computed by using the original scaling to transform $R(x')$. Details of this procedure are presented below.

Early in the development of gradient rescaling we devised a measure of beta diversity which is an alternative to the half-change. This we name the gleason, in honor of H. A. Gleason, who first pointed out the importance of continuous compositional change along environmental gradients (Gleason, 1926). One gleason (G) is the amount of compositional turnover which would occur if all changes were concentrated into a single species whose abundance changed 100%. For example, a 1 G coenocline might consist of three species, two of which change from 50% to 0% relative abundance as the third species changes from 0% to 100% relative abundance. There is no general function which can interconvert gleasons and half-changes. In the following sections we develop computational procedures for both measures.

Computations

We start with field observations of species abun-

dances within a sequence of samples, X, along an ecological gradient. This gradient can be either a direct environmental gradient or an indirect, compositional gradient derived by ordination. We consider two measures of species turnover or compositional change: half-changes and gleasons. Computations for both measures are performed by a FORTRAN program, GRADBETA, which is available with documentation from Cornell Ecology Programs, Ecology and Systematics, Cornell University, Ithaca, NY 14853 (Wilson & Mohler, 1980).

For either gleasons or half-changes species importance values should be standardized so that sample totals are equal. After this standardization, measures of compositional change are measures of proportional turnover. In this way, the influence of trends in total biomass are minimized in the interpretation of compositional turnover.

Field data always include variation due to sample error or the effects of chance and uninvestigated ecological factors. Such noise causes the observed rate of compositional change to be higher than the change attributable to species distributions along recognized gradients. Noise is often ignored in ecologial analyses, but this practice may lead to misinterpretation of ecological patterns.

A convenient index of noise is internal association (IA), the expected similarity of replicate samples. IA decreases as the average noise level of data increases. Any index of compositional similarity can be used to estimate IA; our technique uses percentage similarity (Whittaker, 1975; Goodall, 1978). Replicate samples are equivalent to samples whose separation along the gradient is zero. Therefore, in a plot of sample percentage similarity of each pair of samples versus the gradient separation of each pair, an estimate of IA may be obtained by extrapolating percentage similarity to zero separation (Whittaker, 1960). Whittaker accomplished this by fitting a straight line to the central portion of a plot of the logarithm of percentage similarity versus gradient separation. We make two refinements. First, extrapolation is accomplished by linear least-squares regression instead of by a hand-fitted line. Second, instead of an arbitrary spacing between pairs of samples, we use the ecologically more meaningful gradient distance derived by the gleasons or half-changes procedure.

Half-changes

Whittaker (1960) defined the half-change as the ecological distance between two samples with similarity of species composition 50% that of two replicate samples. The Whittaker method for measuring the beta diversity of an entire gradient is based on the posited relationship:

$$HC(a,b) = \frac{[\log(IA) - \log(PS(a,\,b))]}{\log 2}$$

where $PS(a,b)$ is the percentage similarity of samples a and b, IA is the expected similarity of replicate samples, and $HC(a,b)$ is the beta diversity of the gradient segment from a to b, in half-changes.

This half-change formula implies two basic assumptions. The first assumption is that sample similarity is negatively exponentially related to ecological distance as measured in half-changes, so that as ecological distance increases the rate at which sample similarity decreases is constant (cf. Gauch, 1973). Whittaker (1960, 1972) presents data sets that support this assumption. The calculations of half-changes in our technique are also based on this exponential relationship. The second assumption is that samples are evenly spaced along the ecological gradient. This assumption, which is not met by most field data, is avoided by our technique.

In outline our method is as follows. First, an index of compositional turnover rate is computed for each sample point on the gradient and the among-sample variance in this index is determined. The turnover index used in the procedure is basically the average of $\dfrac{HC(a,b)}{\Delta x(a,b)}$ for all samples b in some vicinity of a, where $HC(a,b)$ is the number of half-changes between samples a and b, and $\Delta x(a,b)$ is the gradient distance between a and b. The ratio $\dfrac{HC(a,b)}{\Delta x(a,b)}$ is the slope of a vector in a 2-space whose axes are gradient distance and compositional distance (in half-changes). The appropriate average for the ratios is thus the slope of the vector sum. Second, using a path-of-steepest-descent algorithm (Berington, 1969), the sample positions are adjusted iteratively until the variance in turnover rate has reached a local minimum. This produces the new scaling, X'. Third, the final turnover rate and new sample positions are used to compute a set of

corrected turnover rates for the original scaling and a set of partial beta diversity values corresponding to the intervals around each sample. Total beta diversity is then computed as the sum of the partial beta diversities. See Wilson & Mohler (1980) for details of the numerical application of these procedures.

To compute the partial and total beta diversity values and the turnover rates with respect to the original gradient we first define an interval, I, about each of the final sample positions:

$$I(j) = \begin{cases} (x'(2) - x'(1))/2 & j = 1 \\ (x'(j+1) - x'(j-1))/2 & j = 2,\ldots,n-1 \\ (x'(n) - x'(n-1))/2 & j = n \end{cases} \quad (2)$$

Partial beta diversities, $B_{HC}(j)$, are then computed as:

$$B_{HC}(j) = R(x'(j)) \cdot I(j)$$

Thus the beta diversity associated with a sample is simply the turnover rate at that point on the rescaled gradient times the gradient interval associated with the sample. Total beta diversity, β_{HC}, of the gradient is then the sum of the partial beta diversities. If var $R = 0$ exactly, then $R(x'(j)) = k$ for all j, and $\beta_{HC} = k \cdot [x'(n) - x'(1)]$.

To compute the turnover rates for the original scaling of the gradient we divide the above partial beta diversities by the intervals for the original scaling. These corrected rates, $R^*(x(j))$, are:

$$R^*(x(j)) = \begin{cases} \dfrac{2B_{HC}(1)}{(x(2) - x(1))} & j = 1 \\[2mm] \dfrac{2B_{HC}(j)}{(x(j+1) - x(j-1))} & j = 2,\ldots,n-1 \\[2mm] \dfrac{2B_{HC}(n)}{(x(n) - x(n-1))} & j = n \end{cases} \quad (3)$$

Equation (3) gives rates of compositional turnover at the position on the gradient of each sample.

Gleasons

A second natural measure of compositional change is the amount that species importance changes from one point on a gradient to another, summed over all species. The unit of change measured in this manner we call the gleason. Rate of change in terms of gleasons is defined as:

$$R(x) = \sum_i \left| \frac{d\hat{Y}_i}{dX} \right| = (1/dX)\sum_i |d\hat{Y}_i|$$

where $\hat{Y}_i$ is the expected abundance of species i as a function of gradient position, x. Substituting $R(x)$ into equation (1) yields

$$G(a,b) = \int_a^b \sum_i |d\hat{Y}_i| \qquad (4)$$

where $G(a,b)$ is the beta diversity in gleasons of the gradient interval between points a and b.

When a and b are near each other, the integral in (4) may be approximated as:

$$G(a,b) = \sum_i |\Delta\hat{Y}_i(a,b)|$$

where $\Delta\hat{Y}_i(a,b) = \hat{Y}_i(b) - \hat{Y}_i(a)$. When the two points are farther apart, however, some species may rise and fall in abundance within the interval. In this case $\sum_i |\Delta\hat{Y}_i(a,b)|$ does not describe the full change in the abundance of species and thus $G(a,b) > \sum_i |\Delta\hat{Y}_i(a,b)|$.

In practice, the expected change in importance of species i, $\Delta\hat{Y}_i$, is estimated from observations at discrete samples along the gradient. To minimize the underestimate of G caused by non-monotonic species response, G is first computed for pairs of adjacent samples and then the beta diversity of longer intervals is obtained by summing these small elements. In general, species show some random variation in abundance not related to the gradient. Thus, the compositional change from one sample to the next which is attributable to the gradient may be viewed as the total difference in species abundance in the two samples minus that component of the difference which is due to noise. That is,

$$G(j,j+1) =$$

$$\sum_i |\Delta Y_i(j,j+1)|_{observed} - \sum_i |\Delta Y_i(j,j+1)|_{noise}$$

where $\Delta Y_i(j,j+1)$ is the difference in abundance of species i in samples j and $j+1$ and the samples are numbered consecutively with respect to the gradient. Since we use species abundance values, Y_i, that have been standardized so that sample totals are equal, percentage similarity (PS) can be calculated as $PS(j,j+1) = 1 - \frac{1}{2}\sum_i |Y_i(j,j+1)|$. This leads to:

$$\sum_i |Y_i(j,j+1)|_{observed} = 2 - 2 \cdot PS(j,j+1).$$

Moreover, since j and $j+1$ are nearby on the gradient it is reasonable to assume that

$$\sum_i |\Delta Y_i(j,j+1)|_{noise} = 2 - 2(IA),$$

and thus $G(j,j+1)$ can be estimated as

$$G(j,j+1) = 2[IA - PS(j,j+1)].$$

The beta diversity of the whole gradient in gleasons, β_G, is then

$$\beta_G = \sum_{j=1}^{n-1} G(j,j+1) = 2\sum_{j=1}^{n-1} [IA - PS(j,j+1)].$$

Rescaled sample positions, x', are derived from knowledge that the rate of species turnover along the rescaled axis is constant:

$$R(x') = \frac{G(j,j+1)}{[x'(j+1) - x'(j)]} = k$$

or, equivalently,

$$x'(j+1) = (\frac{2}{k})[IA - PS(j,j+1)] + x'(j). \qquad (5)$$

To compute $x(j)$ we first set the beginning of the axis equal to zero, $x(1) = 0$, or some other convenient value, and choose an arbitrary value for k. In practice we set $k=1$ since this causes $x'(n) - x'(1) = \beta_G$, and thus sample positions are in units of gleasons. The sample positions, $x'(j)$ for $j = 2, 3, \ldots, n$, are then computed recursively using equation (5).

The partial beta diversity, $B_G(j)$, attributed to the neighborhood around sample j is calculated as,

$$B_G(j) = \begin{cases} G(1,2)/2 & j=1 \\ [G(j-1,j) + G(j,j+1)]/2 & j=2,\ldots,n-1 \\ G(n-1,n)/2 & j=n \end{cases}$$

Note that $\sum_j B_G(j) = \beta_G$. Corrected rates of change, measured as gleasons per original gradient unit, are then calculated as

$$R^*(x(j)) = \frac{B_G(j)}{I(j)} \quad j=1, 2, \ldots, n,$$

where the intervals, $I(j)$, of neighborhoods around the samples, are defined by equation (2).

Thus, β_G, X' and R^* can be calculated simply and directly, without the iteration step required by the half-changes procedure.

212

Evaluation

Gradient rescaling was developed for the analysis of ecological communities. In the field it is often difficult to determine how many samples should be taken and how well samples have been allocated to different segments of ecological gradients. Sampling schemes and characteristics of the study system also influence the noise level of data. Any successful method of community analysis, including gradient rescaling, should be robust under a variety of field conditions. Although we have applied the procedure to many field data sets the following evaluation relies primarily on simulated data, because only with a simulation can the true values of estimated parameters be specified for purposes of comparison.

Evaluation methods

There were four steps to the evaluation procedures. First, we constructed simulated data sets of species distributions along gradients, both with and without noise. Second, using *a priori* knowledge of species distributions from the simulations, we calculated exact values of the rescaled sample positions, the corrected rates of change, and beta diversity. Third, we 'sampled' the simulated data at different points along the gradient, and analyzed the resulting data using our gradient rescaling technique. Fourth, the results of the gradient rescaling of the sampled data were compared to the true values derived from the original simulations.

Gradients of species distributions were simulated using a modification of CEP-1 of Gauch & Whittaker (1972). This program constructs species distributions as Gaussian curves along a gradient. Species curves were drawn at random from a wide range of peak abundance and breadths of distribution, and were placed randomly along the gradients. Modifications included a noise algorithm that, for each species distribution curve, produced noise proportional to abundance, and a routine that added species until total abundance along the gradient was constant. Gradients were simulated with different ecological lengths and with different noise levels. For our main experiment, 18 gradients were simulated in a factorial design, as follows: three ecological lengths (short, medium and long; approximately 1.5, 3.0 and 6.0 half-changes), two

noise levels (IA values of 90% and 70%), and three replicates. For each of the 18 gradients the corresponding noise-free data set represented the underlying ecological conditions of the gradient.

For our evaluation of the effects of sampling intensity nine gradients were simulated: three ecological lengths (approximately 0.7, 3.0 and 10.0 half-changes) with three replicates each. These nine gradients were simulated in noise-free and moderate noise versions (*IA* values of 100% and 80%).

Each of the simulated gradients was 'sampled' by computer at certain locations along the gradient to mimic extremes of sample placement that could be encountered in field work. In the main experiment (Fig. 1), 15 samples were placed in four schemes: evenly spaced, grouped at both ends, grouped in the center, and grouped at one end only. These schemes were the basis for the true or underlying sample spacings. In each analysis by gradient rescaling this sample scheme information was treated as unknown, just as the true relationships among samples would be unknown in a field study. Thus, the naive view that samples were evenly spread along the gradient was assumed for the sake of evaluating the technique. In the intensity experiment the intervals between samples was constant but the intensity was set at 3, 8, 15, 29 or 57 samples per gradient. The data on species abundance within samples were then analyzed by our gradient rescaling technique, using the parameter values and procedures recommended in Wilson & Mohler (1980). Results from these analyses of the simulated field data were then compared with the true values as determined from knowledge of the noise-free species distributions.

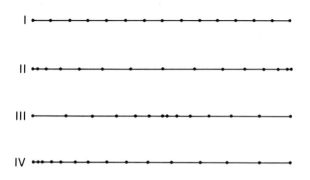

Fig. 1. Sample placement schemes along the simulated gradients for analysis by gradient rescaling. I – evenly spaced, II – clustered at both ends, III – clustered at the center, IV – clustered at one end. Each scheme has 15 samples.

Evaluation results

Gradient rescaling gave accurate estimations of beta diversity under most combinations of gradient length and noise level. Table 1 presents the average proportional error of estimation, for both half-changes and gleasons, when sampling was from 15 evenly spaced locations along the simulated gradient. Gleason estimations are nearly 100% accurate for the noise-free gradients. In contrast, the estimation of half-changes for the low-noise gradients was better than for the noise-free gradients. Very poor estimations in gleasons or half-change occurred only for the short gradients with high noise. Evidently the signal, in this case the patterns of species turnover, was swamped by the noise, the random variations in species abundances. Without noise, estimation for gleasons was better than for half-changes, but with either low or high noise estimation was better for half-changes. These contrasting results may reflect the development of gleasons from theoretical considerations of compositional gradients compared to the lineage of half-changes as an empirical technique.

The ability to estimate beta diversity is affected by the accuracy of assumptions concerning underlying sample spacing. Table 1 also shows accuracy of estimation of beta diversity by the gradient rescaling technique when the underlying sample positions followed three extreme distributions (see Fig. 1). Patterns of estimation accuracy of both half-changes and gleasons followed those seen for evenly spaced samples, with best estimations of gleasons for noise-free gradients and for longer gradients, and best estimations of half-changes with low noise gradients (Table 1). Overall, differences among the four underlying sample distributions did not greatly influence estimation of beta diversity as either half-changes or gleasons. That is, poor knowledge of true sample spacing would not normally be an impediment to the accurate estimation of beta diversity by the technique of gradient rescaling. A possible exception to this general robustness involved data sets with noisy samples clustered toward the center of short gradients. Again, the signal to noise ratio is low for gradients with high noise. Center-clustered sampling schemes are also least useful in ordination (Mohler, 1981) and for extracting information about species distributions (Mohler, 1983).

Table 2 reports the ability of gradient rescaling to recover the three types of underlying sample distri-

Table 1. Accuracy of estimation of beta diversity by the gradient rescaling technique. The index is the average proportional error: average of |estimation-true|/true. Short, medium and long refer to the relative lengths of the simulated gradients, approx. 1.5, 3.0, and 6.0 half-changes, resp. Noise levels correspond to internal associations of 100%, 90%, and 70%. Sampling was at 15 locations in four distributions: evenly spaced, end-clustered, center-clustered, and clustered at one end (see Fig. 1).

Gradient length	Noise level	n	Index of accuracy							
			Half-changes				Gleasons			
			Sample distribution				Sample distribution			
			Even	Ends	Center	One end	Even	Ends	Center	One end
short	none (l)*	3	0.14	0.15	0.14	0.14	0.00	0.00	0.00	0.00
	none (h)*	3	0.09	0.09	0.10	0.09	0.01	0.01	0.01	0.01
	low	3	0.09	0.06	0.32	0.09	0.41	0.61	0.39	0.55
	high	3	0.47	0.09	0.66	0.22	0.83	0.63	0.85	0.65
medium	none (l)*	3	0.05	0.07	0.02	0.06	0.01	0.02	0.01	0.02
	none (h)*	3	0.05	0.07	0.03	0.05	0.00	0.01	0.01	0.01
	low	3	0.02	0.02	0.02	0.01	0.22	0.21	0.23	0.27
	high	3	0.21	0.25	0.34	0.26	0.19	0.26	0.47	0.09
long	none (l)*	3	0.03	0.03	0.02	0.03	0.02	0.04	0.04	0.04
	none (h)*	3	0.03	0.04	0.02	0.02	0.03	0.06	0.03	0.05
	low	3	0.02	0.02	0.04	0.03	0.06	0.09	0.07	0.08
	high	3	0.09	0.11	0.07	0.07	0.16	0.33	0.11	0.12

* The first noise-free gradient was used to simulate the low noise gradient; the second was used for the high noise gradient.

Table 2. Ability of gradient rescaling to recover three types of underlying sample spacings: end-clustered, center-clustered, and clustered at one end (see Fig. 1). A low ratio of convergence means good recovery, a high ratio means bad recovery. The simulated gradients are the same as in Table 1. See text for more explanation.

Gradient length	Noise level	n	Ratio of convergence					
			Half-changes			Gleasons		
			Sample distribution			Sample distributions		
			Ends	Center	One end	Ends	Center	One end
short	none (l)*	3	0.15	0.33	0.10	0.01	0.02	0.01
	none (h)*	3	0.17	0.22	0.10	0.01	0.02	0.01
	low	3	0.56	0.69	0.39	0.79	0.64	0.91
	high	3	1.59	1.48	0.67	1.47	1.03	0.88
medium	none (l)*	3	0.18	0.23	0.12	0.03	0.05	0.01
	none (h)*	3	0.10	0.36	0.08	0.02	0.01	0.01
	low	3	0.32	0.28	0.17	0.50	0.38	0.45
	high	3	1.02	1.18	0.54	0.92	0.91	0.96
long	none (l)*	3	0.08	0.15	0.08	0.04	0.07	0.03
	none (h)*	3	0.07	0.16	0.05	0.07	0.07	0.04
	low	3	0.14	0.27	0.08	0.32	0.28	0.25
	high	3	0.48	0.41	0.19	0.75	0.71	0.67

* The first noise-free gradient was used to simulate the low noise gradient; the second was used for the high noise gradient.

butions discussed, given the initial assumption that samples are evenly spaced along the gradient. Successful recovery, or convergence, was measured as the ratio of: 1) the mean absolute difference between locations of the samples after rescaling and their true, underlying positions to 2) the mean absolute difference between locations of the distorted sample positions and the true positions. Thus, a ratio of convergence of 0.0 indicates that true sample positions have been perfectly revealed by gradient rescaling; a ratio of 1.0 indicates the gradient rescaling did not offer any advantage over the naive, even spacing. As with estimating beta diversity, the gleasons procedure did well on the noise-free gradient of all lengths and with all underlying sample distributions. Sample positions were recovered less well with the low and high noise gradients, although ratios of convergence were lower for low noise and longer gradients. Performance on the short, noisy gradients was unsatisfactory; in these cases gradient rescaling produced little or no improvement on the initial even spacing. Underlying sample distribution seemed to have little overall effect on the ratio of convergence with the gleasons procedure.

The half-changes procedure on the noise-free gradients adequately recovered the true sample positions, but still did not do as well as did the gleasons procedure. With all low noise gradients and the high noise, long gradients, however, the half-changes did well or very well at converging to the underlying sample distribution. Convergence was better with the longer gradients. In contrast to the results with gleasons, the particular form of the underlying sample distribution did make a significant difference in the ability of the half-changes procedure to recover sample positions. In general, the distribution with samples clustered at one end had the best ratios of convergence, especially for the high noise gradients.

The number of samples collected to represent a study system should have some effect on the success of analysis. Many workers attempt to collect as many samples as can reasonably be accomplished. For estimating beta diversity, however, intermediate sampling ($n = 15$) gave the best overall results (Table 3). The sparsest sampling ($n = 3$) did the worst at estimating beta diversity and intensive sampling ($n = 57$) was not advantageous. With noisy gradients best estimation of beta diversity was accomplished by sampling short gradients less intensively and by sampling long gradients more in-

Table 3. Accuracy of estimation of beta diversity by the gradient rescaling technique with different numbers of samples. The accuracy index is the average proportional error: average of estimate-true/true. Very short, medium and very long refer to the relative lengths of the simulated gradients (approx. 0.7, 3.0, and 10.0 half-changes, resp.). Noise levels correspond to internal associations of 100% and 80%. For each sampling intensity samples were spaced evenly.

Gradient length	Noise level	n	Index of accuracy				
			Number of samples				
			3	8	15	29	57
			Half-changes				
very short	none	3	0.22	0.18	0.19	0.19	0.20
	moderate	3	0.20	0.31	0.51	0.69	0.67
medium	none	3	0.20	0.15	0.14	0.13	0.18
	moderate	3	0.36	0.13	0.12	0.32	0.34
very long	none	3	0.19	0.13	0.08	0.10	0.14
	moderate	3	0.24	0.15	0.10	0.10	0.12
			Gleasons				
very short	none	3	0.10	0.02	0.01	0.01	0.01
	moderate	3	0.37	0.72	0.85	0.68	1.13
medium	none	3	0.35	0.12	0.07	0.01	0.17
	moderate	3	0.45	0.15	0.26	0.37	0.89
very long	none	3	0.75	0.24	0.15	0.16	0.28
	moderate	3	0.75	0.24	0.23	0.37	0.83

tensively, indicating an optimal sampling intensity of perhaps 3 samples per beta diversity unit.

Of the gradient characteristics varied in the evaluation – beta diversity, noise level, and sample spacing – noise level had the largest effect on the performance of gradient rescaling, indicating the importance in field studies of using sampling techniques and methods of data summarization that minimize noise. Gradient length was not an important factor except when high noise levels occurred in short gradients; this combination produced poor results. All but extremes of sampling intensity resulted in good performances by gradient rescaling. From these evaluation results we conclude that beta diversity and underlying sample positions can be accurately measured by the technique of gradient rescaling.

Comparisons with other methods

During this study we converted Whittaker's (1960) original method of measuring beta diversity to a more objective linear regression technique (see Computations), which successfully captured the rationale of the original (R. H. Whittaker, personal communication). Like our technique of gradient rescaling, Whittaker's method measures beta diversity, but unlike gradient rescaling it was not designed to give information on proper sample positions with respect to the gradient or on rates of ecological change at particular sample positions. Our formalization of Whittaker's method was applied to each of the simulated gradients discussed above, and the beta diversity results compared to the performance of the half-changes and gleasons procedures of gradient rescaling. As expected, Whittaker's method performed best with those gradients in which the samples were evenly spaced and less well with distorted sample distributions. In general, the half-changes procedure of gradient rescaling was superior to Whittaker's method except when sampling intensity was very high. Whittaker's method usually did better than the gleasons procedure at estimating beta diversity for gradients with noise but did appreciably less well than the gleasons procedure with noise-free gradients.

In gradient rescaling, the effects of errors of estimation of internal association are compounded because the half-changes and gleasons procedures both calculate overall turnover by accumulating individual turnover values between samples. This compounding of errors is responsible for the reduced accuracy of gradient rescaling at the highest sampling intensity (Table 3). Whittaker's method, in contrast, examines only overall turnover and should be less sensitive to errors in estimating *IA*. Despite this sensitivity of gradient rescaling, the half-changes procedure, in particular, calculates the true total turnover better than does Whittaker's method when sample spacing is not uniform.

Detrended correspondence analysis (DCA; Hill & Gauch, 1980) is a recently developed ordination technique, derived from reciprocal averaging (Hill, 1973). Some of the issues which we have addressed in this paper – calculating gradient length and rescaling gradients – are also dealt with by DCA. In a subsequent paper we will evaluate the performance of DCA in these two areas, and will compare DCA with our gradient rescaling technique.

Field example

We applied the gradient rescaling technique to field data on the distribution of tree species along an elevation gradient on quartz diorite substrate in the Siskiyou Mountains of Oregon, U.S.A. (Whittaker, 1960). The data were densities of stems greater than 1 cm dbh (diameter at breast height) within 6 elevation belts, which were centered at 610 m, 915 m, 1 220 m, 1 525 m, 1 800 m, and 2 025 m. Thirty-seven tree species were recorded.

These data of tree densities within elevation belts were analyzed using the gradient rescaling procedures described above. Values for beta diversity, rescaled sample positions and corrected rates of ecological change were obtained for both gleason and half-change units. Results for gleasons and half-changes closely paralleled one another and only the half-change results will be reported here.

When measured by our gradient rescaling method the Siskiyou elevation gradient has a beta diversity of 4.0 *HC*. In contrast, Whittaker's (1960) original methods yield beta diversity values of ca 6.9 *HC*, an apparently large overestimation. Species turnover of 4.0 *HC* indicates a significant amount of ecological change over the elevation change of 1 415 m. The comparison of beta diversity along the Siskiyou elevation gradient with that of other systems must await the availability of accu-

rate beta diversity figures from the analyses of other community gradients.

The interpretation of species distribution curves depends on the scaling used for the gradient. Gradient rescaling of the Siskiyou data produced small but important shifts in sample positions (Table 4). Shifts for samples 3 and 4 were the largest, both approximately 10% of the total gradient length. The mean absolute displacement of the half-change scale from the elevation scale was 6.5%. In Figure 2 the density of stems of *Abies concolor* and *Pseudotsuga menziesii*, the most abundant tree species in

Table 4. Positions of samples along the Siskiyou Mountains elevation gradient with respect to elevation (m) and, after rescaling, with respect to half-changes. The mean absolute displacements for the two sets of sample positions is 6.5%.

Elevation belt	Sample positions			
	Elevation (m)	Half-changes	Elevation (%)	Half-changes (%)
1	610	0.0	0.0	0.0
2	915	0.68	21.6	16.8
3	1220	1.37	43.1	33.8
4	1525	2.20	64.7	54.4
5	1800	3.34	84.1	82.6
6	2025	4.05	100.0	100.0

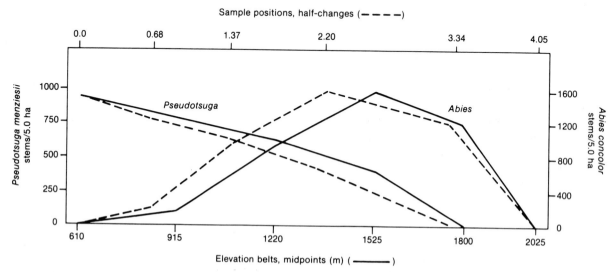

Fig. 2. Distribution curves of *Abies concolor* and *Pseudotsuga menziesii* along an elevation gradient in the Siskiyou Mountains, Oregon, U.S.A., with respect to meters of elevation (——) and to half-changes (---).

the Siskiyou data set, are plotted against both elevation and half-changes. The curves against elevation are skewed to higher elevations; the curve of *Abies concolor* against half-changes is more symmetrical and slightly more broad. Niche breadth of *Abies concolor,* measured as the standard deviation of the distribution curve, is 13% wider with the half-change scaling than with the elevation scaling.

Perhaps the most interesting and significant results of this analysis of the Siskiyou elevation gradient are the calculated rates of eological change through the elevation belts (Fig. 3). Rates of change range between 2.2 *HC*/1000 m and 3.7 *HC*/ 1000 m. Rates are generally higher through the higher elevation belts, indicating that environmental change with elevation is more biologically significant at higher elevations than at lower. The greatest rate of change is through the 1800 m elevation belt. This elevation corresponds to the broad subalpine-montane border as indicated by Whittaker (1960, p. 303). Our gradient rescaling technique has given a quantitative corroboration for Whittaker's field observations.

Discussion

We expect the technique of gradient rescaling presented here will improve a variety of investigations into the nature of biotic communities. Gradient rescaling furnishes accurate and robust estimates of beta diversity, which can be used in the comparison of ecosystems from different geographical or biological regions. Rates of ecological change computed in gradient rescaling can be used to indicate the presence (or absence) of ecotones, either in space or through time. Gradient rescaling should be helpful also in analyses of species distributions. Specifically, measures of niche breadth and overlap can now be calculated with respect to axes rescaled to constant rates of species turnover, in effect weighting for the ecological separation between samples. Because breadth would then be measured in objective and universal units of beta diversity, direct comparisons of species distributions among systems are possible.

Although gradient rescaling is particularly well suited to the analysis of vegetation (e.g., Marks & Harcombe, 1981; Wilson, 1982), the technique is relevant to research in other ecological systems as well. Trends in faunal composition through space (e.g., James, 1971; Terborgh, 1971; Cody, 1975), animal niche relations along resource axes (e.g., Pianka, 1973; Cody, 1974), and patterns of community change through time, including phenology of flowering and pollination (e.g., Mosquin, 1971;

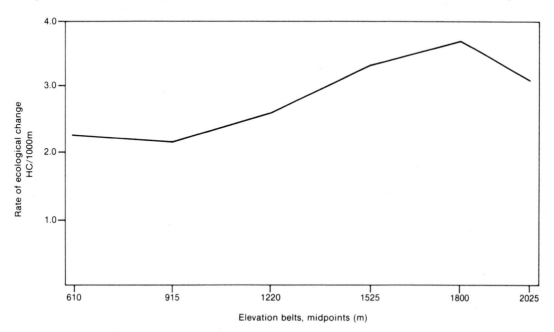

Fig. 3. Rate of ecological change through 6 elevation belts in the Siskiyou Mountains, Oregon, U.S.A.

Reader, 1975; Bratton, 1976) and succession after disturbance (Austin, 1977; Christensen & Peet, 1983) may all be examined with the tools of gradient rescaling.

Some differences were apparent in the use of gleasons and half-changes as units of beta diversity. Gleasons as units of compositional change have several appealing advantages. The measure itself (equation 4) is simple and straightforward. Derived directly from a notion of species turnover rates, gleasons are well suited for examining rates of compositional change along gradients. Using gleasons has the disadvantage of retaining only information of the similarity of samples adjacent on the gradient; using half-changes as units of compositional change utilizes information about the similarities of some non-adjacent samples as well. In general, the half-changes procedure should be superior when moderate or high levels of field noise are present.

Gradient rescaling can be applied to data in which several factors are important in determining species abundances. Each important axis must be identified, using multivariate techniques, such as factor analysis or ordination. Then data dimensionality is reduced by taking the average abundance of each species along segments of single axes, one axis at a time. This averaging is analogous to marginal distributions of multivariate statistics. Gradient rescaling can then be applied to each reduced axis, and results combined into a multi-dimensional representation. The ecological lengths, or beta diversities, of individual axes, for example, would indicate the relative importance of those axes to overall ecological structure. As with most approaches to multivariate data, care must be taken to identify axes of variation that are mutually orthogonal.

Gradient rescaling is not useful for the analysis of all forms of community data. Several assumptions and prerequisites are necessary for its proper use. Samples must be orderable along axes of compositional or environmental variation. These axes can be obtained either by direct observations (e.g., along gradients of elevation, soil nutrient status, etc.), or by indirect analysis (e.g., by ordination). When a gradient structure is inappropriate – for example, when biogeographical phenomena (e.g., migration) or historical effects (e.g., disturbance by fire) create discontinuities in species distributions –

methods of analysis other than gradient rescaling must be employed.

Samples must also be of sufficient size and number to adequately represent gradient conditions. Multi-dimensional data require larger numbers of samples. Poor sampling procedures can contribute to high levels of noise and significantly reduce the accuracy of gradient rescaling results. In particular, when field noise is high and gradient length is short, gradient rescaling (and probably any other technique) is not successful. Because the technique performs best at intermediate sampling intensities, it is often possible to reduce noise for purposes of the analysis by pooling replicate samples. For most studies with adequate sampling along major gradients of environmental or compositional variation, gradient rescaling can be a powerful tool for ecological analysis.

References

Austin, M. P., 1976. On non-linear response models in ordination. Vegetatio 33: 33–41.

Austin, M. P., 1977. Use of ordination and other multivariate descriptive methods to study succession. Vegetatio 35: 165–175.

Beals, E. W., 1969. Vegetation change along altitudinal gradients. Science 165: 981–985.

Berington, P. R., 1969. Data Reduction and Error Analysis for the Physical Sciences. McGraw-Hill, New York.

Bratton, S. P., 1975. A comparison of the beta diversity functions of the overstory and herbaceous understory of a deciduous forest. Bull. of the Torrey Bot. Club 102: 55–60.

Bratton, S. P., 1976. Resource division in an understory herb community: responses to temporal and microtopographic gradients. Am. Nat. 110: 679–693.

Christensen, N. L. & Peet, R. K., 1983. Convergence during secondary forest succession. J. Ecol. (in press).

Cody, M. L., 1974. Competition and the Structure of Bird Communities. Princeton University Press, Princeton. 318 pp.

Cody, M. L., 1975. Towards a theory of continental species diversity. In: Cody, M. L. & Diamond, J. M., (eds.), Ecology and Evolution of Communities. pp. 214–257. Harvard Univ. Press, Cambridge.

Colwell, R. K. & Futuyma, D. J., 1971. On the measurement of niche breadth and overlap. Ecology 52: 567–576.

Gauch, H. G., Jr., 1973. The relationship between sample similarity and ecological distance. Ecology 54: 618–622.

Gauch, H. G., Jr., Chase, G. B. & Whittaker, R. H., 1974. Ordination of vegetation samples by Gaussian species distributions. Ecology 55: 1382–1390.

Gauch, H. G., Jr. & Whittaker, R. H., 1972. Coenocline simulation. Ecology 53: 446–451.

Gleason, H. A., 1926. The individualistic concept of the plant association. Bull. of the Torrey Bot. Club 53: 7–26.

Goodall, D. W., 1978. Sample similarity and species correlation. In: Whittaker, R. H., (ed.), Ordination of Plant Communities Junk, The Jague.

Hill, M. O., 1973. Reciprocal averaging: an eigenvector method of ordination. J. Ecol. 61: 237–249.

Hill, M. O. & Gauch, H. G., Jr., 1980. Detrended correspondence analysis, an improved ordination technique. Vegetatio 42: 47–58.

Hutchinson, G. E., 1958. Concluding remarks. Cold Spring Harbor Symposium on Quantitative Biology 22: 415–427.

Ihm, P. & Groenewoud, H. van, 1975. A multivariate ordering of vegetation data based on Gaussian type gradient response curves. J. Ecol. 63: 767–777.

James, F. C., 1971. Ordinations of habitat relationships among breeding birds. Wilson Bull. 83: 215–236.

Janzen, D. H., 1967. Why mountain passes are higher in the tropics. Am. Nat. 101: 233–249.

Levins, R., 1968. Evolution in Changing Environments. Mon. in Population Biology 2, Princeton University Press, Princeton. 120 pp.

Marks, P. L. & Harcombe, P. A., 1981. Forest vegetation of the Big Thicket, southeast Texas. Ecol. Monogr. 51: 287–305.

McNaughton, S. J. & Wolf, L. L., 1970. Dominance and the niche in ecological systems. Science 167: 131–139.

Mohler, C. L., 1981. Effects of sample distribution along gradients on eigenvector ordination. Vegetatio 45: 141–145.

Mohler, C. L., 1983. Effects of sampling pattern on estimation of species distributions along gradients. Vegetatio, in press.

Mosquin, T., 1971. Competition for pollinators as a stimulus for the evolution of flowering time. Oikos 22: 398–402.

Peet, R. K., 1978. Forest vegetation of the Colorado front range: patterns of species diversity. Vegetatio 37: 65–78.

Pianka, E. R., 1973. The structure of lizard communities. Ann. Rev. of Ecology and Systematics 4: 53–74.

Pielou, E. C., 1972. Niche width and overlap: a method for measuring them. Ecology 53: 687–692.

Reader, R. J., 1975. Competitive relationships of some ericads for major insect pollinators. Can. J. Bot. 53: 1300–1305.

Terborgh, J., 1971. Distribution on environmental gradients: theory and a preliminary interpretation of distributional patterns in the avifauna of the Cordillera Villcabamba, Peru. Ecology 52: 23–40.

Whittaker, R. H., 1960. Vegetation of the Siskiyou Mountains, Oregon and California. Ecol. Monogr. 30: 279–338.

Whittaker, R. H., 1965. Dominance and diversity in land plant communities. Science 147: 250–260.

Whittaker, R. H., 1967. Gradient analysis of vegetation. Biol. Rev. 42: 207–264.

Whittaker, R. H., 1972. Evolution and measurement of species diversity. Taxon 21: 213–251.

Whittaker, R. H., 1975. Communities and Ecosystems, 2nd ed. Macmillan, New York. 385 pp.

Whittaker, R. H., Levin, S. A. & Root, R. B., 1973. Niche, habitat, and ecotope. Am. Nat. 107: 321–338.

Wilson, M. V., 1982. Microhabitat influences on species distributions and community dynamics in the conifer woodland of the Siskiyou Mountains, Oregon. Ph.D. Thesis, Cornell University, Ithaca, N.Y.

Wilson, M. V. & Mohler, C. L., 1980. GRADBETA – A FORTRAN program for measuring compositional change along gradients. Ecology and Systematics, Cornell University, Ithaca, N.Y. 51 pp.

Accepted 30.3.1983.

A new approach to the minimal area of a plant community*

P. Dietvorst, E. van der Maarel & H. van der Putten**
*Institute of Ecological Botany, University of Uppsala, Box 559 S 75122 Uppsala, Sweden****

Keywords: Coenotic molecule, Dominance-diversity relation, Grassland, Heathland, Minimal area, Salt marsh, Similarity, Species richness

Abstract

Current methods for the determination of the minimal area of a phytocoenosis are compared. A new method is described. It is based on the increase in average quantitative similarity between plots in series of nested plots of increasing size in relation to species richness and dominance relations. A simulation model composed of coenotic molecules sensu Moravec is developed from which these relations can be deduced. Both a quantitative and a qualitative minimal area are derived in this way. The analytical minimal area is then defined in relation to the largest of the two minimal area values, because in this way both the species richness and the dominance relations will be represented.

Some examples of grasslands and heathlands are presented. Some further investigations are suggested to improve the approach and to link it with pattern and dominance-diversity studies.

Introduction

The concept of minimal area is important in phytosociology. It is described in classical and modern textbooks (e.g. Braun-Blanquet, 1964; Westhoff & van der Maarel, 1978) usually as the smallest area on which the species composition of a plant community is adequately represented (Mueller-Dombois & Ellenberg, 1974). This approach starts from the recognition of vegetation units in the field (cf. Westhoff & van der Maarel, 1978), and thus the concept is bound to the typological approach to

vegetation. However, there is still no fully satisfactory way to determine the minimal area, whereas the concept itself remains ambiguous on two points:
a. whether it should refer to the abstract plant community or to the concrete stand, and
b. whether it should refer to the qualitative species composition.

Westhoff (1951) recognised a clear distinction between the concrete and the abstract plant community. A parallel distinction should be made between analytical minimal area and synthetical minimal area (van der Maarel, 1966, 1970). Westhoff & van der Maarel (1978) define the analytic minimal area as the area that 'should be defined for a stand under study as a representative area, e.g. as an adequate sample of species of regular occurrence in the stand'. Clearly the analytical minimal area has to be determined on the basis of actual structural and compositional properties in the field. As Moravec (1973) stated: 'A certain minimum part of the biocoenosis contains the basic species which build the "body" of the biocoenosis repeating more or less

* Nomenclature of plant species follows Heukels-van Oost-stroom (1977), of syntaxa Westhoff & den Held (1969) and de Smidt (1975)

** We thank Dr. Jaroslav Moravec, Průhonice, for discussions, Drs. Onno van Tongeren, Nijmegen for help with calculations and Cyril Liebrand and Leo Mertens, Nijmegen for help with salt marsh analyses and discussions on dominance-diversity relations.

*** Research was carried out when the authors were at the Division of Geobotany, University of Nijmegen.

regularly throughout a certain environment as a "coenotic molecule'".

The synthetical minimal area could then be determined as the (average) area on which all species occur which were found to be a member of the characteristic species combination as derived from the table rearrangement and subsequent syntaxonomical comparison (Westhoff, 1951; Westhoff & van der Maarel, 1978). Of course the relevés used for the elaboration of the characteristic species combination should each satisfy an analytical minimal area criterion. Serious problems arise with this approach because the real character-species (i.e. those appearing exclusively in the community type under investigation) need not to be constant and they may be very much scattered at any site within their distribution area, and thus escape from being stored within the analytical minimal area plot.

A more realistic approach would be to take the average (or some other characteristic value e.g. median or maximum value) of the analytical minimal area determinations for sufficiently developed stands which could be assigned to one phytocoenon. One could also think of a minimal number of relevés needed to describe a phytocoenon. This matter has been discussed by Tüxen (1970a) and will be treated in a companion paper. In this paper we will concentrate on the analytical minimal area.

The distinction between qualitative and quantitative minimal area was suggested by Meyer Drees (1954) for forestry purposes and taken over by Beeftink (cf van der Maarel, 1966) for general phytosociological use. In connection with the definition presented above we could give two different definitions by adding the words qualitative and quantitative to 'species combination'. Because the characterization of a phytocoenon is usually based on the quantitative occurrence of species, the quantitative minimal area concept should be central. As most minimal area approaches are based on the qualitative species combination, this represents a re-orientation.

The purpose of this paper is to compare and evaluate current methods of minimal area determination against the background of the foregoing considerations. A new approach is suggested which avoids some of the difficulties inherent in previous methods.

Some current methods for the determination of the minimal area

Five methods of minimal area determination can be mentioned:
(1) Interpretation of species-area curves, (2) Interpretation of frequency-area curves, (3) Species representation, (4) Similarity analysis and (5) Pattern representation.

Interpretation of species-area curves (Braun-Blanquet & Jenny, 1926; Braun-Blanquet, 1964; Cain, 1938; Cain & Castro, 1959) is the original and still frequently recommended method (e.g. Mueller-Dombois & Ellenberg, 1974). The main problem with the method is that the species-number is seldom found to reach a level of saturation (e.g. Hopkins, 1955; van der Maarel, 1970; Peet, 1974), notwithstanding evidence and views collected by Tüxen (1970a).

Methods such as Cain's tangent method (see e.g. Mueller-Dombois & Ellenberg, 1974) are too subjective to be acceptable (Barkman, 1968; van der Maarel, 1966, 1970; Moravec, 1973).

Frequency-area curves have been taken as a basis for minimal area determination in the Uppsala approach (Du Rietz et al., 1920; Du Rietz, 1930). Although these curves tend to flatten somewhat more strongly than normal species-area curves, they remain difficult to interpret (van der Maarel, 1966). Barkman (1968) suggested the use of a 'differential curve' in which the increase in species number, $\triangle S$ is plotted in stead of S in the area. This method can, as Barkman demonstrated, be used for the detection of separate minimal areas for different synusiae which may occur in a phytocoenosis. The problem here is that the significance of drops in the curve is difficult to test (van der Maarel, 1970).

Van der Maarel (1966) observed that there is a relation between the minimal area and the size of the entire homogeneous stand under study. On the basis of observations in a dune grassland he proposed as a pragmatical solution to take as minimal area the area necessary to record 80% of the species found on the total site. He also applied this 80% criterion to frequency data (cf Du Rietz's approach). Werger (1972) followed a similar approach for communities with large areal extension and suggested an area with a proportion of 50–55% of the species found on one ha of the phytocoenosis.

Goodall (1954) seems to have been the first to

attempt application of statistical procedures for minimal area determination. He analyzed plots of increasing size and at different distance to each other. He supposed that if the sample plot reaches a certain size, the variance of cover values of individual species will no longer depend on the distance of individual sample plots in a homogeneous stand. However, he could not find such a sample size. This made Goodall to be doubtful as to the reality of the minimal area concept (see also Moravec, 1973).

Similarity analysis as a basis for minimal area determination was suggested by Gounot & Calléja (1962), who analyzed series of randomly located plots of increasing size. They defined the minimal area as the sample plot size for which the mean pairwise floristic similarity (Sørensen coefficient) between four sample plots exceeds a certain value; they took 80% in the case of the phytocoenosis investigated, a *Brachypodium ramosum* grassland. The value of 80% is realistic to the extent that similarity values of over 80% are not likely to be expected in samples from one and the same community (Curtis, 1959; Kortekaas *et al.*, 1976; see also Gauch & Whittaker, 1976). However, for other communities lower levels could be more realistic and the question

is how to determine the level in a less arbitrary way.

An alternative approach by Gounot & Calléja (1962) is based on the distribution of values of similarity coefficients between sample plots of the same size. The minimal area is said to be reached if this distribution tends to become normal. This method is also arbitrary because the normality of the distribution depends on the number of sample plots investigated. Moreover, from a statistical point-of-view it is not possible to determine the normality of a distribution when the variables are not independent from each other, and floristic similarity coefficients as obtained here are dependent variables.

Moravec (1973) also investigated the use of similarity measurements. He supposed that when the minimal area is reached, the basic set of species of the phytocoenosis is present in all sample plots; however species of lower frequency will also be present and it is due to them that the mean floristic coefficient will not reach 100%. By further increasing the sample plot size the number of species common to all plots increases but there is a simultaneous increase in the number of species of low frequency, so that the mean floristic similarity will not continue to rise. Moravec found that the mean floristic similarity still increases or oscillates after the minimal area has been reached. Hence the same problem arises as with the species area curve viz. 'what is the exact point where the curve flattens'.

Both Gounot & Calléja and Moravec used presence-absence data for their similarity analyses and thus their results are relevant only as far as the qualitative minimal area is concerned. In conclusion we may state that of the approaches discussed so far the similarity method is the only realistic one. Hence the 'similarity-area' approach will be a central element in our approach. However the method should be based on quantitative data, i.e. cover-abundance estimates, instead of on presence-absence data. (After we decided to adopt this approach we learned that Roux & Rieux, 1981 had applied a similar approach, but had followed Gounot & Calléja in adopting the arbitrary 80% similarity level).

As to method (5); pattern representation, van der Maarel (1966, 1970) and Werger (1972) suggested that structural characteristics of the phytocoenosis should be taken into account. Barkman (1968) suggested a relation between minimal area and the mean area occupied by one plant-individual.

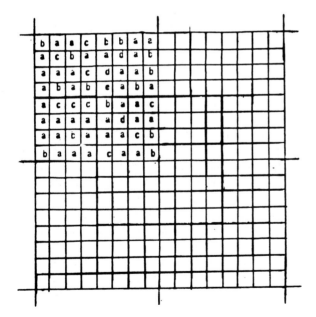

Fig. 1. Phytocoenosis model with coenotic molecules consisting of 64 atoms representing species or open spaces. The example shows 5 species a, with 37, b with 14, c with 8, d with 4 and e with 1 elements. Four coenotic molecules are shown of which one is filled in. Configuration within molecules is at random.

Another possible relation may exist between minimal area and the sociability of the species involved. Van der Maarel (1966, 1970) developed an exclusion index which is calculated as the sum of products of species cover-abundance and sociability according to the Braun-Blanquet scale. This index appeared to be highly correlated with minimal area. More generally, minimal area seems to increase with increasing vertical complexity (e.g. Westhoff & van der Maarel, 1978). Stimulated by these observations we start with a more systematic treatment of the relation between minimal area and structural parameters.

A phytocoenosis model for a similarity based minimal area approach

Outline of the model

In order to relate the similarity – area curve to structural parameters, notably quantitative relations between species, we adopt a structural model of the phytocoenosis, based on the coenotic molecule in the sense of Moravec (1973). Each coenotic molecule consists of a number of atoms, each atom representing a plant unit of one particular species. A coenotic atom can be related to an 'atomic area', the mean area on which one and not more than one plant unit occurs. This area is identical with the sum of the mean area of a plant unit and the mean bare space around it. It is related to the mean area in the sense of reciprocal density for randomly distributed plant units (Greig-Smith, 1964). The model we start with contains 81 coenotic molecules each with 64 randomly filled atoms (Fig. 1).

The phytocoenosis is then viewed as a macro-molecule composed of coenotic molecules, in which the structure, i.e. the quantitative relations between the species, will be the same, but in which the composition of the less abundant species may be different. Minimal area is related to coenotic molecule as follows: it is an area comprising at least one coenotic molecule. If for each structure, i.e. type of quantitative relation between species, a level of similarity between coenotic molecules can be found, we can determine minimal area as the area where the similarity level is reached that belongs to the coenotic molecule of the model with the same structure as the phytocoenosis under investigation. Of course

natural phytocoenoses are not composed of such discontinuous units. As a first adaptation to this reality we perform a similarity – area analysis on the model by starting with one atom area and building nested plots around each of them. As a result of the random placing of atoms the similarity between plots of small sizes will be much lower than 100% and the 100% level may be reached only beyond the area corresponding to one coenotic molecule.

If one atom type would be dominating in the molecule the similarity level will increase more rapidly; in the case of one species with complete dominance the minimal area would be reached at the atom size. To examine the dependence of the similarity area curve on the dominance relations within the molecule, while keeping the random placing of atoms, we varied both the number of species and the dominance relations. In total 17 variants of the model were used with species numbers ranging from 1 to 21 and dominance relations ranging from strong dominance of one species to equal proportions. To simulate sparse vegetation randomly placed empty atoms were introduced in a number of models. Dominance values in the models are based on the % of occupied fields in the molecule. In order to relate the model to the field situation in which usually Braun-Blanquet figures are recorded, the % occupancy figures were transformed into the ordinal transform scale of van der Maarel (1979) as follows:
number of atoms representing a species

| 1 | 2 | 3 | 4–7 | 8–15 | 16–31 | 32–47 | 48–64 |

corresponding ordinal transform value

| 2 | 3 | 4 | 5 | 6 | 7 | 8 | 9 |

Note that the lowest possible occurrence in the model, 1 out of 64, is considered equivalent to Braun-Blanquet symbol +. This relates to a practical simplification in the field analyses to be discussed later: similarity calculations are based on Braun-Blanquet values of + and higher and the very rare occurrences are not considered.

Minimal area determination in the phytocoenosis models

Standard minimal area determinations were made for each of the 17 models. Each determination started in the middle of one coenotic molecule. A series of areas was then analyzed corresponding with resp. 1, 2, 4, 8, 16, 32, 64, 128, 256 and 384

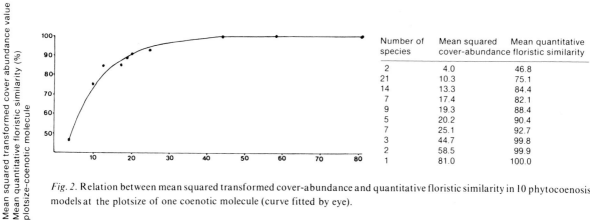

Fig. 2. Relation between mean squared transformed cover-abundance and quantitative floristic similarity in 10 phytocoenosis models at the plotsize of one coenotic molecule (curve fitted by eye).

Number of species	Mean squared cover-abundance	Mean quantitative floristic similarity
2	4.0	46.8
21	10.3	75.1
14	13.3	84.4
7	17.4	82.1
9	19.3	88.4
5	20.2	90.4
7	25.1	92.7
3	44.7	99.8
2	58.5	99.9
1	81.0	100.0

atoms. In each model 10 series were analyzed.

Similarities between all pairs of samples of all sizes involved were calculated both for quantitative and for presence-absence data, using the similarity-ratio SR (see e.g. Westhoff & van der Maarel, 1978), which reduces to Jaccard's formula for presence-absence data. The resulting curves all reached the maximum possible average similarity of 100%, but at different areas. If average similarity on the size of the coenotic molecule was less than 100% this is explained as resulting from differences between coenotic molecules due to the different distribution of atoms of one species respectively empty atoms. We then wished to correlate the different similarity values reached at the coenotic molecule area with some measure expressing the quantitative relations between the species. Amongst the measures we tested were Simpson's diversity index

(which is originally a dominance index) and the Berger-Parker dominance index (being the proportion of the most dominant species; see Southwood, 1978 for references and discussion). The best correlation was obtained with the mean squared ordinal transform value, to be called MSO.

We also tested the qualitative similarity relations and here the best correlation was found with the mean ordinal transform value, MO. Figures 2 and 3 present the respective relationships. It follows from Figure 2 that quantitative similarity values reach 100% already at the coenotic molecule size for values of MSO ≥ 40. Such MSO values are produced by any set of species with a low number of species each having a high score, i.e. at least ordinal transform 6. For the qualitative similarity level at coenotic molecule size the same holds: only with

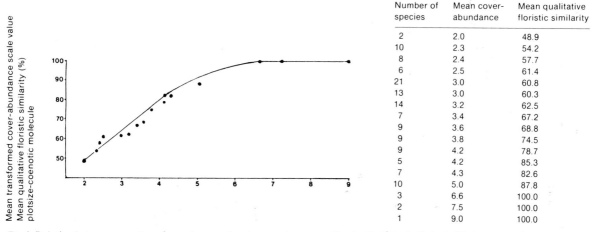

Number of species	Mean cover-abundance	Mean qualitative floristic similarity
2	2.0	48.9
10	2.3	54.2
8	2.4	57.7
6	2.5	61.4
21	3.0	60.8
13	3.0	60.3
14	3.2	62.5
7	3.4	67.2
9	3.6	68.8
9	3.8	74.5
9	4.2	78.7
5	4.2	85.3
7	4.3	82.6
10	5.0	87.8
3	6.6	100.0
2	7.5	100.0
1	9.0	100.0

Fig. 3. Relation between mean transformed cover-abundance and mean qualitative floristic similarity in 17 phytocoenosis models at the plotsize of one coenotic molecule (curve fitted by eye).

Table 1. Species composition of investigated phytocoenoses; transformed Braun-Blanquet values 1 are deleted.

Type number	1					2			3						4		5		6	7	8	9
Location / Species	Otersum	Oeffelt 4	Oeffelt 3	Oeffelt 2	Oeffelt 1	Nieuw Bergen 3	Nieuw Bergen 2	Nieuw Bergen 1	Mook 4	Mook 3	Molenhoek 2	Molenhoek 1	Mook 2	Mook 1	Molenhoek 3	Mook 5	Terschelling 2	Terschelling 1	Terschelling 3	Terschelling 4	Terschelling 5	Erlecom
Festuca rubra ssp. rubra	7	8	9	5	7																	
Anthoxanthum odoratum	6	6	4	4	6														2			
Galium verum	3	5	5	4	5																	
Luzula campestris	2	3	4	2	3																	
Carex arenaria	3	2		4		3														3		
Hieracium pilosella	3	3	3	4	3												2		2			
Hypochaeris radicata	2	2	2	3																		
Rumex acetosella	3	2		4					2				2									
Ononis repens	2			2																		
Teesdalia nudicaulis	2																					
Aira praecox	3																					
Achillea millefolia	2	3	3	2	3																	
Agrostis tenuis	2	2	2	4	5						3											
Poa pratensis	2	2	3	2																		
Cerastium arvense	2	3	3	3	3																	
Plantago lanceolata		2	2	2																		
Lotus corniculatus			2		2																	
Ranunculus bulbosus			2																			
Ornithopus perpusillus				3																		
Erodium cicutarium				2																		
Trifolium repens				2																		
Campanula rotundifolia					4																	
Carex caryophyllea					2																	
Spergularia morisonii						3	3	3														
Corynephorus canescens						2	2	2														
Calluna vulgaris									7	8	8	8	3	3	2	3	8	7	5			
Molinea caerulea									3	3	2	6	4	3								
Carex pilulifera									2	3	3		2	3						9		
Deschampsia flexuosa									3	3	3		4	5							9	
Betula pendula									2	2				2								
Sieglingia decumbens										2			2	2								
Cuscuta epithymum										3												
Sarothamnus scoparius											2		3									
Genista anglica										2					3							

Rubus fruticosus									
Frangula alnus	2								
Empetrum nigrum	2	2							
Erica tetralix			7	7	8	8			
Salix repens			5	4	7	6			
Festuca ovina ssp. tenuifolia			5	4	2				
Carex trinervis			2	2	2				
Oxycoccus macrocarpos			2	3	4	3			
Phragmites australis			2		2	6			
Genista tinctoria			2						
Luzula multiflora			2						
Potentilla erecta			2						
Carex nigra			2						
Myrica gale					7	7			
Calamagrostis epigejos					3	3			
Hydrocotyle vulgaris					3	3			
Limonium vulgare						3			
Plantago maritima						5			
Juncus gerardii						6			
Festuca rubra f. litoralis						7			
Armeria maritima						4			
Glaux maritima						6			
Triglochin maritima						3			
Carex extensa						2			
Agrostis stolonifera						5			
Salicornia europaea						3			
Suaeda maritima						3			
Puccinellia maritima						2			
Halimione portulacoides						2			
Elytrigia repens							9		
Poa trivialis							7		
Alopecurus geniculatus							5		
Rumex obtusifolius							2		

♦ For the description see text

few species and high average ordinal transform is the maximum of 100% reached. Here for lower average ordinal transform values the similarity level remains considerably below 100%

Field data

Description of sites

Data were collected in some grasslands and heathlands near Nijmegen, and in dune heathlands and salt marshes on the Wadden island of Terschelling, (the Netherlands). These were all one- or two layer vegetation types, usually without a moss layer (if present the species were recorded, but not included in the analysis). Extension of the method to multilayered communities will follow. The species composition of the phytocoenoses is shown in Table 1. They are classified syntaxonomically as follows:

1) Dry grassland phytocoenoses on riverdunes, order *Festuco-Sedetalia* R. Tüxen 1951. The stands Oeffelt 1. and 4. are meadows grazed by cattle. Oeffelt 2. and 3. are not grazed. Ottersum is grazed only slightly by horses.
2) Inland sanddune phytocoenoses of the association *Spergulo-Corynephoretum* (R. Tüxen, 1928) Libbert 1932 em., Passarge, 1960.
3) Heathland phytocoenoses of the association *Genisto anglicae-Callunetum* R. Tüxen, 1937. The heathlands are of different ages. The youngest heather is found in Mook 3 and 4. The cover of the vegetation is low because the heather has recently been burned and mowed. Moreover it is heavily grazed by sheep. The heather in Mook 1 is about two years old, in Mook 2 and Molen-

hoek 1 about three years and in Molenhoek 3 about ten years old.
4) Grassland phytocoenoses with *Deschampsia flexuosa* as the dominant species. This type comes close to the association *Genisto anglicae-Callunetum* of which it is a degenerated form. (Mook 5 and Molenhoek 3)
5) Heathland phytocoenoses of the association *Carici arenariae-Empetretum* R. Tüxen et Kawamura, 1975. (Terschelling 1 and 2)
6) Heathland phytocoenosis of the association *Empetro-Genistetum tinctoriae* Westhoff (1947) 1968 nom. nov. According to de Smidt (1975) this association should be included in the association *Carici arenariae-Empetretum*. (Terschelling 3)
7) Wet heathland phytocoenosis of the association *Empetro-Ericetum* Westhoff (1943) 1947. (Terschelling 4)
8) Saltmeadow phytocoenosis of the association *Juncetum gerardii* Warming 1906. (Terschelling 5)
9) Wet grassland phytocoenosis of the association *Rumici-Alopecuretum geniculati* R. Tüxen (1973) 1950. (Erlecom)

Minimal area analysis

A homogeneous stand of 1 000 m² or in some cases 500 m² was located within each phytocoenosis. In each 10 separate series of relevés were collected. Each series consisted of successively enlarged nested plots (Fig. 4). The series were systematically layed out in each phytocoenosis and they never overlapped. The distance between the starting relevés of adjacent series was kept constant (Fig. 4). The species were recorded using a transformed

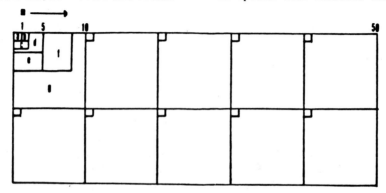

Fig. 4. Location of the sample plots in the standard minimal area determination.

Braun-Blanquet scale for cover abundance (Westhoff & van der Maarel, 1978) as follows:

Braun-Blanquet: r + 1 2m 2a 2b 3 4 5
transformation: 1 2 3 4 5 6 7 8 9

The size of the smallest plots investigated depended on the species pattern within the phytocoenosis. If the phytocoenosis had a fine-grained dispersion of species the sample plot series started with a relevé of 1 dm². If it had a coarse-grained dispersion of species the size of the smallest relevé was 1 m². In subsequent calculations the occurrences with Braun-Blanquet r were left out of consideration. Since they are especially species with an r-occurrence which enter the analysis at greater plot sizes our dominance measures, viz. mean squared transformed resp. mean transformed Braun-Blanquet values, are more stable.

Similarity analysis

Both the quantitative and the qualitative similarity values were calculated for each plot size. So for each plot size of a phytocoenosis we obtained two series of 45 pairwise similarity coefficients.

Results

Table 2 presents the similarity levels for different plot sizes in the various communities analyzed. Most series show a more or less regular increase in similarity with a maximum approached at a plot size well below the maximum plot size investigated. There are exceptions, e.g. Mook 4-qualitative. In general the quantitative series reach a maximum clearer and earlier, but they never reach 100% as many qualitative series do. In some phytocoenoses the 80% similarity level (Gounot & Calléja criterion) is already reached at the smallest plot size, in others it is reached only at 10–20 m². This illustrates the difficulties in using this criterion.

The similarity levels corresponding with the model indicated coenotic molecule size, as derived from interpreting Figures 2 and 3, vary from 0.02 m² to 40 m². In some cases the quantitative and qualitative levels differ considerably. Because of the low levels of MSO and MO, some of the coenotic molecule similarity levels are low, the lowest value being 46.1% in N. Bergen-quantitative. Such low values imply that plot descriptions referring to that size in the same phytocoenosis would be very dissimilar and hard to use. Apparently the similarity level drops if the vegetation is very open and moreover the MSO value is low (see also Fig. 2). In cases where a much higher similarity level is reached at somewhat larger plot sizes, as in the N. Bergen case, it would be unrealistic to adopt the MSO or MO based low similarity value. On the other hand, a low similarity level also on larger areas may be an indication for a heterogeneous phytocoenosis. However, this is not the case in our analyses.

The quantitative similarity levels corresponding with the model indicated molecule size are generally higher than the corresponding qualitative values, but the model indicated molecule sizes on a quantitative basis are generally larger than those on a qualitative basis, in the Erlecom grassland and some Terschelling dune heaths even much larger. Apparently all species are present on a relatively small area, but show differences in local density. In some phytocoenoses, notably the salt marsh the qualitative coenotic molecule is much larger. Here we must have a relatively even density of the dominant species.

A pragmatic minimal area determination.

It would seem realistic to take both the quantitative and the qualitative results into account and base a minimal area on the largest model indicated coenotic molecule size. According to what was said above we could take this size as an approximation. If on that size the average similarity would be less than 80%, we take the smallest area on which 80% is reached as minimal area. For instance, the Oeffelt 1 grassland has a quantitative molecule size of 1–2 m², a qualitative one of 2 m², but the qualitative similarity is only 63.8%. Thus we adopt 20 m² as analytic minimal area because at that size the qualitative similarity has increased to over 80%. The Erlecom grassland has a quantitative molecule size of 20 m² and a qualitative one of 0.1 m². The quantitative similarity is 98.1%, thus we adopt 20 m² as minimal area. Table 3 presents values arrived at in this way for all phytocoenoses studied. We take these values as minimal area values. These values are accompanied by the two coenotic molecule sizes as well as the results of determinations according to Moravec (1973), Gounot & Calléja (1962) and van der Maarel (1966).

Discussion

Range of minimal area values

The minimal area values range from 1 m² in the salt marsh to 40 m² in some of the heathlands. The model-indicated coenotic molecules sizes vary much more. Our values correspond rather well to the general values presented by Westhoff & van der Maarel (1978) for similar plant community types. (see Table 4).

Variation within one syntaxon

The phytocoenoses assigned to one particularly syntaxon (Table 1) may vary in minimal area size: in the *Genisto anglicae-Callunetum* from 2–40 m². This is easy to understand: The delimitation of syntaxa is largely based on character- and differential species and, if structural characteristics are used at all, on vertical structure. On the other hand, the criteria for the minimal area determination are related to the horizontal structure. This includes species richness, dominance relations and sociability.

Table 2. Mean floristic similarity (quantitative and qualitative) of sample plots of the same size for each phytocoenosis investigated. The areas on which the similarity levels corresponding with the model are reached are indicated with -. Mean squared ordinal transform, MSO, and mean ordinal transform, MO, values are added.

Typenumber♦♦	Mean quantitative floristic similarity				1				2		
Location	Oeffelt 1	Oeffelt 2	Oeffelt 3	Oeffelt 4	Ottersum'	N. Bergen 1	N. Bergen 2	N. Bergen' 3	Mook 1	Mook 2
Size (m²)										
0.01					58.2				51.8	51.6
0.02					69.7				85.4	65.2
0.04					72.1				93.9	60.7
0.1		41.6		36.4	75.7	26.9	27.4	49.4	91.2	68.6
0.2/0.25♦		59.1		42.7	80.3	45.4	41.8	59.0	90.7	77.1
0.4/0.5♦		51.6		48.0	80.0	54.6	46.6	63.7	90.3	76.6
1	67.2	79.4	80.0	55.8	84.7	59.9	46.1	70.9	89.1	78.5
2	76.5	85.4⌐	82.7	66.7	87.6⌐	66.8	62.8	73.3	90.4	78.2
4	76.9	88.3	84.0⌐	68.3	92.3	82.6⌐	68.3	84.0⌐	92.0	78.7
10	79.9	91.4	85.0	76.5	94.0	81.0	85.1	87.2	89.0	83.0
20	85.6	94.5	86.5	81.8⌐	95.3	88.5	89.8	85.3	88.2	85.4
40	85.4	95.6	89.5	86.3	95.7	93.3	91.8	89.9	91.6	87.4
100/50♦	91.0	94.7	90.3	♦86.3	95.5	96.0	96.0	91.1	95.4	92.7
MSO	10.8	15.1	14.9	9.9	17.7	7.3	6.5	6.5	14.0	15.4
	Mean qualitative floristic similarity									
Size (m²)										
0.01					53.5				54.4	49.1
0.02					58.1				90.0	51.7
0.04					60.4			-	73.7	51.7
0.1		47.6		42.3	62.9	34.1	36.4	71.1	61.3	51.7
0.2/0.25♦		59.4		53.0	65.6	57.8	59.1	73.3	57.0	51.7
0.4/0.5♦		63.9		54.1	67.8	71.8	57.6	73.3	56.7	47.3
1	59.2	72.8	66.3	66.1	82.8	73.7	60.6	76.7	63.7	46.3
2	63.8	83.7	71.8	73.0	89.4	74.4	71.2	82.2	66.3	41.6
4	67.2	90.3	77.5	74.7	88.2	84.4	71.2	82.2	65.9	46.5
10	75.2	96.7	79.6	77.8	95.8	93.3	100.0⌐	90.0	73.3	65.2
20	81.9⌐	96.7	84.8	85.2	98.2	93.3	100.0	90.0	80.0⌐	66.2
40	86.6	100.0	85.0	94.8	98.2	100.0	100.0	100.0	89.3	73.4⌐
100/50♦	95.0	100.0	86.7	♦95.2	100.0	100.0	100.0	100.0	92.9	89.2
MO	2.9	3.4	3.4	3.0	3.0	2.7	2.5	2.5	3.2	3.4

♦♦ for description see text

— similarity level corresponding to that of coenotic molecule with same MSO resp. MO value

⌐ suggested minimal area

Minimal area and sociability

Some phytocoenoses have the same cover-abundance index but different minimal area sizes. This is related to differences in the sociability of the prevailing species. For example: Mook 2: MSO value 15.4, minimal area 20 m², Oeffelt 2: MSO value 15.1, minimal area only 2 m². Mook 2 is a heathland with a coarse grained pattern mainly of *Calluna vulgaris,* Oeffelt 2 is a grassland with a fine grained pattern.

There are also phytocoenoses with the same minimal area but different index values. Two examples are Oeffelt 3 and Nieuw Bergen 1, having a minimal area of 4 m² and MSO values of 14.9 and 7.3. Mook 5, Molenhoek 3 and Erlecom all have 20 m² as minimal area, but their MSO values are 25.0, 45.0 and 39.7. Oeffelt 1 and Nieuw Bergen 1 are both grasslands, but their species richness is different. Oeffelt 1 has a high number of species and a small grain size, whereas Nieuw Bergen 1 has few species and a relatively large grain size.

The same phenomenon is shown in the second series of phytocoenoses with a relatively large minimal area. Molenhoek 3, an old inland heathland and Erlecom, a wet grassland, are phytocoenoses composed by few species with a comparatively large grainsize and Mook 2, a heathland, is a species-rich phytocoenosis with a relatively small grainsize.

Table 2. (Continued).

	3				4	5		6	7	8	9
Molen-hoek 1	Molen-hoek 2	Mook 3	Mook 4	Mook 5	Molen-hoek 3	Terschel-ling 1	Terschel-ling 2	Terschel-ling 3	Terschel-ling 4	Terschel-ling 5	Erlecom
67.0	57.4			72.1	82.4				68.5	81.6	84.6
73.1	65.9			74.8	81.6				67.3	87.0	87.8
75.6	72.7	15.7		77.9	84.9				74.4	92.3	88.9
84.6	83.3	20.0		79.3	84.5				79.3	94.6	91.4
74.7	87.6	22.8	11.2	83.6	85.5	♦67.9	♦69.4	♦72.0	78.9	94.8	91.1
−81.9	92.4	36.8	30.6	85.4	84.1	♦73.3	♦69.8	♦79.1	83.9	95.8	92.0
83.0	95.8	51.5	45.8	92.0	87.1	79.2	79.1	81.4	87.8	96.6	93.1
86.3	−98.9	60.7	68.4	91.9	90.0	89.3	81.2	84.0	91.4	95.6	95.3
86.8	98.5	64.4	72.4	91.9	95.2	89.0	87.6	87.4	93.8	97.6	96.0
86.7	98.8	78.4	74.1	92.2	96.0	89.6	90.8	87.8	−95.3	97.8	96.3
89.0	99.4	86.0	84.1	93.6	−99.0	89.9	92.6	88.1	96.5	97.9	−98.1
91.0	99.2	86.8	89.7	97.2	99.1	−90.9	−91.3	88.3	97.2	98.0	98.2
♦92.5	99.7	89.4	93.5	98.6	99.2	90.6	93.3	91.2	♦97.8	♦97.9	98.4
73.4	50.0	9.9	9.1	25.0	45.0	20.3	23.8	14.6	27.6	18.1	39.7
45.4	76.7			80.7	90.0				53.4	83.5	88.2
62.5	90.0			80.7	82.2				63.8	81.8	93.3
57.3	90.0	20.0		72.2	82.2				63.2	80.6	93.3
59.4	−100.0	29.7		61.8	76.7				67.6	82.4	−93.3
62.1	100.0	33.8	21.1	61.8	76.7	♦68.1	♦71.8	♦54.2	68.5	82.7	100.0
69.0	100.0	48.9	45.1	60.0	73.3	♦68.3	♦76.3	♦59.3	76.4	83.6	100.0
69.8	100.0	64.0	55.4	56.1	72.2	78.3	80.1	61.3	83.3	88.2	100.0
71.6	100.0	66.6	72.5	52.8	76.7	−76.6	88.9	59.8	85.9	88.9	100.0
66.4	100.0	67.9	78.4	53.5	82.2	78.3	87.8	64.0	91.7	100.0	100.0
76.4	100.0	79.8	78.4	68.0	90.0	80.7	92.2	69.7	95.6	100.0	100.0
81.1	100.0	93.3	83.9	72.0	100.0	82.1	96.7	70.9	100.0	100.0	100.0
84.0	100.0	96.7	85.4	83.3	100.0	79.0	96.7	74.8	100.0	100.0	100.0
♦90.6	100.0	100.0	97.1	100.0	100.0	79.0	100.0	84.8	♦100.0	♦100.0	100.0
3.1	7.0	3.0	2.9	4.0	6.0	3.9	4.5	3.2	4.9	4.6	5.8

Table 3. Minimal area values (m²) estimated according to different approaches. Values for the quantitative and qualitative coenotic molecule sizes and values for squared mean cover-abundance and mean cover-abundance (in ordinal transform) are given as well.

Type Location	Mean cover-abundance of species	Mean squared cover-abundance of species	Coenotic molecule quantitative	qualitative	Analytic minimal area	Minimal area qualitative (Moravec) (4)	Minimal area qualitative (Gounet & Calléja) (5)	Minimal area qualitative van der Maarel)
Festuco-Sedetalia dry grassland								
Oefelt 1	2.9	10.8	1 2	2	20	>100	10 20	10 20
Oefelt 2	3.4	15.1	1 2	0.4 1	2	40	1 2	0.4 1
Oefelt 3	3.4	14.9	4	1 2	4	>100	10 20	4 10
Oefelt 4	3.0	9.9	4 10	0.4 1	20	>100	10 20	4 10
Ottersum	3.9	17.7	2 4	0.4 1	2	100	0.4 1	0.4 1
Spergulo-Corynephoretum dry grassland								
Nieuw Bergen 1	2.7	7.3	1 2	0.2 0.4	4	40	1 2	1 2
Nieuw Bergen 2	2.5	6.5	1 2	0.4	10	10	4 10	1 2
Nieuw Bergen 3	2.5	6.5	0.2	>0.1	4	40	2 4	0.4
Genisto anglicae-Callunetum heathland								
Mook 1	3.2	14.0	0.02 0.04	1	20	>100	20	10 20
Mook 2	3.4	15.4	20	20 40	40	>100	40 100	20 40
Molenhoek 1	3.1	13.4	0.4	2 4	20	>100	10 20	4 10
Molenhoek 2	7.0	50.0	2	0.04 0.1	2	0.1	0.01 0.02	<0.01
Mook 3	3.0	9.9	4 10	0.4 1	10	100	10 20	4 10
Mook 4	2.9	9.1	2 4	1 2	20	>100	10 20	2 4
Deschampsia flexuosa grassland								
Mook 5	4.0	25.0	20 40	20 40	40	100	20 40	10 20
Molenhoek 3	6.0	45.0	20	10 20	20	20	2 4	1 2
Carici arenariae-Empetretum heathland								
Terschelling 1	3.9	20.3	40	2	40	>100	4 10	40 100
Terschelling 2	4.5	23.8	40	1 2	40	100	1	0.5
Empetro-Genistetum tinctoriae heathland								
Terschelling 3	3.2	14.6	1 2	4 10	40	>100	40 100	4 10
Empetro-Ericetum wet heathland								
Terschelling 4	4.9	27.6	10	2 4	10	20	0.4 1	0.2 0.4
Juncetum gerardii saltmeadow								
Terschelling 5	4.6	18.1	0.02 0.04	0.4 1	1	4	0.01	0.04 0.1
Rumici-Alopecuretum genticulati wet grassland								
Erlecom	5.8	39.7	20	0.1 0.2	20	0.2	0.01	1

Table 4. Comparison between minimal area values found with the newly proposed method and those summarized by Westhoff & van der Maarel (1978).

minimal area size of	Westhoff & van der Maarel (1978)	range of values determined with the newly proposed method			number of stands investigated
		quantitative coenotic molecule	qualitative coenotic molecule	analytical minimal area	
Dune grassland *Koelerio-Corynephoretea*	1–10 m²	0.2–10 m²	0.1–2 m²	2–20 m²	8
Heathlands *Nardo-Callunetea*	10–50 m²	0.4–40 m²	0.04–40 m²	10–40 m²	10
Salt marshes *Asteretea tripolii*	2–10 m²	0.02–0.04 m²	0.4–1 m²	1 m²	1

Apparently one next step in the minimal area analysis is the development of an index for sociability, or plant patch size. Van der Maarel's (1966) exclusion index, measured as sum of products of cover-abundance and sociability, which was found to be correlated with minimal area, could be a start here. However, sociability will have to be measured more accurately.

Deviations due to micro-heterogenity

The model-indicated quantitative coenotic molecule found in the heathland phytocoenosis Mook 1: 0.02–0.04 m², appears to be unrealistically low. The reason for this is that the mean floristic similarity keeps fluctuating after reaching the similarity level corresponding to the coenotic molecule. These fluctuations are apparently due to the heterogeneity of the phytocoenosis, i.e. a big variation in plant patch size of the dominating *Calluna vulgaris*.

Comparison with other determination methods

Three other methods are compared. It should be noted that they are applied to our data excluding the very rare occurrences. (Application to the full data learned that almost all minimal area values would be very high, even near to the maximum area analyzed).

Minimal area sizes according to Moravec's method are equal to or larger than the values found in this study (see Table 3). In some phytocoenoses Moravec's minimal area is undetermined, because the similarity limit of 100% is not reached. Apparently an insufficient number of species reach 100%

frequency. With the similarity limit of 80%, as proposed by Gounot & Calléja we find again generally larger areas. Only if the cover-abundance index is high the sizes are smaller than those found with our method. Still, as the model shows, the mean floristic similarity varies per phytocoenosis.

The correspondence with van der Maarel's method (based on the criterion that 80% of the total species must be reached) is good, with a few exceptions. This is remarkable because of the rather different approach involved.

It is difficult to say much more on this comparison, until we would have a more elaborated model of the dominance-diversity relations in a plant community (see below).

Quantitative and qualitative minimal area

Although we decided to adopt one overall analytic minimal area, it is still realistic and informative to consider quantitative and qualitative minimal area as two different expressions of the combined effect of species richness and dominance relations on the size of the coenotic molecule in a phytocoenosis. If a phytocoenosis is dominated by one or a few species, high quantitative similarity values will already be found for relatively small plots. If at the same time one or more species have a scattered distribution, the qualitative minimal area will be larger. Only at this larger size will both dominance relations and species richness be adequately sampled. If on the other hand the species are evenly distributed but with different densities the qualitative species composition may be reached at smaller plot sizes than the quantitative one.

In cases of high dominance and few species both

minimal areas will be small whereas in cases of high species diversity with even distribution both values will be high. This can be summarized in the following scheme.

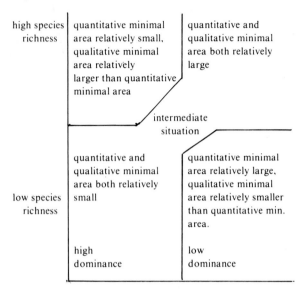

High similarity values as a check of overall homogeneity

In all cases the average similarity values on larger plot sizes exceeded 80%. This means that the entire plots of 1 000 or 500 m² were sufficiently homogeneous. Indeed (cf Curtis, 1959) the similarity approach can at the same time be used as an indication of the homogeneity level of the phytocoenosis under investigation and possible heterogeneities can be detected in this way.

If the similarity level of 80% would not be reached at all, as is to be expected in species rich communities with many species occurring only once or twice, e.g. in tropical rain forest, we may conclude that there is no minimal area in such cases. Then the entire stand recognized as distinct from the surroundings should be investigated.

Dominance relations

First we would emphasize that species performance should be measured as dominance, which can be realized through estimation of cover %. The scale values 1, 2 and 3 are abundance values. They are not entirely independent of the size of the sample plot. For this reason it would be better to use

density estimates (cf Barkman, Doing & Segal, 1964) or cover % values throughout. In the model the cover of species, recorded in number of elements can be expressed in cover %. So it will be better to calculate the mean cover in % and to use their relation with the mean floristic similarity.

Our attempt to relate the minimal area to some dominance character should be elaborated. Instead of the measures empirically found in this study, viz. mean (squared) transformed cover-abundance value an index characterizing the dominance-diversity curve (Whittaker, 1965, 1977) may be found. We have indications that the dominance relations between the main dominants are realized on a relatively small area. If we could find a suitable dominance index (which is in fact now being developed) we could plot the value of this index against area and probably find a saturation curve.

Conclusion

Minimal area determination through similarity analysis is possible, though tedious. In comparison with the original similarity approach by Gounot & Calléja (1962) two elaborations are suggested: extension of the similarity calculations to quantitative data, and adoption of a dominance index with which the 'minimal area similarity level' can be indicated. Instead of the 80% level proposed by Gounot & Calléja we suggest three different situations to be recognized: (1) according to the adopted model and its dominance measure a higher level must be adopted in cases of strong dominance; (2) if the similarity level indicated by the dominance measure does not reach 80% the nearest bigger area where the 80% level is reached is chosen as minimal area; (3) if the similarity level never reaches 80% no minimal area can be determined.

If a minimal area approach could be developed on the basis of a dominance index derived from the dominance – diversity relation this approach should be preferred to the similarity approach.

References

Barkman, J. J., 1968. Das synsystematische Problem der Mikrogesellschaften innerhalb der Biozönosen. In R. Tüxen (ed.) Pflanzensoziologische Systematik Ber. Int. Symp. Stolzenau/Weser 1964: 21–53. Junk, Den Haag.

Barkman, J. J., Doing, H. & Segal, S., 1964. Kritische Bemerkungen und Vorschläge zur quantitativen Vegetationsanalyse. Acta Bot. Neerl. 13: 394–419.

Beeftink, W.G., 1965. De zoutvegetatie van ZW-Nederland beschouwd in Europees verband. (with English summary). Diss. Wageningen Meded. Landbouwhogeschool Wageningen 65 (1): 1–167.

Braun-Blanquet, J., 1913. Die Vegetationsverhältnisse der Schneestufe in den Rätisch-Lepontischen Alpen. N. Denkschr. Schweiz. Nat. Ges. 48.

Braun-Blanquet, J., 1964. Pflanzensoziologie. Springer, Wien, New York. 3. Aufl. XIV + 865 pp.

Braun-Blanquet, J. & Jenny, H., 1926. Vegetationsentwicklung und Bodenbildung in der alpinen Stufe der Zentralalpen. Schweiz. Naturf. Ges. 63(2): 181–349.

Cain, S. A., 1938. The species-area curve. Amer. Midland Nat. 9: 573–581.

Cain, S. A. & de Oliveira Castro, G. M., 1959. Manual of vegetation analysis. XVI + 325 pp. Harper & Brothers, New York.

Curtis, J. T., 1959. The vegetation of Wisconsin. XI + 657 pp. Univ. of Wisconsin Press, Madison.

Du Rietz, G. E., 1930. Vegetationsforschung auf soziations-analytischer Grundlage. Handb. Biol. Arbeitsmeth. XI(5): 293–480.

Du Rietz, G. E., Fries, Th. C. E., Osvald, H. & Tengwall, T. A., 1920. Gesetze der Konstitution natürlicher Pflanzengesellschaften. Vetensk. Prakt. Unders. Lappland. Flora och Fauna 7: 1–47. Uppsala/Stockholm.

Gauch, Jr. H. G. & Whittaker, R. H., 1976. Simulation of Community patterns. Vegetatio 33: 13–16.

Goodall, D. W., 1952. Quantitative aspects of plant distribution. Biol. Rev. 27: 194–245. Cambridge.

Goodall, D. W., 1954. Minimal area: a new approach. Rapp. Int. Bot. Congr. Paris, Sect. 7–8: 19–21.

Goodall, D. W., 1961. Objective methods for the classification of vegetation. IV. Pattern and minimal area. Austr. J. Bot. 9: 162–196.

Gounot, M. & Calléja, M., 1962. Coefficient de communauté, homogénéité et aire minimale. Bull. Serv. Carte Phytogeogr., 7B: 181–210.

Greig-Smith, P., 1964. Quantitative plant ecology. 2nd ed. Butterworths, London.

Heukels, H. & Ooststroom, S. J., 1977. Flora van Nederland. 19e ed. Wolters-Noordhoff Groningen.

Hopkins, B., 1955. The species-area relations of plant communities. J. Ecol. 43: 409–426.

Hopkins, B., 1957. The concept of minimal area. J. Ecol. 45: 441–449.

Kortekaas, W., Maarel, E. van der & Beeftink, W. G., 1976. A numerical classification of European Spartina communities, Vegetatio 33: 51–60.

Maarel, E. van der, 1966. Over vegetatiestructuren, -relaties en -systemen in het bijzonder in de duingraslanden van Voorne. (with English summary) 170 pp. Thesis Utrecht.

Maarel, E. van der, 1970. Vegetationsstruktur und Minimum-Areal in einem Dünen-Trockenrasen. In: Tüxen, R. (ed.): Gesellschaftmorphologie. Ber.Int. Symp. Rinteln 1966: 218–239. Junk, Den Haag.

Maarel, E. van der, 1979. Transformation of cover-abundance values in phytosociology and its effects on community similarity. Vegetatio 39: 97–114.

Meijer Drees, E., 1954. The minimum area in tropical rain forest with special reference to some types in Bangka (Indonesia). Vegetatio 5–6: 517–523.

Moravec, J., 1973. The determination of the minimal area of phytocenoses. Folia Geobot. Phytotax. 8: 23–47.

Mueller-Dombois, D. & Ellenberg, H., 1974. Aims and methods of vegetation ecology. XX + 547 pp. Wiley, New York.

Peet, R. K., 1974. The measurement of species diversity. Ann. Rev. Ecol. Syst. 5: 285–307.

Raunkiaer, C., 1934. The life forms of plants and statistical plant geography; being the Collected Papers of C. Raunkiaer. 632 pp. Oxford.

Roux, C. & Rieux, R., 1981. L'aire minimale des peuplements lichéniques saxicoles-calcicoles. Vegetatio 44: 65–76.

Smidt, J.T. de, 1975. Nederlandse heidevegetaties. (with English summary) Thesis. 98 pp. + tables. Utrecht.

Sørensen, Th. A., 1948. A method of establishing groups of equal amplitude in plant sociology based on similarity of species content. Biol. Skr. K. Danske Vidensk. Selsk. 5(4): 1–34.

Southwood, T. R. E., 1978. Ecological methods. With particular reference to the study of insect populations. XXIV + 524 pp. Chapman and Hall, London, New York.

Tüxen, R., 1970a. Einige Bestandes- und Typenmerkmale in der Struktur der Pflanzengesellschaften. In R. Tüxen (ed.) Gesellschaftsmorphologie, Ber. Int. Symp. Rinteln 1966. p. 76–98. Junk, Den Haag.

Tüxen, R., 1970b. Bibliographie zum Problem des Minimi-Areals und der Art-Areal-Kurve. Exc. Bot. B 10: 291–314.

Werger, M. J.A., 1972. Species-area relationship and plot size with some examples from South African vegetation. Bothalia 10: 583–594.

Westhoff, V., 1951. An analysis of some concepts and terms in vegetation study or phytocenology. Synthese 8: 194–206.

Westhoff, V. & Held, A. J. den, 1969. Plantengemeenschappen in Nederland. 324 pp. Thieme, Zutphen.

Westhoff, V. & Maarel, E. van der, 1978. The Braun-Blanquet approach. 2nd ed. In: R. H. Whittaker (ed.). Classification of plant communities, p. 287–399. Junk, The Hague.

Whittaker, R. H. Dominance and diversity in land plant communities. Science 147: 250–260.

Whittaker, R. H., 1977. Evolution and measurement of species diversity. Taxon 21: 213–251.

Williams, C. B., 1964. Patterns in the balance of nature and related problems in quantitative ecology. 324 pp. Academic Press, London.

Accepted 10.7.82.

Plant species richness at the 0.1 hectare scale in Australian vegetation compared to other continents

B. Rice[1] & M. Westoby[2]*
[1] *12 Douglas St., Putney, NSW 2112, Australia*
[2] *School of Biological Sciences, Macquarie University, North Ryde, NSW 2113, Australia*

Keywords: Australia, Convergence, Diversity:plant species, Richness:plant species, Species-area curves

Abstract

New data are reported, and literature data compiled, for species richness in 0.1 ha plots in Australian vegetation. We conclude that on present evidence the same vegetation types are rich, and the same types poor, at a 0.1 ha scale, in Australia as elsewhere. Tropical rainforest averages 140 species per 0.1 ha in permanently humid types. Temperate sclerophyll shrub-dominated types on low-nutrient soils are generally in the range 50–100 species, with open woodlands somewhat richer than scrublands. Warm semi-desert shrublands can have 50–80 species, counting ephemerals both of summer and of winter. Temperate closed forests generally have fewer than 50 species per 0.1 ha. For none of these types is there clear evidence that they are richer or poorer in species at a 0.1 ha scale than types in similar environments with similar growth-form mixes on other continents. We give data for grassy woodlands and sclerophyll scrublands in the monsoonal tropics; the fragments of data on such types available from other continents suggest there may be a wide range of species richness in sub-types of this very broad grouping. Generally, available data do not support the idea that floristic evolutionary history is a strong influence on the species richness of vegetation at the 0.1 ha scale, relative to the influence of the present-day climatic and soil environment.

Introduction

One hypothesis regards species richness as at equilibrium in evolutionary time, and to be interpreted as the outcome of processes of community organization such as competition, modulated by the physical environment in which the community is assembled. Another hypothesis regards species richness as not in equilibrium over evolutionary time, and as determined by how fast and how long speciation has been going on relative to species extinction. To understand processes operating over evolutionary time we rely on natural experiments; these experiments consist of comparisons between areas on different continents which have similar present-day physical conditions, but where floras have very different evolutionary histories. In recent reviews summarizing such comparisons, Whittaker (1977), Shmida & Whittaker (1979), Naveh & Whittaker (1979) and Shmida (1981) conclude that there are substantial differences in species richness between analogous vegetation types on different continents. They attribute this nonconvergence to the differing evolutionary histories of the floras. This conclusion implies that communities are not sub-

* The National Parks and Wildlife Service of N.S.W. gave permission to work in National Parks. Getty Pty. Ltd., Central Coast Mining N.L., Mount Isa Mines Ltd. and Denison Australia Pty. Ltd. supported work on their mining leases. The staff of Fowlers Gap Research Station were always hospitable. The Herbarium Australiense and the herbaria of N.S.W., Sydney University, the Northern Territory and Queensland helped with identifications. We thank D. J. Parsons, C. Zammit, R. K. Peet, W. E. Westman and R. H. Whittaker for comments and unpublished data. We wish particularly to record our debt to the late Professor Whittaker, who encouraged this work.

stantially organized, with respect to the coexistence of species, by processes operating in ecological time – that at least some community properties cannot be predicted solely from present-day ecological conditions.

In this paper we present data on species richness (some from the literature, mostly new) from a range of vegetation types in Australia; the Australian flora's evolutionary history has long been separate from that of most of the rest of the world. We compare these data with available records from other continents.

The interpretation of species richness is bedevilled by problems of scale. Several processes affect the slope of the species-area curve, the rate at which the species number increases as plots of larger area are examined. Among these are: (a) the increase of the species list as more individuals are examined; (b) the increasing variety of physical environments, at all scales from across a pebble to across a mountain range, in progressively larger areas; (c) how similar the members of the flora are in the physical environments they occupy (equivalent to 'species packing', the aspect of species richness with which most ecological theory deals); (d) an increasing variety of successional stages, in a landscape which is a successional mosaic; and (e) biogeographic separation (one species would outcompete the other within a small area, but until recently they have been isolated). These processes differ in what part of the species-area curve they affect; in consequence:

(1) We expect the absolute number of species found per unit area to be affected by the processes operating over the orders of scale below the sample size being considered. For instance, we expect the numbers of species in 1 m^2 quadrats to be affected by how many individual plants are present and so by any factors which affect plant size, and by the micro-heterogeneity of the environment. At 10^3 m^2 (our sample size and fairly typical of what most ecologists would consider to represent a 'community') we expect to be looking mainly at the effects of meso-heterogeneity and of the limiting similarity for species coexistence ('species packing'), but there may still be a serious effect of increasing numbers of individuals for trees (May, 1975). At areas up to 10^9 m^2 (typical of areas an ecologist might consider a 'study region', within which a collection of sample sites might be examined) we expect to be dealing with the effects of macro-heterogeneity if there is

any, succession if relevant (Pickett & Thompson, 1978; White, 1979), meso-heterogeneity and limiting similarity. At a plant geographer's scale such as 10^{11} m^2 we are looking mainly at the effects of macro-heterogeneity and of geographical isolation.

(2) Whatever size of sample one chooses, the number of species one sees is affected by more than one process. We therefore believe it is not helpful to try to distinguish within-community processes and sampling scales from between-community processes and sampling scales. This distinction has customarily been made and follows naturally from the body of theory aiming to predict local species-richness as the outcome of processes such as competition. This formulation of the problem was lucidly summarized by Pielou (1975): 'given a stable many-species community, how have its constituent species come to share the same habitat, and how do they maintain themselves and interact with one another? Hence, what determines the number of species which can live together and their relative proportions?' But two individual plants could be 'together' for the purposes of some of the biological processes listed above when separated by 10 km; equally, two plants could be 'apart' for the purposes of other processes when separated by 10 cm. As a corollary, we do not believe Whittaker's (1960) conceptual partitioning of gamma (landscape) diversity into alpha ('within-community') and beta ('between-community') components leads to a satisfactory procedure for analysing real data.

(3) The species richnesses observed at different scales are not necessarily highly correlated.

The immediate problem is to collect data in a standardized way from different continents. With these data we can ask empirically whether there is convergence of species richness or not with one particular sampling method. Only once it is established whether there is convergence, and if so at what spatial scales, can further investigation be devised to ask what processes are responsible for convergence or divergence between different pairs of continents and at different spatial scales. Sampling methods could conceivably standardize the number of individuals examined, the size of area examined, or the range of physical heterogeneity examined. Most workers have standardized the plot size, area being the most practical scaling measure to standardize in the field. We have collected data from a plot size of 0.1 ha, chosen to allow comparison with the large published body of such data from other parts of the world collected by Whittaker and several others.

Methods

In the Results section data are summarized for other continents (entirely from the literature) and

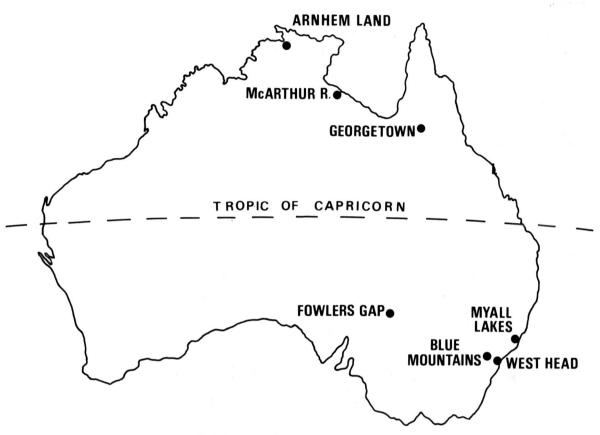

Fig. 1. Locations of our study areas in Australia.

Table 1. Properties of the study areas.

Area	Latitude (°S)	Number of 0.1 ha samples taken (* resamples at another date)	Sampling dates	Soil types	Major vegetation types
Georgetown	18	7	5/77	Yellow earths	Grassy woodlands
McArthur River	16	17	2/76	Yellow earths and lithosols	Grassy woodlands and sclerophyll scrubs
Arnhem Land	13	74 (20*)	4/78, 4/81	Red earths and lithosols	Grassy woodlands and sclerophyll scrubs
Fowlers Gap	31	42 (21*)	1/76, 8/78	Red clay loams	Chenopod shrublands
Myall Lakes	32	10	9/76	Podsol on Holocene sand	Eucalypt forest
Blue Mountains	33	21	1–8/76	Lithosol on sandstone	Eucalypt forest
West Head	34	39	9/76–11/77	Lithosol on sandstone	Eucalypt forest, woodland and sclerophyll scrub

for Australia (partly from the literature but largely from our measurements). Here we describe study areas and methods only for our data, not for those from the literature.

Regions studied

We have collected data from seven study areas within Australia (Fig. 1). Table 1 summarizes the essential properties of each study area. Figure 2 gives climate diagrams for the areas.

Georgetown, McArthur River and Arnhem Land are all tropical. They include vegetation of similar structure, an open savanna woodland over grass in a monsoonal climate with summer rainfall in the 700–900 mm range (data for Georgetown and Darwin; Fig. 2c, d). At McArthur River and in Arnhem Land these woodlands are found on lowlands below a broken sandstone plateau which falls by a sharp escarpment to the lowlands. The vegetation on the sandstone has much in common with the sclerophyll vegetation of the Australian temperate zone, with many *Myrtaceae* and *Fabaceae*. A large proportion of the lowlands is flooded during the wet season, much by sheet flooding only, and different parts of the country remain wet for different periods into the dry season. The tall grass left after the wet season is burnt off during the dry season in most years, either by pastoralists or by Aboriginal hunter-gatherers.

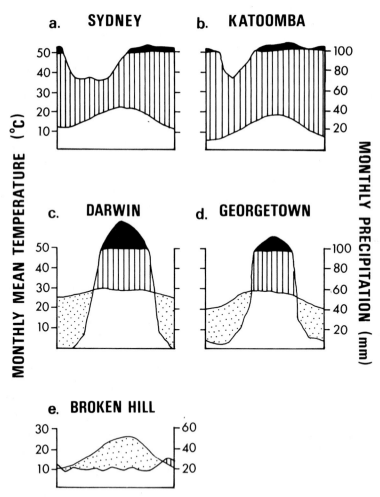

Fig. 2. Climate diagrams (Walter & Lieth, 1960) representative of our study areas. The horizontal axis represents the annual sequence of months from July to June. Dotted areas show periods of water shortage, vertical hatching periods of water surplus, black shows rainfall over 100 mm per month on a 10:1 compressed scale. Sydney (a) represents West Head and Myall Lakes, Katoomba (b) represents Blue Mountains, Darwin (c) represents Arnhem Land, McArthur River would be similar to Darwin and Georgetown (d), Broken Hill (e) represents Fowlers Gap.

Fowlers Gap Research Station is semi-arid with annual rainfall about 200 mm, which is not strongly seasonal on the average, although very unpredictable from any given season to another (data for Broken Hill; Fig. 2e). Vegetation is mostly chenopod shrublands of various kinds on red clay loams. We looked at sites at Fowlers Gap on two occasions, after heavy summer rains in the summer of 1975–1976, and then after mid-winter rains in 1978. After summer rains an annual flora from many families but with sizeable representation of *Poaceae*, *Chenopodiaceae* and *Malvaceae* was added to the perennial chenopods and grasses. After winter rains the annual flora was predominantly of *Asteraceae* and *Brassicaceae*.

The final three areas lie on sandstone originally deposited around the continental margins, like the sandstone plateau sites in the north, or on sands eroded from such sandstones. They are all very low in nutrients, e.g. total phosphorus with typical values of 5–100 ppm (Clements 1980; M. D. Fox, 1981). Clay bands in the sandstones produce local variation in drainage. The eucalypt overstorey varies from about 60% projective foliage cover to absence; the understorey is sclerophyll. In eastern Australia the distribution of sclerophylly is determined more by soil nutrients (Beadle, 1954, 1966; Groves & Specht, 1965; Specht, 1969) than by climate. Figure 2 (a, b) shows that these sites do not have mediterranean climates, although there can be droughts of up to two months in summer.

In the Myall lakes area we have a few observations from a eucalypt forest established on a sand mass produced by the transgression of a coastal dune during the Holocene. These sites have an understorey of shrubs in the *Fabaceae* and *Mimosaceae* to 2–3 m, with lower shrubs of other families.

The remaining two sites are on massive, deeply dissected sandstone plateaus. The Blue Mountains lie about 100 km from the coast at about 1 000 m elevation. Dissection is on a relatively large scale; our plots of 0.1 ha were commonly on flat plateau tops. The vegetation is almost entirely a continuous canopy of eucalypts with a rather varied understorey. In the West Head of Ku-Ring-Gai Chase National Park the sandstone plateau is dissected on a finer scale. West Head is surrounded on three sides by sea, and has a maximum elevation of about 200 m. Scrub, woodland and forest communities occur on broken Hawkesbury Sandstone parent mate-

rials. The West Head data are described in more detail in Rice & Westoby (in press).

Procedure at each sample site

We searched an area 50 × 20 m and recorded the species for all vascular plants found within it; other species may have been present as seeds or tubers only. We recorded life-forms and growth-forms of the species, and present growth-form spectra using a system of categories which allows comparison with Naveh & Whittaker (1979).

Choice of sites within a region

In most of the regions from which we collected data, an important aim was to study variation in the vegetation within the region. The plots were placed subjectively in such a way as to sample a cross section of the major types of vegetation present, and so are not a random sample of all the vegetation in the given geographical region. The types of vegetation which cover a large portion of each region tend to be under-represented. This sort of distortion of fully random sampling is also found in the literature data with which we compare our results. Formal statistics, such as t-tests to compare two populations of plots, cannot safely be applied.

Comparison of results

Whittaker and various colleagues established the first large body of data on species richness on 0.1 ha in a series of studies in North America, in the Great Smoky Mountains (Whittaker, 1956), the Siskiyou Mountains (Whittaker, 1960), the Santa Catalina mountains (Whittaker & Niering, 1965, 1975), and in oak-pine forest at Brookhaven (Whittaker & Woodwell, 1969). These studies were summarized by Whittaker (1965, 1972, 1977). He contrasted the great majority of US vegetation, which on the available evidence he thought had fewer than 50 species per 0.1 ha (1 in Fig. 3), with exceptions of two types: mesic 'cove forests' in the Great Smoky Mountains (3 in Fig. 3), and some semi-arid types at low elevation in the Santa Catalina Mountains, which rise out of the Sonoran desert in Arizona (4 and 6 in Fig. 3). Subsequent North American work (Glenn-Lewin, 1975, 1977; Peet, 1978a, 1981) has confirmed that much vegetation has fewer than 50 species per 0.1 ha, and that semi-arid woodlands can exceed 50. Peet & Christensen (1980) found that forests on particularly fertile soils on the North Carolina Piedmont averaged 75 and ranged up to 120. Since Peet, Glenn-Lewin & Wolf (1983) found 60–85 species per 625 m² in savannas on the southern coastal plain, it may be that values well above 50 per 0.1 ha in southeastern North America are not

restricted to mesic forests but are more widespread than is currently recognised.

Naveh & Whittaker (1979), summarized patterns of species richness in mediterranean-climate shrublands and woodlands in California and Israel. In Californian chaparral and woodlands richness is in the range 24–64. Closed maquis shrublands in Israel are in the same range (10 in Fig. 3), but where shrublands or woodlands are open and grazed, a rich flora of annual herbs can raise total species richness per 0.1 ha as high as 130–140 (11 in Fig. 3).

Naveh & Whittaker (1979) pointed out the sharp distinction between the vegetation of mediterranean climates in California and around the Mediterranean proper, and that of South Africa and Australia. For South African fynbos their data ranged from about 50 to 130 and averaged 75 per 0.1 ha, and they summarized other data indicating a similar range (12 in Fig. 3). The structure is totally different from that of open Israeli woodlands, containing no annuals and being made up predominantly of microphyllous shrubs (Cowling & Campbell, 1980).

It has long been known that tropical rainforest is exceptionally species-rich. Some available data are presented in Figure 4. The only datum including all species is point 1 in Figure 4, which shows that tropical rainforest can be exceptionally species-rich. In most of the available studies only trees above some minimum size are recorded. Under

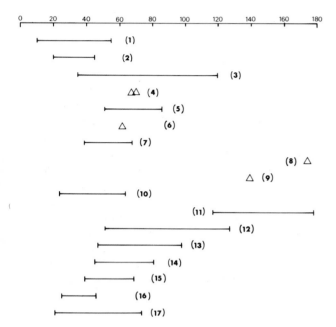

Fig. 3. Species richness in 0.1 ha plots; horizontal bars show ranges of values, triangles single values or means. Where sample is of fewer plots than indicated in Table 1, this is because some plots with very different structure or on extreme soils have been excluded. Notes: (1) most temperate North American forests: from Whittaker (1956, 1960), Whittaker & Woodwell (1969), Whittaker & Niering (1975), Glenn-Lewin (1975), Peet (1978a). (2) Some east Australian forests at Myall Lakes (10 plots, mean 32, range 21–43), Blue Mountains (18 plots, mean 36, range 23–50), and on volcanic soils on the West Head (2 plots, 34 and 37). Forests on sandstone on the West Head had a wider range (10 plots, mean 56, range 35–82, Rice & Westoby, in press), but are not different from those shown in 2 in the same way that those in 3 are different from those in 1. (3) Forests of fertile soils in the Great Smoky Mountains (Whittaker, 1956, 1965) and on the North Carolina Piedmont (Peet & Christensen, 1980). (4) Semi-arid types in foothills of Santa Catalina Mountains (Whittaker & Niering, 1975): *Bouteloua curtipendula* desert grassland (mean 70), spinose-suffrutescent desert shrub (mean 68); these means include 20–24 winter annuals, as well as summer annuals. (5) Semi-arid types at Fowlers Gap, 20 sites, mean 65, range 51–86, including annuals after both summer and winter rains. (6) Open oak woodland (Whittaker & Niering, 1975), mean 62 excluding 18 winter annuals. (7) Mallee woodland (Whittaker *et*

al., 1979), 6 sites, mean 49, range 39–68, not including summer annuals. (8) 175 species in 400 m² of tropical rainforest in Puerto Rico (Smith, 1970). (9) Permanently humid rainforests, mean 140 (Webb *et al.,* 1967); Figure 4 shows a larger selection of data from tropical rainforests. (10) Californian chaparral and woodlands, and closed maquis shrublands in Israel (Naveh & Whittaker, 1979). (11) Open shrublands and grazed woodlands in Israel (Naveh & Whittaker, 1979). (12) South African fynbos (Naveh & Whittaker, 1979). (13) Woodlands on the West Head, 12 sites, mean 77, range 60–93 (Rice & Westoby, in press). (14) Southwest Australian heath (Naveh & Whittaker, 1979). (15) Scrub on the West Head, 14 sites, means 48 for wet scrub and 56 for dry scrub, ranges 43–68 (Rice & Westoby, in press). (16) Scrub on sandstone escarpment in Arnhem Land, in 1978 4 sites, mean 35, range 28–44, in 1981 9 sites, mean 38, range 26–47. (17) Grassy woodlands on lowlands in tropical Australia: around Georgetown 7 sites, mean 39, range 29–47; around MacArthur River, mean 37, range 22–52; Arnhem Land in 1978, 8 sites, mean 44, range 32–51, Arnhem Land in 1981, 16 sites, mean 62, range 44–75.

In the text the following types are considered as cross-continent comparisons (Australian type second): (1) with (2), (4) with (5), (6) with (7), (8) with (9), (12) with (13).

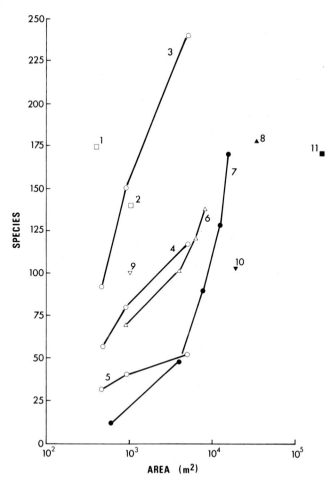

Fig. 4. Species richness in tropical rainforests on plots of known area. Where points are connected by lines, the data are for continuous samples on increasing areas. (1) About 175 species of all kinds in 400 m² in Puerto Rico, of which about 20 were climbers, 4 stranglers, 42 epiphytes, 31 herbs, and 11 saprophytes (Smith, 1970). (2) Average of 140 spp in 'permanently humid' rainforests in Queensland (Webb *et al.*, 1967). Monsoonal and seasonally dry forests have fewer. Includes epiphytes 'where accessible'. (3, 4, 5). Sample plots 38, 30 and 17 of Brünig (1973) from Sarawak and Brunei; all tree species >1 cm diameter. (6) Plot 4 of Paijmans (1976) in Papua; trees >10 cm d.b.h. (7) Wyatt-Smith (1966) from Bukit Lagong and Sungei Menyala forests in Malaya; trees >10 cm d.b.h. (8) Trees >10 cm d.b.h. at Para, Brazil (Pires, Dobzhansky & Black, 1953). (9) Tree species only on Barro Colorado, Panama (Knight, 1975). (10) Tree species >10 cm d.b.h. in the southern Cameroons (Richards, 1963). (11) Tree species >10 cm d.b.h. in southern Nigeria (Jones, 1955).

these circumstances on a plot of 0.1 ha only 40–70 individuals are usually recorded (see Paijmans, 1976), so species richness values seem relatively low. However up to about 2 ha, or 1 000 individuals, the

number of species found increases sharply, as expected from the work of May (1975). The data points 6 and 7 in Figure 4 are of this kind. The importance of including smaller individuals can be seen from the data of Brünig (1973), which include all trees above 1 cm diameter. Curves from sites rich, average and poor in species (points labelled 3, 4 and 5 resp.) are shown in Figure 4. Our picture of total vascular plant species richness in moist tropical forests is as yet fragmentary. Within a region there can be great variation in richness (e.g. the data of Brünig in Fig. 4). Of Smith's 175 species in 0.04 ha, fewer than 70 were trees. Life-forms such as epiphytes (42 species in Smith's data) can add substantially to tree species richness. However the contribution of life-forms other than trees is unlikely to rise as rapidly on larger plots as the number of tree species.

In Australian tropical rainforests Webb *et al.* (1967) found a mean of 140 species per 0.1 ha in permanently humid sites (point 2 in Fig. 4; 9 in Fig. 3). Monsoonal and seasonally dry forests had fewer, ranging down to 54. This range appears similar to that observed in other tropical rainforests (8 in Fig. 3), and is similarly richer than that consistently or reliably observed in any other type of vegetation.

In mallee vegetation in western New South Wales Whittaker, Niering & Crisp (1979) found 49 species per 0.1 ha, counting annuals of spring only (7 in Fig. 3). Mallee is a semi-arid woodland dominated by eucalypts with multiple stems. The stems are killed by fire and the plants regrow from basal lignotubers. The understory is a variable mixture of xerophytic shrubs, perennial grasses and seasonal ephemerals. Probably the nearest non-Australian analogue for which species richness data are available is the open *Quercus* woodland on the lower slopes of the Santa Catalina mountains in Arizona (6 in Fig. 3), which has a similar growth-form mix in the understory, and similar richness (mean 56 per 0.1 ha) if annuals of only one season are counted (Whittaker & Niering, 1975).

Naveh & Whittaker (1979) gave data from seven plots in heath in southwest Australia, an area with a true mediterranean climate; these had a mean of 65. George, Hopkins & Marchant (1979) gave species-area curves for 8 southwestern heathlands, where richness at a 0.1 ha is in the range 45–80 species (14 in Fig. 3). They tabulated data at 500 m² for a larger set of 25 sites; these ranged from 37 to 92 species. Other data reported by Westman (1975), Lamont *et al.* (1977), Cockburn (1978), Braithwaite & Gullan (1978) and Parsons & Cameron (1974) are not for 0.1 ha plots, but suggest a generally similar range of richness for scrubs and heath on a range of substrates, only one of which (Lamont *et al.*, 1977) was in a genuinely mediterranean climate.

In north Australia the vegetation we studied on broken sandstone plateaus overlooking the lowlands at Arnhem Land had 26–47 species per 0.1 ha (16 in Fig. 3) in both 1978 and 1981; the two similar sites at MacArthur River were in the same range. The growth-form mix (Table 2) differed from that of sclerophyllous scrub on sandstone or sand in Australia (13 and 15 in Fig. 3) in having some annuals but fewer shrubs, with lower overall richness. In grassy woodlands, the most common type on lowlands, richness was in the range 22–52 at Georgetown, at MacArthur River and in Arnhem Land in 1978. However the sites sampled in 1981 in Arnhem Land had richness in the range 44–75 (17 in Fig. 3). This might indicate that our measurements in the monsoonal north are susceptible to between-year differences in what species emerge, or in the exact stage of the season at which sites are sampled. On the other hand the two Arnhem Land grassy woodland sites which were sampled in both 1978 and 1981 barely changed in species richness.

Few data are available for vegetation which might correspond to the north Australian grassy woodlands we have studied. Ramsay & de Leeuw (1965) generally found fewer than 30 species in 0.2 ha plots in Nigerian savannah. On the other hand Eiten (1978) commented that Brazilian cerrado can be distinguished from nearby structurally similar types by its extraordinary floristic richness. Eiten gave two reports of about 300 species in 1 ha, and one of more than 230 species in 0.1 ha. Cerrado is found on infertile, usually deep latosols in an intermediate rainfall region with a distinct dry season; other types in a similar climatic range but on different soils are much less rich in species (Eiten, 1978).

At Fowlers Gap the total richness of most sites was in the range 50–80 species (5 in Fig. 3), counting both winter and summer ephemerals. These figures are similar to those for desert grassland and desert scrub at lower elevations in the Santa Catalina Mountains (4 in Fig. 3). Relatively few of the Fowlers Gap species are fully perennial, surviving droughts such as the 26-month one which intervened between our two sampling periods, and few grow as ephemerals during both seasons. In North America the shrublands dominated by chenopods tend to occur in rather different climates. For instance, in Curlew Valley, Utah (Rice & Westoby, 1978) chenopod shrublands are found at low elevations on saline soils in an area where precipitation falls largely as winter snow. The flora of these che-

Table 2. Number of species per 0.1 ha in different growth forms. See text for further explanation of vegetation types, Table 1 for details on areas, footnotes to Figure 3 for replication and ranges of species richness values.

	Trees & shrubs	Climbers & epiphytes	Perennial graminoids	Geophytes	Other perennial herbs	Annual graminoids	Other annual herbs	Total
Monsoon grassy woodland								
Georgetown	11.9	1.1	5.6	0	5.6	5.1	7.0	36.3
MacArthur River	11.0	1.8	4.3	0.6	5.2	3.6	10.7	37.2
Arnhem Land 1978	17.5	1.9	6.9	2.3	3.5	5.0	8.5	45.6
Arnhem Land 1981	22.3	1.0	6.9	3.1	8.5	8.6	11.6	62.0
Monsoon sandstone scrub								
Arnhem Land 1978	22.0	0.5	4.0	1.0	1.0	2.3	3.8	34.6
Arnhem Land 1981	18.4	1.2	5.7	1.3	2.6	3.0	5.8	38.0
Temperate forest								
Blue Mountains	21.4	0.3	5.1	0.5	7.0	0.1	1.6	36.0
Myall Lake	21.8	0.1	1.8	3.9	4.0	0	0.4	32.0
West Head								
Woodland over scrub	55.2	1.0	11.1	0.2	9.2	0.1	0	76.8
Dry scrub	41.9	0.7	6.8	0	5.9	0	0	55.3
Desert shrublands	8.7	0	5.1	0.7	6.3	6.0	37.9	64.7

nopod shrublands is very poor, although a larger flora of herbaceous plants is found on the non-saline soils on foothills typically dominated by *Artemisia* species.

Of the remaining three sites, the Myall Lakes area and the Blue Mountains are occupied by forests with continuous canopy (usually 40–60% projective foliage cover) and the terrain is typically fairly homogeneous across the 0.1 ha plot. Richness in these sites was generally in the range 20–45 species per 0.1 ha (2 in Fig. 3). For the most part then, these Australian sclerophyllous forests on non-patchy surfaces had richness values comparable to those found for most temperate forests in North America.

The West Head data are described in detail by Rice & Westoby (in press). Forests on fertile soils on a volcanic intrusion with a nonsclerophyll understorey had about 35 species per 0.1 ha (2 in Fig. 3). On sandstone, there are two main floristic types of scrub, 'dry scrub' on skeletal soils on ridgetops, and 'wet scrub' on moderate slopes where a clay layer restricts the available soil volume. Wet scrub averaged 48 and dry scrub 56 species per 0.1 ha, with a range for both from 43 to 68 (15 in Fig. 3). The richest sites were woodlands on benched slopes, with mean 77 species per 0.1 ha and some sites ranging up to nearly 100 (13 in Fig. 3). In both growth-form mix (Table 2) and richness, the scrubs match closely the southwest Australian heath, and the woodland the South African fynbos, reported by Naveh & Whittaker (1979).

Discussion

We have found no evidence that at the 0.1 ha scale Australian vegetation types have notably different species richness from equivalent vegetation types on other continents. This applies to temperate forests with closed canopy (usually <50 species), semi-deserts (50–80 species counting ephemerals of both seasons), shrub-dominated vegetation on very infertile soils (about 75 species for shrubs under woodlands) and permanently humid tropical rainforest (> 130 species per 0.1 ha).

Because none of the data available (neither ours nor that in the literature) have been collected from a randomized set of plots within a given landscape or vegetation type, we think it inappropriate to apply formal statistical tests for differences in species

richness between continents or vegetation types. In the future, it will be desirable to collect fully randomized data from specified vegetation types, and test formally for convergence or divergence. However, this problem takes second place to the even more intractable difficulty of defining vegetation types which might be expected to converge, i.e. which occur in similar environments, on different continents. The conclusions we draw from our data depend very much on which vegetation we treat as comparable. For example, Whittaker (1977) and Naveh & Whittaker (1979) treated the different mediterranean-climate shrublands as having comparable environments, and so concluded that their species richness is divergent, while we treat the Australian and South African types as not comparable to the others because of the infertile soils which characterize them. Conversely we have provisionally compared eucalypt forest to US temperate forests, and Australian chenopod shrublands to US Sonoran desert. In both cases it would have been easy, *a posteriori,* to explain away nonconvergence of species richness as due to differences in the environment.

There are three possible approaches to this problem. The most obvious is to match physical environments closely. Generally climates have been matched. Even here, though, there can be problems; for example Australia has a higher incidence of large rainfall events, and conversely of very long droughts (Low, 1978; Westoby, 1980), than other regions with similar mean annual rainfall and evaporative climate. Milewski (1979) attributed to this factor the differences in tree incidence in Australia compared to South Africa. The Australian sclerophyll scrub example shows the importance of matching soils, but it remains unclear what features of soils are important. Is it sufficient to match a general fertility level, or are depth, texture, specific nutritional properties etc. important influences on species richness?

The second approach is to look at variation in species richness within continents, and compare this with variation between continents (Peet, 1978a). Our conclusion here is that on present evidence, in Australia the same sorts of vegetation are relatively rich in species as elsewhere, and the same sorts relatively poor. Since the same pool of taxa has been available to colonize different environments within Australia, this suggests that the variation in species richness at the 0.1 ha scale we can at present distinguish is much more attributable to environmental differences than to floristic history. More generally, variation within a continent can be used as one test for the hypothesis that any given difference between continents is due to present environments rather than to floristic history. For example, the hypothesis that Californian chaparral is poorer in species than Australian or South African temperate scrub due to its more fertile soils would be supported by finding that any Australian scrub on fertile soils was relatively poor in species.

The third approach is to use the growth-form mix and life-form mix of vegetation as an indicator of its present environment. It was the growth-form mix that first drew attention to the differences in soil environment between Gondwanan and other mediterranean-climate shrublands. This approach involves assuming that growth-forms are not substantially limited by the evolutionary history of the floras. There is certainly some growth-form convergence in similar environments, as shown both by general studies (e.g. Cowling & Campbell, 1980; Box, 1981) and by many particular examples. However, Shmida & Whittaker (1978), Peet (1978b), Shmida (1981) and Westman (1981) emphasize cases which have failed to converge in supposedly similar environments, and attribute the differences to evolutionary history. We are inclined to attribute such divergences to more subtle environmental differences, e.g. Westoby (1980) for deserts. Perhaps the most that can be said at present is that where two vegetation types in similar climates have different growth-form spectra, there must be a suspicion that some other environmental factor besides climate has a strong influence on them, and so perhaps on their species richness.

Thus far we have generally followed previous authors in drawing a simple contrast between past floristic history and present physical environment as factors which might control species richness. If species richness converges in similar present-day environments, the reasonable explanation is that species richness values are at equilibrium over evolutionary time and that the equilibria are controlled by the present-day environment. Chance could explain some convergence, of course. If species richness does not converge, the usual explanation (Whittaker, 1977) is that speciation or colonization is proceeding faster than extinction, so that richness

is not at equilibrium over evolutionary time. Richness values would then depend on how long and how fast species accumulation had been going on. However, non-convergence need not imply non-equilibrium. Conceivably, at some time in the distant past two floras could have taken different evolutionary paths (e.g. with respect to herbivore defence) which subsequently produced different degrees of species packing at equilibrium.

A further complication has been introduced by the recent proliferation of allegedly non-equilibrium models for the influence of the physical environment on species richness (e.g. Connell, 1978; Yodzis, 1978; Huston, 1979; Chesson & Warner, 1981). These models explain species richness in terms of the disturbance regime – e.g. the frequencies of fires, hurricanes or landslips. While they depict richness within any single patch as non-equilibrium over ecological time, they are in fact equilibrium models so far as mean species richness and evolutionary time are concerned. However, the models bring into focus a further difficulty – where to draw the line between past history and present conditions. The family composition of a continent's flora, determined during the Tertiary, is clearly past history; an undisturbed soil type is clearly present conditions. But for other kinds of present conditions, their effect might depend on how long they have been in place. For instance Naveh & Whittaker (1979) attribute the richness of grazed Israeli woodlands to several millennia of intense grazing, while explaining the failure of Californian woodlands to be more species-rich when grazed as due to the recent introduction of domestic stock.

Generally, our assessment of the present state of knowledge is that we still need a clear empirical picture of world variation in species richness. Only given a sharper picture can we identify comparisons which are good tests of the importance of the factors discussed above. For these tests we will need to collect data from sites chosen according to proper randomization procedures, within vegetation types circumscribed *a priori*. Most particularly, the tests will need to specify the range of spatial scales over which a given factor is expected to affect species richness. Discussions of species richness continue to suffer from insensitivity to scale. For example many authors (reviewed by Hopper, 1979) have emphasized that southwest Australia is rich in species compared to southeast Australia, but data summarized here shows that 0.1 ha plots are no richer; the greater richness appears at landscape scale. Similarly the richness of the Californian flora is often emphasized (Raven, 1977), but is at landscape-scale rather than at stand-scale. At the other extreme, literature on the effects of disturbance or stress on species diversity often generalizes from plots as small as 1 m² (Grime, 1973a, b; J. F. Fox, 1981). Not only are such generalizations unlikely to hold for other scales – on the West Head vegetation types with different species richness values above 100 m² are commonly no different at 1 m² (Rice & Westoby, unpublished) – but such fine scales are particularly likely to be misleading in the study of disturbance. Species richness is powerfully affected by the numbers of individuals examined up to 200 individuals (May, 1975). Factors which make for smaller individuals, so that more can be fitted into a square metre, can hardly help increasing richness at the 1 m² scale. Conversely, if removing grazing pressure allows a herbaceous sward to grow and self-thin from 200 to 20 individuals per m², it would be surprising if the number of species in 1 m² did not decline. It would be more helpful to know if the number of species in 200 individuals had changed.

References

Beadle, N. C. W., 1954. Soil phosphates and the delimitation of plant communities in eastern Australia. Ecology 35: 370–375.

Beadle, N. C. W., 1966. Soil phosphate and its role in molding segments of the Australian flora and vegetation, with special reference to xeromorphy and sclerophylly. Ecology 47: 992–1007.

Box, E. O., 1981. Predicting physiognomic vegetation types with climate variables. Vegetatio 45: 127–139.

Braithwaite, R. W. & Gullan, P. K., 1978. Habitat selection by small mammals in a Victorian heathland. Aust. J. Ecol. 3: 109–127.

Brünig, E. F., 1973. Species richness and stand diversity in relation to site and succession of forests in Sarawak and Brunei (Borneo). Amazoniana 4: 293–320.

Chesson, P. L. & Warner, R. R., 1981. Environmental variability promotes coexistence in lottery competitive system. Amer. Nat. 117: 923–943.

Clements, A., 1980. The vegetation of bushland in the northern Sydney area. M.Sc. thesis, Macquarie Univ.

Cockburn, A., 1978. The distribution of Pseudomys shortridgei (Muridae: Rodentia) and its relevance to that of other heathland Pseudomys. Aust. Wildl. Res. 5: 213–219.

Connell, J. H., 1978. Diversity in tropical rain forests and coral reefs. Science 199: 1302–1310.

Cowling, R. M. & Campbell, B. M., 1980. Convergence in vegetation structure in the Mediterranean communities of California, Chile, and South Africa. Vegetatio 43: 191–197.

Eiten, G., 1978. Delimitation of the cerrado concept. Vegetatio 36: 169–178.

Fox, J. F., 1981. Intermediate levels of soil disturbance maximize alpine plant diversity. Nature 293: 564–565.

Fox, M. D., 1981. Coexistence between eucalypts in a coastal open forest. Ph.D. thesis, Macquarie Univ.

George, A. S., Hopkins, A. J. M. & Marchant, N. G., 1979. The heathlands of Western Australia. In: R. L. Specht (ed.), Heathlands and Related Shrublands: Descriptive Studies, pp. 211–230. Elsevier, Amsterdam.

Glenn-Lewin, D. C., 1975. Plant species diversity in ravines of the southern Finger Lakes region, New York. Can. J. Bot. 53: 1465–1472.

Glenn-Lewin, D. C., 1977. Species diversity in North American temperate forests. Vegetatio 33: 153–162.

Grime, J. P., 1973a. Competitive exclusion in herbaceous vegetation. Nature 242: 344–347.

Grime, J. P., 1973b. Control of species density in herbaceous vegetation. J. Envir. Manage. 1: 151–167.

Groves, R. H. & Specht, R. L., 1965. Growth of heath vegetation I Annual growth curves of two heath ecosystems in Australia. Aust. J. Bot. 13: 261–280.

Hopper, S. D., 1979. Biogeographical aspects of speciation in the southwest Australian flora. Ann. Rev. Ecol. Syst. 10: 399–422.

Huston, M., 1979. A general hypothesis of species diversity. Amer. Nat. 113: 81–101.

Jones, E. W., 1955. Ecological studies in the rain forest of southern Nigeria IV. The plateau forest of the Okumu forest reserve. J. Ecol. 43: 564–594.

Knight, D. H., 1975. A phytosociological analysis of species-rich tropical forest on Barro Colorado Island, Panama. Ecol. Monogr. 45: 259–284.

Lamont, B. B., Downes, S. & Fox, J. E. D., 1977. Importance-value curves and diversity indices applied to a species-rich heathland in Western Australia. Nature 265: 438–441.

Low, B. S., 1978. Environmental uncertainty and the parental strategies of marsupials and placentals. Amer. Nat. 112: 197–213.

May, R. M., 1975. Patterns of species abundance and diversity. In: M. L. Cody & J. M. Diamond (eds.) Ecology and Evolution of Communities, pp. 81–120. Harvard Univ. Press, Cambridge, Mass.

Milewski, J., 1979. A climatic basis for the study of convergence of vegetation structure in mediterranean Australia and southern Africa. J. Biogeog. 6: 293–299.

Naveh, Z. & Whittaker, R. H., 1979. Structural and floristic diversity of shrublands and woodlands in northern Israel and other Mediterranean areas. Vegetatio 41: 171–190.

Paijmans, K. (ed.), 1976. New Guinea Vegetation. CSIRO and Australian National University Press, Canberra, 213 pp.

Parsons, R. F. & Cameron, D. G., 1974. Maximum plant species diversity in terrestrial communities. Biotropica 6: 202–203.

Peet, R. K., 1978a. Forest vegetation of the Colorado Front Range: patterns of species diversity. Vegetatio 37: 65–78.

Peet, R. K., 1978b. Ecosystem convergence. Amer. Nat. 112: 441–444.

Peet, R. K., 1981. Forest vegetation of the Colorado Front Range: composition and dynamics. Vegetatio 45: 3–75.

Peet, R. K. & Christensen, N. L., 1980. Hardwood forest vegetation of the North Carolina Piedmont. Veröff. Geobot. Inst. ETH, Stiftung Rübel, Zurich 69: 14–39.

Peet, R. K., Glenn-Lewin, D. C. & Wolf, J. W., 1983. Prediction of man's impact on plant species diversity: a challenge for vegetation science. In: W. Holzner, M. J. A. Werger & I. Ikusima (eds.), Man's Impact on Vegetation, pp. 41–54, Junk, The Hague.

Pickett, S. T. A. & Thompson, J. N., 1978. Patch dynamics and the design of nature reserves. Biol. Cons. 13: 27–38.

Pielou, E. C., 1975. Ecological Diversity. Wiley-Interscience, N.Y.

Pires, J. M., Dobzhansky, T. & Black, G. A., 1953. An estimate of the numbers of species of trees in an Amazonian forest community. Bot. Gaz. 114: 467–477.

Ramsay, D. McC. & Leeuw, P. N. de, 1965. An analysis of Nigerian savanna. IV Ordination of vegetation developed on different parent materials. J. Ecol. 53: 661–667.

Raven, J. P., 1977. The California flora. In: M. G. Barbour & J. Major (eds.), Terrestrial Vegetation of California, pp. 109–137. Wiley-Interscience, N.Y.

Rice, B. L. & Westoby, M., 1978. Vegetative responses of some Great Basin shrub communities protected against jackrabbits or domestic stock. J. Ranges Manage. 31: 28–34.

Rice, B. & Westoby, M. (in press). Species richness in vascular vegetation of the West Head, N.S.W. Aust. J. Ecol.

Richards, P. W., 1963. Ecological notes on West African vegetation. II Lowland forest of the southern Bakundu forest reserve. J. Ecol. 51: 123–149.

Shmida, A., 1981. Mediterranean vegetation in California and Israel: similarities and differences. Israel J. Bot. 30: 105–123.

Shmida, A. & Whittaker, R. H., 1979. Convergent evolution of deserts in the old and new worlds. In: O. Wilmanns & R. Tüxen (eds.) Werden und Vergehen von Pflanzengesellschaften, pp. 437–450. J. Cramer, Vaduz.

Smith, R. F., 1970. The vegetation structure of a Puerto Rican rain forest before and after short-term gamma irradiation. In: H. T. Odum (ed.), A Tropical Rain Forest, pp. D103–140. U.S. Atomic Energy Commission, Oak Ridge, Tennessee.

Specht, R. L., 1969. A comparison of the sclerophyllous vegetation characteristic of mediterranean type climates in France, California and southern Australia. I. Structure, morphology and succession. Aust. J. Bot. 17: 277–292.

Walter, H. & Lieth, H., 1960. Klimadiagram – Weltatlas. Fischer-Verlag, Jena.

Webb, L. J., Tracey, J. G., Williams, W. T. & Lance, G. N., 1967. Studies in the numerical analysis of complex rain-forest communities, I. A comparison of methods applicable to site/species data. J. Ecol. 55: 171–191.

Westman, W. E., 1975. Pattern and diversity in swamp and dune vegetation, North Stradbroke Island. Aust. J. Bot. 23: 339–354.

Westman, W. E., 1981. Diversity relations and succession in Californian coastal sage scrub. Ecology 62: 170–184.

Westoby, M., 1980. Elements of a theory of vegetation dynamics in arid rangelands. Israel J. Bot. 28: 169–194.

White, P. S., 1979. Pattern, process and natural disturbance in vegetation. Bot. Rev. 45: 229–299.

Whittaker, R. H., 1956. Vegetation of the Great Smoky Mountains. Ecol. Monogr. 26: 1–80.

Whittaker, R. H., 1960. Vegetation of the Siskiyou Mountains, Oregon and California. Ecol. Monogr. 30: 279–338.

Whittaker, R. H., 1965. Dominance and diversity in land plant communities. Science 147: 250–260.

Whittaker, R. H., 1972. Evolution and measurement of species diversity. Taxon 21: 213–251.

Whittaker, R. H., 1977. Evolution of species diversity in land communities. Evol. Biol. 10: 1–67.

Whittaker, R. H. & Niering, W. A., 1965. Vegetation of the Santa Catalina Mountains, Arizona. II. A gradient analysis of the south slope. Ecology 46: 429–452.

Whittaker, R. H. & Niering, W. A., 1975. Vegetation of the Santa Catalina Mountains, Arizona. V. Biomass, production and diversity along the elevation gradient. Ecology 56: 771–790.

Whittaker, R. H., Niering, W. A. & Crisp, M. D., 1979. Structure, pattern, and diversity of a mallee community in New South Wales. Vegetatio 39: 65–76.

Whittaker, R. H. & Woodwell, G. M., 1969. Structure, production and diversity of the oak-pine forest at Brookhaven, New York. J. Ecol. 57: 155–174.

Wyatt-Smith, J., 1966. Ecological studies on Malayan forests. I. Composition of and dynamic studies in lowland evergreen rainforest in two 5-acre plots in Bukit Lagong and Sungei Menyala Forest Reserve and in two half-acre plots in Sungei Menyala Forest Reserve, 1947–1959. Forest Research Institute, Malaya, Research Pamphlet No. 52.

Yodzis, P., 1978. Competition for Space and the Structure of Ecological Communities. Springer-Verlag, Berlin.

Accepted 12.11.1982.

Diversity relations in Cape shrublands and other vegetation in the southeastern Cape, South Africa*

R. M. Cowling**
Botany Department, University of Cape Town, Private Bag, Rondebosch, 7700, South Africa

Keywords: Disturbance, Fynbos, Renosterveld, Soil nutrients, Species diversity, Tension zone, Vegetation dynamics, Vegetation history.

Abstract

This paper investigates, and seeks explanations for, the diversity relations of Cape shrublands (fynbos and renosterveld), subtropical thicket and Afromontane forest, in the biogeographically complex SE Cape. Global comparisons of richness at the 0.1 hectare scale, of communities in the study area and elsewhere in South Africa with analogous vegetation on other continents, were largely inconclusive. Reasons for this are the unexplained variability of richness within vegetation types, problems associated with the scale of diversity used, and difficulties in defining analogous vegetation types. Diversity comparisons within the Cape Region and within the study area communities showed that alpha diversity of fynbos was not consistently higher than other vegetation types. In the study area highest richness was recorded in renosterveld and highest equitability in subtropical thicket; the most species-poor communities were Mountain Fynbos and Afromontane forest. The results of a correlation analysis showed that an index of phytochorological diversity was the factor most strongly correlated with richness in all vegetation types. Soil nutrients did not emerge as significant correlates of diversity except in fynbos where low levels of available nutrients were associated with low values of phytochorological diversity and low species richness. The diversity of fire-prone and grazed communities could be partly explained by non-equilibrium models of species diversity. Ecological and historical hypotheses were presented as explanations for the richness of communities having island-like distributions in the study area. It was generally concluded that historical and ecological factors should be given equal weight in descriptive studies which seek regional and global explanations of the evolution and maintenance of species diversity.

Introduction

Species diversity is a perennial topic in biology which has long intrigued theoreticians and empiri-

* Nomenclature follows the Albany Museum Herbarium (GRA), Grahamstown.
** I thank W. Bond, B. Campbell, H. P. Linder, E. Moll, R. Peet, S. Pierce, W. Westman and the late R. H. Whittaker for valuable comment and discussion. G. Thompson of the Department of Agriculture and Fisheries analyzed the soils. This study forms part of the Fynbos Biome Project and was funded by the C.S.I.R. A special thanks to M. Jarman of C.S.P.: C.S.I.R., for liaison and other supporting facilities.

cists alike. An understanding of the factors which regulate diversity is central to the science of conservation. However, there appears to be no theory of plant species diversity which is consistently corroborated by data from varied communities on different parts of the globe. Even within the relatively well circumscribed and intensively studied North American temperate forests, the prediction of species diversity remains elusive (Glenn-Lewin, 1977; Peet, 1978).

Cape fynbos vegetation has acquired a reputation for high diversity at all levels (Taylor, 1978; Kruger, 1979) although few data have been published. Campbell & van der Meulen (1980) have investigated patterns of alpha and beta diversity along a successional gradient in southwestern (SW) Cape fynbos. Kruger & Taylor (1979) have demonstrated spectacularly high levels of gamma and delta diversity in the Mountain Fynbos of the SW Cape. It is not known whether these patterns are generalizable for the remainder of the Fynbos Biome (*sensu* Kruger, 1978). There are no published data dealing with diversity relations of the fynbos communities in the southern and southeastern regions of the Fynbos Biome.

This paper reports on patterns of species diversity in Cape shrublands (fynbos and renosterveld), subtropical thicket and Afromontane forest in the southeastern (SE) Cape at the eastern boundary of the Fynbos Biome. The aim of the study was to generate hypotheses to explain the evolution and maintenance of diversity in the communities studied. Due to our poor understanding of continental plant species diversity relations, there is no general deductive model from which test implications can be deduced. The usual approach is to use descriptive techniques such as correlation analysis to explore diversity patterns and derive inductive predictions of diversity based on the available data (Glenn-Lewin, 1977).

Study area

The study area is located in the Humansdorp (34°02′S, 24°47′E) region of the SE Cape. Details of the environment and vegetation of the study area are given elsewhere (Cowling, 1983a).

Physiography, geology and climate

Most of the region comprises a coastal plain (0–300 m) which cuts across sandstones and quartzites of the Table Mountain Group, shales of the Bokkeveld Group, and conglomerates of the Uitenhage Group. Along the coast there are recent deposits of calcareous dune sands and calcrete. An important feature in the eastern part of the area is the deep and wide alluvial valley of the Gamtoos River which is incised into the rocks of the Uitenhage Group. The southwestern wall of the valley is composed of conglomerates while the lower eastern wall comprises soft sandstones and mudstones with coarse-grained sandstones and conglomerates

higher up. Above these last-mentioned Uitenhage deposits is a faulted block of limestones and phyllites of the Malmesbury Supergroup. Resting on these are quartzites of the Table Mountain Group which comprise the Elandsberg Range (700–1 000 m), an east trending axis of the Cape Folded Belt which towers above the Gamtoos Valley.

The climate is warm temperate and transitional between Köppen Csb and Cfb climates (Schulze & McGee, 1978). I recognized three climatic types in the study area (Cowling, 1983a).

(i) Coastal plain subhumid climate

The climate is mild with low diurnal and annual temperature ranges. The average annual temperature at Cape St. Francis on the coast is 17.0 °C and the average annual rainfall is 666 mm. The rainfall at Humansdorp, also on the coastal plain, is 667 mm. Rain can fall at any time of the year but the summer months (December, January and February) are always driest.

(ii) Semi-arid valley climate

The river valleys of the SE Cape have a warmer, drier and more variable climate than the adjacent mountains and interfluves: temperature extremes are great and rainfall variability is high (Anon., 1942; Louw, 1976). At Uitenhage in the nearby Swartkops valley the average annual temperature is 18.1 °C and at Hankey, in the Gamtoos Valley, the average annual rainfall is 432 mm. Rainfall distribution is bimodal with spring and autumn peaks.

(iii) Humid coastal mountain climate

In the SE Cape, where rainfall is under strong orographic control, the highest precipitation is recorded on the coastal axis of the Cape Folded Belt. In the study area the upper seaward slopes of the Elandsberg range receive more than 1 000 mm/yr. Rainfall peaks are in autumn and spring. Temperatures are equable but lower than the surrounding lowlands and light frosts and snow are occasionally recorded in winter. I consider this climate most favourable for plant growth, or mesic, in a universal sense (cf. Peet, 1978).

Vegetation and soils

The vegetation of the study area has been classified hierarchically into a series of classes, orders and communities (Cowling, 1983a; see Table 3 for syntaxonomic hierarchy and nomenclature). Detailed soil data are given in Cowling (1983a).

The class Cape Fynbos Shrublands comprises the shrublands and heathlands of the Cape Floristic Region which are confined to acid infertile sands and alkaline calcareous sands (see Taylor, 1978; Kruger, 1979 for reviews). I recognized three orders in the study area. SE Mountain Fynbos is a chorologically pure fynbos confined to uppermost slopes of the Elandsberg and other ranges in the SE Cape. Soils are extremely acid and highly infertile loamy sands. Grassy Fynbos is a chorologically complex fynbos type which occurs on the north and lower slopes, and planed surfaces of the Cape Folded Belt in the SE Cape. It occupies drier, warmer and more fertile sites than the Mountain Fynbos. Grassy Fynbos has a high cover of subtropical C4 grasses which largely replace Restionaceae. South Coast Dune Fynbos is restricted to calcareous dune sands along the southern (S) and SE Cape coasts. It differs from other fynbos types in lacking Proteaceae and having a strong subtropical thicket shrub component.

The class Cape Transitional Small-leaved Shrublands comprise the non-fynbos small-leaved shrublands of the Cape Region and occur on the fine-grained and moderately fertile soils of the coastal forelands and intermontane valleys. Although many of the constituent species are confined to Cape Region, most are chorological transgressors. I recognized one order in the study area – South Coast Renosterveld. This shrubland type is restricted to the S and SE Cape coastal forelands. The dominant shrub is *Elytropappus rhinocerotis*. Grasses and geophytes are conspicuous in the understorey. Much of the South Coast Renosterveld has been derived from grassland in historical times (Cowling, 1983a).

A non-Cape class of shrublands is the Subtropical Transitional Thicket. Structurally these thickets are best described as a tangle of evergreen, spiny, sclerophyllous shrubs and vines with a high cover of succulents in drier regions. Typical tropical-subtropical shrub and tree genera are *Euclea, Diospyros Sideroxylon, Rhus* and *Cassine* and karroid succulent genera include *Crassula, Euphorbia, Delosperma* and *Zygophyllum*. In the Cape, subtropical thicket is confined to deep, well-drained fertile soils of the coastal forelands and intermontane valleys. I recognized two orders in the study area.

Kaffrarian Thicket is a non-succulent type with fairly strong Afromontane affinities. Kaffrarian Succulent Thicket is confined to the hot, dry river valleys of the S and SE Cape. It is characterized by a strong incidence of succulents including many species of karroid affinity. Unlike the fynbos and renosterveld communities, thickets are not fire-prone and, in the study area, are ungrazed.

Afromontane Forest is a class of temperate African mountain forests distributed in an archipelago of mountain 'islands' from Somalia to the Cape Peninsula (White, 1978). Character taxa include *Podocarpus* spp., *Ocotea bullata, Rapanea melanophloeos,* and *Curtisia dentata.* In the study area this forest type is confined to small patches on deep, colluvial soils in the Elandsberg Mountains.

Methods

Data collection

I sampled a total of 194 plots selectively located in the study area to cover the fullest possible range of vegetational diversity and disturbance regimes. It was necessary to standardize plot size since a comparison of species richness requires equal plot sizes. I therefore established an optimal plot size for all vegetation types which would give a compromise between effort expended and information obtained. 19 standard $1-1\,000$ m² samples (Naveh & Whittaker, 1979; Whittaker *et al.,* 1979) were located in a wide range of vegetation types in the study area (Table 2). I used the approach of Werger (1972) which determines optimal plot size as having 50–55% content of a hectare information (number of species in one hectare). Using the standard species area relationship to compute the number of species in one hectare and assuming one hectare as having 100% information content, I was able to determine the information content for plot sizes of 1, 5, 10, 100 and 1 000 m². 100 m² plots gave an information content ranging from 39% to 64% with a mean and standard deviation of 55.1% ± 7.5% for all vegetation types including grasslands, shrublands and forests. The information content of 14 of the 19 samples fell within the range of 55%–65%. The two samples falling well below this range (38.8% and 42.6%) were on coastal dunes where a complex successional mosaic made it difficult to

select homogeneous 0.1 ha stands. I used 10 m ×
10 m plots in all vegetation types of the study area.
The 0.1 ha plots sampled in the study area and
elsewhere in South Africa (Table 2), provided the
data for global comparisons of species richness at
this scale (cf. Naveh & Whittaker, 1979; Rice &
Westoby, 1983a).

In each 100 m² plot I subjectively estimated the
projected canopy cover of each species and recorded a range of site variables including aspects of
community structure, site topography, climate, soil
and biotic factors including post-disturbance vegetation age and grazing intensity. Table 1 lists these
variables with some information on classes and
methods.

Data analysis

Alpha diversity

Alpha diversity is defined as the within-habitat or
intra-community diversity (Whittaker, 1960, 1972)
and is the major focus of this paper. Species rich-

ness (S) is a biologically appropriate measure (Peet,
1974; Whittaker, 1972, 1977) defined simply as the
number of species per site. Heterogeneity or mixed-
diversity measures incorporate both species richness and species evenness (Peet, 1974). I used both
Simpson's index (C) and the Shannon-Wiener
function (H') (Whittaker, 1972; Peet, 1974). Where
it was necessary to show graphical comparisons of
these indices, the reciprocal of C and the exponential form of H' were used (Hill, 1973). Percentage
canopy cover was used to give importance values
for species.

Values of species numbers and area for the 0.1 ha
samples were fitted by the semi-log form of the least
squares regression $S = b + d \log A$ in which S is
species number, and A is area in m². The coefficients b (mean number of plant species in a 1 m²
plot) and d (rate of increase in species numbers with
increasing plot area) are then diversity expressions
that may be compared with S for 1 000 m² (Naveh &
Whittaker, 1979; Whittaker *et al.,* 1979). Although
the log-log form is often superior for species-rich

Table 1. Environmental and biotic variables recorded in plots and used as independent variables in the correlation analyses. Soil
chemical data are from the A horizon only and for a subset of 97 plots. Some details on classes of variables and methods are shown.
Abbreviations are those used in Tables 4–7.

Variable	Abbreviation	Classes of variables and methods
Vegetation age (yr)	VAG	Post-fire, post-bush cut. Estimates based on information from landowners
Grazing intensity	GRZ	Classes: ungrazed = 1, light = 2, moderate = 3, heavy = 4, over-grazed = 5. Scale based on current stocking rate, past grazing and the effects of grazing (cf. Roberts *et al.,* 1975)
Annual rainfall (mm/yr)	RAI	Data from 1:250 000 isohyet maps and local weather stations
Altitude (m)	ALT	From 1:50 000 topographic sheets
Aspect	ASP	Classes: 1 = SE, 2 = S, 3 = SW, 4 = E, 5 = W, 6 = NE, 7 = N, 8 = NW: a cool to hot gradient estimated from aspect-radiation flux data (Schulze, 1975)
Slope inclination	SLO	Slope angle in degrees
Litter cover (%)	LIT	Subjective estimate
Rock cover (%)	RCO	Subjective estimate
Soil depth (m)	SDE	Estimated from augerings in each plot or, for stony soils, from roadside cuttings and other disturbances
Sand content (%)	SAN	Estimated by feel for all plots and tested using the results of a tex-tural analysis (pipette method) of samples from 97 plots
Soil pH	PHH	1 N KCl 1:2.5 solution
Exchangeable calcium (ppm)	CAC	
S value (sum of exchangeable bases) (meq %)	BAS	1 NH₄ acetate leachate
Oxidizable carbon (%)	CAR	Walkley-Black method
Total nitrogen (%)	NIT	Kjeldahl
Available phosphorus (ppm)	PHO	Bray No. 2 (acid extraction, pH 3) Modified Olsen (alkali extraction, pH 8)

vegetation I used the semi-log form since it is widely used in the literature (see Table 2) and was thus of more value in comparing my data to other data sets.

Phytochorological diversity (PHD)

I determined the phytochorological affinities of each species in terms of established phytochoria (Werger, 1978) and computed an index of phytochorological diversity (referred to as PHD in Tables 4–7) for all plots. A phytochorion is a biogeographical unit of any rank. In SE Cape vegetation there are species characteristic of the Cape, Karoo-Namib, Tongaland-Pondoland and Zambezian phytochoria (Werger, 1978). The distribution of taxa were established from locality records in the Albany Museum Herbarium (GRA) and the Bolus Herbarium (BOL) and also from relevant taxonomic revisions. Species were classified as: (a) endemic to a particular phytochorion; (b) linking two (usually) adjacent phytochoria; (c) widely distributed, common in tropical and subtropical phytochoria; (d) widely distributed, occurring in temperate and tropical phytochoria and often having extra-African distributions.

Categories (b)–(d) are ecological and chorological transgressor species (White, 1978). Details on the distribution tracks and examples of species distributions are given in Cowling (1983b). A list showing the classification of species into the phytochorological groups is available on request.

I used the Shannon-Wiener index to compute phytochorological diversity PHD where

$$PHD = \Sigma \, p_i \log p_i$$

in which p_i is the number of species in phytochorological group i expressed as a fraction of the total number of species in the plot.

Correlation analysis

I used simple correlation analysis using the factors shown in Table 1 and PHD as independent variables against each of the diversity variables. All variables were untransformed. Because the fynbos-renosterveld and forest-thicket communities are fundamentally different in their biogeographical affinities (Cowling, 1983b), structure, and disturbance regimes (Cowling, 1983a), these data sets were analyzed separately, in addition to an analysis of the full data set. Only those plots with the full complement of measured soil variables were used in the correlation analysis, thus reducing the sample to 97 plots.

Results

Global comparisons at the 0.1 ha scale

Table 2 shows species diversity and growth form richness at the 0.1 ha scale from 19 sites in the study area and 11 sites elsewhere in South Africa. The data are depicted in a way that makes them directly comparable to other published sources (e.g. Naveh & Whittaker, 1979).

The Cape Fynbos Shrublands in the study area (samples 1–9) showed variable richness with a mean of 66.4 species and a range of 41–93. Most of the species are shrubs and perennial herbs, particularly graminoids; geophytes are common but annuals are rare. This growth form spectrum seems typical of southern hemisphere heathlands on infertile soils (Kruger, 1979; Naveh & Whittaker, 1979; Specht, 1979; Rice & Westoby, 1983b). Dominance concentration (C) was generally low and equitability (H') high for all samples (cf. Westman, 1983). Kruger (1979) gives a mean of 65.0 and a range of 31–126 species per 0.1 ha for eight fynbos samples from the SW Cape, and Naveh & Whittaker (1979) report a mean of 75.0 and a range of 52–128 species from 10 SW Cape fynbos samples.

In common with Cape fynbos, Australian heathlands are also reputed to have high alpha diversity (Marchant, 1973; Lamont et al., 1977; Whittaker, 1977; George et al., 1979; Naveh & Whittaker, 1979; Rice & Westoby, 1983a, b; Hnatiuk & Hopkins, 1981). Naveh & Whittaker (1979) report a mean of 65.0 with a range of 46–82 species, and a mean value of C of 0.22, from seven Western Australian kwongan (heath) samples. George et al. (1979) give species–area data for kwongan which show that at the 0.1 ha scale, eight samples had a mean of 60.6 and a range of 49–81 species. Hnatiuk & Hopkins (1981) report species numbers per 0.1 ha of 46.3 ± 18.3 ($n = 11$), 82.3 ± 16.1 ($n = 42$) and 91.1 ± 13.6 ($n = 32$) from three kwongan communities on a relatively dry (550 mm yr) lateritic plateau in Western Australia. Rice & Westoby (1983b) present data for standard 0.1 ha samples from woodlands ($\bar{X} = 77.9 \pm 9$; $n = 12$), wet scrub ($\bar{X} = 48.8 \pm 8$; $n = 2$), and dry scrub ($\bar{X} = 56 \pm 7$; $n = 12$) on infertile sandstone-derived

Table 2. Comparative diversity measures and growth form representation in 0.1 ha samples from the study area and elsewhere in South Africa.

| Vegetation[a] | S/0.1 ha | S/m² | Coefficients[b] | | H' | C | Woody | | |
			b	d			Tree & shrub	Sub-shrub	Vine
Cape Fynbos Shrublands									
1 Thamnus-Tetraria	41	12.8	13.3	9.5	1.23	0.08	18	7	
2 Erica-Trachypogon	60	13.1	10.2	14.8	1.00	0.08	14	18	
3 Erica-Trachypogon	83	17.4	15.8	21.6	1.37	0.08	26	16	
4 Protea-Clutia	87	24.5	26.4	20.6	1.61	0.08	9	37	
5 Thamnochortus-Erica	45	15.7	15.7	9.7	1.24	0.10	6	11	
6 Thamnochortus-Tristachya	74	22.0	21.4	16.9	1.18	0.10	1	17	
7 Themeda-Stenotaphrum	93	15.5	7.4	24.6	1.91	0.18	4	21	
8 Restio-Agathosma	60	15.8	14.7	14.6	1.17	0.13	10	19	
9 Restio-Maytenus	55	6.7	2.3	15.8	1.08	0.12	32	1	3
Cape Transitional Small-leaved Shrublands									
10 Themeda-Cliffortia	57	5.1	10.0	16.4	0.99	0.22		13	
11 Themeda-Cliffortia	74	22.6	23.6	17.0	1.25	0.25	2	22	
12 Elytropappus-Eustachys	95	19.4	21.5	25.6	1.41	0.07		40	
13 Elytropappus-Eustachys	87	21.3	23.6	21.8	1.26	0.08	11	29	
14 Elytropappus-Relhania	85	8.6	12.7	25.1	1.03	0.19	10	27	
15 W Coast Renosterveld Cape Town, SW Cape	99	21.0	19.3	26.1	1.34	0.13		28	
16 W Coast Renosterveld Tygerberg, SW Cape	103	20.9	18.7	26.8	1.26	0.24	21	9	
17 S Coast Renosterveld Swellendam, S Cape	60	13.4	11.7	16.5	1.04	0.14	5	15	
18 S Coast Renosterveld Grahamstown, SE Cape	95	18.7	19.6	25.4	1.30	0.09	11	22	
Karroid Shrublands									
19 Karroid Broken Veld Worcester, SW Cape[f]	33	4.6	3.1	10.0	0.75	0.27		6	5
Subtropical Transitional Thicket									
20 Pterocelastrus-Euclea	52	9.7	10.3	14.1	1.32	0.06	25		11
21 Pterocelastrus-Gonioma	71	14.9	14.4	18.3	1.53	0.05	40		11
22 Cassine-Cussonia	37	9.0	9.1	9.4	1.22	0.10	18	1	10
23 Sideroxylon-Euphorbia	61	13.4	10.9	15.5	1.31	0.07	13		17
24 Euclea-Brachylaena	98	4.5	4.1	30.1	1.17	0.12	26	10	14
25 Dune thicket Kinklebosch. SE Cape	48	11.0	10.2	12.5	1.26	0.08	14		17
26 Dune thicket Koeberg, SW Cape[g]	35	6.9	7.3	8.7	1.32	0.12	14	3	3
Subtropical Forest									
27 Coastal forest Durban, Natal	73	14.5	13.6	19.2	1.22	0.08	43	8	18
Subtropical Grassland Savanna									
28 Aristida junciformis grassland, Durban	76	15.0	14.5	20.0	1.05	0.21	2	11	
29 Aristida-Protea roupelliae savanna, Durban	97	14.0	9.8	26.5	1.21	0.11	21	8	
Afromontane Forest									
30 Rapana-Canthium	53	11.0	7.6	16.9	1.44	0.06	31		8

[a] Unless otherwise indicated, samples are from communities in the study area.

[b] From least squares regressions, $S = b + d \log A$ (see methods).

[c] Forbs exclude geophytes as well as graminoids under perennial herbs.

[d] Graminoids include Cyperaceae, Restionaceae (mainly in fynbos communities) and Poaceae.

[e] According to Campbell *et al.* (1981).

[f] Data of R. H. Whittaker, L. Olsvig, R. M. Cowling and S. Milton.

[g] Data of R. H. Whittaker, L. Olsvig and B. Low.

* Data collected in mid-summer and excludes most geophytes and annuals.

Succulent			Herbs-perennial			Annual		Structural characterization[e]
Tree & shrub	Sub-shrub	Vine	Forb[c]	Grami-noid[d]	Geo-phyte	Forb	Grami-noid	
–	–	–	1	9	6	–	–	Mid-high mid-dense proteoid shrubland with a closed restioid understorey
–	6	–	4	12	6	–	–	Mid-high open proteoid shrubland with a low closed ericoid shrub understorey
–	7	–	9	14	10	1	–	Mid-high mid-dense proteoid shrubland with a low mid-dense ericoid shrub understorey
–	1	–	16	17	6	1	–	Low mid-dense ericoid and grassy/restioid shrubland
–	1	–	5	15	6	1	–	Low closed restioid heathland
–	2	–	17	24	9	1	3	Closed restioid grassland
–	3	–	27	16	12	9	1	Closed restioid grassland with a low open ericoid shrub overstorey
–	–	–	13	10	8	–	–	Low mid-dense ericoid and restioid shrubland
–	4	–	5	5	–	–	–	Mid-high mid-dense large-leaved shrubland/low open ericoid shrubland
–	1	–	11	16	15	–	1	Closed grassland
–	3	–	15	12	21	–	–	Low mid-dense small-leaved grassy shrubland
–	7	–	16	17	15	–	–	Closed grassland
–	5	–	14	15	13	–	–	Mid-high mid-dense small-leaved grassy shrubland
–	1	–	20	14	11	–	–	Low mid-dense small-leaved grassy shrubland
–	7	–	16	9	20	4	15	Mid-dense grassland with a low open small-leaved shrub overstorey
–	1	–	14	11	39	1	7	Low closed small-leaved grassy shrubland
–	3	–	7	10	18	–	2	Low mid-dense small-leaved grassy shrubland
–	16	–	16	13	17	–	–	Low closed small-leaved grassy shrubland
–	21	–	2	–	1*	–*	–*	Low sparse small-leaved shrubland with a sparse succulent dwarf shrub understorey
–	3	3	2	4	5	1	–	Tall closed large-leaved shrubland
1	3	3	4	4	5	–	–	Tall mid-dense large-leaved shrubland with a mid-dense low tree overstorey
–	–	–	2	4	–	2	–	Tall closed large-leaved shrubland
3	10	3	5	4	6	–	–	Tall closed large-leaved succulent shrubland with a low sparse succulent tree overstorey
12	16	4	4	8	4	–	–	Mid-high mid-dense large-leaved and succulent shrubland
2	–	3	6	3	2	–	–	Tall closed large-leaved shrubland
3	6	–	–	3	–*	1*	–*	Mid-high open large-leaved shrubland with a sparse dwarf succulent shrub understorey
–	–	–	1	1	2	–	–	Low forest
–	–	–	31	17	13	2	–	Closed grassland
–	2	–	36	18	9	3	–	Closed grassland with a tall mid-dense proteoid shrub overstorey
–	–	–	5	6	3	–	–	Tall forest

soils near Sydney. All communities had strong heathland (*sensu* Specht, 1979) characteristics.

Samples 10–18 in Table 2 are Cape Transitional Small-leaved Shrublands. All samples belong to the Coast Renosterveld (Acocks, 1953; Taylor, 1978; Boucher & Moll, 1980) and while samples 10–14 are confined to the study area, the remaining samples span the full range of this vegetation type from the winter rainfall region of the SW Cape to the transitional summer rainfall region of the SE Cape. These communities occur on moderately fertile loams to clay-loams and are therefore somewhat comparable to mediterranean-type shrublands (*sensu* Di Castri, 1980). In the renosterveld samples richness was higher than all fynbos and most kwongan data presented above, with a mean of 83.8 and a range of 57–103 species; values of C and H' were higher and lower respectively than the fynbos data shown in Table 2. Of particular interest in the renosterveld communities is the exceptional wealth of geophytes.

Renosterveld appears to be more diverse than its putative analogues in other mediterranean-type regions including the mesic, evergreen, sclerophyllous types (e.g. maquis, chaparral, mallee; Naveh & Whittaker, 1979) and the xeric, drought-deciduous, mesophyllous types (e.g. coastal sage scrub, phrygana; Westman, 1981). However, it is poorer than grazed open shrublands and woodlands of Israel where Naveh & Whittaker (1979) found up to 197 species per 0.1 ha in disturbed, annual-rich communities.

Table 2 shows a succulent karroid shrubland (Acocks, 1953) sample on fertile clay-loam in a xerothermic mediterranean climate (average rainfall of 213 mm/yr) in the SW Cape. Overall richness is low but there is a high diversity of succulent sub-shrubs. Werger (1972) gives species–area data for nine non-succulent karroid communities from the semi-arid regions of the northern Cape; he recorded a mean of 46.8 species per 0.12 ha with a range of 25–61. Rice & Westoby (1983a) report a mean of 64.7 species from semi-arid chenopod shrublands in New South Wales where soils are red clay-loams and rainfall is non-seasonal 200 mm/yr. About 60% of the species in the Australian shrublands are annuals. Due to the time of sampling the full complement of annuals and geophytes were not recorded at the South African karoo site. Whittaker & Niering (1965, 1968) record a range of 57–70 species in Sonoran Desert scrub and desert grassland on limestone in the semi-arid, near subtropical climate of the lower slopes of the Santa Catalina Mountains in Arizona. These figures include both summer and winter annuals.

Samples from subtropical thicket are mostly from the study area (20–24) but include a dune thicket sample from elsewhere in the SE Cape (25) and from the SW Cape (26) (Table 2). These thickets, especially the drier ones, have a wealth of growth forms including large-leaved sclerophyll trees, shrubs and vines, arborescent stem and leaf succulents, succulent shrubs and vines, forbs, graminoids and geophytes. Subtropical thicket is moderately rich with the highest species numbers recorded in the dry succulent types (samples 23 and 24) and lowest numbers in the dune thickets (samples 22, 25 and 26) especially the depauperate sample from the SW Cape. Equitability is higher than fynbos samples. These thickets are a subtropical formation which have recently penetrated into the temperate Cape (Cowling, 1983b) and have no clear analogue in other mediterranean and sub-mediterranean regions. Boucher & Moll (1980) and Specht & Moll (1983) call the southwertern Cape thickets mediterranean shrublands (*sensu* Di Castri, 1980), a clearly erroneous interpretation since affinities are strongly tropical and there are very few species endemic to the Cape mediterranean-climate region (Cowling, 1983b). Samples 23 and 24 have similar climatic and soil conditions to the Australian mallee community studied by Whittaker *et al.* (1979) who recorded a mean of 52.7 and a range of 49–62 species in 10 0.1 ha samples. However unlike the South African thicket, the mallee community is a fire-prone, open shrubland with few vines and succulents, and numerous annuals.

The two forest samples in Table 2 include a subtropical coastal forest (Acocks, 1953) from Natal (sample 27) and a temperate Afromontane forest from the study area (sample 30). Rice & Westoby (1983a) quote data on the richness of Australian tropical rain forests which are considerably more diverse than the Natal forest sample. There is a large body of data on species richness in 0.1 ha plots from North American temperate forests (see Whittaker, 1969, 1972, 1977 for reviews; Peet, 1978), where species numbers average about 50 but values as low as 4 and as high as 115 have been recorded (Rice & Westoby, 1983a). Whittaker (1977) gives a mean of 47.7 and a range of 29–105 species from 28 0.1 ha samples of temperate Australian forests and woodlands. Rice & Westoby (1983a) noted similar richness values for their Australian temperate forest samples and the North American forests.

The two remaining samples are from species-rich coastal subtropical grassland and savanna on acid, sandy, infertile lithosols derived from T. M. G. sandstones. The communities are dominated by tropical C_4 grasses but include species of typical Cape fynbos genera (e.g. *Protea, Agathosma, Watsonia, Aristea*). The geophytic flora is relatively rich. Very few data are available from analogous communities on other continents. Cerrado, a tropical Brazilian savanna confined to infertile sands is well noted for its floristic richness where about 230 species were reported in 0.1 ha (Eiten, 1978). Rice & Westoby (1983a) give a range of 25–50 species in north Australian tropical grassy woodlands.

Diversity patterns in the study area

Alpha diversity

Table 3 shows patterns of alpha diversity in 100 m^2 plots from communities in the study area. Within the Cape shrublands, SE Mountain Fynbos had the lowest richness (*S*) but values of dominance concentration (*C*) and equitability (*H'*) were comparable to other fynbos communities. Mountain Fynbos was poorer than all forest and thicket except for the *Rapanea-Ocotea* Afromontane forest community. A comparison with richness data from S and SW Cape Mountain Fynbos communities stresses the depauperate nature of the SE Mountain Fynbos. Mean number of species recorded in 50 m^2 samples from three Mountain Fynbos communities at Jonkershoek, SW Cape, were 39, 39 and 50 (Werger *et al.*, 1972). Mean species numbers recorded in 124 50 m^2 Mountain Fynbos samples (excluding azonal communities) from Cape Hangklip, SW Cape, was 32.2 ± 8.9 (Boucher, 1978). More precisely comparable data come from Bond's (1981) study in the southern Cape mountains. In the Outeniqua Mountains, some 260 km west of the study area, he recognized Mesic Proteoid Shrublands which share with the SE Mountain Fynbos, similar climate, soils and dominant proteoid and graminoid species. In three communities comparable to those in the study area, Bond (1981) recorded mean species richness in 50 m^2 samples of 30.2 (range = 21–37), 41.8 (35–48) and 33.2 (29–39).

Grassy Fynbos communities showed variable values of *S, H'* and *C* (Table 3). The least diverse community (*Thamnochortus-Erica*) had the lowest values of phytochorological diversity (PHD) and the poorest soils. The most diverse community (*Thamnochortus-Tristachya*) is a restioid grassland (*sensu* Campbell *et al.*, 1981) which is frequently burnt or mowed and moderately grazed by domestic livestock.

Dune fynbos communities were poorer than Grassy Fynbos and had lower and higher values of *H'* and *C* respectively. Highest richness but lowest equitability were recorded for the *Themeda-Stenotaphrum* community which is mowed and grazed. Van der Merwe (1976) reported a richness per 100 m^2 plots of 30.9 (16–42) and 33.3 (11–54) in two Dune Fynbos communities 200 km west of the study area.

In the study area *S* was generally highest in the renosterveld communities but *C* was consistently high reflecting the greater dominance by one or few species. Lowest diversity was in the most mesic community (*Themeda-Cliffortia*) on seasonally waterlogged soils while highest diversity was in the drier *Elytropappus-Eustachys* community on well-drained, sandy, stony soils.

Within the forest and thicket communities, Afromontane forest was least diverse. The drier *Rapanea-Canthium* community with its strong admixture of subtropical forest trees and vines was richer than the cooler, wetter and chorologically purer *Rapanea-Ocotea* community. Data from Campbell & Moll (1977) and McKenzie *et al.* (1977) indicate a mean of about 18 species in 100 m^2 plots from chorologically pure Afromontane forests on the Cape Peninsula. McKenzie's (1978) data for SW Cape Afromontane forests show a mean of about 19 species in 100 m^2 plots. However there is a distinct tendency for alpha richness to increase in forest patches closest to the Knysna forest 'source area' (cf. Cody, 1983).

Diversity, particularly heterogeneity diversity, was highest in the dry succulent thickets and lowest in the dune thickets (*Cassine-Cussonia, Cassine-Schotia*). Van der Merwe (1976) recorded a similar richness in 100 m^2 samples ($\bar{X}$: 24.7; range : 16–41) for a community almost identical to the *Cassine-Cussonia* community in Table 3. In related communities on calcareous substrates in the SW Cape, Van der Merwe (1977) recorded mean species richness of 11.3 (6–22) and 19.9 (6–32) in 100 m^2 sam-

Table 3. Alpha diversities in 100 m² plots in communities of the study area, expressed as species richness (*S*) and heterogeneity-diversity (*H'* = Shannon-Wiener index, *C* = Simpsons index). Also shown is the total number of species recorded in combined samples from each community.

Syntaxa[a]	n[b]	n[c]	S			H'			C			Total species numbers in combined samples
			$\bar{X}$	SD	range	$\bar{X}$	SD	range	$\bar{X}$	SD	range	
Cape Fynbos Shrublands												
SE Mountain Fynbos												
Tetraria-Thamnus community	8	4	26.4	5.3	22–36	1.03	0.13	0.83–1.22	0.13	0.05	0.08–0.22	87
Leucospermum-Tetraria	5	2	26.6	4.6	22–34	1.00	0.07	0.90–1.10	0.14	0.02	0.13–0.18	56
Grassy Fynbos												
Thamnochortus-Erica	5	3	33.8	9.2	24–45	1.10	0.14	0.93–1.27	0.14	0.05	0.07–0.20	69
Protea-Clutia	9	4	43.5	9.8	35–66	1.09	0.13	0.85–1.30	0.13	0.04	0.09–0.21	128
Erica-Trachypogon	16	9	40.1	8.4	28–56	1.08	0.17	0.56–1.30	0.14	0.10	0.08–0.51	212
Themeda-Passerina	4	2	36.3	4.5	30–40	1.00	0.15	0.82–1.20	0.17	0.06	0.09–0.24	92
Thamnochortus-Tristachya	7	5	46.6	4.7	38–51	1.20	0.04	1.16–1.27	0.10	0.02	0.07–0.13	133
South Coast Dune Fynbos												
Restio-Agathosma	13	6	32.4	11.2	14–58	0.97	0.18	0.54–1.16	0.17	0.10	0.11–0.48	144
Restio-Maytenus	12	4	33.5	8.4	25–51	1.04	0.19	0.83–1.36	0.17	0.09	0.06–0.35	114
Themeda-Stenotaphrum	5	3	40.2	5.0	32–45	0.81	0.07	0.72–0.89	0.27	0.04	0.25–0.34	104
Cape Transitional Small-leaved Shrublands												
South Coast Renosterveld												
Themeda-Cliffortia	11	6	33.6	8.4	21–48	0.72	0.24	0.44–1.17	0.39	0.17	0.09–0.65	120
Elytropappus-Eustachys	10	5	51.5	14.7	36–73	1.11	0.19	0.87–1.34	0.19	0.13	0.07–0.51	162
Elytropappus-Metalasia	12	7	40.8	7.8	26–52	0.94	0.19	0.69–1.34	0.23	0.11	0.07–0.46	152
Elytropappus-Relhania	4	1	42.5	11.9	35–60	0.81	0.28	0.53–1.17	0.36	0.21	0.11–0.57	97
Afromontane Forest												
Knysna Afromontane Forest												
Rapanea-Canthium	5	2	32.8	2.3	30–36	1.12	0.12	0.90–1.21	0.12	0.07	0.09–0.24	71
Rapanea-Ocotea	4	2	21.2	2.1	19–24	0.97	0.07	0.87–1.04	0.16	0.04	0.12–0.21	34
Subtropical Transitional Thicket												
Kaffrarian Thicket												
Pterocelastrus-Gonioma	11	4	36.9	7.8	23–46	1.24	0.13	1.02–1.39	0.08	0.03	0.06–0.14	107
Pterocelastrus-Euclea	18	8	37.3	8.1	26–54	1.19	0.19	0.80–1.44	0.11	0.07	0.04–0.33	130
Cassine-Cussonia	12	5	24.6	6.1	13–37	0.92	0.20	0.61–1.21	0.21	0.09	0.11–0.39	93
Cassine-Schotia	5	2	29.6	7.2	21–37	1.14	0.13	1.01–1.31	0.12	0.04	0.06–0.15	67
Kaffrarian Succulent Thicket												
Sideroxylon-Euphorbia	15	7	35.8	10.9	20–52	1.21	0.10	1.01–1.37	0.09	0.02	0.06–0.12	120
Euclea-Brachylaena	5	3	45.0	7.6	39–56	1.26	0.08	1.16–1.35	0.09	0.02	0.07–0.11	108

[a] Hierarchy of classes, orders and communities (Cowling, 1983a).

[b] Total number of samples used to compute diversity indices.

[c] Samples with soil data used in correlation analyses.

ples. These data further illustrate the westwards depauperization of subtropical thicket in the Cape Region (Cowling, 1983b).

Community richness, expressed as the total number of species in combined samples (Table 3) showed interesting patterns which did not always correlate with sample means. Comparing communities with similar sample sizes, it can be inferred that highest community richness was recorded in grassy fynbos and renosterveld communities. Thicket communities with similar means for *S* had a lower overall community richness which could mean a lower beta diversity.

Correlation analysis

The three measures of alpha diversity were subjected to simple correlation analysis in relation to 15 environmental and other variables. Matrices of simple correlation coefficients between all variables are given in Tables 4, 5 and 6 for the full data set, fynbos-renosterveld and forest-thicket communities respectively.

Predictably many independent variables were highly co-linear in all data sets. In the full data set (Table 4) species richness (S) showed strong positive relationships with phytochorological diversity (PHD), grazing intensity (GRZ), and negative relationships with litter (LIT) and average annual rainfall (RAI). Similar results were obtained for the fynbos and renosterveld communities (Table 5) where vegetation age (VAG) was also highly correlated with S. In the forest and thicket communities (Table 6) S was also strongly correlated with PHD and RAI and showed a weaker negative relationship with percentage sand (SAN) and a positive relationship with rock cover (RCO). In all data sets there were no significant relationships between S and variables reflecting soil fertility (cf. Grime, 1973; Huston, 1979, 1980; Peet & Christensen, 1980).

Figure 1 shows the relationship of S to PHD. The correlation is stronger for the forest and thicket than the fynbos and renosterveld communities. Figure 2 shows the relationship of S to the two disturbance variables, VAG and GRZ, for grazed and fire-prone fynbos and renosterveld communities. When the renosterveld was treated separately, S was not significantly correlated with both VAG and GRZ. In the fynbos communities disturbance is clearly implicated in the regulation of species rich-

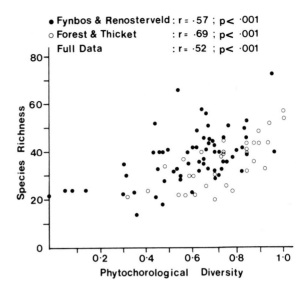

Fig. 1. Relationship of species richness per 100 m² to phytochorological diversity for fynbos and renosterveld ($n = 64$), forest and thicket ($n = 33$), and the full data set.

ness (Campbell & van der Meulen, 1980; Van Wilgen, 1980; Kruger, 1983). The relationship between S and GRZ is curvilinear and the overall pattern supports the intermediate disturbance hypothesis (Connell, 1978) for which there is considerable empirical support from different ecosystems (Harper, 1969; Zeevalking & Fresco, 1977; Lubchenco, 1978; Naveh & Whittaker, 1979; Peet, 1978). The possibility of a spurious relationship between S and both

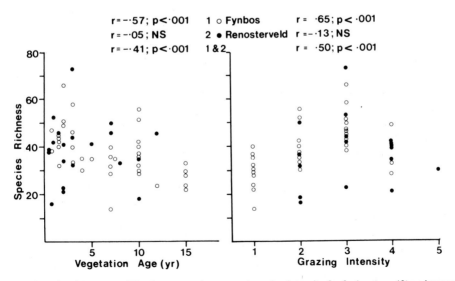

Fig. 2. Relationship of species richness per 100 m² to vegetation age and grazing intensity for fynbos ($n = 43$) and renosterveld ($n = 21$) communities.

Table 4. Correlation matrix for species diversity variables and various environmental and other variables. Full data set ($n = 97$).

	RAI	ALT	SLO	ASP	LIT	SDE	RCO	SAN	PHH
Annual rainfall	1.00								
Altitude	0.49***	1.00							
Slope inclination	0.21*	0.32***	1.00						
Aspect	-0.13	-0.09	-0.11	1.00					
Litter cover	0.00	-0.16	0.30**	-0.03	1.00				
Soil depth	0.22*	-0.15	-0.03	0.03	0.20*	1.00			
Rock cover	-0.14	0.29***	0.09	0.02	-0.24*	-0.58***	1.00		
% sand content	0.29**	-0.15	-0.11	-0.19	-0.24*	0.19	0.09	1.00	
Soil pH	-0.16	-0.57***	-0.14	-0.05	0.10	0.42***	-0.40***	0.24*	1.00
Exchangeable calcium	-0.21*	-0.34***	0.05	-0.14	0.35***	0.31***	-0.30**	-0.03	0.75***
S value	-0.24*	-0.39***	0.19	-0.10	0.45***	0.23*	-0.31**	-0.20*	0.65***
Oxidizable carbon	0.02	-0.06	0.41***	-0.03	0.40***	-0.06	-0.06	-0.30***	0.09
Total nitrogen	-0.02	-0.13	0.28**	0.03	0.49***	0.08	-0.15	-0.38***	0.21*
Available phosphorus	-0.10	-0.27**	-0.06	0.06	0.28**	0.37***	-0.24*	0.05	0.49***
Vegetation age	-0.12	0.03	-0.40	-0.03	0.83***	0.30**	-0.24*	-0.37***	0.16
Grazing intensity	-0.05	-0.08	-0.38***	0.04	-0.80***	-0.27**	0.11	0.11	-0.19
Phytochorological diversity	-0.68***	-0.37***	-0.04	0.14	0.02	0.07	-0.09	-0.24*	0.25*
Species richness	-0.37***	-0.08	-0.10	-0.04	-0.31**	-0.13	0.01	-0.02	-0.11
Shannon-Wiener	-0.25*	0.19	0.22*	-0.04	0.27**	0.03	0.07	-0.04	-0.11
Simpsons	0.13	-0.25*	-0.22*	0.04	-0.29**	-0.05	-0.16	-0.06	0.12

* $p < 0.05$; ** $p < 0.01$; *** $p < 0.001$

VAG and GRZ cannot be overlooked. For example there were no recently burnt or grazed plots in the SE Mountain Fynbos communities. I do not know what the effects of grazing or frequent burning would be on these communities. There are indications in SW Cape Mountain Fynbos that burning mature vegetation causes an increase in S (Campbell & van der Meulen, 1980) and Van Wilgen (1980) has reported richer stands in vegetation burnt on a short (6 yr) rotation than those with longer fire cycles. Much of the ambiguity results from combining different communities in the correlation analysis. I therefore examine later, for single fynbos and renosterveld communities, the response of diversity to different fire and grazing regimes.

In the full data set (Table 4) equitability (H') was highly correlated with S and showed similar relationships to independent variables. H' increased with increasing vegetation age (VAG) but this does not necessarily contradict the generalization that equitability decreases with successional time (Auclair & Goff, 1971; Whittaker, 1969, 1972, 1977;

Shafi & Yarranton, 1973; Houssard et al., 1980; Westman, 1981). This relationship reflects the relatively higher values of H' for non fire-prone forest and thicket communities when compared to the frequently disturbed fynbos and renosterveld communities. Measures of soil fertility such as sum of exchangeable cations (BAS) and total nitrogen (NIT), both highly intercorrelated, emerged as significant negative correlates with H' for the fynbos-renosterveld data; there was also a significant positive correlation with SAN (Table 5). These relationships draw attention to the lower equitability of renosterveld communities on the heavier, more fertile soils. In the forest and thicket data, H' and S were highly co-linear ($r = 0.83$; $p < 0.001$) and were correlated with similar suites of independent variables.

Dominance concentration (C) and H' were highly intercorrelated in all data sets ($r = -0.92$; $r = -0.93$; $r = -0.89$ for full data, fynbos-renosterveld and forest thicket data sets respectively). Relationships are therefore the inverse of those with H'.

1.00										
0.91***	1.00									
0.46***	0.68***	1.00								
0.55***	0.73***	0.85***	1.00							
0.40***	0.35***	0.06	0.11	1.00						
0.45***	0.59***	0.51***	0.59***	0.31***	1.00					
-0.40***	-0.47***	-0.42***	-0.42***	-0.31***	-0.83***	1.00				
0.34***	0.34***	0.09	0.15	0.10	0.24*	-0.09	1.00			
-0.06	0.09	-0.14	-0.16	-0.16	-0.15	0.34***	0.52***	1.00		
0.12	0.14	0.13	0.13	0.06	0.40***	-0.26**	0.33***	0.52***	1.00	
-0.09	-1.00	-0.11	-0.08	-0.11	-0.35***	0.27**	-0.17	-0.33**	-0.92***	1.00
CAC	BAS	CAR	NIT	PHO	VAG	GRZ	PHD	S	H'	C

Dynamics and disturbance

In this section I investigate the role of dynamics (succession) and disturbance (burning, mowing and grazing) in regulating species diversity in selected communities. Since different communities may respond to similar perturbations in different ways I determined the effects of grazing and burning/mowing on the diversity relations in a Grassy Fynbos and a renosterveld community.

A structural classification of these communities showed that they could be divided into two major structural cover states (cf. Slatyer, 1976). The *Erica-Trachypogon* Grassy Fynbos community comprises an ericoid shrub/grassy cover state which is burnt on a 3–4 yr rotation and is moderately grazed; and a proteoid shrub cover state which is burnt on a 10–12 yr rotation and is lightly grazed (Table 7). The former cover state had significantly more species per 100 m² plots than the latter; it also had higher values of H' and lower values of C although these differences were not significant. Frequent burning and moderate grazing appear to increase species diversity in this and other Grassy Fynbos communities in the study area. There are certain compositional changes which are not reflected in the diversity measures (Table 7). Frequent burning in fynbos communities often eliminates obligate reseeding species many of which are serotinous and cannot produce new propagules in the short intervals between fires (Bond, 1980, 1983; Campbell & van der Meulen, 1980; Van Wilgen, 1980; Kruger, 1983). In the study area these include the proteoid shrubs, *Protea neriifolia* and *P. repens*. However, the elimination of these shrubs under a short fire cycle is more than compensated for by the post-fire invasion of many shade intolerant forbs and C_4 grasses which are suppressed under dense proteoid shrub canopy.

The *Themeda-Cliffortia* community is a wet renosterveld on seasonally waterlogged duplex soils and exists in a small-leaved shrub or grassland cover states (Table 7). The shrub state is derived from the grassland as a consequence of severe and prolonged overgrazing which reduces grass vigour and

Table 5. Correlation matrix for species diversity variables and various environmental and other variables. Fynbos and renosterveld communities ($n = 64$).

	RAI	ALT	SLO	ASP	LIT	SDE	RCO	SAN	PHH
Annual rainfall	1.00								
Altitude	0.43***	1.00							
Slope inclination	0.36**	0.47***	1.00						
Aspect	-0.07	0.17	-0.12	1.00					
Litter cover	0.29*	0.44***	0.33**	-0.07	1.00				
Soil depth	0.39**	-0.24	-0.03	0.05	-0.21	1.00			
Rock cover	-0.27	0.34**	0.25*	-0.01	0.36**	-0.57***	1.00		
% sand content	0.31*	0.06	0.04	-0.24	0.11	0.43***	0.04	1.00	
Soil pH	0.02	-0.51***	-0.17	0.05	-0.17	0.47***	-0.40***	0.25*	1.00
Exchangeable calcium	-0.02	-0.43***	-0.05	0.05	-0.05	0.40***	-0.24	0.24	0.91***
S value	-0.12	-0.51***	-0.07	0.14	-0.12	0.31*	-0.23	-0.02	0.82***
Oxidizable carbon	-0.16	-0.17	0.15	0.23	0.06	-0.14	0.20	-0.24	0.12
Total nitrogen	-0.26*	-0.39**	-0.01	0.27*	-0.13	-0.02	0.02	-0.34**	0.35**
Available phosphorus	0.16	-0.31*	0.00	0.01	-0.07	0.48***	-0.19	0.35**	0.72***
Vegetation age	0.27*	0.50***	0.43***	-0.07	0.75***	-0.25*	0.36**	0.02	-0.25*
Grazing intensity	-0.33**	-0.25*	-0.32**	0.04	-0.64***	0.01	-0.21	-0.31*	-0.03
Phytochorological diversity	-0.59***	-0.40**	-0.27*	0.10	-0.38***	0.03	-0.01	-0.06	0.21
Species richness	-0.32*	-0.16	-0.24	-0.22	-0.42***	0.01	-0.09	0.11	-0.05
Shannon-Wiener	-0.10	0.29*	0.05	-0.16	-0.02	-0.04	0.20	0.38**	-0.24
Simpsons	0.05	-0.34**	-0.09	0.08	-0.08	0.03	-0.29*	-0.39**	0.22

* $p < 0.05$; ** $p < 0.01$; *** $p < 0.001$.

basal cover, and results in soil capping, increased run-off and eventually the disturbance of water table dynamics (Cowling, 1983a). There ensues a thickening up of relatively deep-rooted small-leaved shrub species (e.g. *Cliffortia linearifolia, Elytropappus rhinocerotis*) which are able to exploit subsoil moisture, unavailable to the shallow rooted grasses. Table 7 shows that there is a slight, but insignificant reduction of S in the shrubland cover state relative to the grassland; the latter had significantly lower values of H' and higher values of C'. The grassland is overwhelmingly dominated by a single grass species, *Themeda triandra,* but has many other grasses, forbs, geophytes and small-leaved shrubs (including those which dominate in the shrubland state), which contribute little to overall cover. Disturbance of the grassland results in reduced dominance concentration and increased equitability as species importance is more evenly distributed as the shrubs cover increases. However overall composition and richness do not change much although certain grasses are eliminated in the shrubland state.

Next I investigated diversity trends along a successional and grazing intensity gradient in communities on coastal dunes (Fig. 3). The successional communities are contemporary and therefore not strictly sequential. I was careful to select only plots on deep, well-drained sand from a climatically homogeneous area where vegetation ultimately develops into a dense dune thicket. However, there is no certainty that pre-climax stages will actually

1.00										
0.94***	1.00									
0.38**	0.57***	1.00								
0.56***	0.77***	0.87***	1.00							
0.62***	0.53***	0.06	0.20	1.00						
-0.18	-0.23	0.05	-0.17	-0.07	1.00					
0.23	0.01	-0.14	-0.10	-0.17	-0.70***	1.00				
-0.08	0.28*	0.16	0.28*	0.19	-0.37**	0.28*	1.00			
-0.11	-0.15	-0.19	-0.12	-0.02	-0.41***	0.50***	0.57***	1.00		
-0.22	-0.33**	-0.19	-0.29*	-0.07	-0.05	0.09	0.07	0.43***	1.00	
0.19	0.30*	0.19	0.30*	0.04	-0.05	0.01	0.01	-0.34**	-0.93***	1.00
CAC	BAS	CAR	NIT	PHO	VAG	GRZ	PHD	*S*	*H'*	*C'*

follow this sequence. The assumption that spatially separate vegetation represents states of the same system at different stages of development has been criticized (Drury & Nisbet, 1973; Goodall, 1977). On the dunes and in the renosterveld and fynbos communities mentioned below, this assumption is borne out by fence-line contrasts separating different (structural) communities on identical sites (Cowling, 1983a). On the dunes the only physical site variable to show a consistent trend along the gradient was an increase in soil organic content due largely to a build-up in decomposition products from litter accumulated in the more mature communities.

The successional communities, ranged along the gradient of increasing vegetation age and decreasing grazing intensity, were a restioid grassland (*Themeda-Stenotaphrum*), dune fynbos (*Restio-Agathosma*), mixed dune fynbos-thicket (*Restio-Maytenus*) and the climax dune thicket (*Cassine-Cussonia*) (Fig. 3). The relative importance of graminoids decreased, and large-leaved shrubs increased, along the gradient while small-leaved shrubs showed a mid-successional peak. Richness decreased monotonically along the gradient and equitability increased to a maximum in the pre-climax stage and then declined. Dominance concentration showed exactly the inverse trend of H'; when expressed as $1/C$ and $e^{H'}$, these diversity measures changed in parallel (Fig. 3).

Many studies show a mid-successional peak in diversity followed by a reduction associated with

Table 6. Correlation matrix for species diversity variables and various environmental and other variables. Forest and thicket communities ($n = 33$).

	RAI	ALT	SLO	ASP	LIT	SDE	RCO	SAN
Annual rainfall	1.00							
Altitude	0.57***	1.00						
Slope inclination	0.20	0.24	1.00					
Aspect	-0.24	0.06	0.11	1.00				
Litter cover	-0.02	0.05	-0.27	0.06	1.00			
Soil depth	0.15	0.06	-0.35*	0.03	0.05	1.00		
Rock cover	-0.03	0.12	0.33	0.13	-0.10	-0.59	1.00	
% sand content	0.20	-0.22	-0.04	-0.16	-0.11	0.21	-0.21	1.00
Soil pH	-0.36	-0.67***	-0.33	-0.22	-0.06	0.20	-0.35*	0.51**
Exchangeable calcium	-0.31	-0.31	-0.19	-0.35*	-0.03	-0.07	-0.18	0.09
S value	-0.26	-0.42*	0.01	-0.37*	0.02	-0.31	0.06	0.04
Oxidizable carbon	0.31	0.08	0.40*	-0.26	0.00	-0.49**	0.04	-0.10
Total nitrogen	0.37*	0.16	0.19	-0.17	0.28	-0.39*	0.23	-0.14
Available phosphorus	-0.22	-0.26	-0.31	0.12	0.18	0.13	-0.29	0.04
Phytochorological diversity	-0.78***	-0.32	-0.03	0.24	-0.24	-0.14	0.09	-0.31
Species richness	-0.50**	0.05	0.16	0.30	-0.25	-0.32	0.39*	-0.39*
Shannon-Wiener	-0.43*	0.11	0.18	0.24	-0.13	-0.37*	0.52**	-0.46**
Simpsons	0.27	-0.15	-0.27	-0.13	0.11	0.33	-0.41*	0.47**

* $p < 0.05$; ** $p < 0.01$; *** $p < 0.001$

the competitive elimination of non-climax species in the climax stage. The mid-successional peak is attributed to the co-existence of early and late successional species in a single stand (Loucks, 1970; Auclair & Goff, 1971; Shafi & Yarranton, 1973; Whittaker, 1972, 1977; Westman, 1975b) and that in many landscapes the mid-successional stage is most frequent within the successional mosaic (Loucks, 1970; Denslow, 1980). Increase of species diversity from successional stages to the climax is often observed (Monk, 1967; Reiners et al., 1971; Nicholson & Monk, 1974) and has been stated as a generalization (Margalef, 1963; Odum, 1969). At least two studies have reported a steady decrease in diversity after initial post-fire establishment (Habeck, 1968; Westman, 1981).

The decline in diversity of the dune communities towards the terminal climax is probably associated with the dominance of the community by large-leaved shrubs and the closure of the canopy (cf. Auclair & Goff, 1971; Peet, 1978; Houssard et al., 1980; Westman, 1981). The mid-successional peak in H' coincides with the roughly equal representation of graminoids, small- and large-leaved shrubs (Fig. 3). The high richness of the grassland community could be the result of a higher grazing intensity and frequent mowing allowing the establishment and persistence of certain species which are not found in the other communities (van der Maarel, 1971). High C is explained by the dominance of Stenotaphrum secundatum, a grass which forms an extensive, dense turf under conditions of frequent mowing and grazing.

Discussion

Global comparisons

Our understanding of intercontinental diversity patterns is still in a rudimentary phase and is

1.00									
0.63***	1.00								
0.54**	0.86***	1.00							
-0.16	0.23	0.54**	1.00						
-0.18	0.26	0.46**	0.76***	1.00					
0.25	0.10	0.01	-0.22	-0.28	1.00				
0.17	0.26	0.17	-0.25	-0.28	-0.14	1.00			
-0.15	0.09	0.09	-0.03	-0.14	-0.28	0.69***	1.00		
-0.20	0.06	0.07	0.04	0.01	-0.12	0.58***	0.83***	1.00	
0.25	-0.06	-0.07	-0.12	-0.06	-0.06	-0.41*	-0.62***	-0.89***	1.00
PHH	CAC	BAS	CAR	NIT	PHO	PHD	S	H'	C

plagued by a number of conceptual problems, most of which have been discussed succinctly by Rice & Westoby (1983a). Convergence theory predicts that in genetically isolated habitats sharing similar environments, overall community structure would be similar (Cody & Mooney, 1978). This theory assumes that species richness is at equilibrium in evolutionary time and can be interpreted largely in terms of ecological processes. Comparisons of species richness in mediterranean climate regions of the world (Whittaker, 1977; Naveh & Whittaker, 1979; Cowling & Campbell, 1980; Westman, 1981) and other globally comparable biome types (Shmida & Whittaker, 1979; Rice & Westoby, 1983a) have shown important differences in species richness among analogous vegetation types on different continents. Historical differences are usually invoked to explain this non-convergence: for example Whittaker (1977) and Naveh & Whittaker (1979) attribute the higher richness of the southern

hemisphere 'mediterranean' heathlands, when compared to their shrubland (*sensu* Di Castri, 1980) 'analogues', to the longer evolutionary histories of the heathland floras. Clearly this hypothesis refutes the notion of a global equilibrium or saturation of species richness (Rice & Westoby, 1983a).

I concur with Rice & Westoby (1983a) that the level or scale of diversity is of paramount importance for global comparisons. Naveh & Whittaker (1979) cite data from 0.1 ha samples (alpha richness) as evidence for the high diversity of Cape fynbos. My data indicates that at this scale (Table 2) fynbos is no richer than other shrublands, some of which (e.g. subtropical thicket) certainly have not had a long history in their present area (Cowling, 1983b). The similarly high richness of certain fynbos communities and subtropical grassland on infertile soils (Table 2) indicates that ecological factors such as low levels of available soil nutrients could be implicated in the regulation of diversity at

Table 7. A comparison of diversity in 100 m² samples of two structural cover states in a Grassy Fynbos and a South Coast Renosterveld community in the study area.

	Erica-Trachypogon fynbos community		*Themeda-Cliffortia* renosterveld community	
	Ericoid shrub/grassy cover state $n = 9$	Proteoid shrub cover state $n = 7$	Mismanaged shrubland cover state $n = 4$	Well managed grassland cover state $n = 7$
Diversity variables				
S	42.8 ± 6.4	36.6 ± 9.8*	31.3 ± 9.8	35.0 ± 7.9
H'	1.11 ± 0.08	1.04 ± 2.3	0.94 ± 0.23	0.59 ± 0.16**
C	0.12 ± 0.03	0.17 ± 0.17	0.18 ± 0.08	0.51 ± 0.10***
Disturbance variables				
VAG[a]	3.6 ± 2.4	11.4 ± 2.4	1.9 ± 1.0	2.2 ± 0.75
GRZ[a]	2.3 ± 0.71	1.7 ± 0.49	4.0 ± 0.0	3.1 ± 0.38
Structural variables (relative % cover)				
Proteoid shrubs[b]	5.2 ± 6.7	16.2 ± 9.2	0.4 ± 0.4	0.0
Small-leaved shrubs[c]	32.2 ± 14.8	47.5 ± 14.8	32.3 ± 11.4	4.5 ± 3.8
Grasses	38.8 ± 14.6	14.9 ± 11.4	37.1 ± 10.3	85.5 ± 6.4
Restioids[d]	9.2 ± 7.8	10.1 ± 6.6	5.6 ± 5.5	1.3 ± 2.7

* $p < 0.1$; ** $p < 0.01$; *** $p < 0.001$

[a] See Table 1 for explanations of, and classes within, these variables.

[b] Shrubs having isobilateral sclerophyllous leaves characteristic of the Cape Proteaceae.

[c] Shrubs having leaves < 25 mm².

[d] Restionaceae.

this scale (Campbell & Cowling, 1980; Rice & Westoby, 1983a). Kruger & Taylor (1979) observe that in sites of about 1.0 km² and less, South African grassland and savanna have approximately as many species as do Cape fynbos sites of equal size. However they recorded exceptionally high levels of delta diversity. Perhaps the turnover in species composition along analogous landscape gradients or within analogous habitats (Cody, 1983) on different continents are more appropriate diversity scales for global comparisons.

Perhaps the situation is not so bad that global variation in floristic richness appears best explained by historical events specific to each region. That historic events act as constraints to convergence cannot be ignored (Cody & Mooney, 1978). However a deeper understanding of global patterns of plant species richness will emerge once we understand the factors controlling the different levels of diversity in each region. The remainder of this paper is devoted towards this end for the SE Cape.

Diversity patterns in the study area: within-Cape comparisons

Both Taylor (1978) and Kruger (1979) have stated that strikingly high alpha richness and equitability are general features of fynbos communities. My data do not support these generalizations. The highest richness in Tables 2 and 3 were recorded in renosterveld communities and highest equitability in thicket. There is a trend for renosterveld to have lowest values of *H'* and highest of *C* although high values of *C* (e.g. 0.51) were recorded in some fynbos plots.

Phytochorological complexity and tension zones

Biogeographic relations can have a significant influence on species diversity as they determine historic differences in the availability of species from different areas (Whittaker, 1972; Danin, 1978). Past climatic changes result in shifts of the boundaries of phytochoria causing contacts with different floras which may increase the number of species in certain environments.

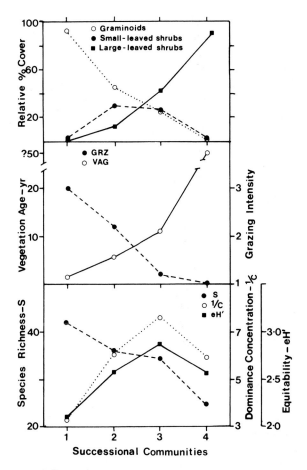

Fig. 3. Trends in diversity measures per 100 m² plots in dune communities along a gradient of decreasing grazing intensity and increasing vegetation age. Successional communities are:
1. *Themeda – Stenotaphrum* (*n* = 2)
2. *Restio – Agathosma* (*n* = 6)
3. *Restio – Maytenus* (*n* = 9)
4. *Cassine – Cussonia* (*n* = 8)

In the study area an index of phytochorological diversity (PHD) emerged as the variable most significantly correlated with species richness in all data sets (Tables 4–6). The SE Cape is a complex tension zone where species characteristic of at least five phytochoria contribute to the richness of phytochorologically mixed communities. Danin (1978) and Naveh & Whittaker (1979) recorded high species richness in the most biogeographically complex regions of their study areas in the Middle East. Westman (1981) noted the highest increase in diversity in California coastal sage communities where a mixing of southern and northern floras occurs. Bond (1981) found that the richest communities

(waboomveld) in the southern Cape mountains occurred on the transition between renosterveld and fynbos and have a diverse assemblage of fynbos shrubs and graminoids, karroid shrubs, and subtropical grasses.

The SE Cape comprises a transition zone between the warm temperate winter rainfall region and the subtropical summer rainfall region. Generally the rainfall distribution is bimodal with spring and autumn maxima although it is highly variable with a winter peak one year and a summer peak the next (Schulze, 1965; Gibbs Russel & Robinson, 1981). It has been shown (Roux, 1966) that in the mixed grass-karroid shrub communities of the region, grass growth is favoured by summer rains and shrubs by winter rains: farmers speak of 'grass' years and 'bush' years. This transitional and variable rainfall regime allows species of diverse biogeographical affinities to co-exist in single stands and facilitates high alpha richness since there are a wide range of conditions for germination, establishment and growth (Grubb, 1977; Pierce & Cowling, 1983).

Generally there is a greater phytochorological mixing in drier communities with more fertile soils; mesic communities which have an adequate and predictable rainfall, and, in fynbos, have highly infertile soils, are phytochorologically pure and species-poor. I comment further on the effect of soil nutrients on phytochorological mixing and species richness in a later section.

The discussion of the relationship between phytochorological diversity and species richness has the potential to appear circular. Phytochorological groups are subjectively defined on the basis of groups of species which show similar distribution patterns. Thus high species richness could be a trivial consequence of high phytochorological diversity. However the reverse does not hold since certain species-rich SW Cape fynbos communities have low phytochorological diversity.

Equilibrium biogeography versus history

In the SE Cape, Mountain Fynbos and Afromontane forests have patchy or island-like distributions within a continental setting. It is therefore possible to invoke island biogeography theory to explain the diversity patterns of these vegetation types (Cody, 1983). In this section I present, as alternatives, an ecological and a historical hypothesis to explain the richness of Mountain Fynbos and Afromontane forest communities in the study area.

In the SE Cape, Mountain Fynbos is pinched into an eastwardly narrowing peninsula on the upper slopes of the higher mountains, which is separated from the 'mainland source area' of the mesic coastal mountains by a wide belt of inhospitable terrain. Afromontane forest is patchily distributed in an archipelago of mountain islands throughout the African continent (White, 1978). In the study are these forests are confined to small patches on deep colluvial soils in the cool, wet parts of the Elandsberg Mountains and are isolated from the large contiguous block of forests in the Knysna region to the west. Cody (1983) has demonstrated the significance for bird diversity of the patchy distribution of Afromontane forests west of the Knysna region.

Central to the theory of island biogeography is the assumption that present day ecosystems are in an equilibrium state (MacArthur & Wilson, 1967; MacArthur, 1972). If one assumes ecological saturation it is permissible to ignore largely historical events (Pianka, 1966; Vuilleumier & Simberloff, 1980) and explore the great potential for modelling, predicting and interpreting diversity (MacArthur, 1972). Assuming a dynamic equilibrium between species immigration and extinction, mediated by competitive interactions in ecological time, it is possible to predict the species numbers on a given island by means of a conveniently small number of easily measured environmental variables such as distance from the mainland source and island area (Johnson et al., 1968; Johnson & Raven, 1968; Simpson, 1974; Nilsson & Nilsson, 1978).

The predictions of island biogeography theory afford a reasonable explanation for the low richness of Mountain Fynbos and Afromontane forest patches in the study area. In the small areas occupied by these vegetation types there may be few habitat types and a tendency for higher extinction rates due to smaller population sizes than on the larger contiguous areas that constitute the 'mainland' source. I have shown that alpha richness of Mountain Fynbos communities of the Outeniqua mountains to the west (part of the 'mainland') is higher than in the study area. It would be desirable to have richness data from the contiguous Knysna Afromontane forests.

In recent years island biogeography theory has been criticized both from a theoretical (Whittaker, 1977; Connor & McCoy, 1979; Connor & Simberloff, 1979) and an empirical basis (Vuilleumier & Simberloff, 1980; Weaver & Kellman, 1981). There has been much criticism of the equilibrium assumption for plant communities: the difficulty of explaining competitive niche diversification in species-rich plant communities and also the notion that disturbance in the long term (climatic change) and the short term (fire, grazing, landslides, etc.) results in continuous changes in the competitive hierarchy (Connell, 1978; Huston, 1979). Geographical ecology has been criticized for not fully appreciating the impact of historical factors (Sieb, 1980; Vuilleumier & Simberloff, 1980) even though these are often poorly understood and not amenable to conclusive tests (Pianka, 1966; Cody, 1983). I present below an alternative historical hypothesis which does not assume equilibrial conditions, to explain the richness of Afromontane forest and Mountain Fynbos in the study area.

There is a growing body of data to suggest that Pleistocene glacial climates were both colder and drier in the Cape coastal region than present Holocene interglacial conditions (Deacon, 1983). Therefore during the last glacial, mesic Afromontane forest would have been largely eliminated from the study area and it is possible that only a few relic patches survived in the Knysna region (Cowling, 1983b). Mountain Fynbos would have been restricted to even smaller patches than present, and the predominant vegetation on the mountains of the study area would have been an admixture of dry fynbos, C_3 grasses and karoo (Cowling, 1983b). With the onset of warmer wetter conditions after about 12 000 B.P. (Deacon, 1983), the areal extent of both vegetation types would have expanded and communities would have been enriched by the immigration of new species from relics. It could be argued that the relatively low richness of Mountain Fynbos and Afromontane forest in the study area is because too little time has elapsed since climatic amelioration for many species to disperse from the source areas to the expanding 'islands' (cf. Taylor & Regal, 1978; Vuilleumier & Simberloff, 1980). It is hard to envisage the establishment of equilibrial conditions in plant communities under the continuously changing climate of a glacial-interglacial sequence.

There are problems with both hypotheses. The

ecological theory is certainly more testable but it would be very difficult in practice to evaluate the respective roles of the absence of required ecological habitat and competition as factors regulating the richness of 'island' floras – we simply lack the required autecological information. Historical hypotheses are difficult to test and the model of climatic change used to predict vegetation changes in the study area is very rudimentary (Deacon, 1983).

Dynamics and disturbance

There has been a surge of interest in non-equilibrium models of species diversity (e.g. Caswell, 1978; Connell, 1978; Huston, 1979) arising from the notion that the 'frequencies of natural disturbance ... are often much faster than the rate of recovery from perturbations' (Connell, 1978). In this section I discuss the role of short-term disturbances (fire, grazing) as factors which regulate species diversity in the study area.

Connel (1978), Huston (1979) and Peet *et al.* (1983) postulate a dynamic model of species diversity: in the absence of disturbance-mediated population reduction, a competitive equilibrium results with one or few species dominating and diversity is low; at high frequencies of population reduction only those species capable of reproducing under the heavy disturbance regime will persist, and diversity will also be low. Species diversity will be highest at an intermediate disturbance level, at which most species can co-exist.

Figure 2 supports the intermediate disturbance hypothesis for the relationship between grazing intensity and species richness in fynbos communities. Disturbance, in the form of frequent fires, enhances species richness in fire-prone fynbos shrublands (Fig. 2, Table 7). In fynbos, much of the patch area at any given time is in the form of large-scale clearings after fire. The high richness in the youthful post-fire stage could be because most species are adapted for establishment in open, high-light intensity post-fire conditions and many are suppressed and do not reproduce successfully as the community matures and the canopy closes (cf. Loucks, 1970; Whittaker & Levin, 1977; Denslow, 1980). I did not observe similar patterns in the equally fire-prone renosterveld communities.

On the coastal dunes I demonstrated a trend of decreasing richness associated with gradients of decreasing grazing intensity and increasing vegetation age and thicket development (Fig. 3). However, certain thicket communities which have remained relatively undisturbed for decades or even centuries, are very rich and have the highest equitability in the study area (Table 3). It seems that recurrent large-scale disturbances such as fire are not implicated in the regulation of diversity in these communities. In the drier thicket a variable soil moisture supply has allowed for the co-existence of a diverse flora of shallow-rooted succulents with deep-rooted, evergreen, sclerophyllous subtropical shrubs and trees. The latter component comprises species which nearly all produce fleshy, bird-dispersed fruits. Fruiting periods are highly variable within and among species (Pierce & Cowling, 1983) and therefore the proportion of viable seed of different species available to recolonize small disturbance patches (e.g. landslide, occasional shrub death, or elephant damage in pre-colonial times) will show appreciable temporal and spatial variability. This asynchronism in fruit production and dispersal may represent a way to limit competition (Grubb, 1977; Houssard *et al.,* 1980) and thus account for the relatively high alpha diversity in thicket.

The role of soil nutrients

Huston (1979) has presented a non-equilibrium hypothesis of species diversity that predicts, amongst other things, high diversity on moderately infertile substrates. He argues that slow growth rates on these stressed sites would 'reduce the rate of competitive displacement allowing a longer period of co-existence among competitors and thus the maintenance of diversity' (Huston, 1979; see also McNaughton, 1967; Grime, 1973, 1977; Newman, 1973). There is much experimental evidence to show that the fertilization of pastures on infertile soils results in an increase of standing crop and a reduction in diversity (e.g. Willis, 1963; Jeffrey & Pigott, 1973). Huston (1980) found strong correlations between species richness of Costa Rican forests and low levels of soil nutrient availability and other factors which would reduce growth rates. Some studies show an increase in plant diversity with increasing soil fertility (e.g. Monk, 1967; Westman, 1975a; Whittaker, 1977).

My data would appear to falsify Huston's (1979) hypothesis. Fynbos communities on the least fertile soils (Mountain Fynbos, *Thamnochortus-Erica* community) have low diversity whereas thicket on the most fertile soils can have very high diversity (Tables 2 and 3). In the correlation analysis measures of soil fertility were not significantly correlated with species richness (*S*) in all data sets and there were some significant negative correlations between soil nutrients and equitability (*H'*) (Tables 4–6). It could be argued that the highly infertile fynbos soils represent a case of extreme deficiency and would therefore support species-poor communities (Grime, 1973; Huston, 1979, 1980). However these soils would be considered only moderately infertile when compared to the fynbos soils of the SW Cape (Campbell, 1983) which support species-rich communities. The relatively high diversity of thicket, which grows on fertile soils, is unexplained by Huston's (1979) theory.

In the SE Cape fynbos, highly leached infertile soils are an edaphic barrier for the migration of species from non-Cape phytochoria (e.g. subtropical grasses), and they support communities of low phytochorological diversity and species richness (see below).

When the 27 samples of Grassy Fynbos and Mountain Fynbos on T.M.G. quartzites were treated separately, there were significant positive correlations between *S* and available phosphorus (*r* = 0.56; *p* < 0.01), total nitrogen (*r* = 0.42; *p* < 0.02), sum of exchangeable bases (*r* = 0.47; *p* < 0.02) and pH (*r* = 0.48; *p* < 0.02). The generally highest diversities in the study area were recorded in dry renosterveld communities on moderately fertile soils where conditions appeared suitable for the establishment, growth and persistence of species from a wide range of phytochorological groups.

Other hypotheses

My data are consistent with other studies (e.g. McNaughton, 1967; Risser & Rice, 1971; Whittaker, 1972, 1977; Grime, 1973; Whittaker & Niering, 1975; Glenn-Lewin, 1977; Naveh & Whittaker, 1979) in giving no support to the producivity hypothesis (Connell & Orias, 1964), if annual average rainfall can be taken as a rough index of productivity potential. Maximum diversity was not recorded in the most mesic or favourable environments (cf. Terborgh, 1973; Richardson & Lum, 1980).

Environmental instability, expressed as a greater temporal heterogeneity of water supply and temperature regime in the valley climate of the study area, has promoted species diversity in thicket communities by permitting the co-existence of a variety of growth forms specialized in different ways to the irregular availability of water and great diurnal and annual temperature ranges (cf. Whittaker & Niering, 1975; Whittaker, 1977).

Environmental heterogeneity is frequently correlated with plant species diversity (Pianka, 1966; van der Maarel, 1971; Harner & Harper, 1976; Grubb, 1977; Rickleffs, 1977; Houssard *et al.,* 1980). Richness (*S*) and equitability (*H'*) showed a weak positive correlation with rock cover (RCO) for the forest-thicket communities (Table 6) but was not significantly correlated in the other data sets.

Concluding remarks

No single hypothesis discussed in this paper could adequately explain diversity relations in all vegetation types studied. Phytochorological diversity emerged as a consistently significant correlate in all data sets. However this factor is probably an unimportant predictor of plant species diversity outside tension zones. Perhaps it is too ambitious to seek unifying patterns in very heterogeneous data sets.

It is difficult to untangle the complex of factors which have influenced the evolution and maintenance of diversity in any given region, let alone make generalizations on a global scale. It is clear to me that historical factors should not be merely treated as 'noise' or the substance of *ad hoc* explanations when ecological theories fail. Although ecological theories often yield more, and better defined, predictions than historical explanations, the latter should be considered at all stages of a study. In many cases ecological and evolutionary time cannot be realistically separated in their effects on community structure (Whittaker, 1977; Vuilleumier & Simberloff, 1980). In my study area historical events such as climate change and phytochorological mixing, and ecological factors such as fire, grazing, soil nutrients, and climate as well as biological factors have all contributed to the evolution and maintenance of diversity in one or more communities at all times.

References

Acocks, J. P. H., 1953. Veld types of South Africa. Mem. Bot. Surv. S. Afr. 28.

Anon., 1942. Weather on the coasts of southern Africa. Part 3. Union of South Africa from Mossel Bay to East London. Government Printer, Pretoria. 52 pp.

Auclair, A. N. & Goff, F. G., 1971. Diversity relations of upland forests in the western Great Lakes area. Am. Nat. 105: 499 528.

Axelrod, D. I., 1978. Origin of coastal sage vegetation, Alta and Baja California. Am. J. Bot. 65: 1117 1131.

Bond, W. J., 1980. Fire and senescent fynbos in the Swartberg, southern Cape. S. Afr. For. J. 114: 68 71.

Bond, W. J., 1981. Vegetation gradients in the southern Cape mountains. Unpubl. thesis, University of Cape Town.

Bond, W. J., 1983. Fire and serotiny in Cape Proteaceae. Unpubl. ms.

Boucher, C., 1978. Cape Hangklip area. II. The vegetation. Bothalia 12: 455 497.

Boucher, C. & Moll, E. J., 1980. South African mediterranean shrublands. In: Castri, F. di, Goodall, D. W. & Specht, R. L. (eds.), Mediterranean-Type Shrublands, pp. 233 248. Elsevier, Amsterdam.

Campbell, B. M., 1983. Montane plant environments of the Fynbos Biome. Bothalia (in press).

Campbell, B. M. & Moll, E. J., 1977. The forest communities of Table Mountain, South Africa. Vegetatio 34: 105 115.

Campbell, B. M. & van der Meulen, F., 1980. Patterns of plant species diversity in fynbos vegetation, South Africa. Vegetatio 43: 43 47.

Campbell, B. M., Cowling, R. M., Bond, W. J. & Kruger, F. J., 1981. Structural characterization of vegetation in the Fynbos Biome. S. Afr. Nat. Sci. Prog. Rep. 53. C.S.I.R., Pretoria.

Caswell, H., 1978. Predator-mediated coexistence: a nonequilibrium model. Am. Nat. 112: 127 154.

Cody, M. L., 1983. Continental diversity patterns and convergent evolution in bird communities. In: Kruger, F. J., Mitchell, D. T. & Jarvis, J. U. M. (eds.), Mediterranean-Type Ecosystems. The role of nutrients. Springer, Berlin, Heidelberg, New York (in press).

Cody, M. L. & Mooney, H. A., 1978. Convergence versus non-convergence in mediterranean-climate ecosystems. Ann. Rev. Ecol. Syst. 9: 265 321.

Connell, J. H., 1978. Diversity in tropical rain forests and coral reefs. Science 199: 1302 1310.

Connell, J. H. & Orias, E., 1964. The ecological regulation of species diversity. Am. Nat. 98: 399 414.

Connor, E. F. & McCoy, E., 1979. The statistics and biology of the species-area relationship. Am. Nat. 113: 791 833.

Connor, E. F. & Simberloff, D., 1979. The assembly of species communities: chance or competition? Ecology 60: 1132 1140.

Cowling, R. M., 1983a. A syntaxonomic and synecological study in the Humansdorp region of the Fynbos Biome. Mem. Bot. Surv. S. Afr. (in press).

Cowling, R. M., 1983b. Phytochorology and vegetation history in the south eastern Cape, South Africa. J. Biogr. (in press).

Cowling, R. M. & Campbell, B. M., 1980. Convergence in vegetation structure in the mediterranean communities of California, Chile and South Africa. Vegetatio 43: 191 197.

Danin, A., 1978. Plant species diversity and ecological districts of the Sinai Desert. Vegetatio 36: 83 93.

Deacon, H. J., 1983. The comparative evolution of mediterranean-type ecosystems: a southern perspective. In: Kruger, F. J., Mitchell, D. T. & Jarvis, J. U. M. (eds.), Mediterranean-Type Ecosystems. The role of nutrients. Springer, Berlin, Heidelberg, New York (in press).

Denslow, J. L., 1980. Patterns of plant species diversity during succession under different disturbance regimes. Oecologia 46: 18 21.

Di Castri, F., 1980. Mediterranean-type shrublands of the world. In: Castri, F. di, Goodall, D. W. & Specht, R. L. (eds.), Mediterranean-Type Shrublands, pp. 1 52. Elsevier, Amsterdam.

Drury, W. H. & Nisbet, I. C. T., 1973. Succession. J. Arnold Arboretum 54: 331 368.

Eiten, G., 1978. Delimitation of the cerrado concept. Vegetatio 36: 169 178.

George, A. S., Hopkins, A. J. & Marchant, N. G., 1979. The heathlands of Western Australia. In: Specht, R. L. (ed.), Heathlands and Related Shrublands of the World, A. Descriptive studies, pp. 211 230. Elsevier, Amsterdam.

Gibbs Russel, G. E. & Robinson, E. R., 1981. Phytogeography and speciation in the vegetation of the eastern Cape. Bothalia 13: 467 472.

Glenn-Lewin, D. C., 1977. Species diversity in North American temperate forests. Vegetatio 33: 153 162.

Goodall, D. W., 1977. Dynamic changes in ecosystems and their study: the roles of induction and deduction. J. Environ. Manage. 5: 309 317.

Grime, J. P., 1973. Competition and diversity in herbaceous vegetation. Nature 244: 311.

Grime, J. P., 1977. Evidence for the existence of three primary strategies in plants and its relevance to ecological and evolutionary theory. Am. Nat. 111: 1169 1194.

Grubb, P. J., 1977. The maintenance of species richness in plant communities: the importance of the regeneration niche. Biol. Rev. 52: 107 145.

Harner, R. F. & Harper, K. T., 1976. The role of area, heterogeneity and favourability in plant species diversity on Pinyon-Juniper ecosystems. Ecology 57: 1254 1263.

Harper, J. L., 1969. The role of predation in vegetational diversity. Brookhaven Symp. Biol. 22: 48 62.

Hill, M. O., 1973. Diversity and evenness: a unifying notation and its consequences. Ecology 54: 427 432.

Hnatiuk, R. J. & Hopkins, A. J. M., 1981. An ecological analysis of kwongan vegetation south of Eneabba, Western Australia. Austr. J. Ecol. 6: 423 438.

Houssard, C., Escarré, J. & Romane, F., 1980. Development of species diversity in some Mediterranean plant communities. Vegetatio 43: 59 72.

Huston, M., 1979. A general hypothesis of species diversity. Am. Nat. 133: 81 101.

Huston, M., 1980. Soil nutrients and tree species richness in Costa Rican forests. J. Biogeogr. 7: 147 158.

Jeffrey, D. W. & Pigott, C. D., 1973. The response of grasslands on sugar-limestone in Teesdale to application of phosphorus and nitrogen. J. Ecol. 61: 85 92.

Johnson, M. P. & Raven, P. H., 1970. Natural regulation of plant species diversity. Evol. Biol. 4: 127 162.

Johnson, M. P., Mason, L. G. & Raven, P. H., 1968. Ecological parameters and plant species diversity. Am. Nat. 102: 297–306.

Kruger, F. J., 1978. A description of the Fynbos Biome Project. S. Afr. Nat. Sci. Progr. Rep. 28. C.S.I.R. Pretoria.

Kruger, F. J., 1979. South African heathlands. In: Specht, R. L. (ed.), Heathlands of the World. A. Descriptive studies, pp. 19–80. Elsevier, Amsterdam.

Kruger, F. J., 1983. Plant community diversity and dynamics in relation to fire. In: Kruger, F. J., Mitchell, D. T. & Jarvis, J. U. M. (eds.), Mediterranean-Type Ecosystems. The role of nutrients. Springer, Berlin, Heidelberg, New York (in press).

Kruger, F. J. & Taylor, H. C., 1979. Plant species diversity in Cape fynbos: gamma and delta diversity. Vegetatio 47: 85–93.

Lamont, B., Downes, S. & Fox, J. E. D., 1977. Importance value curves and diversity indices applied to species-rich heathland in Western Australia. Nature 265: 438–441.

Loucks, O. L., 1970. Evolution of diversity, efficiency and community stability. Amer. Zool. 10: 17–25.

Louw, W. J., 1976. Mesoclimate of the Port Elizabeth-Uitenhage metropolitan area. Pretoria Weather Bureau Technical Paper No. 4. Government Printer, Pretoria.

Lubchenco, J., 1978. Plant species diversity in marine intertidal community: importance of herbivore food preference and algal competitive abilities. Am. Nat. 112: 23–29.

Maarel, E. van der, 1971. Plant species diversity in relation to management. In: Duffey, E. & Watt, A. S. (eds.), The Scientific Management of Animal and Plant Communities, pp. 45–63. Blackwell, London.

Marchant, N. G., 1973. Species diversity in the south western flora. J. Roy. Soc. Western Australia 56: 23–30.

Margalef, R., 1963. On certain unifying concepts in ecology. Am. Nat. 97: 357–374.

MacArthur, R. H., 1972. Geographical Ecology: patterns in the distribution of species. Harper & Row, New York. 269 pp.

MacArthur, R. H. & Wilson, E. O., 1967. The Theory of Island Biogeography. Princeton University Press, Princeton, N.J. 203 pp.

McKenzie, B., 1978. A quantitative and qualitative study of the indigenous forests of the south-western Cape. Unpubl. thesis, University of Cape Town.

McKenzie, B., Moll, E. J. & Campbell, B. M., 1977. A phytosociological study of Orange Kloof, Table Mountain, South Africa. Vegetatio 34: 41–53.

McNaughton, S. J., 1967. Relationships amongst functional properties of Californian grassland. Nature 216: 168–169.

Monk, C. D., 1967. Tree species diversity in the eastern deciduous forest with particular reference to north central Florida. Am. Nat. 101: 173–187.

Naveh, Z. & Whittaker, R. H., 1979. Structural and floristic diversity of shrublands and woodlands in northern Israel and other mediterranean areas. Vegetatio 41: 171–190.

Newman, E. I., 1973. Competition and diversity in herbaceous vegetation. Nature 244: 310.

Nicholson, S. A. & Monk, C. D., 1974. Plant species diversity in old field succession on the Georgia Piedmont. Ecology 55: 1075–1085.

Nilsson, S. G. & Nilsson, I. N., 1978. Species richness and dispersal of vascular plants to islands in Lake Möckeln, southern Sweden. Ecology 59: 473–482.

Odum, E. P., 1969. The strategy of ecosystem development. Science 164: 267–270.

Peet, R. K., 1974. The measurement of species diversity. Ann. Rev. Ecol. Syst. 5: 285–307.

Peet, R. K., 1978. Forest vegetation of the Colorado Front Range: patterns of species diversity. Vegetatio 37: 65–78.

Peet, R. K. & Christensen, N. L., 1980. Hardwood forest vegetation of the North Carolina piedmont. Veröff. Geobot. Inst. ETH, Stiftung Rübel, Zürich 69: 14–39.

Peet, R. K., Glenn-Lewin, D. C. & Wolf, J. W., 1983. Prediction of man's impact on plant species diversity: a challenge for vegetation science. In: Holzner, W., Werger, M. J. A. & Ikusima, I. (eds.), Man's Impact on Vegetation. Junk, The Hague.

Pianka, E. R., 1966. Latitudinal gradients of species diversity: a review of concepts. Am. Nat. 100: 33–46.

Pierce, S. M. & Cowling, R. M., 1983. Phenology of fynbos and non-fynbos communities in the south eastern Cape. S. Afr. J. Bot. (in press).

Reiners, W. A., Worler, I. A. & Lawrence, D. B., 1971. Plant diversity in a chronosequence at Glacier Bay, Alaska. Ecology 52: 55–69.

Rice, B. & Westoby, M., 1983a. Plant species richness at tenth-hectare scale in Australian vegetation compared to other continents. Vegetatio 52: 129–140.

Rice, B. & Westoby, M., 1983b. Species richness in vascular vegetation of the West Head, N.S.W. Submitted to Austr. J. Ecol.

Richardson, P. J. & Lum, K., 1980. Patterns of plant species diversity in California: relation to weather and topography. Am. Nat. 116: 504–536.

Rickleffs, R. E., 1977. Environmental heterogeneity and plant species diversity: a hypothesis. Am. Nat. 111: 376–381.

Risser, P. G. & Rice, E. L., 1971. Diversity in tree species in Oklahoma upland forests. Ecology 52: 876–880.

Roberts, B. R., Anderson, E. R. & Fourie, J. H., 1975. Evaluation of natural pastures: quantitative criteria for assessing conditions in the Themeda veld of the Orange Free State. Proc. Grassld. Soc. sth. Afr. 10: 133–140.

Roux, P. W., 1966. Die uitwerking van seisoenreënval en beweiding op gemengde karooveld. Proc. Grassld. Soc. sth. Afr. 1: 103–110.

Schulze, B. R., 1965. Climate of South Africa. Part 8. General survey. South African Weather Bureau 28: Government Printer, Pretoria. 330 pp.

Schulze, R. E., 1975. Incoming radiation on sloping terrain: a general model for use in southern Africa. Agrochemophysika 7: 55–61.

Schulze, R. E. & McGee, O. S., 1978. Climatic indices and classification in relation to biogeography of southern Africa. In: Werger, M. J. A. (ed.), Biogeography and Ecology of Southern Africa, pp. 19–52. Junk, The Hague.

Shafi, M. I. & Yarranton, G. A., 1973. Diversity, floristic richness and species eveness during a secondary (post-fire) succession. Ecology 54: 879–902.

Shmida, A. & Whittaker, R. H., 1979. Convergent evolution of deserts in the old and new worlds. In: Wilmanns, O. & Tuxen, R. (eds.), Werden und Vergehen von Pflanzengesellschaften, pp. 437–450. J. Cramer, Vaduz.

Sieb, R. L., 1980. Baja California: a peninsula for rodents but not for reptiles. Am. Nat. 115: 613–620.

Slatyer, R. O. (ed.), 1976. Dynamic changes in terrestrial ecosystems: patterns of change, techniques for study and applications to management. M.A.B. Technical Notes No. 4. UNESCO, Paris.

Specht, R. L., 1979. Heathland and related shrublands of the world. In: Specht, R. L. (ed.), Heathlands and Related Shrublands of the World. A. Descriptive studies, pp. 1–18. Elsevier, Amsterdam.

Specht, R. L. & Moll, E. J., 1983. Heathlands and sclerophyllous shrublands – an overview. In: Kruger, F. J., Mitchell, D. T. & Jarvis, J. U. M. (eds.), Mediterranean-Type Ecosystems. The role of nutrients. Springer, Berlin, Heidelberg, New York (in press).

Taylor, H. C., 1978. Capensis. In: Werger, M. J. A. (ed.), Biogeography and Ecology of Southern Africa, pp. 171–229. Junk, The Hague.

Taylor, R. J. & Regal, P. J., 1978. The peninsula effect on species diversity and the biogeography of Baja, California. Am. Nat. 112: 583–593.

Terborgh, J., 1973. On the notion of favourableness in plant ecology. Am. Nat. 107: 481–501.

Van der Merwe, C. V., 1976. Die plantekologiese aspekte en bestuursprobleme van die Goukamma-Natuurresevaat. Unpubl. thesis, University of Stellenbosch.

Van der Merwe, C. V., 1977. 'n Plantegroei opname van die De Hoop Natuurresevaat. Bontebok 1: 1–29.

Van Wilgen, B., 1980. Some effects of fire frequency on fynbos at Jonkershoek, Stellenbosch. Unpubl. thesis, University of Cape Town.

Vuilleumier, F. & Simberloff, D., 1980. Ecology versus history as determinants of patchy and insular distribution in high Andean birds. Evol. Biol. 13: 235–379.

Weaver, M. & Kellman, M., 1981. The effects of forest fragmentation on woodlot tree biotas in Southern Ontario. J. Biogeogr. 8: 199–210.

Werger, M. J. A., 1972. Species-area relationships and plot size: with some examples from South African vegetation. Bothalia 10: 583–594.

Werger, M. J. A., 1978. Biogeographical division of southern Africa. In: Werger, M. J. A. (ed.), Biogeography and Ecology of Southern Africa, pp. 145–170. Junk, The Hague.

Werger, M. J. A., Kruger, F. J. & Taylor, H. C., 1972. A phytosociological survey of Cape fynbos and other vegetation at Jonkershoek, Stellenbosch. Bothalia 10: 599–614.

Westman, W. E., 1975a. Edaphic climax pattern of the pygmy forest region of California. Ecol. Monogr. 22: 1–44.

Westman, W. E., 1975b. Pattern and diversity in swamp and dune vegetation, North Stradbroke Island. Austr. J. Bot. 23: 339–354.

Westman, W. E., 1981. Diversity relations and succession in Californian coastal sage scrub. Ecology 62: 170–184.

Westman, W. E., 1983. Spatial partitioning of resources. In: Kruger, F. J., Mitchell, D. T. & Jarvis, J. U. M. (eds.), Mediterranean-Type Ecosystems. The role of nutrients (in press).

White, F., 1978. The Afromontane Region. In: Werger, M. J. A. (ed.), Biogeography and Ecology of Southern Africa, pp. 463–513. Junk, The Hague.

Whittaker, R. H., 1960. Vegetation of the Siskiyou Mountains, Oregon and California. Ecol. Monogr. 30: 279–338.

Whittaker, R. H., 1969. Evolution of diversity in plant communities. Brookhaven Symp. Biol. 22: 178–196.

Whittaker, R. H., 1972. Evolution and measurement of species diversity. Taxon 21: 213–251.

Whittaker, R. H., 1977. Evolution of species diversity in land communities. Evol. Biol. 10: 1–67.

Whittaker, R. H. & Levin, S. A., 1977. The role of mosaic phenomena in natural communities. Theor. Pop. Biol. 12: 117–139.

Whittaker, R. H. & Niering, W. A., 1965. Vegetation of the Santa Catalina Mountains. II. A gradient analysis of the south slope. Ecology 46: 429–452.

Whittaker, R. H. & Niering, W. A., 1968. Vegetation of the Santa Catalina Mountains, Arizona. IV. Limestone and acid soils. J. Ecol. 56: 523–544.

Whittaker, R. H. & Niering, W. A., 1975. Vegetation of the Santa Catalina Mountains, Arizona. V. Biomass, production and diversity along the elevation gradient. Ecology 56: 771–790.

Whittaker, R. H., Niering, W. A. & Crisp, M. D., 1979. Structure, pattern and diversity of a mallee community in New South Wales. Vegetatio 39: 65–76.

Willis, A. J., 1963. Braunton Burrows: the effects on the vegetation of the addition of mineral nutrients to dune sands. J. Ecol. 51: 353–374.

Zeevalking, H. J. & Fresco, L. F. M., 1977. Rabbit grazing and species diversity in a dune area. Vegetatio 35: 193–196.

Accepted 30.3.1983.

Coexistence of plant species with similar niches*

A. Shmida[1]** & S. Ellner[2]**
[1] *Department of Botany, The Hebrew University, Jerusalem, Israel;*
[2] *Department of Mathematics and Graduate Program in Ecology, University of Tennessee, Knoxville, TN 37996, U.S.A.*

Keywords: Coexistence, Competition models, Diversity, Habitat differentiation, Niche differentiation, Non-equilibrium communities, Species richness

Abstract

In the context of a simple mathematical model, we derive several mechanisms whereby plant species can coexist in a community without differing in their trophic niches (their relations with habitats, resources and exploiters). The model is based on the dynamics of species turnover in microsites, and incorporates localized competition, non-uniform seed dispersal and aspects of spatiotemporal environmental heterogeneity. These factors, which are not included in most standard competition models, allow stable coexistence of trophically equivalent species due to:
  (a) Differences in life-history 'strategy'.
  (b) Input of seeds from nearby habitats (spatial Mass Effect).
  (c) Differences in demographic responses to environmental fluctuations (temporal Mass Effect).
  (d) Turnover in species composition between different habitat patches.
Quantitative descriptive studies are presented, demonstrating the occurrence of vegetation patterns predicted on the basis of the hypothesized mechanisms. We also review previously proposed mechanisms that would allow trophically equivalent species to coexist, and explore the theoretical and methodological implications of recognizing coexistence mechanisms independent of trophic niche differentiation. In particular, we propose that these mechanisms contribute to the dissimilarity of within-community replicate samples and the maintenance of many rare species in plant communities.

* Nomenclature follows Zohary & Feinbrun-Dothan (1966 →).
** Acknowledgements. We are grateful to D. Cohen, R. H. Whittaker, and S. A. Levin for establishing the Cornell/Hebrew University association that enabled us to work together, and for discussions of the ideas presented here. Innumerable staff from the field schools of the Society for the Protection of Nature in Israel assisted with the field work. We have freely incorporated ideas and improvements suggested by volunteer reviewers, including I. Noy-Meir, S. Levin, M. Rosenzweig, P. Chesson, R. K. Peet, and especially P. J. Grubb. Support came in part from NSF Grant DEB 78-09340 to R. H. Whittaker (AS), NSF Grants MCS 77-01076 and MCS 80-01618 to S. A. Levin (SPE) and NSF and Lady Davis Trust (Jerusalem) Graduate Fellowships to SPE. Typing was done graciously under pressure by B. Marks and C. Blair; the figures are by M. Shmida and W. Brown. This paper is dedicated to the memory of R. H. Whittaker.

Introduction

Identification of the factors governing the species richness of plant communities has been a major theme of plant ecology (Whittaker, 1965, 1969, 1972, 1977). Classical ecological theory offers two principal explanations for the coexistence of species in a community (Hutchinson, 1958, 1959; MacArthur, 1968, 1972; Whittaker, 1969, 1975, 1977):

1. *Habitat differentiation* – species utilize different portions of the available habitats, or differ in the range of habitats wherein they have a competitive advantage over other species.

2. *Resource differentiation* – species partition the limiting resources (nutrients, water, light, etc.) in

such a way that each is limited by a different component of the available resources.

Historically, plant ecologists have tended to emphasize interspecific differentiation in microhabitat requirements, whether edaphic, climatic, or biotic. Consequently, much effort has been devoted to identifying the adaptations of plants for the habitats in which they are found (e.g. Whittaker, 1972, 1975; Mueller-Dombois & Ellenberg, 1974; Harper, 1977; Barbour *et al.*, 1980) and to identifying the conditions yielding optimal growth of particular species (e.g. Major, 1951; Billings, 1952; Daubenmire, 1974). Complex patterns of plant distribution are often interpreted as reflecting a multiplicity of interacting plant/habitat relations. These are analyzed by statistical techniques of community ordination, classification, or pattern analysis.

The classical framework has recently been expanded in several directions. More attention is being paid to differentiation in regeneration (Grubb, 1977). Regeneration-niche factors identified as particularly important include: the density and composition of surrounding or overtopping vegetation (Watt, 1947, 1964; Forcier, 1975; Fox, 1977; Grubb, 1977; Turkington & Harper, 1979; Shmida & Whittaker, 1981); host-specific predation (Harper, 1969; Connell, 1970, 1978, 1979; Janzen, 1970); and, in general, the microhabitat requirements of seedlings (Harper *et al.*, 1961, 1965; Whittaker, 1969; Grubb, 1977; Harper, 1977). Disturbances initiating local successions are being recognized as an important aspect of community structure and dynamics. The concept of 'regional coexistence' (Levins & Culver, 1971; Slatkin, 1974) expands the classical framework by viewing the community as a mosaic of patches in different successional stages (Whittaker & Levin, 1977) and emphasizes patch successional stage as a fundamental axis of habitat-niche differentiation (e.g. Whittaker, 1969; Levin & Paine, 1974; Platt, 1975; Platt & Weis, 1977; Caswell, 1978; Connell, 1978, 1979; Werner, 1979; Pickett, 1980; Paine & Levin, 1981).

The classical and modern approaches are all elaborations on the principle (variously called Gause's hypothesis or the competitive exclusion principle) that coexisting species must differ in their Eltonian or trophic niche, their 'relations to food and enemies' (Elton, 1927). Our thesis in this paper is that plant species can coexist without trophic niche differentiation; that is, without differences in their habitat preferences, their resource utilization or their exploitation as resources. We will refer to such species as *trophically equivalent* (*TE*) species. This coinage is intended to emphasize that *TE* species may still differ in ecologically significant characteristics such as life-form, reproductive strategy, and seed-dispersal mechanism. Our goal is to develop, and to support with quantitative descriptive studies, several mechanisms whereby these non-trophic factors may allow *TE* species to coexist in a community.

Our approach is to construct a simple mathematical model of competing trophically equivalent plant species, based on the dynamics of species turnover in microsites. The model incorporates spatially localized competitive interactions, non-uniform seed dispersal, and aspects of the spatial distribution of microsites and environmental fluctuations. Spatial effects, independent of the microhabitat and trophic relations of species, have long been recognized as important aspects of plant population dynamics and community structure, but typically have not been incorporated into reverse order mathematical theories of plant species coexistence (Schaffer & Leigh, 1976; Fowler & Antonovics, 1981).

Validation of our proposals will require detailed demographic and experimental studies of community dynamics. In support of our ideas we present quantitative descriptive studies (most of them previously unpublished) demonstrating the occurrence of vegetation patterns predicted on the basis of our proposed mechanisms. Although we recognize that they fall short of the demographic desideratum, we offer them in the hope that students of vegetation confronted by similar phenomena might be prompted to test our hypotheses and models.

The model: competition for microsites in a spatially structured habitat

The model we present is intended to capture some essential features of plant competition and environmental heterogeneity in a way that is convenient for theory formulation and mathematical analysis. The principal features of the model are:
1. Competition is age-dependent, and localized within microsites: adults cannot be dislodged from a microsite by competing juveniles, and

competition for possession of vacant microsites occurs among only those juveniles established in the microsite.

2. Seed dispersal is non-uniform, with the seed-rain from an adult concentrated in its own microsite. Seeds disperse passively, without the ability to seek preferred or open microsites.

3. Large-scale spatial heterogeneity is modelled by considering a system of disjunct 'patches' of many contiguous microsites (microsites that are close relative to the scale of seed dispersal), with only weak seed dispersal between patches. Temporal variability is introduced by allowing the model parameters to vary randomly from year to year.

4. In order to model trophically equivalent species, we assume that at any given time there are no microsite-to-microsite variations in either the characteristics of adults or the process of competition among juveniles. Each microsite is occupied by a different adult, but there are no other differences. The *TE* species have identical microsite requirements and neither has a decisive advantage in competition for a particular microsite (equations 1–3 below).

Our assumption of exact trophic equivalence is certainly unrealistic; however, by analyzing this limiting case we identify mechanisms whose efficacy does not depend on any trophic niche differentiation whatsoever.

The remaining model assumptions are made solely for the sake of tractability. Competition among juveniles for occupancy of a microsite is by 'lottery' (Sale, 1977, 1978; Hubbell, 1979; Chesson & Warner, 1981). That is, among all juveniles on a vacant site, one of them is chosen in a (possibly biassed) random draw to occupy that site as an adult. We also assume that establishment in vacant microsites is entirely by seed rather than by vegetative spread; that all seeds germinate in the year after their formation; that all sites are always occupied; and that juveniles are reproductively mature at age one.

The general properties of model communities composed of weakly connected patches have been extensively analyzed (e.g. Cohen, 1970; Karlin & McGregor, 1972a; Levin, 1974, 1976; Yodzis, 1978; De Angelis *et al.*, 1979). These results allow us to extrapolate from the within-patch dynamics to mechanisms for coexistence in the system as a whole.

Notation and governing equations

Each habitat patch is taken to be a collection of identical microsites capable of supporting a single adult of either species 1 or species 2. The number of microsites is assumed to be large enough that

1. the fractions $f_i(t)$ of sites occupied by species i in year t, may be treated as continuous variables;

2. stochastic effects, other than those due to year-to-year environmental fluctuations, are negligible.

In particular, our models do not include any effects arising from the fact that a population actually consists of a finite number of individuals whose behavior varies about the population mean (demographic stochasticity *sensu* Keiding, 1975).

The life-cycle of the model plants is diagrammed in Figure 1, and the notation is summarized in Table 1. Each year, each adult of species i produces m_i seeds. A fraction b_i of these are dispersed uniformly over all microsites in the patch, with survivorship p_i en route. The remaining $v_i = m_i(1 - b_i)$ seeds are deposited in the parent's microsite. In this simplification of a leptokurtic dispersal curve the near-parent 'peak' is represented by v_i and the 'tail' by the $w_i = m_i b_i p_i$ dispersers. Mortality of seeds between dispersal and germination is absorbed into the m_i. Following reproduction, a fraction d_i of the adults of species i die. Surviving adults retain their site in the next year. In vacant sites all seeds germi-

Table 1. Notation for the model. Subscripted i is the species number.

Symbol	Definition
$f_i(t)$	Fraction of microsites occupied by species i in year t
d_i	Annual death rate of adults.
m_i	Annual seed production per adult.
b_i	Fraction of seed production that is dispersed over the patch.
p_i	Survivorship during dispersal.
v_i	$= m_i(1 - b_i)$, per capita number of seeds deposited on parent's microsite.
w_i	$= m_i b_i p_i$, per capita number of dispersed seeds that survive to establish at some microsite.
λ_i	$= f_i w_i$, total number of dispersed seeds of species i that establish on a given microsite.
F_i	Fraction of those sites vacated by death of a species-i adult which are occupied by species 1 in the following year.
β_i	Invasion coefficient, defined in equation (6).
Z_t	Environment-type in year t.
q_i	Juvenile mortality.

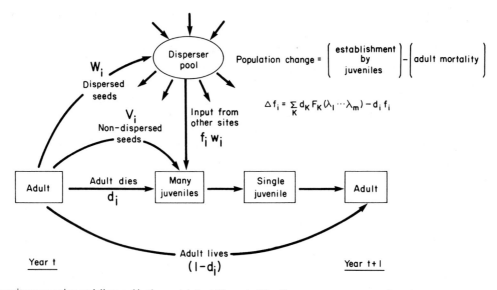

Fig. 1. Microsite occupation and dispersal in the model plant life-cycle. The diagram represents one microsite over the course of one year.

nate, and compete among themselves for possession of the site.

We restrict now to the case of two species. Gains and losses for species 1 occur only in sites where the adult has died. Sites last occupied by an adult of species i each contain v_i non-dispersed seeds from the former occupant, plus $\lambda_1 = f_1 w_1$ dispersed seeds of species 1 and $\lambda_2 = f_2 w_2$ dispersed seeds of species 2. On the assumption of unbiased 'lottery' competition, a species 1 seedling is the eventual victor in a fraction

$$F_1 = \frac{v_1 + \lambda_1}{v_1 + \lambda_1 + \lambda_2} \tag{1}$$

of the sites vacated by an adult of species 1, and in a fraction

$$F_2 = \frac{\lambda_1}{v_2 + \lambda_1 + \lambda_2} \tag{2}$$

of the sites vacated by an adult of species 2. A biassed lottery could be represented by taking instead of (1)

$$F_1 = \frac{Q_1 v_1 + Q_1 \lambda_1}{Q_1 v_1 + Q_1 \lambda_1 + Q_2 \lambda_2} \tag{3}$$

and similarly for F_2, where Q_j is the relative survival rate during competition of species-j seedlings. However, if we define $m_i' = Q_i m_i$ and use m_i' instead of m_i throughout, then (3) reduces to (1). Henceforth we assume that this has been done, and use the 'unbiassed' forms (1) and (2).

We are assuming that all sites are occupied, so $f_2(t) = 1 - f_1(t)$ and the state of the system is specified by $f_1(t)$. Adding up the gains and losses to species 1, we have

$$f_1(t + 1) = (1 - d_1) f_1(t) + d_1 f_1(t) F_1(\lambda_1, \lambda_2) +$$
$$+ d_2 (1 - f_1(t)) F_2(\lambda_1, \lambda_2) = H(f_1(t)) \tag{4}$$

wherein

$$\lambda_1 = f_1(t) w_1 \text{ and } \lambda_2 = f_2(t) w_2 = (1 - f_1(t)) w_2 \tag{5}$$

and the F_i are given by (1) and (2). The first term in (4) is site-retention by surviving adults; the second and third are establishment in microsites vacated by mortality of species 1 and species 2 adults.

Our discussions will be based mostly on the simple model (4). Some of the artificialities of this model could be removed, at the price of much additional complexity. Variations in microsite characteristics (such as the relative competitive abilities of the species) could be incorporated by having separate f's and F's for each microsite type. An extended pre-reproductive juvenile phase would require an age-structured model (similar to a Leslie matrix). Inclusion of extended dormancy would force us to abandon (4) entirely, since the aboveground site occupancies would not completely characterize the state of the system. Instead, we would have to consider the frequency distribution of belowground population densities; for single-species models of this type see e.g. Chesson (1981), Elderkin (1982).

Microsite-based competition models were introduced by Skellam (1951), and have since become popular as models of competition for space, most notably as models of forest dynamics (Horn, 1975, 1976; Acevedo L., 1981). In the forest models (which are quite similar to (4)) the microsites are generally heterogeneous (characterized by the previous occupant, light regime, etc.), and interspecific differences in microsite-type preferences are found to allow stable coexistence. Models similar to (4), but with uniform seed dispersal, have also been studied with regard to the coexistence of *TE* species in constant (Hubbell, 1979, 1980) and randomly varying environments (Fagerström & Ågren, 1979; Chesson & Warner, 1981; Chesson, 1982). One of our main theoretical results here is that non-uniform seed dispersal substantially enhances the potential for stable coexistence of *TE* species. Hastings (1980) analyzed a variant of (4) with a strict competitive hierarchy (high-ranking juveniles being able to displace low-ranking adults) and clearing of sites by disturbances, finding that species richness was maximized at intermediate disturbance rates. Microsite-based models have also been used for theoretical studies of the evolution of dispersal (Hamilton & May, 1977; Comins *et al.*, 1980; Comins, 1982; Motro, 1982). These models generalize (4) in a variety of ways, but the competitors are allowed to differ only in the fraction of seed allocated to dispersal (b_i in our notation).

Model dynamics in a constant environment

The qualitative behavior of our two-species, one-patch model (4), can be described in terms of the 'invasion coefficients' β_1 and β_2, defined by

$$\beta_1 = \frac{w_1 d_2}{v_2 + w_2} - \frac{w_2 d_1}{v_1 + w_2}$$

$$\beta_2 = \frac{w_2 d_1}{v_1 + w_1} - \frac{w_1 d_2}{v_2 + w_1}$$

(6)

The β_i have a simple biological interpretation: $\beta_i + 1$ is the finite growth rate of species i, when it is introduced at low densities into a patch dominated by species j, i.e.

$$1 + \beta_i = \lim_{f_i(t) \to 0} \frac{f_i(t+1)}{f_i(t)}.$$

(7)

Thus if $\beta_i > 0$, species i will increase in numbers when it is rare. In Appendix A we show that there are essentially four possible behaviors of model (4), determined by the signs of the β_i;

    (a) $\beta_1 > 0, \beta_2 < 0$: species 1 excludes species 2.
    (b) $\beta_1 < 0, \beta_2 > 0$: species 2 excludes species 1.
    (c) $\beta_1 < 0, \beta_2 < 0$: contingent exclusion.
    (d) $\beta_1 > 0, \beta_2 > 0$: coexistence.

If competitive exclusion holds (a or b), the excluded species declines to zero from any initial density (unless the other species is absent). In contingent exclusion, stable coexistence is impossible and neither species is able to invade a patch dominated by the other. There is an unstable equilibrium at $f_1 = f^*$, and $f_1(t) \to 0$ if $f_1(0) < f^*$, $f_2(t) \to 0$ if $f_1(0) > f^*$: one or the other goes extinct. In the case of coexistence, there is a stable equilibrium $(f_1^\dagger, f_2^\dagger)$ with both species present, and the system tends to $(f_1^\dagger, f_1^\dagger)$ from any initial densities except $f_1(0) = 0$ or $f_2(0) = 0$.

The possible coexistence mechanisms in our model depend only on the relationship between model parameters and the signs of the β_i. Consequently, the mechanisms are likely to be robust under slight modifications in the model dynamics. For example, the assumption of 100% site occupancy can be relaxed by allowing a fractin c of all sites to become vacant prior to reproduction. The local stability behavior of the model then depends on 'invasion coefficients' $\beta_i(c)$ exactly as above, and the β_i are continuous functions of c (Appendix A). Sufficient conditions for coexistence derived from (4) therefore remain approximately valid so long as the frequency of unoccupied sites is fairly low. Vacant sites also arise if seed dispersal is randomly (rather than deterministically) uniform over the patch, so that each microsite receives a Poisson-distributed number of seeds of each species. We have analyzed this case along the lines of Appendices A and B; all coexistence mechanisms derived from (4) persist, without any qualitative differences in the life-history interpretations of the coexistence criteria.

280

Applications of the model

In applying our model to plant communities we are concerned primarily with the maintenance of species richness at two levels: that of the single, spatially delimited community, and that of its abstraction, the community-type, which is restricted to a single habitat-type in a small enough region that macroenvironmental gradients are negligible, but may occur in disjunct stands.

We will also pay attention to the maintenance of 'rare' (low abundance) species. In communities which we have sampled (Shmida, 1972, 1977, and unpublished) and in published species-abundance or species-presence curves for plants (e.g. Williams, 1964; Kilburn, 1966), a substantial fraction of the species are represented by 1–5 individuals/m² on average. Similarly, in replicate samples from a community-type, many species occur in only a small fraction of the samples (Figure 2 is typical of communities we have sampled). Were the sample-size greatly enlarged, species of intermediate abundance might then be predominant, as in the log-normal species-abundance curve (Preston, 1948, 1962; May, 1975; Whittaker, 1977). Nonetheless, it is apparent that many plant species are locally rare and/or patchily distributed on the scale of competition for individual sites. Several of the mechanisms we discuss would result in plant species that are rare, at least at some times in some parts of their habitat, and we will suggest possible connections between the theory and the occurrence of many rare plant species.

Coexistence of species with alternate life-history strategies

A striking (and entirely unexpected) property of our model is that two trophically equivalent species can coexist stably in a single patch of identical microsites, in an unchanging environment. The mathematical conditions for this mode of coexistence ($\beta_1 > 0$, $\beta_2 > 0$) correspond to a pair of species adopting alternative 'strategies' for utilizing the patch, one emphasizing adult survivorship and the other emphasizing fecundity.

The condition $\beta_1 > 0$, $\beta_2 > 0$ is equivalent to

$$\frac{v_2 + w_2}{v_1 + w_2} < \frac{w_1 \, d_2}{w_2 \, d_1} < \frac{v_2 + w_1}{v_1 + w_1}. \tag{8}$$

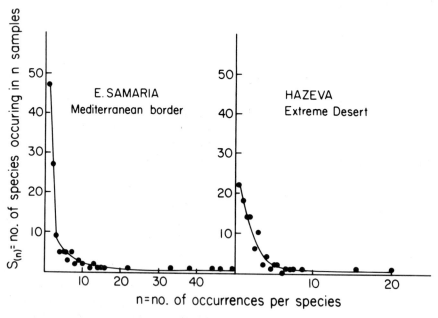

Fig. 2. Species-presence curves (frequency distribution of numbers of 100-m² quadrats in which a species occurs). n = number of quadrats in which the species occurs, $S(n)$ = number of species occurring in exactly n quadrats. (a) Mediterranean border vegetation, East Samaria. (b) Extreme desert vegetation, Hazeva study area.

This can only be satisfied if either (Appendix A)

a) $v_1 > v_2$, $w_1 > w_2$, and $d_1 > d_2$; or
b) $v_1 < v_2$, $w_1 < w_2$, and $d_1 < d_2$.　　　　(9)

In case (a) species 1 has the higher fecundity, and also the higher death rate; species 1 is the better 'space claimer', and species 2 the better 'space holder'. In (b) the roles are reversed. These conditions are necessary for coexistence, but they are not sufficient.

Non-uniform seed dispersal by at least one of the species ($v_i > 0$) is necessary for (8) to hold (otherwise the outer inequality reads $1 < 1$). Essentially, non-uniform seed dispersal generates a clumped seedling distribution in each species, and the clumping reduces interspecific competition to a level where stable coexistence is possible. The same effect also arises if seed dispersal is random. In a similar vein, Atkinson & Shorrocks (1981) simulation-modelled competing insect species with a negative binomial distribution of eggs per microsite, and found that sufficiently high aggregation by a superior competitor allowed an inferior competition to persist. These results emphasize the importance of non-uniform seed dispersal, localized competition and other possible causes of spatially non-uniform competitive interactions, for the coexistence of plant species.

Some life-history restrictions are also implicit in (8). Firstly, two annuals cannot satisfy (8) since they have $d_1 = d_2 = 1$. Secondly, there must be substantial within-patch seed dispersal. If the w_i are small relative to the v_i, the outer inequality in (8) is nearly an equality, leaving only a minuscule range of parameters within which (8) is satisfied. The robustness of these restrictions with respect to modifications of our simplifying assumptions cannot be judged at present.

We are unable to offer empirical support for this mechanism, although there is an analogy in the coexistence of *Ranunculus* species in grasslands (Harper & Sagar, 1953; Sarukhan & Harper, 1973; Sarukhan, 1974; reviewed in Harper, 1977). These species have similar habitat preferences but are sharply differentiated in their reproductive 'strategy', ranging from primarily by seed (*R. bulbosus*) to primarily vegetative (*R. repens*), with *R. acris* intermediate.

Spatial mass effect

Source- and sink-populations

The dispersal of seeds is at least partially random, subject to the vagaries of wind, animal movements, etc. Self-maintaining populations of a species in its primary habitat therefore serve as a source of emigrant seeds, that augment populations in other nearby habitats. Thus, input of seeds from neighboring habitats ('mass effect' *sensu* Shmida & Whittaker, 1981) can maintain 'sink' populations that would otherwise be non-viable. This occurs (trivially) in our model if externally input seeds are added to the disperser pool of a species ($\lambda_j = f_j w_j + I_j$, I_j = external input). It also can occur if $\beta_1 > 0$, $\beta_2 < 0$ in some patches (so species 1 excludes species 2 in the absence of external input) and $\beta_2 > 0$, $\beta_1 < 0$ in others. This is intuitively clear – each patch is the stable 'source' for 'sink' populations elsewhere – and follows formally from our one-patch analysis in Appendix A and the Karlin-McGregor (1972) theorem, so long as interpatch dispersal is weak.

The potential role of the mass effect in augmenting community species richness has been recognized in studies of animal populations (Anderson, 1970; Itzkowitz, 1977) and desert herbs (Shmida & Whittaker, 1981). Smith (1972) argued that source populations in refuges from predation are important in maintaining prey species, and Pielou (1975) discusses spatial mass effect 'blurring' zonation patterns along environmental gradients. Nonetheless enumerations of the factors possibly contributing to plant community species richness (e.g. Whittaker, 1975, 1977; Grubb, 1977; Connell, 1978; Pickett, 1980) usually omit it entirely. The situations in which mass effect is likely to be particularly important are:

1. Habitat mosaics

Suppose that different habitat types, supporting different assemblages of plant species, occur as a mosaic of patches in the landscape (Whittaker & Levin, 1977). Mass effect will then cause species to occur in the primary habitats of other species even if they would otherwise rapidly be competitively excluded by the resident species (Levin, 1974, 1981; Shmida & Whittaker, 1981). The habitat mosaic need not be static, so long as all habitat types are present at any given time (e.g. under-shrub and between-shrub habitats).

2. 'Slow' dynamics of competitive exclusion

Mass effect from their primary habitats can maintain numerous species in 'neutral' habitats that are secondary for all. If no species has a decisive competitive advantage, then small amounts of mass effect can stabilize coexistence over large regions of neutral habitat. This hypothesis was advanced by Dale (1978) for territorial coral reef fishes, which according to Sale (1974, 1977, 1978, 1979) occupy very similar territories and compete for open territories in a nearly even lottery. In plant communities, disturbances may create local reductions in the intensity of competitive intensity, and mass effect could then allow species from diverse habitats to co-occur throughout the disturbed area. This was suggested by Peet *et al.* (1983), who also review supporting studies. Other circumstances producing slow dynamics in our model, and theories relating slow dynamics to species richness, are discussed below.

Experimental verification of coexistence by mass effect will be difficult, because the *cause* of coexistence (source-populations in the primary habitat) may be very far from the *locality* of coexistence (the secondary or neutral habitats). If the source-populations can be identified, and their location is fairly static, they could be screened to block dispersal. Mass effect would be indicated by local extinction of sink-populations. Alternately one could attempt to screen presumed sink-populations against all incoming seeds, without inducing confounding habitat modifications.

Sink-populations far from their source, or in inhospitable habitats, are likely to consist of only a few individuals. Thus, mass effect is potentially a source of rare species, and the following study indicates that mass effect can be a major factor in maintaining rare species.

Spatial patterns of abundance at Har Gilo, Israel

The abandoned terrace farm at Har Gilo, Israel is one of the most species-rich areas of that size in the world known to us. The community is dominated by annuals: 187 annual species are found in the 1-ha study area, and 1-m^2 quadrats often contain 40–60 different annual species. The highest average spe-

cies richness yet reported for North America is 39/m^2, with a maximum of 52, in savannas on the North Carolina coastal plain (Peet *et al.*, 1983), a community-type with very few annual species. At Har Gilo we have a similar average richness, but individual quadrats contained up to 67 species of annuals alone. The spatial patterns of relative abundance at Har Gilo suggest that the associations of species with primary habitats in a dynamic landscape, and the maintenance of sink-populations by mass effect, both contribute to the high species richness of the annual community.

The study area (described in Appendix C) was thoroughly mapped to identify the most abundant species in each square meter, for four years beginning in 1979. Species abundances were estimated visually each year in 148 1-m^2 quadrats placed randomly throughout the area. The abundance patterns were not entirely static: although the 1979, 1980, and 1981 patterns were similar, 1982 was markedly different. Our report here refers to 1980; in that year we also recorded all individuals in 51 1-m^2 quadrats distributed among all dominance-types in the area.

Of the annuals, 33 were the most abundant species in at least one locality (a 'dominance patch'). The dominance was associated with higher per capita fecundity (2 to 15 times the average for quadrats where the species was not the most abundant; visual estimates based on the number of reproductive structures just prior to seed-set). If mass effect is occurring, we would expect the dominance patches to act as source-populations, with smaller sink-populations nearby. This pattern was in fact observed. Almost all (28) of the 33 species with dominance patches were 'rare' (5 or fewer specimens per quadrat) in at least one of the fully censused quadrats, and all of the occurrences as rare species were within 5 m of an identified dominance-patch.

The data from the 51 censused quadrats indicate that these 'rare' species, which we interpret as sink-populations, are an appreciable component of the measured species richness; results from 10 representative quadrats are summarized in Table 2. The apparent sink-populations are also a major component of the large number of 'rare' species observed in each quadrat. About half (83) of the annuals were 'rare' in at least one censused quadrat. If our interpretation of the 28 with dominance patches is correct, mass effect then accounts for one third of the annual species recorded as 'rare' at some location in the study area.

Mass effect is also indicated by the presence, in all 51 quadrats, of at least one species represented by only depauperate non-reproductive populations. These populations must be sustained by viable populations, either in other locations or in other years (held over in the seed bank). They are a component of species richness as it is usually measured, although they probably should be regarded as 'non-resident' species, analogous to non-resident birds that pass through a habitat without nesting there.

Dissimilarity of within-community replicate samples

Replicate samples from a single community rarely contain exactly the same species. Indeed, the percentage of recorded species that appears in both samples (their 'index of similarity', Curtis, 1959; or 'coefficient of community', Whittaker, 1975) is typically 70–90% in samples from a single stand (Whittaker, 1972), and 30–70% in samples from different stands of the same community-type (Curtis, 1959; Shmida, 1972, 1977; Barbour *et al.*, 1980). Dissimilarity of replicate samples can be due to a number of factors, including cryptic microhabitat differences, sampling inadequacies and errors, and the vagaries of dispersal and competition. We present here analyses of within-community replicate samples suggesting that much of the dissimilarity can also be due to mass effect from adjacent habitats.

If mass effect contributes substantially to the dissimilarity of within-community replicate samples, we would predict that

1. many of the species responsible for the dissimilarities would have their centers of abundance in other nearby communities;
2. consequently, distant replicates exposed to mass effect from different source-populations will be more dissimilar than nearby replicates.

We tested these predictions in two community-types in Israel: annual grassland on sandfields near Qesarya (Roman Caesaria) and an *Artemisia* semi-shrub community in the Judean Desert (the study areas are described in Appendix C). Both communities occur as scattered stands in a habitat mosaic, so quadrats were sampled throughout the mosaic and classified into community-types by nodal ordination (Noy-Meir, 1972).

Coefficients of community between 1-m^2 replicate samples from these communities are given in Table 3 for (a) 10 contiguous quadrats in a linear transect and (b) 10 quadrats placed randomly in the community-type, subject to a 5 m minimum distance between quadrats. As predicted, the contiguous quadrats are substantially more similar than the disjunct quadrats.

Regarding prediction 1: for each pair of quadrats, the species may be classified as shared or unshared, unshared species causing the dissimilarities. In both communities studied, most of the species that were unshared by at least one pair of quadrats, were most abundant in a different community (i.e., in samples assigned to a different node in the ordination) – 72% of the unshared species in the sandfields, 64% in the semi-shrub community.

We recognize that cryptic microhabitat gradients could be invoked to explain these patterns, and those observed at Har Gilo. In the absence of exper-

Table 2. Numbers of rare and common species in 10 permanent 1-m^2 quadrats in the Har Gilo study area. (Species with less than 5 individuals in the quadrat were considered rare. Dominance-patches are defined in the text.) Quadrats were chosen to represent all major vegetation types identified by nodal ordination of 88 quadrats; quadrat choice within types was random.

	1	2	3	4	5	6	7	8	9	10	Mean	S.D.	% of total
Number of species in quadrat	28	38	67	36	35	39	34	31	32	49	38.9	11.4	100 %
Number of rare species	17	16	27	13	20	16	13	13	15	22	17.2	4.6	44.2%
Number of rare species which have dominance-patch within 5 m.	8	5	10	8	9	4	4	3	7	9	6.8	2.4	17.5%
Number of rare species which are rare throughout the study area	6	8	13	5	9	10	9	10	8	11	8.9	2.3	22.9%

Table 3. Floristic similarity (= 100 × CC) between 1 m² quadrats at Qesarya (above diagonal) and *Artemisia* (below diagonal) study areas. (a) Contiguous quadrats in linear transect. (b) Quadrats randomly placed, subject to 5 m minimum distance between quadrats. All samples placed within single communities identified by nodal ordination. Qesarya data are from Barbour *et al.* (1982). Diagonals (CC = 1) omitted from summary statistics.

$$CC = \frac{2S_{AB}}{S_A + S_B}$$

where S_A = # of species in sample A, S_B = # of species in Sample B, S_{AB} = # of species in both A and B.

3(a)	1	2	3	4	5	6	7	8	9	10
1	100	68	74	71	66	64	64	74	71	60
2	72	100	72	78	71	69	69	63	50	47
3	66	69	100	80	82	81	75	73	58	55
4	34	50	54	100	82	77	72	87	68	65
5	50	48	60	67	100	82	68	76	62	59
6	45	66	48	63	57	100	71	75	66	62
7	66	48	61	61	52	67	100	65	55	63
8	45	52	66	72	40	66	72	100	82	72
9	57	46	52	78	36	72	78	61	100	75
10	52	66	45	67	70	75	83	66	55	100

3(b)	1	2	3	4	5	6	7	8	9	10
1	100	22	25	12	32	42	35	55	40	33
2	31	100	50	31	31	50	45	52	41	43
3	28	25	100	36	41	21	23	42	40	22
4	26	24	30	100	31	34	47	54	52	45
5	35	32	25	37	100	40	43	50	38	50
6	26	35	29	47	26	100	36	60	50	34
7	22	40	35	50	44	38	100	40	23	20
8	37	33	53	40	25	42	23	100	58	40
9	35	42	50	52	33	50	22	35	100	42
10	25	33	40	33	23	21	33	25	27	100

Qesarya $\overline{CC}$ (contiguous) = 69.2 ± 9.2 (mean ± S.D.)
  $\overline{CC}$ (disjunct)   = 38.7 ± 11.2 (mean ± S.D.)
Artemisia $\overline{CC}$ (contiguous) = 59.68 ± 11.2 (mean ± S.D.)
  $\overline{CC}$ (disjunct)   = 33.71 ± 8.9 (mean ± S.D.)

imental manipulations, or demographic study of apparent sink-populations, we can only say that mass effect is a predictive and parsimonious explanation for the observed abundance patterns.

Non-equilibrium coexistence or 'temporal mass effect'

Hutchinson (1961) suggested that community-wide environmental fluctuations might constantly reverse the direction of competitive exclusion, thus allowing species to recover before they are completely eliminated from the community. Coexistence then derives from 'safe years' rather than 'safe sites' for each species: a temporally-defined refuge. Our model predicts that environmental fluctuations can allow coexistence of trophically equivalent species that could not exist at equilibrium in a constant environment.

Specifically, we consider two species in one patch, and allow the model parameters to vary as functions of the (random) 'environment-type' Z_t in year t. The Z_t are assumed to be independent, identically distributed random variables uninfluenced by the history of the populations. The iteration $f_1(t + 1) = H(f_1(t))$ in model (4) is thereby converted into a stochastic difference equation of the form

$$f_1(t + 1) = H(f_1(t), Z_t) \tag{10}$$

Our criterion for non-equilibrium coexistence in (10) is the 'stochastic boundedness' criterion developed by Chesson (1978, 1982):

$$\lim_{\epsilon \to 0} \sup_{t>0} P[f_i(t) < \epsilon] = 0, \, i = 1, 2. \tag{11}$$

where $P[A]$ is the probability of the event A. (11) can be viewed as requiring that the f_i not spend 'too much time' near 0 (Chesson, 1982). In particular, equation (11) implies that neither species drifts to extinction: $P[f_1(t) \to 0] = P[f_2(t) \to 0] = 0$.

If the environmental fluctuations are not excessively severe (in a sense discussed in Appendix B), two trophically equivalent species modelled by (10) will coexist *sensu* (11) if

$$E[\ln(1 + \beta_1(Z_t))] > 0 \text{ and } E[\ln(1 + \beta_2(Z_t))] > 0 \tag{12}$$

Here $E[\cdot]$ denotes expectation with respect to the distribution of Z_t. In particular, (12) implies coexistence so long as the death rates d_i are both constant and < 1. The species are maintained by a sort of temporal mass effect, in which gains made during 'good years' (when β is large) allow each population to persist through the 'bad years' (when β is small).

As an illustrative example, consider two perennial species whose fecundities fluctuate randomly from year to year, and for simplicity assume there are only two environment types. In environment type 1, species 1 has the higher fecundity ($R_1 = \dfrac{m_1(1)}{m_2(1)} > 1$), and in environment type 2, species 2 has the higher fecundity ($R_2 = \dfrac{m_2(2)}{m_1(2)} > 1$). Then (12) will be satisfied, and the species will coexist, whenever

$$a_1 \ln(1 - d_1 + d_2 \, \gamma_1 \, R_1) + a_2 \ln(1 - d_1) > 0$$
$$a_2 \ln(1 - d_2 + d_1 \, \gamma_2 \, R_2) + a_1 \ln(1 - d_2) > 0 \tag{13}$$

where a_i is the frequency of environment type i and $\gamma_i = (b_i p_i)(1 - b_j + b_j p_j)^{-1}$. The derivation of (13) is in Appendix B. The conditions (13) are quite conservative, and describe only a small portion of the parameter values that would lead to coexistence.

In essence, (13) is a requirement that 'good years' be good enough: if R_1 and R_2 are sufficiently large,

the species coexist. The population consequences of 'bad years' are limited by the survival of adults, so occasional bursts of reproduction are adequate to maintain the population. This logic cannot be applied to annuals without a seed bank, and indeed our model does not seem to allow non-equilibrium coexistence of annuals. This appears to be a general characteristic of 'lottery' competition models in random environments: overlapping generations at some stage in the life-cycle broadens the range of conditions allowing non-equilibrium coexistence (Chesson & Warner, 1981; Chesson, 1982, and personal communication). However, non-equilibrium coexistence of annuals is possible if we allow vacant microsites and random seed dispersal (Appendix B) or (more importantly) if we allow a seed bank. Annuals with delayed germination are functionally long-lived interoparous species, whose population declines in 'bad years' are buffered by the seed bank. Models in which the seed bank allows non-equilibrium coexistence of annuals can be easily manufactured, e.g. by allowing random variation in the parameters of the MacDonald & Watkinson (1981) 'bottleneck' model for annual plants with a seed bank (Ellner, 1983).

Our model predicts that environmental fluctuations can allow non-equilibrium coexistence of trophic equivalents, so long as each species has the advantage in some years. Fluctuations of this sort in the seed production by populations of perennial plants are well documented (see reviews in Grubb, 1977; Harper, 1977; Hubbell, 1980). We have observed the same phenomenon in annual vegetation at Nahal Darga and En-Gedi in the Judean Desert, Israel. Table 4 compares the abundance and fecundities of the most common annuals at each location in 1979 (relatively dry, rainfall 56% of average) and in 1980 (relatively wet, rainfall 123% of average; rainfall measurements from En-Gedi Station, Israel Meterological Service). Evidently different species were favored in each year, and the seed production/m^2 by a given species typically varied by several orders of magnitude between years.

Of course, the occurrence of population fluctuations does not prove that this is the factor allowing coexistence. However, if our 2-year records capture the typical range of variation in the annual flora, the above ground populations during unfavorable years are largely derived from the much higher seed

Table 4. Seed production and abundance of the most common annuals in 1-ha study areas at Nahal Darga and En-Gedi, Judean Desert. Abundance = mean # of individuals in 10 1-m² quadrats. Seed production estimated as average seed production by 30–100 randomly chosen individuals ajacent to the study area.

Species	Abundance		Per capita seed production		Relative seed production per m², 1979/1980
	1979	1980	1979	1980	
Nahal Darga					
Stipa tortilis	374	31	25	4	7.5×10^{1}
Reboudia pinnata	5	75	81	731	7.4×10^{-3}
Limonium thouinii	18	1	31	3	1.9×10^{2}
Asteriscus pygmeus	7	38	7	86	1.5×10^{-2}
Trigonella stellata	2	45	38	435	3.9×10^{-3}
Plantago ovata	12	425	4	33	3.4×10^{-3}
En-Gedi					
Aaronsohnia factorovskii	15	121	12	210	7.1×10^{-3}
Pteranthos dichotomus	3	27	4	76	5.9×10^{-3}
Aizon canariense	1	16	331	$3.5 \cdot 10^{4}$	5.9×10^{-3}
Mesembrianthemum forsskalii	18	3	2351	283	5.0×10^{1}
Trigonella stellata	7	31	48	859	1.3×10^{-2}
Asteriscus pygmeus	13	85	11	125	1.4×10^{-2}
Stipa tortilis	41	7	21	3	4.1×10^{1}

output that occurs in favorable years, and the seed-set during unfavorable years probably makes a negligible contribution to the maintenance of the populations. If this is correct, then differentiation in 'safe years' is an important contributing factor in the species richness of the annual communities at Nahal Darga and En-Gedi.

Note also that all of the annuals recorded as 'rare' in one year (5 or fewer individuals per m²) were not rare in the other. Just as spatial mass effect can maintain sink-populations of locally rare species, temporal mass effect is maintaining bad-year populations of temporarily rare species in these desert annual floras. Any given year is 'bad' for many of the species, which are then recorded as rare even though they may be abundant in the seed bank and will be abundant in other years.

Slow dynamics

It is well known that chronically stressed communities can be quite species-rich, and that enrichment or removal of stresses can reduce species richness. Generalizing from these observations, several recent theories have suggested that slow rates of competitive exclusion are a major factor contributing to high species richness (Grime, 1973, 1979;

Huston, 1979; Peet *et al.*, 1983). Equilibrium and non-equilibrium versions of this idea may be distinguished. In the former, some 'stress' prevents the competitive dominants from monopolizing the available sites or resources, leaving space for stable populations of other species. This seems to be the essence of Paine's (1966) and Grime's (1973, 1979) ideas, as well as Naveh and Whittaker's (1979) hypothesis on the role of grazing in maintaining species richness. The 'patch dynamics' (Pickett, 1980) or 'intermediate disturbance' (Cornell, 1978, 1979) hypotheses might also be placed under this heading. The non-equilibrium version views species richness as a dynamic balance between species establishment and competitive exclusion (Huston, 1979). Species richness is then predicted to be highest at intermediate levels of stress, at which establishment is possible but competition is not so intense that the top competitors rapidly exclude all others. The slow dynamics allow coexistence to be maintained by weak or infrequent forces, such as occasional external inputs or occasional 'good' years for competitively inferior species.

Factors contributing to slow exclusion

In our model for two species in one patch, unless both β_i are > 0, one species will increase steadily at the expense of the other. Their transient coexistence nonetheless may be lengthy. At least five separate factors may lead to a system with imperceptibly slow dynamics: (a) long life of adults, (b) small differences in competitive ability, (c) wide dispersion of microsites and/or patches with low dispersal of propagules between them, (d) many open microsites and/or patches relative to the number of occupied sites, and (e) highly aggregated or clumped distributions, with little interspecific overlap.

In (a) microsite turnover is infrequently because vacant sites are rare. Case (b) is a pseudo-equilibrium in which gains and losses nearly balance. Cases (c), (d), and (e) are non-interaction hypotheses. In (c) and (e), when an adult dies the juveniles competing for the microsite will generally be conspecifics, so the site will be reoccupied without altering the abundance of either species. In (d) interspecific competition is weak because environmental 'stress' or predation keep population densities so low that resources are superabundant; coexistence of some seastars (Menge, 1979) and of folivorous insects (Lawton & Strong, 1981) have recently been explained on this basis.

For a theoretical example of the interplay of factors (a)–(d), we can add (d) to our model by assuming that interspecific competition for microsites by juveniles is negligible. This simplification may be realistic for extreme deserts (Gulmon *et al.*, 1979), the U.S. Northeast rocky intertidal (Menge, 1979) or other harsh environments where vacant sites are common and establishment of juveniles is rare. If few species are dispersed into many sites, the chance that a species-i seed reaches any given site will be a small number μ_i. Dropping terms of order $\mu_i \mu_j$ or smaller, our two-species model then becomes

$$f_i(t+1) = f_i(t)(1-d_i) + f_i(t)d_i(1-q_i) + \\ + (1 - f_1(t) - f_2(t))f_i(t) w_i(1-q_i). \quad (14)$$

where q_i is the mortality of species-i juveniles. Model (14) is thus an approximation to (4) that becomes increasingly accurate as the $\mu_i \to 0$. The first term is survival by adults; the second is re-establishment by offspring of an adult who dies; and the third is colonization of vacant sites (microsites or patches). We assume $w_i(1-q_i) < 1$ to keep juvenile establishment rare: adults successfully establish no more than one offspring per year.

(14) is a discrete-time Lotka-Volterra competition model with parallel isoclines: $f_i(t+1) = f_i(t)$ when $f_1(t) + f_2(t) = f_i^*$, where

$$f_i^* = \frac{w_i(1-q_i) - d_i q_i}{w_i(1-q_i)}$$

is the equilibrium density of species i in the absence of species j. The species with the larger f^* can invade and will then competitively exclude the other species (Appendix A). The 'velocity' of competitive exclusion can be measured by the Euclidean distance between successive population densities:

$$V = \sqrt{(\Delta f_1)^2 + (\Delta f_2)^2} \quad (15)$$

where $\Delta f_i(t) = f_i(t+1) - f_i(t)$. Following an invasion by species 1 (assuming $f_1^* > f_2^*$), the rate of exclusion satisfies (Appendix A)

$$V \leqslant f_1^*(f_1^* - f_2^*) \max_i [w_i(1-q_i)]. \quad (16)$$

From (16) the limits on the rate of exclusion imposed by factors (b)–(d) are apparent. $(f_1^* - f_2^*)$ is a measure of species 1's competitive edge, and a superabundance of open sites is equivalent to f_1^* being small. Wide dispersion of sites and/or high juveniles mortality produces a low effective dispersal $w_i(1-q_i)$.

Longevity also shows exclusion in (14), since $f_2(t+1) \geqslant (1 - d_2 q_2)f_2(t)$. If species 2 is long-lived (d_2 near 0) its decline is necessarily slow; but the relationship between d_i and V is not straightforward. Decreasing d_1 increases species 1's competitive edge and hastens species 2's decline, while a decrease in d_2 would slow or possibly reverse the process of exclusion. .

The effects of (e), highly aggregated distributions, were studied by Atkinson & Shorrocks (1981) in a computer simulation model of competing insect utilizing a patchy, ephemeral habitat. They found that clumping slowed competitive exclusion, and that highly clumped distributions allowed stable coexistence, especially if the clumping increased with population density. The coexistence in their model is Skellam (1951)-type, with the subordinate persisting as a fugitive in sites where the dominant species is temporarily absent.

We can offer two apparent examples of slow

dynamics due to virtual non-interaction of plant species. Thirteen species of shrubs and small trees are found in rock outcrops along the Rift Valley in Israel. We sampled the rock outcrops on a 78 km transect from Jericho to Sedom along the western edge of the Rift Valley. Most of the outcrops (165 of 175) contained none of the species, and 7 of the other outcrops contained only one species. The outcrops containing the shrubs were not noticeably different from the others, nor could we detect any association of particular species to particular types of outcrops. Unoccupied patches of apparently suitable habitat appear to be similarly superabundant throughout the distribution of these species in Israel and the Sinai.

The temperate ferns *Phyllitis scolopendrium* and *P. sagittata* are an even more extreme example. In southern Europe and the northern Mediterranean the species are common. At the southern limit of their distribution in Israel they are found only on rubble piles in ruined cisterns. Despite intensive searches by many naturalists, *P. scolopendrium* has been found (in Israel) in only 7 nearby cisterns (of over 200 examined) in the Judean Mountains and two cisterns (of 57 examined) in Samaria, and *P. sagittata* is known only from 3 cisterns in Samaria. No co-occurrence of the species has been observed: the wide dispersion of cisterns and the rarity of the species have apparently eliminated competition between them in Israel.

Contingent exclusion, and geographical replacement in a patchy habitat

In the case of contingent exclusion ($\beta_1 < 0$, $\beta_2 < 0$), coexistence in a single patch is impossible: whichever species exceeds its (unstable) internal equilibrium eventually excludes the other. Consequently, globally stable coexistence is also impossible in a system of several patches linked by dispersal. Such a system can nonetheless have several locally stable equilibria at which the species coexist. The community dynamics then consist of shifts from one equilibrium to another, occasioned by disturbances, local drift to extinction, or other perturbations. The rate of competitive exclusion is then limited by the frequency of perturbations, rather than the rate at which species are excluded from individual patches.

The locally stable equilibria may be generated as

follows. Consider first a system of isolated patches (no interpatch dispersal), each patch containing only one species. Any initial assignment of species to patches is stable when $\beta_1 < 0$, $\beta_2 < 0$. Intuitively, this should imply a successful invasion of patches dominated by the other species; the system with dispersal should have a locally stable equilibrium at which each patch is dominated by one species, with the other present in small numbers due to dispersal. A theorem of Karlin & McGregor (1972a, Theorem 4.4) implies that this is in fact correct, so long as the fraction of seeds that disperses between patches is sufficiently small. Their theorem applies to a broad class of difference-equation models which includes ours; analogous results also hold for differential equation models (Levin, 1974, 1979; Yodzis, 1978; Matano, 1979).

A community structured by contingent exclusion was termed 'founder controlled' by Yodzis (1978), because 'its structure is almost entirely governed by the colonization process' in which each species comes to dominate some patches. If patches never again become open for colonization by other species, a founder-controlled community can be viewed as an example of stable coexistence (e.g. Karlin & McGregor, 1972b; Yodzis, 1978). If patches do become open for colonization, one species or the other will come to dominate the patch. The resulting pattern would again be a (deterministically) locally stable equilibrium. Community dynamics therefore occur on the *slow* time-scale of patch colonization/extinction rather than the *fast* time-scale of population dynamics. Competitive exclusion from the multi-patch community might then be prevented by forces too weak or infrequent to allow coexistence in one patch, such as long-term climatic changes, or habitat differentiation over very gradual environmental gradients and the resultant weak mass effect from primary habitats outside the community of interest. The role of founder control in these circumstances is to segregate species into different patches, weakening interspecific competition so that slow time-scale forces might prevent competitive exclusion.

Founder control depends in general on the existence of 'alternate stable states' for an isolated patch ($f_1 = 0$ and $f_1 = 1$ in our model). The potential role of alternate stable states in enhancing community species richness has long been recognized (e.g. Karlin & McGregor, 1972b; Levin, 1974, 1979, 1981;

Sutherland, 1974; Yodzis, 1978). In plant communities these could result from biotic site-modifications by dominants that favor establishment by conspecifics rather than competitors (Acevedo L., 1981).

Our main point here is that the spatial effects in our model are adequate to generate alternate stable states, even for trophically equivalent species, so coexistence by founder control is possible. The parameter values producing alternate stable stages (contingent exclusion) are $\beta_1 < 0$, $\beta_2 < 0$, i.e.

$$\frac{v_2 + w_1}{v_1 + w_1} < \frac{w_1 d_2}{w_2 d_1} < \frac{v_2 + w_2}{v_1 + w_2}. \tag{17}$$

This is exactly (8) with the inequalities reversed, so (as discussed relative to (8)) the within-patch spatial structure and non-uniform seed dispersal are essential for coexistence. Satisfying (17) requires that $v_1 > v_2$ and $w_1 < w_2$, or the reverse: one must be a relatively 'high dispersal' strategist (larger w), the other a 'lower dispersal' strategist (larger v).

Some large-scale habitat 'patchness' is also necessary for coexistence mechanisms based on alternate stable states (Levin, 1981). In our model, if all microsites are coalesced into a single homogeneous habitat-patch, one is left with contingent competition in which one or the other species will be eliminated. Thus, coexistence depends on the spatial patterning of suitable microsites, not just the intrinsic characteristics of species and microsites.

Founder control without habitat differentiation requires very low inter-patch dispersal rates. This is most likely to occur as a consequence of rarity: either rarity of patches in the landscape, or rarity of suitable microsites in each patch. It therefore seems likely that founder-control should be more common in rare species than abundant ones.

In summary, founder control refers to disjunctive, patchy habitats wherein the species that numerically dominates a patch is able to resist invasion by competitors, This situation could result from microsite-level spatial effects (producing β_1, $\beta_2 < 0$ in our model), or from patch-modification by the dominant (litter, allelochemics, shade, etc.) favoring establishment by conspecifics. Species coexist then by geographical replacement, each being dominant in some patches. These configurations are locally stable, but not stable against major perturbations reversing dominance in a patch. Stabilization against major perturbations must result from other factors, which may operate on the timescale of the major perturbations.

Experimental verification of coexistence by founder control and geographical replacement would involve demonstrating that the outcome of within-patch competition depends on the initial abundances of the competitors. This might be accomplished by establishing populations over a range of initial abundances, or perturbing natural populations, and monitoring the subsequent population dynamics.

Examples of geographical replacement

Coexistence by geographical replacement is suggested by the high turnover in species composition between cliffs (large vertical rocky areas) reported anecdotally by Runemark (1969, 1971) and Snogerup (1971). If many cliff species are coexisting by geographical replacement, we would expect that species turnover among replicate samples in cliffs would be higher than that among equally spaced samples on the surrounding 'slope' habitat.

We tested this prediction by sampling disjunct cliffs separated by 1 km or more in two areas in Israel: the Mediterranean phytogeographic region (*sensu* Zohary, 1973), and a transect along the western edge of the Rift Valley between Fazael and Sedom (a distance of 120 km). The exposed bedrock and slope were limestone at all sites. At each cliff we compiled complete species lists for two 2m × 25 m quadrats: one running down the cliff, the other on the adjacent slope.

Despite the identical procedure and locality of cliff and slope samples the cliff samples had markedly higher species-turnover (Fig. 3). The higher turnover gives the cliff habitat a higher species richness than the slope, even though single slope-quadrats contained more species than single cliff-quadrats.

Similarly high between-patch turnover has also been reported among cays composing a Caribbean atoll (Linhart, 1980), and California vernal pools (Holland & Jain, 1981). Linhart (1980) observed between-cay CC's of 0.4 to 0.8 (average ca. 0.6), and judged that the local distributions reflected historical accidents rather than intrinsic differences between the cays. Holland & Jain (1981) concluded

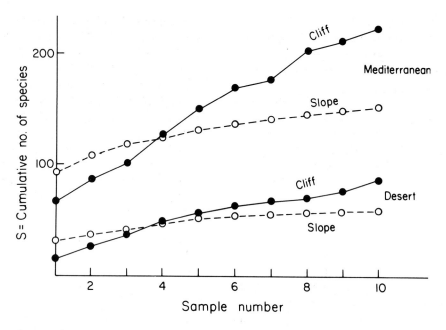

Fig. 3. Species-sample curves for cliff (●) and adjacent slope (○) samples at disjunct cliffs in the Mediterranean and desert regions of Israel. Samples are complete species lists for 25 × 25 m strips in each cliff and slope, arranged in random order for the curves.

that niche partitioning along habitat gradients within California vernal pools occurred largely at the level of genera, with congeners found 'in different pools but under similar ecological conditions'. Between-pool turnover within genera appeared to have a large random component. If Holland & Jain's (1981) interpretations are correct, the congeners are trophic equivalents coexisting in the pool archipelago by geographical replacement. These congruent findings in a variety of communities suggest that, in general, coexistence by geographical replacement may be an important component of the species richness of disjunctive habitats.

Other alternatives to trophic niche differentiation

In the preceding sections we have considered coexistence mechanisms for *TE* species in relation to a simple mathematical model. Some other coexistence mechanisms have been proposed, based on factors not included in the model. In this section we will briefly review these hypotheses, and discuss the relationships among the various coexistence mechanisms. The principal extant hypotheses may be classified as follows.

1. Predator pressure, or compensatory mortality

Limitation of different species by different predators, or predator limitation of a competitively superior species are trophic niche differences which may be adequate to allow coexistence (Paine, 1966; Harper, 1969). Generalized predation can also allow coexistence if the most abundant species is most heavily predated (Harper, 1969; Murdoch, 1969; Murdoch & Oaten, 1975; Roughgarden & Feldman, 1975) or if predators are attracted to monospecific clumps (Connell, 1970; Janzen, 1970). The predator may be an actual herbivore, or any other environmental 'stress' that most severely affects the competitive dominant (Grime, 1979) or whichever species happens to be most abundant (Connell, 1978). Predation could also act to keep all species rare, resulting in slow dynamics (Huston, 1979).

2. Genetic feedback ('advantage of the rare')

As a species becomes abundant, natural selection will favor intraspecific rather than interspecific competitive ability, while the reverse holds for a rare species; the rarer species avoids extinction by evolving into the superior competitor (Park & Lloyd, 1955; Pimentel *et al.*, 1965; Pimentel, 1968). This process has been observed in experimental fly populations (Pimentel *et al.*, 1965; Ayala, 1966) but we know of no natural examples.

3. Equal-chance 'lottery' competition

This hypothesis assumes that 'all species are equal in their abilities to colonize empty spaces [and] hold them against invaders' (Connell, 1978). In the original version (put forth by Sale (1974, 1975), for coral reef fish) the composition of the juvenile pool is independent of the adult population. Hubbell (1979) removed this assumption, and showed that the resulting model mimicked empirical species-abundance curves. In Hubbell's (1979) model, all species are random-walking to extinction, but not yet gone: a 'slow dynamics' hypothesis that competitive exclusion has not yet occurred. Our analysis shows that non-uniform seed dispersal and localized competition allow deterministically stable coexistence in an equal-chance lottery, and Chesson & Warner (1981) showed that random environmental fluctuations could produce coexistence in the sense of stochastic boundedness. Sale's lottery hypothesis remains controversial among coral reef ecologists (Talbot *et al.*, 1978; Sale, 1979; Sousa, 1979), and to our knowledge there are no studies supporting his hypothesis in plant communities.

4. Alternate life-history strategies

One mechanism for coexistence by alternate life-history strategies was outlined above. In addition, a number of models (see Levin, 1976 for a critical review) allow for coexistence of a superior competitor and a superior disperser. The prerequisite is a supply of places where the disperser flourishes at least briefly without competition from the other species; the disperser is a fugitive, which survives by avoiding the competitor. In Skellam's (1951) seminal model for annual plants, vacant microsites arise each year from random seed dispersal. In this model, as in ours and in the Atkinson & Shorrocks (1981) model, the competitor and disperser are not habitat-differentiated. Marshall & Jain (1967) argued that two coexisting species of *Avena* conformed to Skellam's model. Life-history differences

can also be a component of regional coexistence, enabling some species to enter and exploit patches in early successional stages (see below).

Several well-known 'alternatives' to trophic niche differentiation posit that coexistence results from habitat differentiation in response to biotic factors, especially in the regeneration phase. We would therefore regard these as extensions of the trophic niche hypothesis to account for responses to the existing pattern of vegetation, rather than fundamentally different mechanisms. There are essentially two such proposals, although numerous minor variants of each have appeared.

5. Cyclical regeneration

Suppose that the adults in each of two species modify their local environment in a way that favors establishment by the other species. A reduction in the number of adults in either species would then be followed by increased recruitment, so the species would coexist. Replacement sequences of greater length and complexity ('circular networks' *sensu* Jackson & Buss, 1975) are also possible. This phenomenon is documented in some temperate tree species (Fox, 1977; Whittaker & Levin, 1977; Woods, 1979) and grassland herbs (Lieth, 1960; Turkington & Harper, 1979) and Watt (1947, 1964) describes examples from a variety of plant communities.

6. Patch dynamics and regional coexistence

This is essentially the same as cyclical regeneration, save that the cycle includes exogeneous disturbances or other events initiating local secondary successions (Slatkin, 1974; Levin & Paine, 1974). Numerous species can then coexist by habitat differentiation with respect to successional stage (the 'successional niche' of Whittaker, 1969). Patch dynamics is now recognized as an important factor in the species richness of many plant communities (reviewed e.g. by Whittaker & Levin, 1977; Connell, 1979; Werner, 1979; Pickett, 1980; Paine & Levin, 1981). Patch-dynamic mechanisms have been classed as 'non-equilibrium' coexistence (Caswell, 1978; Connell, 1978; Sousa, 1979; Pickett, 1980). However, as Connell (1979) noted, they are only non-equilibrial at the local level of individual patches or sites. The community as a whole is at a global steady-state maintained by asynchronous local events (Connell, 1979; Paine & Levin, 1981), analogous to a chemical equilibrium.

All in all, we have discussed seven different coexistence mechanisms for trophically equivalent species. While they are biologically distinct, all are constructed from three basic components: Spatial variations in the competitive interaction parameters; temporal variations in the competitive interaction parameters, and niche differentiation with respect to non-trophic factors.

Most of the coexistence mechanisms derived in the context of our model (all except temporal mass effect) cannot operate without spatial habitat structure at the level of microsites or patches: localized competition, non-uniform distributions of seeds and/or adults, or spatial variations in habitat characteristics. Similarly, the compensatory mortality, cyclical regeneration, and patch dynamics hypotheses are based on spatial variations in the biotic environment. These spatial effects are particularly important in plants because of their sedentary habit, site retention by adults, and the limitations on dispersal and habitat selection by seeds. We cannot yet claim these mechanisms to be major factors in the coexistence of highly mobile species such as birds.

Discussion

Theoretical and experimental analyses of plant species coexistence have emphasized trophic niche differentiation: differences in habitat or resource utilization, and (to a lesser extent) differences in exploitation by higher trophic levels. Coexisting plant species often differ profoundly in their trophic niches (e.g. Bratton, 1976; Parrish & Bazzaz, 1976; Pickett & Bazzaz, 1976; Werner & Platt, 1976; Harper, 1977; Braakhekke, 1980; Watson, 1981; Tilman, 1982), and it is undeniable that these differences are important in maintaining the species richness of plant communities. However, any two plant species will inevitably differ in some aspect of their responses to resource availability, microhabitats, predators, pollinators, etc. Quantitative elaboration of the differences can be highly suggestive, but the actual problem is to determine if these differences are in fact the proximate cause of coexistence. The question, 'how different is different enough?'

has no general answer. Theoretical 'limiting similarities' (which predict the minimal degree of resource-use differentiation consistent with coexistence on that basis (e.g. May & MacArthur, 1972) are now known to be highly model-dependent (e.g. Abrams, 1975; Rappoldt & Hogeweg, 1980; Turelli, 1981). In addition, it is always possible to explain cases of high overlap by positing differentiation with respect to factors perceived by the organisms but not by the biologist.

A more fruitful approach is to recognize alternative hypotheses, and to design studies that can discriminate among them. This paper has taken the form of an anthology of explicitly formulated, testable, and hopefully plausible alternatives to trophic niche differentiation. We have developed several mechanisms whereby trophic-niche equivalent (*TE*) species can coexist as a consequence of spatial heterogeneity at the level of microsites and patches, and have reviewed coexistence mechanisms involving environmental fluctuations, predation, stress, and coevolution between competitors.

We believe that the mechanisms discussed in this paper have both theoretical and practical implications for plant ecology:

1. The mechanisms offer possible explanations for the dissimilarity of within-community replicate samples; for the species richness of disjunctive habitats; and for the predominance of rare species in species-abundance curves.

2. Studies of species richness and community structure need to consider habitat pattern in their choice of study sites, and interpretations of community patterns must recognize mass effect and other *TE* mechanisms. In particular, recognition and enumeration of apparent mass effect species, and a partition of the species into guilds of trophic equivalents, may allow meaningful intercommunity comparisons that are obscured if all sources of diversity are lumped together.

These considerations are particularly important in the development and testing of predictive theories of community structure. The compositions of nearby communities, and local patterns of geographical replacement, are fundamentally unpredictable from the intrinsic characteristics of a single community. Significant amounts of trophic equivalence or near-equivalence, particularly as produced by mass effects preclude accurate 'assembly rules' (Diamond, 1975) for plant communities and limit the convergence between distant communities in similar environments.

3. We believe that the body of theory predicting the maximum overlap in resource utilization ('limiting similarity' or 'species packing') of coexisting competitors, is not applicable to plant communities. Theories of this sort (e.g. MacArthur & Levins, 1967; May & MacArthur, 1972; Yoshiyama & Roughgarden, 1977; Rappoldt & Hogeweg, 1980) begin by positing some monotonic relationship between the overlap in resource utilization ('niche overlap') and the magnitude of interspecific competition coefficients (α_{ij}) in some competition model (usually Lotka-Volterra) that exhibits competitive exclusion for large α_{ij}. The maximum possible overlap is set by the limits on the α_{ij} that permit coexistence.

However, if competition is localized, refuges may allow a species to persist despite high niche overlap and competition with other species in the system. Measures of the average intensity of competition (such as α_{ij}) do not reflect the presence of refuges and their contributions to coexistence. For example, in the theoretical habitat-utilization curves in Figure 4, the areas exclusive to one species (shaded) could allow coexistence regardless of the amount of overlap. The refuge areas must be large enough to maintain stable populations, but they could be quite small compared to the non-refuge areas (unshaded). In a theory based on limits to α_{ij}, a suffi-

ciently high overlap would be predicted to force exclusion of one or more species.

In addition, spatiotemporal environmental variability allows coexistence of species using different 'strategies' for exploiting a single set of habitats and resources (as discussed in the previous section), despite 100% overlap in resource utilization. Of course, these objections might not apply to highly mobile organisms, or organisms in a spatially uniform habitat, for which models based on spatially uniform competition for resources (e.g. Tilman, 1982) may be appropriate.

4. The possibility of coexistence by temporal mass effect merits careful empirical study. Fluctuations in the vital rates of plant populations have been observed in many field studies (reviewed e.g. by Grubb, 1977; Hubbell, 1980; Pickett, 1980). In many cases these fluctuations conform to our model, to the extent that different species are favored in different years (Grubb, 1977; Loria & Noy-Meir, 1979/80; and above). Thus, explanations of plant coexistence and plant community diversity should take into account the possibility of non-equilibrium coexistence due to environmental fluctuations. Careful demographic studies of non-equilibrium communities, and development of the mathematical theory to the stage where it can aid in designing and interpreting field studies, will be important steps towards determining the importance of non-equilibrium coexistence in plant communities.

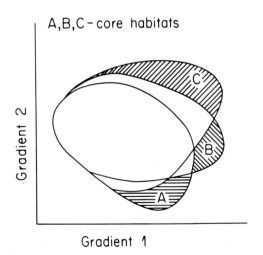

Fig. 4. Hypothetical microsite utilizations by three species along a pair of environmental gradients. Regions A, B, C are primary habitats exclusive to one species.

References

Abrams, P., 1975. Limiting similarity and the form of the competition coefficient. Theor. Pop. Biol. 8: 356–375.

Acevedo, L., M. F., 1981. On Horn's Markovian model of forest dynamics with particular reference to tropical forests. Theor. Pop. Biol. 19: 230–250.

Anderson, P. K., 1970. Ecological structure and gene flow in small mammals. Symp. Zool. Soc. London 25: 299–325.

Atkinson, W. P. & Shorrocks, B., 1981. Competition on a divided and ephemeral resource: a simulation model. J. Anim. Ecol. 50: 461–471.

Ayala, F., 1966. Reversal of dominance in competing species of Drosophila. Am. Nat. 100: 81–83.

Barbour, M. H., Burk, J. H. & Pitts, W. D., 1980. Terrestrial Plant Ecology. Benjamin Cummings, Menlow Park, California, U.S.A.

Barbour, M. G., Shmida, A. & Johnson, A, 1982. A comparison of coastal scrub vegetation in California and Israel: physiognomy, associations, and spatial diversity. Israel J. Bot. (in press).

294

Billings, D., 1952. The environmental complex in relation to plant growth and distribution. Rev. Biol. 27: 251–265.

Braakhekke, W. G., 1980. On Coexistence: a causal approach to diversity and stability in grassland vegetation. Agr. Res. Rep. 902, Pudoc, Wageningen, The Netherlands.

Bratton, S. P., 1976. Resource division in an understory herb community: responses to temporal and microtopographic gradients. Am. Nat. 110: 679–693.

Caswell, H., 1978. Predator mediated coexistence: a non-equilibrium model. Am. Nat. 112: 127–154.

Chesson, P. L., 1978. Predator-prey theory and variability. Ann. Rev. Ecol. Syst. 9: 323–347.

Chesson, P. L., 1981. Models for spatially distributed populations: the effect of within-patch variability. Theor. Pop. Biol. 19: 288–325.

Chesson, P. L., 1982. The stabilizing effect of a random environment. J. Math. Biol. 15: 1–36.

Chesson, P. L. & Warner, R. R., 1981. Environmental variability promotes coexistence in lottery competitive systems. Am. Nat. 117: 923–943.

Cohen, J. E., 1970. A Markov contingency table model for replicated Lotka-Volterra systems near equilibrium. Am. Nat. 104: 547–559.

Comins, H. M., 1982. Evolutionarily stable strategies for localized dispersal in two dimensions. J. Theor. Biol. 94: 579–606.

Comins, H. N., Hamilton, W. D., May, R. M., 1980. Evolutionarily stable dispersal strategies. J. Theor. Biol. 32: 205–230.

Connell, J. H., 1970. On the role of natural enemies in preventing competitive exclusion in some marine animals and in rain forest trees. In P. J. den Boer & G. R. Gradwell (eds.), Dynamics of Populations. P. 298–312, Pudoc, Wageningen.

Connell, J. H., 1978. Diversity in tropical rain forests and coral reefs. Science 199: 1302–1310.

Connell, J. H., 1979. Tropical rain forests and coral reefs as open non-equilibrium systems. In R. M. Anderson, Turner, B. D. & Taylor, L. R., (eds.), Population Dynamics, P. 141–163. Symp. Brit. Ecol. Soc. 20. Blackwell, Oxford-London-Edinburgh.

Curtis, J. T., 1959. The Vegetation of Wisconsin. University of Wisconsin Press, Madison, Wisconsin.

Dale, G., 1978. Money-in-the-bank; a model for coral reef fish coexistence. Env. Biol. Fish. 3: 103–108.

Danin, A., Orshan, G. & Zohary, M., 1975. The vegetation of the northern Negev and the Judean desert of Israel. Israel J. Bot. 24: 118–172.

Daubenmire, R. F., 1974. Plants and Environment. A textbook of plant autecology. John Wiley and Sons, New York, New York, U.S.A.

DeAngelis, D. L., Travis, C. C. & Post, W. M., 1979. Persistence and stability of seed-dispersed species in a patchy environment. Theor. Pop. Biol. 16: 107–125.

Diamond, J. M., 1975. Assembly of species communities. In M. L. Cody & Diamond, J. M., (eds.), Ecology and Evolution of Communities, P. 342–444. Belknap Press, Cambridge, Massachusetts, U.S.A.

Elderkin, R. H., 1982. Seed dispersal in a patchy environment with global age dependence. J. Math. Biol. 13: 283–303.

Ellner, S., 1984. Asymptotic behavior of some stochastic difference equation population models. J. Math. Biol. (in press).

Elton, C. S., 1927. Animal Ecology. Sidgwick and Jackson, London.

Fagerström, T. & Ågren, G. I., 1979. Theory for coexistence of species differing in regeneration properties. Oikos 33: 1–10.

Forcier, L. K., 1975. Reproductive strategies and co-occurrence of climax tree species. Science 189: 808–810.

Fowler, N. L. & Antonovics, J., 1981. Small-scale variability in the demography of transplants of two herbaceous species. Ecology 62: 1450–1457.

Fox, J. F., 1977. Alternation and coexistence of three species. Am. Nat. 111: 69–89.

Grime, J. P., 1973. Competitive exclusion in herbaceous vegetation. Nature 242: 344–347.

Grime, J. P., 1979. Plant Strategies and Vegetation Processes. Wiley and Sons, New York.

Grubb, P. J., 1977. The maintenance of species richness in plant communities: the importance of the regeneration niche. Biol. Rev. 52: 107–145.

Gulmon, S. L., Rundel, P. W., Ehleringer, J. R. & Mooney, H. A., 1979. Spatial relationships and competition in a Chilean desert cactus. Oecologia 44: 40–43.

Hamilton, W. D. & May, R. M., 1977. Dispersal in stable habitats. Nature 269: 578–581.

Harper, J. L., 1969. The role of predation in vegetation diversity In G. M. Woodwell & Smith, H. (eds.), Diversity and Stability in Ecological Systems. P. 48–61. Brookhaven Symposia in Biology 22.

Harper, J. L., 1977. Population Biology of Plants. Academic Press, New York.

Harper, J. L. & Sagar, G. R., 1953. Some aspects of the ecology of buttercups in permanent grassland. Proc. Br. Weed Control Conf. 1: 256–265.

Harper, J. L., Clatworthy, H. N., McNaughton, S. & Sagar, G. R., 1961. The evolution and ecology of closely related species living in the same area. Evolution 15: 209–227.

Harper, J. L., Williams, J. T. & Sagar, G. R., 1965. The behavior of seeds in soil I. The heterogeneity of soil surface and its role in determining the establishment of plants from seeds. J. Ecol. 53: 273–286.

Hastings, A., 1980. Disturbance, coexistence, history, and competition for space. Theor. Pop. Biol. 18: 363–373.

Holland, R. F. & Jain, S. K., 1981. Insular biogeography of vernal pools in the central valley of California. Am. Nat. 117: 24–37.

Horn, H. S., 1975. Markovian properties of forest succession. In M. Cody & Diamond J. (eds.), Ecology and Evolution of Communities. p. 196–211. Belknap Press, Cambridge, Mass.

Horn, H. S., 1976. Succession. In R. M. May (ed.), Theoretical Ecology: Principles and Applications, p. 187–204. Saunders, Philadelphia.

Hubbell, S. P., 1979. Tree dispersion, abundance, and diversity in a tropical dry forest. Science 203: 1299–1309.

Hubbell, S. P., 1980. Seed predation and the coexistence of tree species in tropical forests. Oikos 35: 214–229.

Huston, M., 1979. A general hypothesis of species diversity. Am. Nat. 113: 81–101.

Hutchinson, G. E., 1958. Concluding remarks. Cold Spring Harbor Symposia on Quantitative Biology 22: 415–427.

Hutchinson, G. E., 1959. Homage to Santa Rosalia, or why are there so many kinds of animals? Am. Nat. 93: 145–159.

Hutchinson, G. E., 1961. The paradox of the plankton. Am. Nat. 95: 137–145.

Itzkowitz, M., 1977. Spatial organization of the Jamaican damselfish community. J. Exp. Mar. Biol. Ecol. 28: 217–241.

Jackson, J. B. C. & Buss, L., 1975. Allelopathy and spatial competition among coral reef invertebrates. Proc. Natl. Acad. Sci. U.S.A. 72: 5160–5163.

Janzen, D. H., 1970. Herbivores and the number of tree species in tropical forests. Am. Nat. 104: 401–428.

Karlin, S. & McGregor, J., 1972a. Polymorphisms for genetic and ecological systems with weak coupling. Theor. Pop. Biol. 3: 210–238.

Karlin, S. & McGregor, J., 1972b. Application of method of small parameters to multi-niche population genetic models. Theor. Pop. Biol. 3: 186–209.

Keiding, N., 1975. Extinction and exponential growth in random environments. Theor. Pop. Biol. 8: 49–63.

Kilburn, P. D., 1966. Analysis of the species area relation. Ecology 47: 831–843.

Lawton, J. H. & Strong, D. R., Jr., 1981. Community patterns and competition in folivorous insects. Am. Nat. 118: 317–338.

Levin, S. A., 1974. Dispersion and population interactions. Am. Nat. 114: 103–114.

Levin, S. A., 1976. Population models and community structure in heterogeneous environments. In S. A. Levin (ed.), Mathematical Association of America. Studies in the Mathematical Biology, II: Populations and communities. P. 439–476.

Levin, S. A., 1979. Non-uniform stable solutions to reaction-diffusion equations: applicatons to ecologial pattern formulation. In H. Haken (ed.), Pattern Formation by Dynamic Systems and Pattern Recognition. p. 210–222. Springer-Verlag, Berlin-Heidelberg-New York.

Levin, S. A., 1981. Mechanisms for the generation and maintenance of diversity in ecological communities. In M. S. Bartlett & Hiorns, R. W. (eds.), The Mathematical Theory of the Dynamics of Biological Populations, 2. P. 173–194. Academic Press, New York.

Levin, S. A. & Paine, R. T., 1974. Disturbance, patch formation and community structure. Proc. Natl. Acad. Sci. U.S.A. 71: 2744–2747.

Levins, R. & Culver, D., 1971. Regional coexistence of species and competition between rare species. Proc. Natl. Acad. Sci. U.S.A. 68: 1246–1248.

Leith, H., 1960. Patterns of change within grassland communities. In J. L. Harper (ed.), The Biology of Weeds, P. 27–39. Blackwell, Oxford.

Linhart, Y., 1980. Local biogeography of plants on a Caribbean atoll. J. Biogeogr. 7: 159–171.

Loria, M. & Noy-Meir, I., 1979/80. Dynamics of some annual species in a desert loess plain. Israel J. Botany 28: 211–225.

MacArthur, R. H., 1968. The theory of the niche. In R. C. Lewontin (ed.), Population Biology and Evolution. P. 159–186. Syracuse University Press, Syracuse, New York, U.S.A.

MacArthur, R. H., 1972. Geographical Ecology. Harper and Row, New York.

MacArthur, R. H. & Levins, R., 1967. The limiting similarity, convergence and divergence of coexisting species. Am. Nat. 101: 377–385.

MacDonald, N. & Watkinson, A. R., 1981. Models of an annual plant population with a seedbank. J. Theor. Biol. 93: 643–653.

Major, J., 1951. A functional, factorial approach to plant ecol-

ogy. Ecology 39: 392–412.

Marshall, D. R. & Jain, S. K., 1967. Cohabitation and relative abundance of two species of wild oats. Ecology 48: 656–659.

Matano, H., 1979. Asymptotic behavior and stability of solutions of semilinear diffusion equations. RIMS Kyoto Univ. 15: 401–451.

May, R. M., 1975. Patterns of species abundance and diversity. In M. L. Cody & Diamond, J. M. (eds.), Ecology and Evolution of Communities. P. 81–120. Belknap Press, Cambridge, Massachusetts, U.S.A.

May, R. M. & MacArthur, R. H., 1972. Niche overlap as a function of environmental variability. Proc. Natl. Acad. Sci. U.S.A. 69: 1109–1113.

Menge, B. A., 1979. Coexistence between the seastars Asterias vulgaris and A. forbesi in a heterogeneous environment: a non-equilibrium explanation. Oecologia 41: 245–272.

Motro, U., 1982. Optimal rates of dispersal. I. Haploid populations. II. Diploid populations. Theor. Pop. Biol. 21: 394–411, 412–429.

Mueller-Dombois, D. & Ellenberg, H., 1974. Aims and Methods of Vegetation Science. John Wiley and Sons, New York, New York, U.S.A.

Murdoch, W. W., 1969. Switching in general predators: experiments on predator specificity and stability of prey populations. Ecol. Monogr. 39: 335–354.

Murdoch, W. W. & Oaten, A., 1975. Predation and population stability. Adv. Ecol. Res. 9: 1–131.

Naveh, Z. & Whittaker, R. H., 1979. Structural and floristic diversity of shrublands and woodlands in Northern Israel and other Mediterranean areas. Vegetatio 41: 171–190.

Noy-Meir, I., 1972. Multivariate analysis of the semi-arid vegetation in southeastern Australia. I. Nodal ordination by component analysis. Proc. Ecol. Soc. Australia 6: 159–193.

Paine, R. T., 1966. Food web complexity and species diversity. Am. Nat. 100: 65–75.

Paine, R. T. & Levin, S. A., 1981. Intertidal landscapes: disturbance and the dynamics of pattern. Ecol. Monogr. 51: 145–178.

Park, T. & Lloyd, M., 1955. Natural selection and the outcome of competition. Am. Nat. 96: 235–240.

Parrish, J. A. D. & Bazzaz, F. A., 1976. Underground niche separation in successional plants. Ecology 57: 1281–1288.

Peet, R. K., Glenn-Lewin, D. C. & Wolf, J. W., 1983. Prediction of man's impact on plant species diversity: a challenge for vegetation science. In: W. Holzner, Werger, M. J. A. & Ikusima, I. (eds.), Man's Impact on Vegetation, Junk, The Hague.

Pickett, S. T. A., 1980. Non-equilibrium coexistence of plants. Bull. Torrey Bot. Club 107: 238–248.

Pickett, S. T. A. & Bazzaz, F. A., 1976. Divergence of two co-occurring successional annuals on a soil moisture gradient. Ecology 57: 169–176.

Pielou, E. C., 1975. Ecological Diversity. Wiley-Interscience, N.Y.

Pimentel, D., 1968. Population regulation and genetic feedback. Science 159: 1432–1437.

Pimentel, D., Feinberg, E. H., Wood, P. W. & Hayes, J. T., 1965. Selection, spatial distribution, and the coexistence of competing fly species. Am. Nat. 99: 97–109.

Platt, W. J., 1975. The colonization and formation of equilibrium plant species associations on badger disturbances in a tall-grass prairie. Ecol. Monogr. 45: 285–305.

Platt, W. J. & Weis, I. M., 1977. Resource partitioning and competition within a guild of fugitive prairie plants. Am. Nat. 111: 479–513.

Preston, F. W., 1948. The commonness and rarity of species. Ecology 41: 611–627.

Preston, F. W., 1962. The canonical distribution of commonness and rarity. Ecology 43: 185–215, 410–432.

Rappoldt, C. & Hogeweg, P., 1980. Niche packing and number of species. Am. Nat. 116: 480–492.

Roughgarden, J. & Feldman, M., 1975. Species packing and predation pressure. Ecology 56: 489–492.

Runemark, H., 1969. Reproductive drift, a neglected principle in reproductive biology. Botaniska Notiser 122: 90–129.

Runemark, H., 1971. Distribution patterns in the Aegean. In P. H. Davis, Harper, P. C. & Hedge, I. C. (eds.), Plant Life of Southwest Asia. P. 3–14. Botanical Society of Edinburgh, Edinburgh.

Sale, P. F., 1974. Mechanisms of coexistence in a guild of territorial fishes at Heron Island. In Proc. Second. Intern. Symp. Coral Reefs 1. P. 193–206.

Sale, P. F., 1977. Maintenance of high diversity of coral reef fish communities. Am. Nat. 111: 337–359.

Sale, P. F., 1978. Coexistence of coral reef fishes – a lottery for living space. Env. Biol. Fish 3: 85–102.

Sale, P. F., 1979. Recruitment, loss and coexistence in the guild of territorial coral reef fishes. Oecologia 42: 159–177.

Sarukhan, J., 1974. Studies on plant demography: Ranunculus repens L., R. bulbosus L. and R. acris. II. Reproductive strategies and seed population dynamics. J. Ecol. 62: 151–177.

Sarukhan, J. & Harper, J. L., 1973. Studies on plant demography: Ranunculus repens L., R. bulbosus L. and R. acris. I. Population flux and survivorship. J. Ecol. 61: 675–716.

Schaffer, W. M. & Leigh, E. G., 1976. The prospective role of mathematical theory in plant ecology. Systematic Botany 1: 209–232.

Shmida, A., 1972. The vegetation of Gebel-Maghara, North Sinai. M. Sc. Thesis, Hebrew University, Jerusalem, Israel. English summary.

Shmida, A., 1977. A quantitative analysis of the tragacanthic vegetation of Mt. Hermon and its relation to environmental factors. Dissertation, Hebrew University, Jerusalem, Israel. English summary.

Shmida, A. & Whittaker, R. H., 1981. Pattern and biological microsite effects in two shrub communities, Southern California. Ecology 62: 234–251.

Skellam, J. G., 1951. Random dispersal in theoretical populations. Biometrika 38: 196–218.

Slatkin, M., 1974. Competition and regional coexistence. Ecology 55: 128–134.

Smith, F. E., 1972. Spatial heterogeneity, stability, and diversity in ecosystems. In E. S. Deevey (ed.), Growth by Intussesception. Ecological Essays in Honor of Evelyn Hutchinson. Trans. Conn. Arts. Sci. 44.

Snogerup, S., 1971. Evolution and plant geography of chasmophytic communities. In P. H. Davis, Harper, P. C. & Hedge, L. C., Plant Life of South-west Asia. P. 157–170. Botanical Society of Edinburgh, Edinburgh.

Sousa, W. P., 1979. Disturbances in marine intertidal boulder fields: the non-equilibrium maintenance of species diversity.

Ecology 60: 1225–1239.

Sutherland, J. P., 1974. Multiple stable points in natural communities. Am. Nat. 108: 859–873.

Talbot, F. H., Russell, B. C. & Armstrong, G. R. V., 1978. Coral-reef fish communities: unstable high-diversity systems? Ecol. Monogr. 48: 425–440.

Tilman, D., 1982. Resource Competition and Community Structure. Monographs in Population Biology 17. Princeton University Press, Princeton, New Jersey.

Turelli, M., 1981. Niche overlap and invasion of competitors in random environments. I. Models without demographic stochasticity. Theor. Pop. Biol. 20: 1–56.

Turkington, R. & Harper, J. L., 1979. The growth, distribution, and neighbour relationships of Trifolium repens in a permanent pasture. I. Ordination, pattern, and contact. IV. Fine-scale biotic differentiation. J. Ecol. 67: 201–218, 245–254.

Watson, M. A., 1981. Patterns of microhabitat occupation of six closely related species of mosses along a complex altitudinal gradient. Ecology 62: 1067–1078.

Watt, A. S., 1947. Pattern and process in the plant community. J. Ecol. 35: 1–22.

Watt, A. S., 1964. The community and the individual. J. Ecol. 52: 203–211.

Werner, P. A., 1979. Competition and coexistence of similar species. In O. T. Solbrig, Jain, S., Johnson, G. B. & Raven, P. H. (eds.), Topics in Plant Population Biology. P. 287–310. Columbia University Press, New York.

Werner, P. A. & Platt, W. J., 1976. Ecological relationships of co-occurring goldenrods. Am. Nat. 111: 959–971.

Whittaker, R. H., 1965. Dominance and diversity in land plant communities. Science 147: 250–260.

Whittaker, R. H., 1969. Evolution of diversity in plant communities. In G. M. Woodwell & Smith, H. (eds.), Diversity and stability in ecological systems. P. 178–195. Brookhaven Symposia in Biology 22.

Whittaker, R. H., 1972. Evolution and measurement of species diversity. Taxon 21(2–3): 213–251.

Whittaker, R. H., 1975. Communities and Ecosystems. MacMillan, New York.

Whittaker, R. H., 1977. Evolution of species diversity in land communities. Evolutionary Biology 10: 1–67.

Whittaker, R. H. & Levin, S. A., 1977. The role of mosaic phenomena in natural communities. Theor. Pop. Biol. 12: 117–139.

Williams, C. B., 1964. Patterns in the Balance of Nature. Academic Press, New York.

Woods, K. D., 1979. Reciprocal replacement and the maintenance of codominance in a beech-maple forest. Oikos 33: 31–39.

Yodzis, P., 1978. Competition for Space and the Structure of Ecological Communities. Lecture Notes in Biomathematics 25. Springer-Verlag, Berlin-Heidelberg-New York.

Yoshiyama, R. M. & Roughgarden, J., 1977. Species packing in two dimensions. Am. Nat. 111: 107–121.

Zohary, M., 1973. Geobotanical Foundations of the Middle East. Gustav-Fischer Verlag, Stuttgart.

Zohary, M. & Feinbrun-Dothan, N., 1966→. Flora Palaestina. Israel Academy of Sciences and Humanities, Jerusalem.

Accepted 3.2.1984.

Appendix A

Stability analyses in a constant environment

1. Equation (4)

In this section we derive the qualitative dynamics of equation (4), modelling two trophically equivalent species in one patch, and their interpretation in terms of life-history parameters. The iteration (4) is a one-dimensional difference equation

$$f_1(t + 1) = H(f_1(t)) \tag{A.1}$$

where

$$H(f) = f(1 - d_1) + d_1 f F_1(\lambda_1, \lambda_2) + d_2(1 - f) F_2(\lambda_1, \lambda_2) \tag{A.2}$$

and the F_i, λ_i are defined as in equations (1) and (2).

We consider first the behavior near the equilibria $f_1 = 0$ and $f_1 = 1$. Differentiating (A.2) and denoting $F_{ij} = \dfrac{\partial F_i}{\partial \lambda_j}$,

$$H'(f) = 1 - d_1 + d_1[F_1 + f(F_{11} w_1 - F_{12} w_2)] + d_2[-F_2 + (1 - f)(F_{21} w_1 \ F_{22} w_2)]. \tag{A.3}$$

Differentiating the F_i and setting $f = 0$ gives

$$H'(0) = 1 - d_1 + \frac{v_1 d_1}{v_1 + w_2} + \frac{w_1 d_2}{v_2 + w_2} = 1 + \beta_1 \tag{A.4}$$

Consequently $H(f) - f$ has the sign of β_1 for f near 0, so long as $\beta_1 \neq 0$. Swapping species-indices gives the corresponding results at $f_1 = 1$ (i.e., $f_2 = 0$): $H(f) - f$ has the sign of $-\beta_2$ for f near 1, so long as $\beta_2 \neq 0$.

Equilibria of (4) are solutions of $H(f) = f$; for $f \neq 0$ or 1, $H(f) = f$ simplifies straightforwardly to

$$G(f) = \frac{v_2 + w_2 + f(w_1 - w_2)}{v_1 + w_2 + f(w_1 - w_2)} = \frac{w_1 d_2}{w_2 d_1} \tag{A.5}$$

$G'(f)$ has the sign of $(v_1 - v_2)(w_1 - w_2)$, so G is either constant or strictly monotonic. If $G(0) = G(1) = \dfrac{w_1 d_2}{w_2 d_1}$, then all f's are neutrally stable equilibria ($H(f) = f$ for all f); this occurs only if $\beta_1 = \beta_2 = 0$. Otherwise there will be an interior equilibrium if and only if G is non-constant and

$$G(0) < \frac{w_1 d_2}{w_2 d_1} < G(1) \tag{A.6a}$$

or

$$G(1) < \frac{w_1 d_2}{w_2 d_1} < G(0) \tag{A.6b}$$

Substituting (A.5) into (A.6) shows that (A.6a) corresponds to $\beta_1 > 0$, $\beta_2 < 0$ and (A.6b) to $\beta_1 < 0$, $\beta_2 > 0$: both boundaries unstable or both stable. The monotonicity of G implies that there is at most one interior equilibrium.

Finally, note that $H(f)$ is increasing, since $H'(f) > 0$. The only negative term in (A.3) is $-d_2 F_2$, and $F_1 > F_2$, so if $d_1 \geqslant d_2$ we have $H'(f) \geqslant d_1 F_1 - d_2 F_2 > 0$. On the other hand, if $d_1 < d_2$,

$$H'(f) \geqslant 1 - d_1 + d_1 F_1 - d_2 F_2 \geqslant 1 - d_2 + d_2 F_1 - d_2 F_2 = 1 - d_2 + d_2(F_1 - F_2) > 0.$$

Thus $H(f)$ and f always lie on the same side of any internal equilibrium f^*.

These results combine to give a complete qualitative description of the behavior of (4):

1) $\beta_1 > 0$, $\beta_2 > 0$. Both boundaries are unstable, and there is a unique interior equilibrium f^* where $H(f) - f$ changes sign from positive to negative. Since H is increasing, this implies $f < H(f) < f^*$ for $0 < f < f^*$ and $f^* < H(f) < f$ for $f^* < f < 1$; i.e., f^* is globally stable on $(0, 1)$, so the species coexist.

2) $\beta_1 < 0$, $\beta_2 < 0$. Both boundaries are stable, and there is again a unique interior equilibrium. In this case however, $H(f) < f$ if $f < f^*$ and $H(f) > f$ if $f > f^*$; so unless $f_1(t) \equiv f^*$, $f_1(t)$ converges to 0 or 1, depending on whether $f_1(0) < f^*$ or $f_1(0) > f^*$.

3) The β_i have opposite signs, or only one $\beta_i = 0$. In these cases there is no interior equilibrium, i.e., $H(f) - f$ does not change sign on

$(0, 1)$. If $\beta_1 > 0$ we know $H(f) > f$ for f near 0, so $H(f) > f$ for all f in $(0, 1)$. Thus from any initial densities (except $f_1(0) = 0$) $f_1(t)$ will increase to 1. Similarly, if $\beta_2 > 0$ species 1 is competitively excluded unless $f_2(0) = 0$.

If we allow a fraction c of all microsites to become vacant prior to reproduction, (4) becomes

$$f_1(t + 1) = (1 - c)[(1 - d_1)f_1(t) + d_1 f_1(t) F_1(\lambda_1, \lambda_2) + d_2(1 - f_1(t)) F_2(\lambda_1, \lambda_2)] + c F_0(\lambda_1, \lambda_2 = \pi(f_1(t)) \tag{A.7}$$

where now $\lambda_1 = (1 - c)f_1(t) w_1$, $\lambda_2 = (1 - c)(1 - f_1(t)) w_2$ and $F_0 = \dfrac{\lambda_1}{\lambda_1 + \lambda_2}$ is the fraction of vacant sites on which a species 1 juvenile

becomes established. Arguing as before, we find that the equilibrium $f_1 = 0$ is stable if $\beta_1(c) < 0$ and unstable if $\beta_1(c) > 0$, where

$$\beta_1(c) = (1 - c)^2 \left[\frac{w_1 d_2}{v_2 + (1 - c)w_2} - \frac{d_1 w_2}{v_1 + (1 - c)w_2} \right] + c \left[\frac{w_1 - w_2}{w_2} \right] \tag{A.8}$$

Reversing indices gives the corresponding results for the equilibrium $f_2 = 0$. $\pi(f)$ is increasing in f (by much the same argument as before), so $\beta_1(c) > 0$, $\beta_2(c) > 0$ still guarantees the existence of a locally stable interior equilibrium at which the species can coexist.

Returning to the case $c = 0$, the condition for coexistence in one patch ($\beta_1, \beta_2 > 0$) can be rearranged to

$$\frac{v_2 + w_2}{v_1 + w_2} < \frac{w_1 d_2}{w_2 d_1} < \frac{v_2 + w_1}{v_1 + w_1}. \tag{A.9}$$

Satisfying (A.9) requires that $g(w_1) > g(w_2)$, where

$$g(w) = \frac{v_2 + w}{v_1 + w} \tag{A.10}$$

$g'(w)$ has the sign of $(v_1 - v_2)$ for $w \geq 0$. Thus $g(w_1) > g(w_2)$ if $v_1 > v_2$ and $w_1 > w_2$, or if $v_1 < v_2$ and $w_1 < w_2$. The condition $\beta_1, \beta_2 < 0$ gives (A.9) with the inequalities reversed; arguing as above, this requires $v_1 > v_2$ and $w_1 < w_2$, or $v_1 < v_2$ and $w_1 > w_2$. The inner inequalities then similarly constrain the relative values of the d_i as stated in the discussions of the cases $\beta_1, \beta_2 > 0$ and $\beta_1, \beta_2 < 0$ in the text.

2. Equation (14)

We turn now to the 'many vacant sites' approximation (14), which may be written as

$$f_i(t + 1) = f_i(t) [1 - d_i q_i + w_i(1 - q_i)(1 - f_1(t) - f_2(t))], \text{ for } 0 \leq f_1 + f_2 \leq 1, i = 1, 2. \tag{A.11}$$

Species i alone has an equilibrium at $f_i(t) = f_i^*$, where

$$f_i^* = \max(0, \frac{w_i(1 - q_i) - d_i q_i}{w_i(1 - q_i)}) \tag{A.12}$$

and the line $f_i(t) + f_2(t) = f_i^*$ is the isocline for species i. We assume $f_1^* > f_2^* > 0$ (if $f_i^* = 0$ then clearly $f_i(t) \to 0$) and $w_i(1 - q_i) < 1$ for $i = 1, 2$, ; in this section we will show that $(f_1(t), f_2(t)) \to (f_1^*, 0)$ so long as $f_1(0) > 0$, and then derive the velocity bound (16). The proof is straightforward but exasperatingly messy, so we begin with three lemmas. Henceforth ϕ will denote the map defined by (A.11) taking $(f_1(t), f_2(t))$ to $(f_1(t + 1), f_2(t + 1))$; note that (A.11) has equilibria only at $(0, 0)$, $(f_1^*, 0)$ and $(0, f_2^*)$.

Lemma 1. The one-species iterations

$$f_1(t + 1) = \phi(f_1(t), 0) \tag{A.13}$$

$$f_2(t + 1) = \phi(0, f_2(t)) \tag{A.14}$$

converge monotonically to their equilibria f_i^* so long as $f_i(0) > 0$.

Proof: By direct substitution into (A.13) and (A.14),

$$(f_i(t + 1) - f_i^*) = [1 - w_i(1 - q_i)f_i(t)](f_i(t) - f_i^*),$$

and we are assuming $w_i(1 - q_i) < 1$.

Lemma 2. The regions $R_1 = \{(f_1, f_2): f_1 + f_2 \leq f_1^*\}$ and $R_2 = \{(f_1, f_2): f_2^* \leq f_1 + f_2\}$ are both invariant for (A.11).

Proof: Let T be the function taking $(f_1(t), f_2(t))$ to $(f_1(t+1) + f_2(t+1))$. R_1 is invariant for (A.11) if $T(x, s - x) \leqslant f_1^*$ whenever $0 \leqslant s \leqslant f_1^*$ and $0 \leqslant x \leqslant s$. The function $T(x, s - x)$ is linear in x. This is so because the only nonlinearity in (A.11) comes from the $f_1(1 - f_1 - f_2)$ term, which equals sx along the line $f_1 = x, f_2 = s - x$. To prove R_1 invariant we therefore need only show that $T(0, s) \leqslant f_1^*$ and $T(s, 0) \leqslant f_1^*$ whenever $s \leqslant f_1^*$. But $T(0, s) = \phi(0, s)$; by the last lemma $\phi(0, s) \leqslant f_2^* < f_1^*$ when $s \leqslant f_2^*$ and $\phi(0, s) < s \leqslant f_1^*$ when $f_2^* \leqslant s \leqslant f_1^*$. Similarly $T(s, 0) = \phi(s, 0) \leqslant f_1^*$ whenever $s \leqslant f_1^*$, so R_1 is invariant. The proof for R_2 is symmetric.

Lemma 3. If $(f_1(0), f_2(0)) \in R_1$ and $f_1(0) > 0$, then $(f_1(t), f_2(t)) \to (f_1^*, 0)$ as $t \to \infty$.

Proof: R_1 and R_2 are invariant, so $(f_1(t), f_2(t))$ remains outside $R_1 \cap R_2$ only if $f_1(t) + f_2(t) < f_2^*$ for all t. If this is true, then f_1 and f_2 both increase monotonically to limits $\bar{f_1}, \bar{f_2}$. $(\bar{f_1}, \bar{f_2})$ is therefore an equilibrium of (A.11) satisfying $0 < \bar{f_1} \leqslant \bar{f_2}^*$. But no such equilibrium exists, so eventually $(f_1(t), f_2(t))$ enters $R_1 \cap R_2$. Thereafter $f_2^* \leqslant f_1(t) + f_2(t) \leqslant f_1^*$ (by Lemma 2), so f_1 increases, and f_2 decreases monotonically to limits $\hat{f_1}, \hat{f_2}$. $(\hat{f_1}, \hat{f_2})$ is therefore an equilibrium of (A.11) satisfying $\hat{f_1} > 0$, so we must have $(\hat{f_1}, \hat{f_2}) = (f_1^*, 0)$ as desired.

Finally, consider arbitrary $(f_1(0), f_2(0))$. If $(f_1(t), f_2(t))$ enters R_1, then by Lemma 3 $(f_1(t), f_2(t)) \to (f_1^*, 0)$. If not, then f_1 and f_2 both decrease monotonically to limits $\bar{f_1}, \bar{f_2}$. Arguing as in Lemma 3, we find that $(\bar{f_1}, \bar{f_2}) = (f_1^*, 0)$. This completes the proof of convergence to $(f_1^*, 0)$.

The velocity bound (16) is stated for 'invasion' of species 2 by species 1, i.e. for initial densities $f_2 = f_2^*$, f_1 positive but small. Along these orbits $f_2^* \leqslant f_1(t) + f_2(t) \leqslant f_1^*$ (Lemma 2), hence

$$|f_1(t+1) - (f_1(t)| = f_1(t)[-d_1 q_1 + w_1(1 - q_1)(1 - f_1(t) - f_2(t))] \leqslant f_1^*[-d_1 q_1 + w_1(1 - q_1)(1 - f_2^*)] = f_1^*[-d_1 q_1 + w_1(1 - q_1)(1 - f_1^*)$$

$$+ w_1(1 - q_1)(f_1^* - f_2^*)] = f_1^*[w_1(1 - q_1)(f_1^* - f_2^*)].$$

A similar argument gives

$$|f_2(t+1) - f_2(t)| \leqslant f_1^*[w_2(1 - q_2)(f_1^* - f_2^*)]$$

and from these (16) follows immediately.

Appendix B

Sufficient conditions for non-equilibrium coexistence

The condition (12) for non-equilibrium coexistence in our model follows from a more general criterion for stochastic-boundedness persistence in stochastic difference equation population models (Ellner, 1983). A simple version of that result, adequate for our purposes here, is as follows.

We consider the stochastic difference equation

$$X_{t+1} = H(X_t, Z_t) \quad t = 0, 1, 2, \ldots \tag{B.1}$$

and assume that
  1) The $\{Z_t\}_{t=0}^{\infty}$ are independent, identically distributed random variables, independent of the population history (i.e., Z_n and X_m are independent whenever $n \geqslant m$).
  2) The functions $H(\cdot, Z_t)$ are (with probability 1) non-negative increasing on $[0, \infty)$, and differentiable at 0.
  3) There exist numbers $a, b, c > 0$ such that $ax < H(x, Z_t) < bx$ for all x in $[0, c]$.
Then if $E[\ln H'(0, Z_t)] > 0$, X_t is persistent in the sense of stochastic-boundedness (as defined in equation (11)).

For our model (10), equation (A.4) gives $H'(0, Z_t) = 1 + \beta_1(Z_t)$; thus $f_1(t)$ is persistent whenever $E[\ln(1 + \beta_1(Z_t))] > 0$. The same applies to species 2, whence (12). The limits on the severity of environmental fluctuations stated above (assumption 3) are unnecessarily restrictive; see Ellner (1984) for a more general (but correspondingly messier) version.

The example presented in the text assumed only two possible values for Z_t, and fluctuations only in the fecundities m_i. From (A.4), we have

$$1 + \beta_1 \geqslant 1 - d_1 + \frac{w_1 d_2}{v_2 + w_2} = 1 - d_1 + d_2 \gamma_1 r_1 \tag{B.2}$$

where $r_1 = \dfrac{m_1}{m_2}$ and $\gamma_1 = (b_1 p_1)(1 - b_2 + b_2 p_2)^{-1}$. Thus

$$E[\ln(1 + \beta_i(Z_t))] \geqslant E[\ln(1 - d_1 + d_2 \gamma_1 r_1(Z_t))] \geqslant a_1 \ln(1 - d_1 + d_2 \gamma_1 R_1) + a_2 \ln(1 - d_1) \tag{B.3}$$

where $R_1 = r_1(1)$ and $a_i = P[Z_t = i]$, as in (13). Interchanging species indices proves the same for species 2, hence (13) implies coexistence in the sense of stochastic boundedness.

Finally, we need to show that random dispersal and vacant microsites can allow non-equilibrium coexistence of annuals. To obtain a simple example, we modify model (A.7) by assuming that

1) $d_1 = d_2 = 1$ (annual species).
2) Each microsite receives a Poisson-distributed number of seeds of species i with mean $\lambda_i = (1 - c)f_i w_i$, $i = 1, 2$.
3) In year-type 1, competition between juveniles is always won by species 1; only in sites where no species 1 seeds are present can a species 2 seedling become established. In year-type 2 the competitive asymmetry is exactly reversed.
4) The transition between year-types occurs immediately after juvenile competition. Thus a 'year' encompasses a reproductive phase and the following competitive phase.

As usual, stochastic effects other than those due to environmental fluctuations are assumed to be negligible. The biological generality of these assumptions is undoubtedly minimal – our aim here is to present a mathematically simple example rather than a general analysis of non-equilibrium coexistence of annuals.

Under the random dispersal assumption $f_1(t) + f_2(t)$ is not a constant, so the one-dimensional persistence criterion above does not apply directly. However, we can bound f_1 and f_2 in terms of de-coupled one-dimensional iterations to which the criterion applies, obtaining thereby sufficient conditions for persistence of both species, as follows:

In year-type 1, species 1 reclaims all sites on which it had a reproducing adult last year, and establishes on all other sites reached by one of its diaspores. Thus in year-type 1

$$f_1(t + 1) = H_1(f_1(t))$$

where

$$H_1(f) = (1 - c)f + (1 - (1 - c)f)(1 - e^{-\lambda_1}) \tag{B.4}$$

and as before c is the fraction of sites on which the adult fails to reproduce.

In year-type 2, species 1 reclaims all sites where it had a reproducing adult last year and no species 2 diaspore lands, and establishes at least on those vacant sites where only species 1 diaspores land. Thus in year-type 2,

$$f_1(t + 1) \geqslant H_2(f_1(t))$$

where

$$H_2(f) = (1 - c)f\,e^{-(1 - c)(1 - f)w_{22}} + c\,e^{-(1 - c)(1 - f)w_2}(1 - e^{-\lambda_1}). \tag{B.4}$$

(B.4) is the underestimate of $f_1(t + 1)$ obtained from the bound $f_2(t) \leqslant 1 - f_1(t)$. The general persistence criterion implies that $f_1(t)$ will persist whenever

$$a_1 \ln H_1'(0) + a_2 \ln H_2'(0) > 0, \tag{B.5}$$

where as before a_1 is the frequency of year-type i. Taking the derivatives, we find that species 1 is persistent whenever

$$a_1 \ln(1 + w_{11}) + a_2 \ln(e^{-(1 - c)w_{22}}(1 + c\,w_{12})) > -\ln(1 - c), \tag{B.6}$$

wherein w_{ij} is the value of w for species i in year-type j. Reversing indices shows that species 2 persists whenever

$$a_1 \ln(e^{-(1 - c)w_{11}}(1 + c\,w_{21})) + a_2 \ln(1 + w_{22}) > -\ln(1 - c) \tag{B.7}$$

and (B.6), (B.7) together imply coexistence in the sense of stochastic boundedness. These conditions can be satisfied by letting $w_{12}, w_{21} \to \infty$. This describes a situation of compensating advantages: in year-type 2 species 1 is competitively inferior but has a high fecundity (large w_{12}), and vice versa for species 2.

Appendix C

Study areas

Artemisia study area is in the Judean Desert on the main Judean-Samarian ridge, 32 km southeast of Jerusalem, elevation 525 m, with 170 mm average rainfall. All samples are places on a gentle soft-limestone slope. The community is a typical chamaephyte semi-desert (steppe) strongly dominated by *Artemisia herba-alba*, with 20 to 30% perennial cover. The areas between the shrubs are rich in geophytes and there is a large flush of annuals in the spring (20 to 80% cover). Most of the species are Irano-Turanian elements.

En-Gedi study area is near the western shore of the Dead Sea, elevation –350 m, with 75 mm average rainfall. Samples are placed on rocky slopes covered by talus debris, with hard limestone bedrock and typical desert lithosol soil. The vegetation is limited to a diffuse, sparse cover of chamaephytes (about 10% cover) and a flush of annuals (1–15% cover) flowering between early February and mid-March, rainfall permitting. The most common species are *Anabasis articulata, A. setifera, Salsola tetranda,* and *Zygophyllum dumosum* (Chamaephytes), and *Aizon canariense, Mesembryanthemum nodiflorm, Aaronsohnia factorovskii,* and *Plantago ovata* (annuals).

Har Gilo study area is on the main ridge of the Judean mountains 10 km south of Jerusalem, 916 m above sea level. The climate is mediterranean – hot, dry summers and cold winters, with 600 mm average rainfall. Samples are placed in abandoned terraces that have not been cultivated for at least 30 years. The terraces are on a limestone slope with Terra Rossa and Redzina soils, dominated by winter annual vegetation. Some chamaephytes and chaparral shrubs occur in the rocky remnants of the retaining walls that border individual terraces.

Hazeva study area is located in the Rift Valley 60 km south of the Dead Sea, elevation 125 m, with 47 mm average rainfall. Aridisol hammada is the predominant soil. The area is a true desert with perennial vegetation almost entirely restricted to depressions, north-facing slopes, and washes where their cover is 1 to 10%. The perennials are mostly Saharo-Arabian element chamaephytes (e.g., *Anabasis articulata, Gymnocarpus decandrum*) occurring in largely monotypic stands. Some Sudanian element trees and shrubs (mostly *Acacia tortilis* and *A. raddiana*) are found in the larger washes. In years of particularly high rainfall (about 1 year in every 6) the slopes support a large flush of annuals that germinate after a major rain and complete their life-cycle in about 60 days. The most common annuals are *Aaronsohnia factorovskii, Plantago ovata, Stipa tortilis, Trigonella stellata, Asteriscus pygmaeus,* and *Anastatica hierochuntia.*

Nahal Darga study area is in the Judean Desert at 112 m elevation, with 90 mm average rainfall. The study area is dissected by three first- to second-order wadis (washes) with limestone slopes in between. All samples were taken on the slopes. The vegetation is a typical *Zygophyllum dumosum* desert chamaephyte community (Danin *et al.,* 1975) with 3% cover by the chamaephytes and an ephemeral flush of winter annuals.

Qesarya study area is in the coastal plain of Israel 3 km east of the coast, 38 km south of Haifa, elevation 25 m. The climate is Mediterranean with 560 mm average rainfall. Samples are placed in an open woodland on sandy soil dominated by *Ceratonia siliqua* and *Pistacia lentiscus.* The patches of annual vegetation between the trees and shrubs are dominated by *Trifolium palaestinum, Plantago sacrophylla,* and *Aegilops ovata.* The vegetation is described in detail by Barbour *et al.* (1982).

Composition and species diversity of pine–wiregrass savannas of the Green Swamp, North Carolina*

Joan Walker & Robert K. Peet**

Department of Biology 010A, University of North Carolina, Chapel Hill, NC 27514, U.S.A.

Keywords: Biomass, Coastal plain, Fire, Grassland, North Carolina, Phenology, *Pinus palustris,* Primary production, Savanna

Abstract

Fire-maintained, species-rich pine–wiregrass savannas in the Green Swamp, North Carolina were sampled over their natural range of environmental conditions and fire frequencies. Species composition, species richness, diversity (Exp *H'*, 1/*C*), and aboveground production were documented and fertilization experiments conducted to assess possible mechanisms for the maintenance of high species diversity in these communities.

Although savanna composition varies continuously, DECORANA ordination and TWINSPAN classification of 21 sites facilitated recognition of 3 community types: dry, mesic, and wet savannas. These savannas are remarkably species-rich with up to 42 species/0.25 m^2 and 84 species/625 m^2. Maximum richness occurred on mesic, annually burned sites. Aboveground production, reported as peak standing crop, was only 293 g · m^{-2} on a frequently burned mesic savanna but was significantly higher (375 g · m^{-2}) on an infrequently burned mesic site. Production values from fertilized high and low fire frequency sites were equivalent. Monthly harvest samples showed that savanna biomass composition by species groups did not vary seasonally, but within groups the relative importance of species showed clear phenological progressions.

The variation in species richness with fire frequency is consistent with non-equilibrium theories of species diversity, while phenological variation in production among similar species and the changing species composition across the moisture gradient suggest the importance of equilibrium processes for maintenance of savanna diversity.

Introduction

The savannas or grass–sedge bogs of the coastal plain of the southeastern United States are characterized by scattered pines with an understory sward of mixed graminoids and forbs. These savannas are

* Botanical nomenclature follows Radford, Ahles & Bell (1968). See footnotes to Appendix for details.

** We thank the Federal Paper Board Co., Inc. and the North Carolina Nature Conservancy for permission to work on their lands. Support for this work was provided by the National Science Foundation and the University of North Carolina. Helpful comments on the manuscript were offered by N. L. Christensen, J. D. Doyle, M. A. Huston & W. H. Schlesinger.

well known for their floristic richness and especially their numerous orchids, insectivorous plants and regional endemics (e.g. Wells, 1928; Wells & Shunk, 1928; Lemon, 1949). The descriptive reports of Wells and others prompted our study of the diversity of these communities in the Green Swamp of North Carolina where the most extensive and best preserved mesic savannas on the Atlantic coast are located.

Initial results showed that vascular plant species richness was often near 40/m^2, a level of small-scale species diversity higher than any previously reported for North America and roughly equivalent to the highest values reported in the world literature

(Whittaker, 1977b; Grime, 1979; Peet *et al.,* 1983). The Green Swamp savannas, however, are unique among communities reported to have high small-scale species diversity. Other equivalently species-rich communities have long histories of chronic grazing or mowing. Although neither grazed nor mowed, these savannas have a long history of regular burning. In addition, many of the species encountered in the savannas are specific to this habitat and many are endemic to southeastern North America. Few annuals and no adventives were encountered. In contrast, previously described areas of high small-scale species diversity often contain many non-habitat-specific species (e.g. many species of the species-rich pastures in Britain and chalk grasslands in Holland are described as also inhabiting roadsides and waste places; see Clapham *et al.,* 1962; Bowen, 1968) and sometimes numerous annuals (Naveh & Whittaker, 1979; Shmida & Ellner, 1984).

Many papers present theories to explain the maintenance of high species diversity (see reviews in Whittaker, 1977a; Tilman, 1982; Peet *et al.,* 1983; Shmida & Ellner, 1984). These theories are often divided into two classes: equilibrium and non-equilibrium. Equilibrium theories derive from the competitive exclusion principle and require that species have significant differences in their ecological niches if they are to co-exist indefinitely (Schoener, 1974, 1982; Grubb, 1977). Non-equilibrium theories invoke external factors which interrupt competitive interactions and thereby allow co-existence of species that might otherwise be unable to persist together (Paine, 1966; Connell, 1978, 1980; Grime, 1979; Huston, 1979; Peet *et al.,* 1983; Shmida & Ellner, 1984).

Published evidence suggests that both types of mechanisms might be important in maintaining high species diversity in coastal plain savannas. Numerous studies have contrasted low diversity unburned savannas with more diverse burned sites (e.g. Heyward & Barnette, 1934; Stoddard, 1931; Lemon, 1949, 1967; Eleutarius & Jones, 1969; Vogl, 1973). The maintenance of species diversity by fire, especially on low production sites, is consistent with non-equilibrium theories (e.g. Grime, 1979; Huston, 1979). Other papers describe a striking phenological variation among savanna species (Wells & Shunk, 1928; Folkerts, 1982; Gaddy, 1982). Such differentiation may represent a form of

niche-partitioning consistent with equilibrium theories.

Our discovery of the extraordinary small-scale diversity of the Green Swamp savannas motivated the research reported here. In this paper we first document the variation in species composition and high species diversity of the Green Swamp savannas. We then examine evidence for two commonly suggested mechanisms for maintenance of high diversity in savannas: high fire frequency coupled with low primary production – a non-equilibrium mechanism, and phenological differentiation – an equilibrium mechanism.

Study area

At the time of European settlement mesic pine–wiregrass savanna was widely distributed on fine, poorly drained, oligotrophic sands of the coastal plain from roughly the James River in Virginia south to Florida and then west to the Mississippi River (Harper, 1906, 1914, 1943; Gano, 1917; Wells, 1932; Fernald, 1937; Wharton, 1978; Folkerts, 1982). Outlying communities floristically similar to savannas could be found north into the New Jersey pine barrens (Olsson, 1979) and west to eastern Texas (Streng & Harcombe, 1982). Regionally, savannas can be viewed as communities occupying the center of a soil moisture gradient along which vegetation ranges from shrub bog through savannas and pine flatwoods to dry sandhills (see Wells, 1932; Christensen, 1979). Small pockets of savanna-like vegetation can be found in depressions and sinks within the flatwoods, and in seeps where clay layers surface in the sandhill regions of the uppermost coastal plain (Wells & Shunk, 1931). All of these savanna habitats have a fire cycle typically of less than eight years (Christensen, 1981).

The Green Swamp, originally an area of roughly 80 000 ha, is primarily an ombrotrophic peatland with elevation ranging from 12 to 18 m above sea level. Emerging from the matrix of peat are low islands of mineral soil on which savanna vegetation frequently occurs. The islands vary in convexity, height, and drainage, though most rise less than 1.5 m above the surrounding peat surface. Island size ranges from 1 to 20 ha. The more isolated islands generally have lower fire frequencies and consequently are often overgrown with shrubs. The more species-rich areas are flat and poorly drained with the water table within a few centimeters of the surface during spring and after heavy rains.

The mineral soils of the savannas developed from Pleistocene deposits of fine to coarse sand and clay. The most frequently encountered savanna soils are of the Leon (Aeric Haplaquod), Rains (Typic Paleaquult), Lynchburg (Aeric Paleaquult), and Foreston and Wrightsboro (Aquic Paleudult) series. All of these soils are notably acidic and low in nutrients, particularly nitrogen and phosphorus (Metz *et al.,* 1961; Plummer, 1963; Buol, 1973).

Climatological records are available for Southport, NC, approximately 35 km southeast of the study area. There the average annual precipitation of 1 318 mm is distributed fairly evenly throughout the year with the highest precipitation occurring from June through September. Mean annual temperature is 16.8 °C with the July mean 26 °C and the January mean 8 °C. Precipitation data from a U.S. Forest Service station 25 km north of the study site (1 600 mm/yr) indicate that annual precipitation in the Green Swamp might be slightly higher than in Southport.

Like virtually all of the coastal plain of the southeastern United States, the Green Swamp has been significantly modified by man. Large areas of original *Taxodium, Nyssa, Chamaecyparis* swamp forest were converted to pocosin by cutting and burning in the late 19th century. Subsequently large areas of pocosins and original savannas have been converted to commercial pine plantations.

The remaining undeveloped savannas have also been influenced by man, who, as archeological evidence suggests, has inhabited the Green Swamp savannas for at least 3 000 yr (Rights, 1947; Ward, 1979). Burning was widely used by aboriginal peoples throughout the southeastern United States to improve hunting (Lawson, 1714; Bartram, 1791; Maxwell, 1910; Vogl, 1973), and it is almost certain that the early inhabitants burned these savannas regularly. Burning was continued by the European settlers who burned the savannas to keep the land open and to improve grazing conditions. (Our limited information about early land use by the European settlers in the study area comes from historical records and anecdotal sources compiled by H. McIver and R. Kologiski.) Although European settlements were already established near the Cape Fear River in the 1660's, it was not until the 19th

century that the Green Swamp savanna lands were variously exploited for grazing and for tar, turpentine and timber production. Today, probably as a result of these operations, few pines over 40 cm dbh remain in the Swamp. Grazing by domestic stock ended by the late 1930's and turpentining by the late 1940's. Fire has continued on a 1 to 4 year cycle as a management tool to control fuel accumulation and to minimize growth of broadleaved trees.

Methods

The multiplicity of the questions to be addressed required that a variety of methods be used. It was necessary to sample both species composition and aboveground production of the savannas over the range of environmental conditions. Sampling with a series of nested quadrats provided data to describe species diversity at different size scales. Harvest data were collected to test the hypothesis that there is phenological variation in aboveground primary production. Biomass data also provide a more effective measure of species importance in these complex communities than does frequency, and they provide a basis for comparison with other studies which relate diversity to standing crop (e.g. Grime, 1973, 1979; Al-Mufti *et al.,* 1977; Willems, 1980). Fertilization experiments were used to examine the interaction between increased site production and diversity. Because fire history has been shown to affect both diversity and site production (Parrot, 1967; Christensen, 1977), harvesting and fertilization experiments were conducted in sites with histories of both annual burning and infrequent burning.

Composition

Twenty-one savanna sites were selected to represent a range of environmental conditions and fire frequencies. At each site species presence was recorded in nested quadrats including 10 contiguous square 0.25 m² quadrats (= 1 × 2.5 m quadrat), a 25 m², and a 625 m² quadrat. In each large quadrat five soil samples were collected from the top 10 cm of mineral soil. Soils were analyzed by the North Carolina State Soils Laboratory for exchangeable phosphate, potassium, calcium and magnesium, organic matter content, cation exchange capacity, percent base saturation and pH. Topographic position scores were assigned, ranging from 1 for dry ridges to 4 for depressions. Soil drainage was scored with 0.5 for well-drained sites and 2.5 for the poorly drained extreme. These two subjective scores were summed to produce a single site moisture scalar. The length of the fire cycle during the last two decades for each savanna was estimated based on fire records of the Nature Conservancy and Federal Paper Board Company, Inc., the degree of isolation of the site, and the age of woody growth since the last fire.

Aboveground production

Six rectangular 12×24 m permanent plots were established for measurement of net aboveground production. Three plots were set up in an annually burned mesic savanna. One of these plots was fertilized two weeks after the burn with 5.4 g/m² N, 1.3 g/m² P and 2.7 g/m² K. Paired control and fertilized plots were established on a mesic savanna with a 3–4 year fire cycle. A final control plot was established on a drier, annually burned site. All sites were burned in February, 6 weeks before data collection was initiated.

Net aboveground production was estimated by harvesting biomass in 10 square 0.5×0.5 m quadrats at 2 to 4 week intervals from April 7 through September 2. Five regularly spaced quadrats were centered in alternating meters along each of two transects. This design provided buffer strips between sampled quadrats, minimized trampling damage, and ensured sampling across any within-plot variation. Aboveground biomass was cut to within 0.5 cm of the soil, placed in plastic bags, and stored at 4 °C until sorting. Samples were sorted into live and dead parts by species, dried at 60–65 °C for 48 hr, and weighed. Production is reported as maximum standing crop biomass, a measure which underestimates site production. In this report we do not account for asynchrony among species biomass peaks, nor do we estimate belowground production.

From mid-1979 through 1981 a major drought occurred on the North Carolina coastal plain (U.S. Dept. Commerce, 1977-83). During this period many of the savanna forbs did not flower, which suggested a decline in vigor and may also have led to a decline in the available seed pool. In a season of normal precipitation immediately following the drought (1982), we resampled a control plot (Plot 3) to assess year to year consistency of our results and the possible impact of the drought.

Data analysis

The 21 stand × 151 species data matrix was ordinated using Detrended Correspondence Analysis (DCA: Hill, 1979a; Hill & Gauch, 1980). DCA ordination of both presence/absence and frequency data, both with and without woody species, produced similar arrangements of stands and species. The effects of fire history on the ordination were minimized by using only herbaceous species and presence/absence data in the final analysis. In addition to helping identify environmental factors highly correlated with major trends in species composition, the ordination provided the species order for a phytosociological summary table (see Appendix).

The stands were divided into three community types based on position on the first DCA axis. TWINSPAN (Hill, 1979b), a polythetic devisive clustering program (see Gauch & Whittaker, 1981; Gauch, 1982) was used to further examine possible groupings of stands.

Among methods for quantifying species diversity, species richness, measured as species per unit area, is probably the most computationally simple and easily interpretable (Peet, 1974; Whittaker, 1977a). The nested quadrat sampling scheme allowed determination of species richness at several size scales. Two dominance-weighted indices of species diversity, Simpson's index ($C = \Sigma p_i^2$) and the Shannon-Wiener index ($H' = -\Sigma p_i \log p_i$), were calculated. Importance values (p_i) were measured as relative aboveground biomass for harvested plots and relative frequency of occurrence in 10 0.5×0.5 m quadrats for nested-quadrat samples. While C is strongly affected by the importance of the most abundant species, H' gives more weight to species of intermediate importance. We use the reciprocal form of C (Hill's N_2) and the exponentiated form of H' (Hill's N_1) which are interpretable as the number of equally common species that will produce the observed heterogeneity (see Hill, 1973; Peet, 1974). Evenness was calculated as $(N_2-1)/(N_1-1)$ which is Alatalo's (1981) revision of Hill's (1973) ratio $E_{1,2}$.

Results

Community composition

The DCA ordination of the twenty-one 625 m² plots is shown in Figure 1. Values of the moisture scalar strongly correlate with the first axis (Spearman rank correlation = 0.80, $p < 0.001$), which can be interpreted as a moisture gradient from wet to dry sites (see ordered species list in Appendix). The second ordination axis varies with fire frequency such that the annually burned sites are consistently located above the infrequently burned sites.

Soil pH is negatively correlated with the first axis score (Spearman rank correlation = 0.44). Among environmental parameters soil pH is negatively correlated with the estimated length of the fire cycle, while percent organic matter, CEC, K, and Ca are positively correlated with fire cycle length. These correlations were significant at the $p < 0.05$ level.

The distributions of some groups of ecologically related species are strongly correlated with the first DCA axis moisture gradient (Table 1). The mean DCA positions of species within a group, which indicate the centers of distribution for the groups along the moisture gradient, vary markedly and corroborate the direct correlations with the axis scores. The percentages of the flora composed of

Table 1. Distributions of species groups along the DCA ordination axes. The first 2 columns show Spearman rank correlations between percentages of all species in the 625 m² quadrats which belong to the group and the position of the stand on the DCA axis (* = $p < 0.05$, ** = $p < 0.001$). The final columns show mean species positions on the axes. The ranges of species positions on the two axes are –205 to 473 and –229 to 418 respectively. Axis 1 is a gradient from wet to dry savannas; Axis 2 is correlated with fire frequency.

	Spearman rank correlation		Species mean DCA position	
DCA axis:	1	2	1	2
Species group				
Grasses	–0.018	0.068	79.79	61.05
Sedges	–0.413*	–0.004	58.14	113.57
Other monocots	–0.612*	–0.318	45.52	66.43
Composites	0.758**	0.599*	196.96	103.50
Legumes	0.855**	0.040	332.80	35.00
Other herbaceous dicots	–0.423*	–0.137	78.45	78.94
Ferns and lycopods	–0.118	–0.140	186.00	84.17
Woody species	–0.024	–0.191	151.23	115.26

sedges, monocots (excluding grasses and sedges), and herbaceous dicots (excluding legumes and composites) decrease from wet to dry sites, whereas composite and legume importances increase along the gradient. The relative importance of composites also increases along the second axis suggesting an increase with frequent burning. Of the species which occurred in at least 2 high and 2 low fire frequency mesic sites, 5 were significantly more abundant (frequency in 0.5 × 0.5 m quadrats; Student's t-test, $p < 0.05$) in annually burned sites than in less frequently burned sites (*Helianthus heterophyllus, Marshallia graminifolia, Polygala hookeri, Tofieldia racemosa, Trilisa paniculata*). No species showed the opposite trend at a statistically significant level. Of the 46 species which occurred in only high or low fire frequency sites, 36 were found in the high frequency type.

We recognize wet, mesic, and dry savanna types. The somewhat arbitrary divisions of the DCA axis are consistent with the TWINSPAN results. The first TWINSPAN division separated 1 very dry site (#11). Subsequent divisions segregated first the wet sites, and then mesic from dry savannas.

All of the Green Swamp savannas have a sparse pine canopy (0–150 stems/ha) and a grass-dominated understory. Shrubs up to 1.5 m tall, the real-

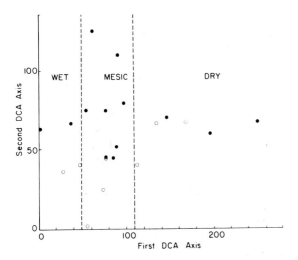

Fig. 1. DCA ordination of presence/absence data for non-woody species in 625 m² quadrats. Open symbols indicate sites with infrequent fires and filled symbols sites with annual fires.

ized height depending on the length of time since the last fire, can be present under all moisture conditions. The bunch-forming grasses grow up to 50 cm in height and are largely responsible for creating microtopographic variation within the communities. Grass clump size varies inversely with fire frequency from 5–15 cm basal diameter in annually burned sites to 35–40 cm. Wet sites along with low fire frequency sites have relatively pronounced vertical relief, up to 10 cm difference between the tops of grass hummocks and the hollows between them.

Wet savannas

Wet savannas are found in shallow depressions and as ecotones between mesic savanna vegetation and the shrub-dominated pocosins. In particular, wet savannas are likely to be found where fires on mesic savannas have spread into the adjacent pocosin eliminating the dominant shrubs.

Dominant grasses of wet savannas include *Ctenium aromaticum, Muhlenbergia expansa* and *Sporobolus teretifolius,* all of which are also dominants in mesic savannas. *Dichromena latifolia* and *Carex verrucosa* are sedges restricted to these communities or found only rarely in wet microsites on drier savannas. Other species which are apparently specific to wet savannas in the Green Swamp include *Drosera intermedia, Coreopsis falcata, Rhynchospora chalarocephala, Oxypolis filiformis, Iris tridentata, Aristida affinis, Anthaenantia rufa,* and the shrubs *Vaccinium corymbosum* and *Cyrilla racemiflora.* In contrast to drier savannas where *Pinus palustris* is the only tree species present, wet savannas support occasional individuals of *Pinus serotina, Taxodium ascendens* and *Nyssa sylvatica* var. *biflora.*

Most species in wet savannas are confined to the grass hummocks; the low depressions are typically devoid of vegetation. The extended period of winter and spring inundation in these microsites, coupled with incomplete combustion of wet litter restricts the growth of small plants and inhibits most seedling establishment.

Dry savannas

Dry savannas typically occur on the high, central portion of the more dome-shaped islands (up to 1.5 m above the surrounding pocosin), or where the soil is coarse textured and well drained. Total plant cover ranges between 50 and 70% compared to over 100% cover in the other savanna communities. The open tree canopy contains only *Pinus palustris,* and the grass *Aristida stricta* dominates the field layer. Although relatively uncommon on mesic and wet sites, legumes are well represented in dry savannas (Table 1, Appendix). Legumes that are apparently confined to drier sites in the study area include *Cassia fasciculata, Lespedeza capitata, Clitoria mariana,* and *Amorpha herbacea.* Other characteristic species (100% constancy) are the woody *Myrica cerifera* and *Smilax glauca,* and the forbs *Euphorbia curtisii* and *Viola septemloba.* Sedges, where present, are restricted to wet microsites.

Mesic savannas

Mesic savannas occupy an intermediate position on the moisture gradient and are especially rich in species. Where present, the tree canopy consists almost exclusively of *Pinus palustris,* though rarely does it exceed 40% cover. The few species that are restricted to mesic savannas include *Polygala cruciata, Pinguicula* spp., *Lachnocaulon anceps, Lilium catesbei* and *Xyris smalliana.* The only species with 100% constancy is the endemic Venus' flytrap, *Dionaea muscipula.* A large group of species with high constancy in mesic savannas can be identified in the Appendix.

Although hummock and hollow microtopography occurs on these mesic sites, the relief is less pronounced than in wet savannas. Graminoids, the most important of which are *Sporobolus teretifolius, Muhlenbergia expansa, Ctenium aromaticum, Andropogon* spp. and *Rhynchospora plumosa,* form the hummocks and dominate the herbaceous layer. Most of the hollows are litter free and are important microhabitats occupied by species such as *Lycopodium carolinianum,* the sedges *Rhynchospora breviseta* and *Rhynchospora chapmanii,* the insectivorous *Dionaea muscipula* and *Drosera capillaris,* and many species of composites. Frequently burned sites are dominated by numerous, small interdigitating hummocks. As litter accumulates in the absence of fire, plants which typically grow between the hummocks and even the small hummocks largely disappear.

Diversity patterns

Species richness

Species richness in savannas ranges up to 42 spe-

Table 2. Species richness statistics for wet, mesic and dry savannas under annual and infrequent fire regimes. Mean (s.d.) species counts are shown for quadrats of 0.25, 1, 2.5, 25 and 625 m².

Moisture class:	Wet		Mesic		Dry	
Fire frequency:	High	Low	High	Low	High	Low
Number of sites:	2	2	8	3	3	3
Quadrat size (m²)						
0.25	19.5	20.8	22.2	14.0	13.2	11.0
	(0.5)	(1.7)	(3.6)	(3.2)	(4.9)	(5.8)
1.0[a]	32.2	33.6	35.2	25.9	24.3	22.4
	(2.3)	(2.4)	(4.8)	(7.3)	(8.8)	(9.1)
2.5	43.5	44.5	46.6	34.7	33.7	31.7
	(4.9)	(2.1)	(6.3)	(3.2)	(12.7)	(12.9)
25	53.5	54.0	57.3	50.3	43.0	43.3
	(2.1)	(5.7)	(8.6)	(1.1)	(14.4)	(11.1)
625	73.5	68.0	79.3	71.2	58.0	54.3
	(3.5)	(5.7)	(5.3)	(2.9)	(13.0)	(8.1)

[a] Calculated from species area curves obtained by regression. Species richness = A + B (log area).

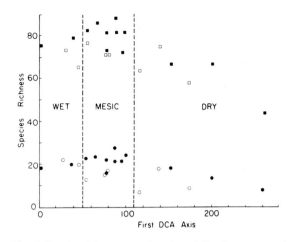

Fig. 2. Species richness as a function of fire frequency and position on the DCA Axis 1. Open symbols indicate infrequently burned sites and filled symbols annually burned sites. Squares indicate richness for 625 m² quadrats and circles the mean richness in ten 0.25 m² quadrats.

cies per 0.25 m² quadrat, to over 50 per m², and from 63 to 84 (43 in site 11) species in 625 m² plots. For all quadrat sizes maximum richness is found in annually burned, mesic savannas (Table 2, Fig. 2). Richness for both 0.25 and 625 m² samples increases with site moisture (decreasing axis 1 score) to a peak in mesic savannas, and levels off or decreases slightly in wet sites. At the 0.25 m² scale the dry sites are only 60% as rich as annually burned, mesic savannas. This in part reflects the smaller plant size in mesic savannas, but at the 625 m² scale richness in dry savannas is still only 73% of that on mesic sites.

Annually burned mesic savannas average 26% more species per m² than less frequently burned savannas. In wet and dry savannas, however, differences in richness with fire frequency are not significant.

Diversity indices

Frequency data were used to calculate dominance-weighted diversity statistics for the 6 classes of savannas (3 moisture classes under 2 fire regimes; Table 3). Consistent with the observations on species richness, the highest diversity values ($1/C$; Exp H') occurred in the frequently burned mesic savannas. The mesic savannas also showed the greatest drop in diversity with a shift to low fire frequency. Among infrequently burned savannas diversity was highest in wet sites. No clear differences were apparent in Alatalo's evenness index. Biomass data were used to calculate diversity indices for the six harvested sample sites (Table 4). Results for 1979 from sites 1 and 3 (selected as replicates) are nearly identical.

Diversity ($1/C$; Exp H') appeared notably higher in 1982 while at the same time the number of species

Table 3. Diversity indices calculated using frequency in ten 0.5 × 0.5 m quadrats per stand in the 21 plots sampled using nested quadrats. Values are means (s.d.).

Number plots	Moisture class	Fire frequency	Spp./0.25 m²	Hill's N_2 ($1/C$)	Hill's N_1 (Exp(H'))	Alatalo's $E_{1,2}$ ((N_2-1)/(N_1-1))
2	wet	high	19.5 (0.5)	28.01 (2.12)	32.14 (2.71)	0.87 (.01)
2	wet	low	20.8 (1.7)	29.95 (0.89)	33.96 (0.08)	0.88 (.03)
8	mesic	high	22.2 (3.6)	31.05 (0.27)	35.44 (4.74)	0.87 (.03)
3	mesic	low	14.0 (3.3)	21.24 (4.58)	25.14 (4.65)	0.83 (.03)
3	dry	high	13.2 (4.9)	20.99 (8.87)	25.00 (10.52)	0.83 (.03)
3	dry	low	11.0 (5.8)	18.73 (9.48)	22.90 (10.62)	0.80 (.04)

Table 4. Diversity indices calculated using aboveground biomass in 6–10 0.25 m² plots. Values for richness are means (s.d.).

Plot/year	Moisture class	Fire frequency	Spp./0.25 m²	Hill's N_2 $(1/C)$	Hill's N_1 $(Exp(H'))$	Alatalo's $E_{1,2}$ $((N_2\text{-}1)/(N_1\text{-}1))$
Control						
1/79	mesic	high	33.1 (4.0)	7.45	13.76	0.51
3/79	mesic	high	31.4 (1.8)	8.18	13.41	0.58
3/82	mesic	high	26.0 (6.1)	10.36	15.59	0.64
2/79	dry	high	24.0 (3.3)	5.00	8.04	0.54
5/79	mesic	low	11.5 (3.2)	2.43	4.51	0.49
Fertilized (1 season)						
4/79	mesic	high	32.9 (4.9)	10.00	14.41	0.67
6/79	mesic	low	13.3 (2.4)	4.02	6.26	0.57
Fertilized (4 seasons)						
4/82	mesic	high	29.0 (5.7)	7.40	12.36	0.56

per 0.25 m² plot had dropped markedly. The data suggest a post-drought decline in the abundance of the typical dominants leading to lower concentration of dominance (note the decline in importance of grasses shown in Table 6), but also a decline in the average number of species per plot.

Fertilization resulted in no decrease in species per 0.25 m² during the first year. After four years the drop in richness was no more than on the control plot where drought apparently caused a modest decline.

Spatial homogeneity

We have observed conspicuous spatial heterogeneity in savannas. On particularly wet or dry sites, or on sites with only infrequent fire, the dominant grasses form large, discrete tussocks. In contrast, frequently burned mesic sites are characterized by small, anastomosing clumps of grasses and sedges. We attempted to quantify this difference in homogeneity by calculating the mean pairwise coefficient of community (see Whittaker & Gauch, 1973) between the ten 0.25 m² quadrats for each plot. As shown in Figure 3, the mesic, frequently burned sites had the highest homogeneities and the dry sites (with one exception notable for its low diversity) had low values. Thus, on mesic sites with frequent fires, the between-patch diversity at the 0.25 m² scale is small relative to that of other savanna types.

Aboveground production

Peak aboveground biomass

Peak live, aboveground biomass for the six harv-

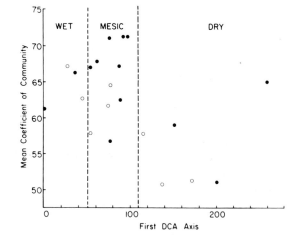

Fig. 3. Within-community homogeneity as a function of fire frequency and position on the DCA Axis 1. Homogeneity is represented by the mean pairwise similarity of ten 0.25 m² quadrats. Open symbols indicate in frequently burned sites and filled symbols annually burned sites.

ested savanna plots is shown in Table 5. Peak standing crop in the two annually burned mesic savannas was estimated at 293 and 236 g/m² for 1979, while the dry site had a lower value of 216 g/m². The mesic, low fire frequency site supported a significantly greater living biomass than the mesic sites that were annually burned.

Grasses contributed about 70% of the biomass on all sites, whereas the importance of other groups varied with moisture and fire frequency (Table 6). With low fire frequency, sedges, other monocots, and composites declined in relative biomass, while other dicot herbs and shrubs increased. Differences

Table 5. Peak aboveground living biomass on control and fertilized sites. Values are means (s.d.) of 6–10 0.25 m² quadrats per site.

| Moisture class: | Mesic | | Mesic | | Dry | |
| Fire frequency: | Annual | | Infrequent | | Annual | |
Year	Plot	Biomass	Plot	Biomass	Plot	Biomass
Control						
1979	1	73.15	5	94.18	2	54.00
		(11.5)		(35.4)		(17.2)
1979	3	58.99				
		(44.9)				
1982	3	75.08				
		(06.7)				
Fertilized (one season)						
1979	4	110.36	6	113.13		
		(27.4)		(24.2)		
Fertilized (four seasons)						
1982	4	93.74				
		(5.2)				

in shrub biomass were particularly dramatic, shifting from 1% to 15% of the standing crop with decreased fire frequency.

Changes following fertilization

Fertilization of an annually burned mesic savanna nearly doubled peak standing crop the following summer (264 vs 441 g/m²), though no further increase was observed after four seasons of fertilization (Table 5). In contrast, fertilization of a low fire frequency mesic site resulted in a much smaller increase in production. Biomass production values from the fertilized, high and low fire frequency sites were almost identical.

Seasonal variation in production

Throughout the sampling season the relative contribution of species groups to standing crop remained nearly constant, but production associated with various species within groups showed considerable seasonal variation. These seasonal differences were examined for three important groups: composites, grasses, and sedges. Figure 4 shows the seasonal progression in standing crop for selected members of these groups as measured for plot 1, an annually burned mesic savanna.

Numerous species of composites were present and they varied markedly in flowering phenology as well as production. In general, peak aboveground biomass coincided with bolting and flowering. Early season species, not shown because of their infrequent occurrence, include *Chaptalia tomentosa* and *Erigeron verna,* both of which flower March through April. Several species flower mid-season while most species flower in late summer or autumn.

Grasses, which form the dominant species group both in biomass and in contribution to community structure, generally accumulate aboveground biomass through the spring and summer and then bloom in late summer or fall. (Figure 4 shows standing live biomass and does not include senescent leaves which begin to accumulate by midsummer.) Although phenologies are similar, periods of maximum growth rates are different among the grass species. For example, *Ctenium aromaticum,* the most abundant species in this plot, grows rapidly as early as 6 weeks after fire, and reaches maximum biomass in July coincident with its maximum flow-

Table 6. Percent contributions of species groups to peak biomass in control and fertilized plots. Group codes: G = grasses, S = sedges, M = other monocots, C = composites, D = other herbaceous dicots, F = ferns and lycopods, W = woody dicots.

Plot	Moisture class	Fire frequency	Year	Species group						
				G	S	M	C	D	F	W
Control										
1	mesic	high	1979	72.0	8.0	5.4	12.4	0.6	0.05	1.6
3	mesic	high	1979	74.4	8.1	1.3	12.8	0.3	0.0	1.3
3	mesic	high	1982	63.8	14.3	3.9	14.2	2.7	0.1	0.9
2	dry	high	1979	74.2	6.9	1.3	16.8	0.4	0.0	0.4
5	mesic	low	1979	75.1	1.8	0.6	1.2	2.3	0.0	19.0
Fertilized (1 season)										
4	mesic	high	1979	72.5	5.9	3.0	15.4	0.5	1.7	1.0
6	mesic	low	1979	62.5	0.8	1.6	0.9	1.5	3.3	29.4
Fertilized (4 seasons)										
4	mesic	high	1982	71.4	8.7	3.5	11.1	3.8	0.02	1.4

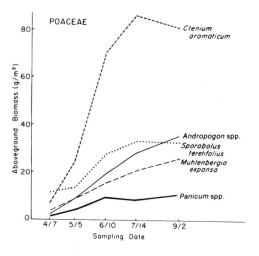

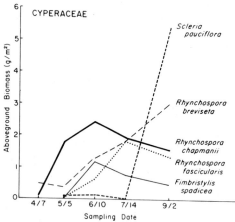

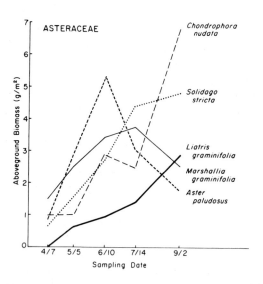

Fig. 4. Seasonal trends in aboveground living biomass of selected individual species of (a) grasses, (b) sedges, and (c) composites. Data show biomass in Plot 1, a mesic annually burned site.

ering but ahead of the peak for most species. *Panicum chamaelonche,* which flowers twice each season, similarly peaks early.

Sedges also show variable seasonal patterns of growth, but generally peak before the dominant grasses. For these species flowering and the associated peak biomass proceeds from the small statured species like *Rhynchospora chapmanii* early in the season to the taller canopy species like *R. breviseta* and *R. plumosa* in the autumn.

Discussion

Savanna composition

Published descriptions of savanna vegetation include the physiognomic sketches provided by early botanical explorers (Bartram, 1791; Elliot, 1821–1824; Nash, 1895), as well as floristic analyses featuring species lists (e.g. Thorne, 1954; Pullen & Plummer, 1964; Eleutarius & Jones, 1969; Gaddy, 1982). Only a few studies systematically describe savannas in the context of regional vegetation or interpret variation in savanna vegetation relative to environmental gradients. Among these few reports, all suggest the importance of a moisture gradient and several also emphasize the significance of fire frequency (e.g. Lemon, 1949, 1967; Vogl, 1973).

Woodwell (1956) described savannas as coastal plain wetland communities. He recognized three distinct herbaceous components which he found associated with *Pinus palustris* or *P. serotina* on wetter sites. His *Pinus/Aristida* communities, which were found on sandy flats and ridges, overlap with the dry savannas described here. He reported herb layers dominated by *Muhlenbergia* and *Andropogon,* and compositionally similar to mesic and wet savannas occurring on clay substrates.

Both Snyder (1980) and Kologiski (1977) described geographically more restricted vegetation units. It is clear that the sites Snyder studied in the Croatan National Forest (150 km northeast of the Green Swamp) were not as species-rich as the Green Swamp counterparts, resulting perhaps from drier conditions at his sites, and in part from the lack of propagules of typical savanna species, many of which reach their northern limits in the area. Kologiski's Green Swamp *P. palustris–Aristida stricta–Rhynchospora* community covers the range of mesic and dry sites in our study. His *P. serotina-Aristida stricta-Rhynchospora* type generally coincides with our wet sites, but the name is somewhat misleading considering that *Aristida stricta* dominates only on dry sites in the study area.

Wells & Shunk (1928) related species composition to habitat factors such as seasonal variation in the water table, soil aeration, and soil nutrient status. They described *Campulosus (Ctenium), Panicum,* and *Andropogon* consocies which represent a gradient from the wet to the dry-mesic savannas of our study. Our dry savannas correspond closely with a hydric or *Aristida* semi-bog phase of Wells' *Quercus-Aristida* associes (Wells & Shunk, 1931).

Species richness

The Green Swamp savannas are remarkable for their high species diversity. At a 625 m² scale, the range of 70 to 84 species places mesic savannas among the more species-rich communities in temperate North America (see Peet, 1978). At the 1 m² and 0.25 m² scales, however, the Green Swamp savannas appear to surpass all other North American plant communities. In over 20 000 1 m² plots examined for temperate North American woodlands and forests, species richness never exceeded 17 species, and in extensive studies of North American tall-grass prairie, richness averaged 18 m⁻² and never exceeded 28 m⁻² (Peet *et al.*, 1983). The average richness of 35 m⁻² in frequently burned mesic savannas of the Green Swamp provides a striking contrast that begs explanation.

Species richness on a 1.0 or 0.25 m² scale equal to that in the Green Swamp savannas has been reported in the world literature only in areas chronically grazed by domestic stock or wildlife, or regularly mowed for long periods. These communities include chalk grasslands in Britain (Rorison, 1971; Grime, 1973; 1979), oligotrophic grasslands in Japan (Itow, 1963) and in the Netherlands (van den Bergh, 1979), the alvars of Sweden (Sjögren, 1971; Rosén & Sjögren, 1973), dune slacks in Holland (van der Maarel, 1971; van der Maarel & Leertouwer, 1967; Thalen, 1971), Mediterranean scrub in Isreal (Whittaker, 1977b; Naveh & Whittaker, 1979; Shmida & Ellner, 1984), calcareous fens in Sweden (Regnéll, 1980), and wildlife-grazed grassland in Sri Lanka (Mueller-Dombois, 1977). All of the examples appear to be from communities where

soil conditions are less than optimal for plant growth.

Non-equilibrium theories of species diversity

Several authors have advanced non-equilibrium theories of species diversity which predict that a moderate amount of disturbance maximizes species diversity (e.g. Loucks, 1970; Grime, 1973; 1979; Whittaker, 1977b; Connell, 1978; Huston, 1979; Abugov, 1982). Grime (1973; 1979) proposed that species richness can be interpreted as a function of 'environmental stress' and/or intensity of management 'such as grazing, mowing, burning and trampling, which tend to prevent potentially competitive species from attaining maximum size and vigor'. He hypothesized that species richness will peak at some intermediate value of these two factors. He attributed low species richness in environmentally stressful or frequently disturbed sites to the small number of species adapted to the imposed stress, and he hypothesized that low richness in stable and/or highly productive sites results from the ability of a few species to pre-empt resources, thereby suppressing richness by competitive exclusion.

A refinement of Grime's model has been proposed by Huston (1979). Huston's model differs in that he explicitly described an interaction between disturbance and rate of competitive exclusion. In Huston's view, the frequency of population reduction or disturbance which maximizes richness in a particular community is dependent on potential growth rate and thus competitive exclusion by the potential dominants. The rate of competitive displacement is correlated with site favorability: good sites yield more rapid growth and consequently have more rapid extinction of competitors. Thus, rather than a single, complex gradient, Huston proposed that two gradients are necessary to determine a richness response surface.

Our observations on species richness in the Green Swamp savannas are consistent with the Grime and Huston models. Mesic, annually burned sites typically have higher species richness than less frequently burned sites (Fig. 2), a result consistent with the prediction that when the disturbance regime is relaxed, richness will drop. As would be predicted from Huston's model, the influence of fire is less pronounced at the extremes of the moisture gradient where the potential for competitive exclusion is lower. None of the savannas burn with less than a 1-year cycle, so we cannot directly assess the impact of more frequent disturbance. The frequency of disturbance in savannas contrasts with the continual biomass reduction typical of most other species-rich communities. The low fertility of savanna soils enforces a slow recovery from disturbance and it is thus likely that any increase in frequency or intensity of disturbance would lead to a reduction in species richness.

Al-Mufti et al. (1977) scaled Grime's one-dimensional gradient of stress and disturbance using aboveground standing crop. Among a variety of herbaceous communities, they found maximum richness to be limited to sites supporting biomass in the range of 350 to 750 g $\cdot$ m^{-2}. In the Green Swamp savannas maximum richness occurs at roughly 280 g $\cdot$ m^{-2} with high values occurring when standing crop is as low as 165 g $\cdot$ m^{-2} (dry sites). Similarly, Willems (1980) reported richness to peak in northwestern European chalk grasslands with biomass values between 150 and 350 g $\cdot$ m^{-2}. Willems attributed this deviation from the Al-Mufti et al. results to his own harvesting at 2 cm rather than ground level.

The high savanna diversity at a somewhat lower biomass than observed by Al-Mufti et al. can be interpreted in an evolutionary context. Evolutionary history should lead to a local pool of species adapted to prevailing ecological conditions (see Terborgh, 1973). The savanna flora evolved under conditions of lower fertility than did the British communities where Al-Mufti et al. worked. Whittaker (1977b; Naveh & Whittaker, 1980) proposed a similar explanation when confronted with differing richness responses to grazing intensity in different ecosystems.

The role of fire

The species-rich savannas of the Green Swamp appear distinctive among communities currently known to have species richness greater than 35 m^{-2} in that the only chronic disturbance is fire. This association with fire is an ancient one. It is likely that man has burned the savannas of southeastern North America for more than 3 000 years (Rights, 1947; Vogl, 1973; Christensen, 1981), and lightning is likely to have ignited savanna fires long before the first human occupation (Vogl, 1973). The large

number of species endemic to these communities attests to the antiquity of the assemblage.

Fire can function in several ways to enhance savanna diversity. The most conspicuous effect is the reduction in woody plants which would otherwise increase in size and density until the savannas were replaced by forest or shrubland (Wells & Shunk, 1928; Lemon, 1949; Eleutarius & Jones, 1969; Wells & Whitford, 1976; Christensen, 1981; Folkerts, 1982). Fire also removes grass and sedge foliage which casts a heavy shade and which can lead to a loss of the smaller grasses and forbs which grow between the clumps. As many savanna species grow under the grass canopy and conduct much of their photosynthesis by means of rosettes of leaves, failure to burn the litter layer for more than one or two seasons may contribute to a decline in the number of species occupying a typical 1 m^2 area. Low light levels as found under heavy litter accumulations have been implicated in the loss of Dionaea muscipula from such communities (Roberts & Oosting, 1958).

In addition to stimulating flower and seed production in many savanna species (Greene, 1935; Lemon, 1949; Parrot, 1967; Vogl, 1973), fire also opens microsites in which new seedlings can become established (Grubb, 1977). The importance of fire for seedling establishment in the Green Swamp savannas has not yet been ascertained, but Lemon (1949) has described a group of readily dispersed species which colonize savanna sites made available by fire.

Species which become established following disturbance are likely to experience less rigourous competition than might otherwise be the case. Consequently, species typically found in adjacent habitats may develop substantial populations in communities which they could not invade without periodic fire. Peet et al. (1983) used this argument to predict that average niche breadth along the moisture gradient should be greater in annually burned savannas than in the less frequently burned savannas. By similar reasoning beta diversity, compositional changes along the gradient, should decrease with increased fire frequency. Our observation that many species common on annually burned savannas appear confined to mesic ecotones between savanna and shrub bog in the absence of fire is consistent with this prediction.

Equilibrium theories of species richness

According to equilibrium theories, species should co-exist in a community only if significant differences exist in their use of resources such that competitive exclusion does not occur. Niche differentiation among species should be the rule.

The phenological patterns in savannas, in both flowering (Wells & Shunk, 1928; Gaddy, 1982) and aboveground growth (Fig. 5), may represent a form of niche differentiation resulting in reduced competition. It has been suggested that in a similar way seasonal partitioning of primary production serves to maintain species richness in North American prairies (Williams & Markley, 1973).

Seasonal differences in aboveground growth are particularly pronounced among the sedges. This observation could be interpreted to be the result of competition, with the remaining sedges having divergent phenologies. However, coupling these observations with species morphological data suggests an alternative interpretation. The sedges appear to have adopted different strategies for co-existence with the dominant grasses. Small-statured sedges which are likely to be overtopped by grasses grow rapidly early in the season thus avoiding competition (e.g. *Rhynchospora chapmanii, R. ciliaris*), whereas larger species which effectively compete with the grasses for canopy space (e.g. *Rhynchospora breviseta*), peak concurrently with the grasses.

Marked phenological variation in production can also be seen among the composites, but this is largely due to timing of bolting and may be unrelated to photosynthesis, much of which could take place during the winter months in the basal rosette leaves.

A second likely form of species differentiation is in response to soil moisture. Our ordination analysis of the Green Swamp savannas shows a compositional gradient correlated with soil moisture. Significant species turnover occurs along this gradient as indicated in the Appendix. In addition, the relative contributions of various species groups show pronounced trends (Table 1).

We have reported only the most prominent mechanisms of potential species differentiation in savannas. Numerous other factors are likely to be important, some perhaps peculiar to one or a few species. For example, a number of woody species appear largely confined to growth on old, rotting pine logs, and *Carduus virginiana* has a conspicuous affinity for ant hills (Wells & Shunk, 1928). Other possible factors such as root distribution and adaptations for uptake of scarce nutrients and differing nutrient requirements (see Tilman, 1982) still need to be examined.

Conclusions

Our investigations confirm that the Green Swamp savannas are remarkably species-rich, especially at small scales, and that this richness is maximal in sites with high fire frequency near the middle of the moisture gradient. Both non-equilibrium and equilibrium processes appear to contribute to the establishment and maintenance of high species diversity in these savannas.

Appendix

Constancy of species in wet, mesic and dry savannas

Herbaceous species which occur in more than one and less than 17 of the 21 stands are ordered by increasing DCA first axis scores.

Savanna type: Number of stands: Species	Wet 4	Mesic 11	Dry 6
Dichromena latifolia	75	09	00
Drosera intermedia	50	00	00
Coreopsis falcata	50	00	00
Eragrostis refracta	25	09	00
Pleea tenuifolia	50	09	00
Sarracenia flava	100	18	00
Sarracenia purpurea	100	36	00
Scleria reticularis	50	09	17
Sarracenia rubra	100	45	00
Eriocaulon decangulare	100	45	00
Spiranthes praecox	00	18	00
Lachnanthes caroliniana	50	45	00
Smilax walteri	25	18	00
Calopogon spp.[1]	100	73	00
Xyris ambigua	75	82	00
Oxypolis ternata	25	64	00
Bartonia virginica	75	73	00
Arundinaria gigantea	75	82	00
Rhexia lutea	100	55	33
Polygala hookeri	75	75	00
Polygala cruciata	75	27	17

Appendix (Continued).

Savanna type:	Wet	Mesic	Dry
Number of stands:	4	11	6
Species			
Rhexia mariana	00	18	00
Eryngium integrifolium	75	91	17
Osmunda cinnamomea	00	27	00
Dionaea muscipula	50	100	17
Carphephorus tomentosus	100	82	33
Cleistes divaricata	25	36	17
Pinguicula spp.[2]	00	45	00
Lachnocaulon anceps	00	75	00
Sabatia difformis	50	36	17
Solidago stricta	75	82	33
Polygala ramosa	25	36	17
Carex verrucosa	100	09	17
Erigeron verna	100	55	50
Seymeria cassioides	25	91	17
Lilium catesbaei	00	18	00
Viola primulaefolia	50	73	33
Smilax laurifolia	100	64	67
Helianthus angustifolius	25	36	17
Aristida virgata	50	45	17
Xyris smalliana	00	45	00
Lycopodium alopecuroides	50	55	17
Woodwardia virginica	50	27	50
Helianthus heterophyllus	100	82	17
Scleria pauciflora	25	82	50
Fimbristylis spadicea	00	36	17
Agalinis aphylla	25	64	17
Polygala lutea	75	73	83
Rhynchospora ciliaris	00	82	33
Sisyrinchium arenicola	100	64	67
Rhynchospora plumosa	75	73	67
Hypoxis micrantha	00	45	50
Carduus virginiana	00	45	50
Lobelia nuttallii	25	55	50
Panicum spp.[3]	50	36	83
Linum virginianum	00	18	17
Liatris graminifolia	100	45	67
Agalinis obtusifolia	00	36	50
Trilisa paniculata	00	82	67
Prenanthes autumnalis	00	36	50
Baptisia tinctoria	00	09	50
Desmodium tenuifolium	00	36	67
Smilax glauca	00	27	50
Gymnopogon breviseta	00	55	33
Andropogon virginicus glom.[4]	50	09	16
Ludwigia virgata	00	27	67
Panicum ciliatum[3]	00	73	100
Desmodium lineatum	00	27	83
Aster squarrosus	00	27	67
Eupatorium rotundifolium	25	55	83
Pteridium aquilinum	25	45	67
Chaptalia tomentosa	25	00	50
Euphorbia curtisii	00	55	100
Amorpha herbacea	00	45	100
Chamaelirium luteum	00	00	33

Appendix (Continued).

Savanna type:	Wet	Mesic	Dry
Number of stands:	4	11	6
Species			
Rhynchospora sp.	00	09	17
Viola septemloba	00	18	100
Stylosanthes biflora	00	18	67
Crotalaria purshii	00	00	67
Tephrosia hispidula	00	18	83
Lespedeza capitata	00	00	50
Cassia fasciculata	00	00	33
Eryngium yuccafolium	00	09	33
Anthaenantia villosa	00	09	17
Pterocaulon pycnostachyum	00	00	33
Heterotheca mariana	00	00	33
Elephantopus nudatus	00	00	33
Aster pilosus	00	00	33

Additional herbaceous species occurring in >80% of all stands (ordered by first DCA axis score): *Rhynchospora chapmanii, Ctenium aromaticum, Tofieldia racemosa, Habernaria* spp.[5], *Drosera capillaris, Rhynchospora fascicularis, Marshallia graminifolia, Lycopodium carolinianum, Rhynchospora breviseta*[6], *Rhexia petiolata, Panicum chamaelonche*[3], *Coreopsis gladiata, Eupatorium leucolepis, Chondrophera nudata, Aster paludosus, Aster dumosus, Aletris farinosa, Sporobolus teretifolius, Rhexia alifanus, Andropogon* spp.[4], *Aristida stricta, Xyris caroliniana, Heterotheca graminifolia.*

Herbaceous species occurring in only one stand (ordered by first DCA axis score): *Utricularia subulata, Rhynchospora chalerocephala, Oxypolis filiformis, Erianthus* sp., *Aristida affinis, Anthaenantia rufa, Iris tridentata, Scleria minor, Paspalum laeve, Sabatia gentiana, Hypoxis hirsuta, Asclepias longifolia, Spiranthes longilabris, Rhynchospora torreyana, Osmunda regalis, Balduina uniflora, Solidago odora, Clitoria mariana.*

Woody species (ordered by first DCA axis score): *Taxodium ascendens, Vaccinium corymbosum, Cyrilla racemiflora, Symplocus tinctoria, Ilex cassine, Pinus serotina, Nyssa sylvatica* var. *biflora, Myrica heterophylla, Ilex coriacea, Rhododendron* sp.[7], *Hypericum stans, Lyonia mariana, Vaccinium atrococcum, Hypericum reductum, Gaylussacia dumosa, Vaccinium crassifolium, Pinus palustris, Vaccinium staminium, Vaccinium tenellum, Diospyros virginiana, Rubus* sp.

[1] *Calopogon* spp. includes *C. barbatus, C. pallidus* and *C. pulchellus.*

[2] *Pinguicula* spp. includes *P. caerulea* and *P. lutea.*

[3] *Panicum* is divided into three groups: *P. chamaelonche* (which includes some *P. portoricense* and *P. longiliigulatum*), *P. ciliatum* (which includes some *P. strigosum*) and *Panicum* sp. (which includes all other small stature *Panicum* species).

[4] Two groups of vegetative *Andropogon* are recognized: *A. virginicus* (which includes some *A. scoparius* and possibly *A. tenarius*) and *A. virginicus* var. *glomeratus* which in the study area is vegetatively and ecologically distinct, but is variously recognized in manuals as a synonym of *A. virginicus* (Radford *et al.,* 1968) and as a separate species (Hitchcock, 1950).

5 *Habenaria* spp. includes, in order of decreasing abundance, *H. cristata, H. integra, H. ciliaris, H. nivea* and possibly *H. blephariglottis.*

6 *Rhynchospora breviseta* includes some *R. oligantha.*

7 *Rhododendron* spp. includes *R. atlanticum* and *R. viscosum.*

References

Abugov, R., 1982. Species diversity and phasing of disturbance. Ecology 63: 289–293.

Alatalo, R. V., 1981. Problems in the measurement of evenness in ecology. Oikos 37: 199–204.

Al-Mufti, M. M., Sydes, C. L., Furness, S. B., Grime, J. P. & Band, S. R., 1977. A quantitative analysis of shoot phenology and dominance in herbaceous vegetation. J. Ecol. 65: 759–791.

Barnes, P. W., Tieszen, L. L. & Ode, D. J., 1983. Distribution, production, and diversity of C_3- and C_4-dominated communities in a mixed prairie. Can. J. Bot. 61: 741–751.

Bartram, W., 1791. Travels through North and South Carolina, Georgia, East and West Florida, etc., Facsimile ed., Dover Publ., Inc., New York.

Bergh, J. P. van den, 1979. Changes in the composition of mixed populations of grassland species. In: M. J. A. Werger (ed.). The Study of Vegetation, pp. 59–80. Junk, The Hague.

Bowen, H. J. M., 1968. The Flora of Berkshire, Holywell Press, Oxford. 389 pp.

Buol, S. W., 1973. Soils of the Southern States and Puerto Rico. USDA South. Coop. Ser. Bull. 174. 105 pp.

Christensen, N. L., 1977. Fire and soil–plant nutrient relations in a pine–wiregrass savanna on the coastal plain of North Carolina. Oecologia 31: 27–44.

Christensen, N. L., 1979. The xeric sandhill and savanna ecosystems of the southeastern Atlantic Coastal Plain, U.S.A. In: H. Lieth & E. Landolt (eds.). Contributions to the Knowledge of Flora and Vegetation in the Carolinas. Vol. 68, pp. 246–262. Veröff. Geobot. Inst. ETH, Stiftung Rübel, Zürich.

Christensen, N. L., 1981. Fire regimes in southeastern ecosystems. In: H. A. Mooney *et al.* (eds.). Fire Regimes and Ecosystem Properties, pp. 112–136. USDA For. Ser. Gen. Tech. Rpt. WO-26.

Clapham, A. R., Tutin, T. G. & Warburg, E. F., 1962. Flora of the British Isles, 2nd ed. Cambridge Univ. Press, Cambridge. 1269 pp.

Connell, J. H., 1978. Diversity in tropical rain forests and coral reefs. Science 199: 1302–1310.

Connell, J. H., 1980. Diversity and the coevolution of competitors, or the ghost of competition past. Oikos 35: 131–138.

Eleutarius, L. N. & Jones, S. B., 1969. A floristic and ecological study of pitcher plant bogs in south Mississippi. Rhodora 71: 29–34.

Elliot, S., 1821–1824. A Sketch of the Botany of South-Carolina and Georgia. J. R. Schenk, Charleston, S. C.

Fernald, M. L., 1937. Local plants of the inner coastal plain of southeastern Virginia. III. Phytogeographical considerations. Rhodora 39: 465–491.

Folkerts, G. W., 1982. The Gulf coast pitcher plant bogs. Am. Sci. 70: 260–267.

Gaddy, L. L., 1982. The floristics of three South Carolina pine savannahs. Castanea 47: 393–402.

Gano, L., 1917. A study in physiographic ecology in northern Florida. Bot. Gaz. 63: 337–372.

Gauch, H. G., 1982. Multivariate Analysis in Community Ecology. Cambridge Univ. Press, Cambridge. 298 pp.

Gauch, H. G. & Whittaker, R. H., 1981. Hierarchical classification of community data. J. Ecol. 69: 537–557.

Greene, S. W., 1935. Relation between winter grass fires and cattle grazing in the longleaf pine belt. J. For. 33: 339–341.

Grime, J. P., 1973. Control of species density in herbaceous vegetation. J. Environ. Manage. 1: 151–167.

Grime, J. P., 1979. Plant Strategies and Vegetation Processes. Wiley, New York. 222 pp.

Grubb, P. J., 1977. The maintenance of species-richness in plant communities: the importance of the regeneration niche. Biol. Rev. 52: 107–145.

Harper, R. M., 1906. A phytogeographical sketch of the Altamaha Grit Region of the Coastal Plain of Georgia. Ann. N.Y. Acad. 17: 1–415.

Harper, R. M., 1914. Geography and vegetation of northern Florida. Ann. Rep. Florida State Geol. Surv. 6: 167–451.

Harper, R. M., 1943. Forests of Alabama. Geol. Survey of Alabama Monogr. 10. Univ. of Alabama.

Heyward, F. & Barnette, R. M., 1934. Effect of frequent fires on chemical composition of forest soils in the longleaf pine region. Florida Ag. Expt. Sta. Tech. Bull. 265.

Hill, M. O., 1973. Diversity and evenness: a unifying notation and its consequences. Ecology 54: 427–432.

Hill, M. O., 1979a. DECORANA. Cornell Ecology Program, Cornell Univ., Ithaca, New York.

Hill, M. O., 1979b. TWINSPAN. Cornell Ecology Program, Cornell Univ., Ithaca, New York.

Hill, M. O. & Gauch, H. G., 1980. Detrended correspondence analysis: an improved ordination technique. Vegetatio 42: 47–58.

Huston, M., 1979. A general hypothesis of species diversity. Am. Nat. 113: 81–101.

Itow, S., 1963. Grassland vegetation in uplands of western Honshu, Japan. Part II. Succession and grazing indicators. Jap. J. Bot. 18: 133–167.

Kologiski, R. L., 1977. The phytosociology of the Green Swamp, North Carolina. N.C. Ag. Expt. Sta. Tech. Bull. No. 250. 100 pp.

Lawson, J., 1714. Lawson's history of North Carolina. Garrett and Massie, Inc., Richmond, VA. 259 pp. (2nd ed. printed 1952.)

Lemon, P. C., 1949. Successional responses of herbs in longleaf-slash pine forest after fire. Ecology 30: 135–145.

Lemon, P. C., 1967. Effects of fire on herbs of the southeastern United States and central Africa. Proc. Tall Timbers Ecol. Conf. 6: 112–127.

Loucks, O. L., 1970. Evolution of diversity, efficiency, and community stability. Am. Zool. 10: 17–25.

Maarel, E. van der, 1971. Plant species diversity in relation to management. In: E. Duffey & A. S. Watt (eds.). The Scientific Management of Animal and Plant Communities for Conservation. Symp. Brit. Ecol. Soc. 11: 45–63.

Maarel, E. van der & Leertouwer, J., 1967. Variation in vegetation and species diversity along a local environmental gra-

318

dient. Acta Bot. Neerl. 16: 211-221.

Maxwell, H., 1910. The use and abuse of forests by the Virginia Indians. William & Mary Coll. Q. Hist. Mag. 19: 73-103.

Metz, L. J., Lotti, T. & Klawitter, R. A., 1961. Some effects of prescribed burning on Coastal Plain forest soil. USDA For. Ser. Res. Pap. SE-113, Asheville, N.C. 10 pp.

Mueller-Dombois, D., 1977. Comment on Whittaker's paper. In: R. Tüxen (ed.). Vegetation und Fauna, pp. 421-422. J. Cramer, Vaduz.

Nash, G. V., 1895. Notes on some Florida plants. Bull. Torrey Bot. Club 2: 141-161.

Naveh, Z. & Whittaker, R. H., 1979. Measurements and relationships of plant species diversity in Mediterranean shrublands and woodlands. In: F. Grassle, G. P. Patil, W. Smith & C. Taillee (eds.). Ecological Diversity in Theory and Practice, Statistical Ecol. Vol. S6, pp. 219-239. Internat. Coop. Publ. House, Fairland, Md.

Naveh, Z. & Whittaker, R. H., 1980. Structural and floristic diversity of shrublands and woodlands in northern Israel and other Mediterranean areas. Vegetatio 41: 171-190.

Ode, D. J.,Tieszen, L. L. & Lerman, J. C., 1980. The seasonal contribution of C_3 and C_4 plant species to primary production in a mixed prairie. Ecology 61: 1304-1311.

Olsson, H., 1979. Vegetation of the New Jersey pine barrens: a phytosociological classification. In: R. T. T. Forman (ed.). Pine Barrens Ecosystem and Landscape, pp. 245-263. Academic Press, New York.

Paine, R. T., 1966. Food web complexity and species diversity. Am. Nat. 100: 65-75.

Parrot, R. T., 1967. A study of wiregrass (Aristida stricta Michx.) with particular reference to fire. M.A. Thesis, Duke Univ., Durham, N.C. 137 pp.

Peet, R. K., 1974. The measurement of species diversity. Ann. Rev. Ecol. Sys. 5: 285-307.

Peet, R. K., 1978. Forest vegetation of the Colorado Front Range: patterns of species diversity. Vegetatio 37: 65-78.

Peet, R. K., Glenn-Lewin, D. C. & Wolf, J. W., 1983. Prediction of man's impact on plant species diversity. In: W. Holzner, M. J. A. Werger & I. Ikusima (eds.). Man's Impact on Vegetation, pp. 41-54. Junk, The Hague.

Plummer, G. L., 1963. Soils of the pitcher plant habitats in the Georgia coastal plain. Ecology 44: 727-734.

Pullen, T. & Plummer, G. L., 1964. Floristic changes within pitcher plant habitats in Georgia. Rhodora 66: 375-381.

Radford, A. E., Ahles, H. E. & Bell, C. R., 1968. Manual of the Vascular Flora of the Carolinas. Univ. N.C. Press, Chapel Hill. 1183 pp.

Regnéll, G., 1980. A numerical study of successions in an abandoned, damp calcareous meadow in South Sweden. Vegetatio 43: 123-130.

Rights, D. L., 1947. The American Indian in North Carolina. Duke Univ. Press, Durham, N.C.

Roberts, P. R. & Oosting, H. J., 1958. Responses of Venus' flytrap (Dionaea muscipula) to factors involved in its endemism. Ecol. Monogr. 28: 193-218.

Rorison, I. H., 1971. The use of nutrients in the control of the floristic composition of grassland. In: E. Duffey & A. S. Watt (eds.). The Scientific Management of Animal and Plant Communities for Conservation. Symp. Brit. Ecol. Soc. 11: 65-78.

Rosén, E. & Sjögren, E., 1973. Sheep grazing and changes of vegetation on the limestone heath of Öland. Zoon, suppl. 1: 137-151.

Schoener, T. W., 1974. Resource partitioning in ecological communities. Science 185: 27-39.

Schoener, T. W., 1982. The controversy over interspecific competition. Am. Sci. 70: 587-595.

Shmida, A. & Ellner, S. P., 1984. Coexistence of plant species with similar niches. Vegetatio (in press).

Sjögren, E., 1971. The influence of sheep grazing on limestone heath vegetation on the Baltic island of Öland. In: E. Duffey & A. S. Watt (eds.). The Scientific Management of Animal and Plant Communities for Conservation. Symp. Brit. Ecol. Soc. 11: 487-495.

Snyder, J. R., 1980. Analysis of coastal plain vegetation, Croatan National Forest, North Carolina. In: H. Lieth & E. Landolt (eds.). Contributions to the Knowledge of Flora and Vegetation in the Carolinas, Vol. 69, pp. 40-113. Veröff. Gebot. Inst. ETH, Stiftung, Rübel, Zürich.

Stoddard, H. L., 1931. The Bobwhite Quail, Its Habits, Preservation and Increase. Charles Scribner's Sons, New York.

Streng, D. R. & Harcombe, P. A., 1982. Why don't Texas savannas grow up to forest? Am. Mid. Nat. 108: 278-294.

Terborgh, J., 1973. On the notion of favorableness in plant ecology. Am. Nat. 107: 481-501.

Thalen, D. C. P., 1971. Variation in some saltmarsh and dune vegetations in the Netherlands with special reference to gradient situations. Acta Bot. Neerl. 20: 327-342.

Thorne, R. F., 1954. The vascular plants of southwestern Georgia. Am. Mid. Nat. 52: 257-327.

Tilman, D., 1982. Resource Competition and Community Structure. Princeton Monogr. Pop. Biol. 17, Princeton Univ. Press, Princeton, New Jersey. 296 pp.

U.S. Dept. Commerce, National Oceanic and Atmospheric Adm., 1977-83. Climatological Data North Carolina, Vol. 82-88. National Climatic Data Center, Asheville, N.C.

Vogl, R. J., 1973. Fire in the southeastern grasslands. Tall Timbers Fire Ecol. Conf. Proc. 12: 175-198.

Ward, T., 1979. Memo to Thomas Massengale, North Carolina Nature Conservancy, Chapel Hill, N.C. April 13, 1979.

Wells, B. W., 1928. Plant communities of the coastal plain of North Carolina and their successional relations. Ecology 9: 230-242.

Wells, B. W., 1932. The Natural Gardens of North Carolina. Univ. of N.C. Press, Chapel Hill. 458 pp.

Wells, B. W. & Shunk, I. V., 1928. A southern upland grass-sedge bog: an ecological study. N.C. Ag. Expt. Sta. Tech. Bull. 32.

Wells, B. W. & Shunk, I. V., 1931. The vegetation and habitat factors of the coarser sands of the North Carolina coastal plain: an ecological study. Ecol. Monogr. 1: 465-520.

Wells, B. W. & Whitford, L. A., 1976. History of stream-head swamp forests, pocosins, and savannahs in the southeast. J. Elisha Mitchell Sci. Soc. 92: 148-150.

Wharton, C. H., 1978. The Natural Environments of Georgia. Off. of Planning and Res., Ga. Dept. Natl. Resources, Atlanta. 227 pp.

Whittaker, R. H., 1977a. Evolution of species diversity in land communities. In: M. K. Hecht, W. C. Steere & B. Wallace (eds.). Evolutionary Biology, Vol. 10, pp. 1-67. Plenum

Publ. Corp., New York.

Whittaker, R. H., 1977b. Animal effects on plant species diversity. In: R. Tüxen (ed.). Vegetation und Fauna, pp. 409–425. J. Cramer, Vaduz.

Whittaker, R. H. & Gauch, H. G., 1973. Evaluation of ordination techniques. In: R. H. Whittaker (ed.). Ordination and Classification of Communities, Hndbk. of Veg. Sci. 5, pp. 287–321. Junk, The Hague.

Willems, J. H., 1980. Observations on north-west European limestone grassland and communities. Proc. K. Ned. Akad. Wet. Series C 83: 279–306.

Williams, G. J. & Markley, J. L., 1973. The photosynthetic pathway type of North American shortgrass prairie species and some ecological implications. Photosynthetica 7: 262–270.

Woodwell, G. M., 1956. Phytosociology of coastal plain wetlands of the Carolinas. M.A. Thesis, Duke Univ., Durham, N.C. 51 pp.

Accepted 17.10.1983.

Diversity models applied to a chalk grassland*

H. J. During & J. H. Willems**
Department of Plant Ecology, University of Utrecht, Lange Nieuwstraat 106, 3512 PN Utrecht, The Netherlands

Keywords: Chalk grassland, Diversity model, Exclosure, Fertilization, Grazing, Management, Mowing, Sod cutting, Species richness

Abstract

In a permanent plot experiment started in 1971, the effects of several management regimes on diversity of a Dutch chalk grassland have been investigated and the results have been compared to existing models predicting general trends in diversity. Treatments included grazing, mowing, and leaving untouched; in the mown plots, the effects of fertilizing and sod cutting were also studied. Grazing resulted in the highest diversity, leaving untouched in the lowest (ca. 42 and 15 spp. per m², respectively). Within the mown plots, fertilizing decreased diversity. The effects of sod cutting disappeared after some years. The results conform best with the response surface model of Huston (1979) relating diversity to frequency of population reduction and rate of competitive displacement, but a slight modification of the shape of the surface is suggested.

Introduction

Diversity patterns in ecosystems cannot be explained in detail by simple rules: 'dominance and diversity form an area of complex and often obscure relationships, not subject to neat, unitary formulation' (Whittaker, 1969). Still diversity and its causation remain among the most intriguing questions in community ecology (e.g., Whittaker, 1972, 1977a; May, 1973). Every step towards more insight into these matters will be of great importance for choice and management of nature reserves (Whittaker, 1969; van Leeuwen, 1966; Westhoff, 1970, 1971; Bakker, 1979).

Limestone grasslands belong to the most diverse grassland communities in NW Europe (Willems, 1973, 1978, 1982). For ages, grazing was a widespread management practice in these grasslands, but in recent years many have become derelict, some are mown, and use of fertilizers has increased enormously. In a 10-year experiment in a Dutch chalk grassland Willems (1979, 1980, 1983) compared the results of several management practices, focusing on the effects of manipulation of the nutrient level on the development of species composition, productivity and diversity. Development of species composition and aboveground phytomass are treated in Willems (1980, 1983). Here we consider the results of experiments involving different frequencies of disturbance and levels of fertilization against the background of some diversity models.

Diversity models

The factors influencing diversity are manifold and it is difficult to make generalizations and predictions. Nevertheless, some models have been proposed to explain general trends in diversity and a number of these have proved fruitful, both in evoking further and more specific studies (e.g.,

* Nomenclature follows van der Meijden *et al.,* 1983.

** The authors are very grateful to Dr J. Miles (Banchory, U.K.) for stimulating remarks and to Prof. Dr M. J. A. Werger, Dr R. K. Peet and two reviewers for many valuable comments and suggestions.

McArthur & Wilson, 1967), and in providing valuable tools for choice and management of nature reserves (van Leeuwen, 1966). Such models necessarily are simplifications, concentrating on the effects of one or a few factors; the suitability of any model will depend on a balance between this simplification and the degree of accurateness necessary for the specific purposes for which it is used.

The diversity response surface of Auclair & Goff (1971) has as the main axes successional stage and 'fertility' or 'favourability' of the environment. This graphical model predicts maximum species density in early successional stages in mesic environments, but at a later stage of succession in intermediate environments (cf. Peet, 1978). This model did not include disturbance (e.g., grazing) as a factor determining diversity. Whittaker (1965, 1969, 1977a) stressed the complexity of diversity causation and usually refrained from simplified models, but once (Whittaker, 1977b) he presented a response surface model with as axes evolutionary time and grazing pressure.

A model relating species density of the vegetation to environmental stress and disturbance (intensity of management: grazing, mowing, or trampling etc.) has been presented by Grime (1973). Starting from the viewpoint of the community as an essentially non-equilibrium phenomenon, Huston (1979) explored the consequences of stress, expressed as the rate of competitive exclusion, and disturbance, expressed as the frequency of population reduction on population dynamics and diversity. He presented as a general model a response surface relating diversity to these factors, predicting maximum diversity at moderate levels of stress and disturbance. Grime's (1979) model may be considered a special case of this model at one level of disturbance (Peet et al., 1983). In a general way, the model is consistent with Sousa's (1979) results on the relationship between dynamics of the substrate and diversity of intertidal algal communities. The absence of a time axis such as that of Whittaker's (1977b) model puts restrictions to its usefulness in comparisons of vegetation from different regions, however (Peet et al., 1983). Also, the degree of 'phasing' of disturbance influences diversity (Abugov, 1982).

The relation theory of van Leeuwen (1966, 1970; cf. van der Maarel, 1980) considers the relation between pattern and process in the vegetation from a more cybernetical point of view. Van Leeuwen relates pattern diversity and species diversity mainly to gradient types and constancy and regularity in time, including constancy and differentiation in management practices and regularity of disturbance (pulse stability, Odum, 1969). Later (Westhoff et al., 1970; see Bakker, 1979; van der Maarel, 1980) he combined the factors of successional time, disturbance and stress in one compound axis of environmental dynamics, ranging from constant situations on dry, humic, nutrient-poor soils to rapidly changing ones in moist, mineral, nutrient-rich soils. He postulated a negative correlation between diversity and environmental dynamics (van Leeuwen, in Westhoff et al., 1970). A similar hypothesis relating diversity primarily to the energy flow in the system has been proposed by Odum (1975). One prediction of the relation theory is that an anthropogenic increase in the rate of succession should eventually lead to lower diversity (van Leeuwen, l.c.).

In conclusion, efforts to model the processes controlling diversity have mainly concentrated on succession, evolutionary time, degree of disturbance, and environmental stress (typically nutrient and water availability).

Study area and methods

In a chalk grassland on a NW-facing slope in the Gerendal nature reserve (SE Netherlands) a series of plots with different management regimes was laid out in 1971. The slope had been heavily fertilized and grazed by cattle up to 1967. Since then, it has been mown annually without application of fertilizers. From 1973 onwards a part of the site has been grazed by Mergelland sheep (stocking rate approx. 6/ha). In this part, 3 plots (1.5 × 1.5 m²) were established in 1971. Another plot (2 × 5 m²) was fenced in 1971 and left untouched ('exclosure'). Ten more plots (1.5 × 1.5 m² each) have been mown annually in August. In 5 of these, sods were cut with a spade in 1971 to a depth of 7–10 cm. The objective was to remove a large proportion of available nutrients and organic matter in the top soil (Tjallingii & Willems, unpubl.; Londo, 1983). In these 10 plots, different fertilization regimes have been applied: NPK in ratios 170:50:50 or 115:355:115 (all in kg/ha; the amounts being roughly equal to those used by the farmer up to 1967); farm yard manure

(100 g/plot annually, corresponding to about 2, 1.5 and 1.5 kg/ha N, P, and K); chalk; or nothing ('control'). Further details concerning site, vegetation, climate, and management can be found in Willems (1980, 1983).

In the 10 mown plots, Braun-Blanquet relevés were made each year, and production was estimated by determining peak aboveground standing crop for all phanerogams together each year since 1973. Occasionally, a plot was not harvested because of excessive wasp (*Vespa* spp.) and/or honey buzzard (*Pernis apivorus*) activity. In 1975 (Voskuilen) and 1981, aboveground dry weight of all phanerogam species was determined using a 0.5 m² subplot.

In 1980, peak standing crop was determined in a temporary exclosure in the grazed area and in some subplots (0.5 × 0.5 m²) in the permanent exclosure. In 1982, relevés were made in 1 m² subplots in the permanent exclosure (2), in temporary exclosures in the grazed area (3) and in the mown plots, using a somewhat more sensitive estimate of abundance (Barkman *et al.*, 1964). The Braun-Blanquet symbols were transformed for calculation of diversity indices following Barkman (pers. comm.) by calculating the geometrical mean of the minimum and maximum cover value in each class. Diversity was measured using Shannon's index *H'* and its derivative *J'* and Simpson's index *C*. Slopes of dominance-diversity curves were calculated from the 1975 and 1981 data on species production of the mown plots using Whittaker's formula $E'_c = \frac{S}{4\sigma_e}$ (Whittaker, 1972) where σ_e is the geometric standard deviation.

Results

The development of phanerogam species numbers and production in the mown plots with different treatments are shown in Figure 1–4. The addition of F.Y.M. or chalk appeared to have no marked effect on production (Fig. 1) nor on diversity (Fig. 3), as compared to the controls. The slight deviation of the F.Y.M. plot without sod cutting from 1979 onwards was due to a large wasp colony and much honey buzzard activity in this plot in 1979 which disturbed the soil heavily. Unfortunately, no production data from prior to 1973 are known; Willems (unpubl.) estimated peak standing crop in 1970 to be ca. 500 g/m² all over the slope.

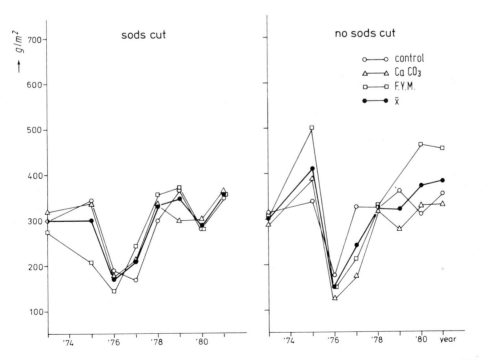

Fig. 1. Peak aboveground standing crop (g/m²) of the mown plots that did not receive artificial fertilizer. Either CaCO3, F.Y.M. or nothing added. Left: sods cut in 1971; right: no sods cut.

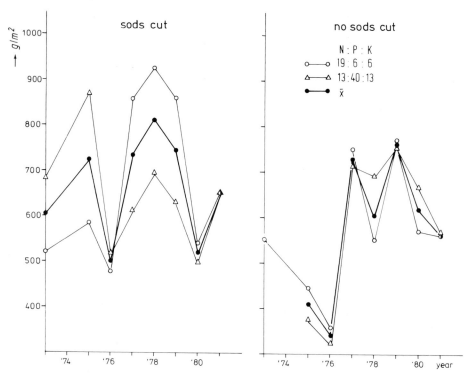

Fig. 2. Peak aboveground standing crop (g/m²) of the mown plots that have received NPK fertilizer annually in the ratios 19:6:6 or 13:40:13.

Apparently, production in these plots had already reached a stable level in 1973, though species numbers continued to rise through 1982. Sod cutting hardly influenced production and species number, apart from the occurrence of some weeds in the first years which is reflected in a slightly higher species number.

The plots receiving artificial fertilizers showed a more differentiated response. In the plots where sods had been cut productivity was at a high level continuously from 1973 onwards, whereas in the plots without sod cutting the reaction to fertilization appears to have been retarded, rising to levels above 500 g/m² only in 1977 (Fig. 2). Species numbers decreased quickly to remain at a level of ca. 18/2.25 m² (Fig. 4). The ratio of NPK in the fertilizer hardly affected productivity in the plots without sod cutting, whereas in the plots with sod cutting initial productivity was higher in the high-P treatment, but after the dry summer of 1976 the reverse was true. In the first years, in the high-P treatment grasses (especially *Dactylis glomerata* and *Festuca rubra*) grew luxuriously, being re-

placed after 1976 by a dense mat of legumes, especially *Lathyrus pratensis*, lying over and 'choking' the other species. In the high-N plot, dicot rosettes were more abundant from the beginning. The high productivity of 1977 and onwards in this plot is mainly due to a few very large individuals of *Heracleum sphondylium* reaching a height of 2 m.

Species numbers were not much affected by the NPK ratio in the plots without sod cutting, whereas in the plots with sod cutting high P lead to a slightly lower number (Fig. 4), maybe related to the difference in vegetation structure.

From 1979 onwards, differences in productivity and species number between series with and without sod cutting and NPK ratios were only slight. Thus, within the mown plots two broad types may be recognized, hereafter called unfertilized and fertilized plots, respectively (Willems, 1980).

The dry summer of 1976 resulted in markedly lower production in all plots, but species number was lower in the fertilized plots only. Apart from the instances mentioned, no persistent floristic changes occurred as a result of this dry summer.

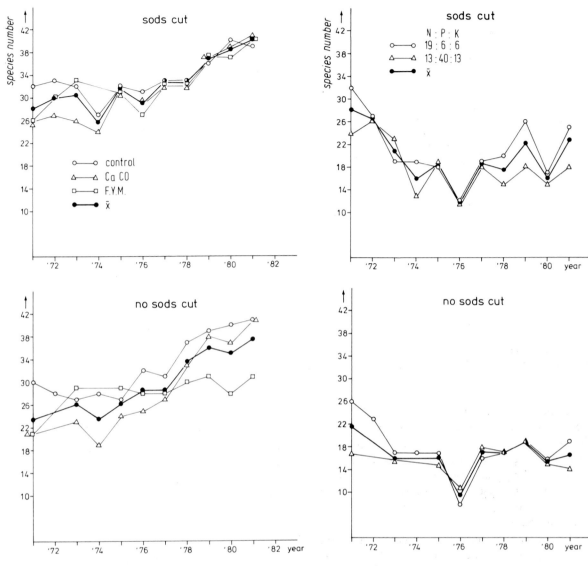

Fig. 3. Phanerogam species number of the mown unfertilized plots in the years 1971–1981.

Fig. 4. Phanerogam species number of the mown fertilized plots in the years 1971–1981.

In 1975 the fertilized plots showed a strong concentration of dominance in one or a few species and very steep, nearly geometric dominance–diversity curves resulting in low E'_c values (Table 1). In the unfertilized plots often several species were co-dominant, especially in the plots receiving F.Y.M., and the number of species per log-cycle was rather high, resulting in curves more nearly lognormal in shape. In 1981, all plots had an increase in number of species in the lowest importance classes leading to a slightly higher E'_c value in the fertilized plots.

Table 1. Slopes of dominance–diversity curves (E'_c) of the mown plots with different treatments in 1975, 1981 and 1982.

	Sods cut in 1971			Sods left intact		
	1975	1981	1982	1975	1981	1982
control	4.25	3.34	3.26	2.81	3.60	3.47
chalk added	3.12	3.43	3.69	3.30	3.70	3.59
F.Y.M. added	3.22	3.21	2.34	3.11	2.48	3.00
NPK 170:50:50	1.01	2.21	2.12	1.04	2.03	1.89
NPK 115:355:115	1.46	1.32	1.62	1.20	1.52	2.09

326

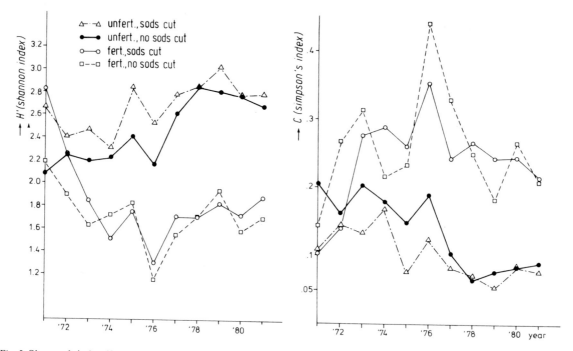

Fig. 5. Shannon's index H' and Simpson's index C of the mown plots in 1971–1981. Data of fertilized and unfertilized plots have been combined into mean values for the series with and without sod cutting.

Differences between fertilized and unfertilized plots, tested with Wilcoxon's sign-rank test, are significant ($r = 2.56$, $p < 0.05$); differences between the plots with and those without sod cutting are not significant. The effect of NPK ratio is complex; in 1975, high-N treatment lead to steeper curves, but in 1981 the reverse was true, whereas in 1982, no consistent pattern emerged.

The pattern in the other diversity measures (Fig. 5) is generally in accordance with these trends. Simpson's index C is high in the fertilized plots, showing large fluctuations with no consistent trend as regards the effect of sod cutting. In the unfertilized plots, it decreased gradually in the first years, more quickly so in the series with sod cutting, stabilizing at a low level after 1976, while S still increased here. Shannon's index H' (Fig. 5) and J' decreased in the fertilized plots stabilizing at ca. 1.7 and 0.6, respectively, and in the unfertilized plots rose to a level of ca. 2.8 and 0.8, respectively. In the latter group, both tend to decrease slightly in the last two years, while S is still increasing. Again, the effect of sod cutting is only visible in the first years in the unfertilized plots.

The dry summer of 1976 markedly affected spe-

cies number in the fertilized plots only. However, the other measures show a sharp increase in dominance and decrease in H' in all plots in this year. The high values of Simpson's index in the fertilized plots are mainly due to a rise in *Festuca rubra* and other grasses, which was not observed in the unfertilized plots (Willems, 1980).

The situation in 1982 in the grazed plots and the exclosure was compared with that in the fertilized and unfertilized, mown plots (Table 2). Production in enclosures in the grazed plot was similar to that of the unfertilized, mown plots. Production of the exclosure in 1982 was ca. 450 g/m². In this plot species number had been lowered drastically in comparison to the situation in 1970, when all plots were more or less equal. Simpson's index C is very high, and the other diversity indices are extremely low. *Brachypodium pinnatum* had become completely dominant, forming a dense litter layer. In the grazed area the species number was distinctly higher than in the mown, unfertilized plots, and all diversity measures indicated a higher diversity in the grazed plots in 1982. In 1970, species number in this plot was 31 (Willems, 1983).

Table 2. Diversity with standard deviations in 1982 of 1 m² plots which have been managed differently since 1971. Abundance estimated in the scale of Braun-Blanquet as refined by Barkman *et al.* (1964). For statistical evaluation of the differences Wilcoxon's signed rank sum test was applied to treatment pairs. The following pairs differed significantly ($p < 0.05$); Mown fert./mown unfert. (*S, H'*); mown unfert./exclosure (*S, C, H', J'*); mown unfert./grazed (*S, H'*).

Management	Exclosure	Mown plots		Grazed
		Fertilized	Unfertilized	
Number of plots	2	4	6	3
Number of species *S*	14.5 ± 2.1	23.0 ± 6.7	35.0 ± 4.1	41.6 ± 1.5
Simpson's index *C*	.75 ± .06	.14 ± .05	.09 ± .01	.07 ± .01
Shannon's index *H'*	.71 ± .18	2.36 ± .34	2.84 ± .08	3.10 ± .08
Evenness *J'*	.27 ± .05	.76 ± .04	.80 ± .02	.83 ± .02

Discussion

Effects of dominance

The decrease in diversity in the exclosure and the fertilized plots correlates markedly with an increase in the degree of dominance in the vegetation. The dominant species were *Brachypodium pinnatum* in the former and *Dactylis glomerata, Festuca rubra* and *Poa angustifolia* in the latter. These species produce numerous overhanging leaves and much litter, which is very persistent in the case of *Brachypodium pinnatum*. According to these characters, they might be considered as competitors *sensu* Grime (1974,1979). In English chalk grasslands, dominance of *Bromus erectus* does not markedly affect the species number under grazing or mowing regimes, even if the species covers more than 75% of the surface (Willems, 1978). This species forms dense clumps with some open space in between. Even if the vegetation cover is nearly closed in summer, this pattern ensures ample opportunities for establishment of many species in spring which may hold out between the clumps in summer. If the vegetation is left untouched, *Bromus erectus* produces copious litter and causes a strong decrease in diversity (Wells, 1971). Thus, the effect of dominance depends partly on the structure of the dominant species, as is also exemplified by the different effects on species number with dominance of *Heracleum* or with grasses or legumes in the fertilized plots. The degree of dominance of the species depends also on the soil structure, however. Thus, on shallow soils *Bromus erectus* rarely reaches complete dominance (Wells, 1971).

One aspect of dominance is a heavy competitive effect on adult plants of subordinate species, for many of these leading to competitive exclusion. Another aspect is the 'smothering' effect on present environmental heterogeneity, which is especially important for the differentiation of the regeneration niche (Grubb, 1977). In our mown plots, many species yet germinate in large numbers, especially in the fertilized plots, but very few or none of these seedlings survive. This would suggest that dominance does not affect very much the differentiation in the regeneration niche in the restricted sense, but prevents the species from becoming established (cf. Newman, 1982). In the exclosure the situation is different; here the thick litter layer of *Brachypodium pinnatum* prevents any germination, presumably because of its strong effect on spectral composition and intensity of light reaching the surface (Willems, 1983).

Interestingly, the fall in species number in the fertilized plots without sod cutting does not coincide with an increase in production. The early disappearing species mainly are small-sized Mesobromion species (Willems, 1980), while *Dactylis glomerata* quickly gains dominance.

Both mowing and grazing without fertilization have led to a steady increase in *S* since 1973. The indices *H'* and *J'* increased as well, but stabilized after 1976 and decreased slowly in 1980 and 1981. Thus even though *H'* and *J'* are strongly dependent on *S* (Peet, 1974; Alatalo, 1981), these measures show a pattern of their own.

Wells (1971) found no significant difference between mowing and grazing after 6 years as regards their effect on diversity in a chalk grassland. Our

328

results show stabilizing H' and J' values in the unfertilized mown plots after 8 years, while in the grazed plot these values have reached a higher level. A nearby chalk grassland with the same exposition and slope, never fertilized, and grazed in the past but only mown since ca. 1960 provides a useful comparison. Here, number of species and evenness are at a lower level (ca. 30 spp/m^2, H' ca. 2.4, J' ca. 0.7), and the bryophyte layer is continuous and rather thick. Though exact data on the situation before 1960 are lacking, from unpublished inventories it appears to have been much richer in species, including several orchids. We regard this as an indication that mowing once a year is not the optimum disturbance level at the lower nutrient level of the unfertilized plots.

The effects of grazing depend on the herbivore species (Wells, 1971; McNaughton, 1977, 1979), and even on the breed (Wells, 1980). A local sheep breed in the Dutch province of Limburg, the 'Mergelland' sheep, was nearly extinct but is now 're-made'. These sheep eat *Brachypodium pinnatum*, especially young shoots, much more readily than, for example, the 'Texel' sheep which are now used widely. Palatability of old shoots appears to be much lower, but at the end of the winter even dead *Brachypodium* will be eaten. Of old, shepherds used to promote consumption of the dead *Brachypodium pinnatum* on roadside verges by now and then throwing a handful of salt on the vegetation! At low stocking rates, grazing results in a distinct patterning of the vegetation cover; some spots are kept low continuously but in others only the more palatable species are picked out, and the dense mat of *Brachypodium* is more or less left intact. This parallels the effect of rabbit grazing in the dunes of the Dutch island of Schiermonnikoog (Zeevalking & Fresco, 1977).

A comparison with diversity models

Species density per m^2 of Dutch chalk grasslands appears to reach a maximum at ca. 54 phanerogams (Boeijen & van Leeuwen, 1980). Among the grasslands reaching this level, considerable differences in pattern diversity are present, the Kunderberg having a particularly fine-grained pattern (Boeijen & van Leeuwen, 1980). Together with the relatively low age of this community type, presumably some 3–4 millennia (Willems, 1978), this suggests that

under continuous grazing (which until recently was almost the only management practiced) a continued rise in species density is possible, in accordance with Whittaker's (1977b) model.

Chalk grasslands are not a clear-cut stage in the succession from bare chalk to woodland (Hope-Simpson, 1940); rather, they must be considered as a plagioclimax (Wells, 1971). Thus, interpretation of their species richness in terms of models wherein successional time is a major axis (Auclair & Goff, 1971) is inappropriate. Thus, the development of the exclosure for 10 years might be considered as succession, but the decrease in diversity in this plot does not fit the existing models.

Our experiments contained different levels of fertilization and disturbance, and therefore our data should conform with models using these parameters (e.g., Grime, 1973; Huston, 1979). The predicted rise in diversity as an effect of disturbance as compared to removal of disturbance (exclosure) is apparent. Also, the higher nutrient level leading to higher production (Grime) and a corresponding higher rate of competitive exclusion (Huston) is in accordance with the models. The slow increase in S in the unfertilized plots is noteworthy. Apparently, the process of establishment of new species in the plots is itself rather slow, which may be related to the low dispersal potential of a number of these species (Verkaar et al., 1983).

Cutting of sods (i.e. removal of all aboveground and a major part of the belowground vegetation, including a considerable part of the seed bank; Schenkeveld & Verkaar, in prep.) must be considered as a very strong disturbance. However, species number in these plots initially is even higher than in the series without sod cutting, partly due to a number of arable weeds growing from seeds buried in deeper soil layers, partly also to a fast regeneration from seeds and belowground parts of most of the species present before (Willems, 1980).

Persistence of the weeds for some years is the main reason for the more gradual loss of species in the fertilized plots with sod cutting as compared to those without sod cutting. The high productivity of the fertilized plots after sod cutting as compared to those without sod cutting was unexpected, since by this treatment a large proportion of the nutrients in the system was removed (Tjallingii & Willems, unpubl.). One explanation might be that with sod removal phytotoxins produced by decaying organisms were also removed (Toai & Linscott, 1979; Wallace & Elliott, 1979). Alternatively, sod removal might accelerate weathering of soil minerals, either by higher temperatures in the soil after sod removal (Jackson & Sherman, 1953) or by greater numbers of rhizosphere bacteria which biochemically weather soil minerals (Jackson & Voigt, 1971; Boyle & Voigt, 1973) after intrusion of many new roots in the mineral zone.

Grime's model (Al-Mufti et al., 1977) predicts high species density at productivity levels of 350–750 g/m². In our plots, the negative effect of a higher productivity level on diversity starts at lower levels, approx. 400 g/m² and in the dry summer of 1976 at 300 g/m² (Willems, 1980). This may be in part due to different harvesting methods, since we clipped at 2 cm above the ground and did not include the (sparse) litter. Differences in growth form on the dominant plant species, notably leaf size and inclination, may be important in this respect, as well.

It is not yet possible to scale the axes of Huston's (1979) response surface model, but it is possible to get an impression of the relative position of grazing and mowing along the axis of frequency of population reduction (F-axis). This axis includes effects of frequency as well as intensity of the disturbance which will have to be combined into the 'impact' of the management regime. Mowing involves a disturbance of great intensity, but low frequency. Grazing usually implies a consistent, low intensity, but a high frequency. In our experiments the stocking rate was sufficient to keep the vegetation continuously low, whereas in the mown plots in summer a tall turf (50–80 cm) developed. Apparently, the impact of grazing here is higher than that of mowing, and placement along Huston's F-axis should shift accordingly (Willems, 1983). Since our grazed plots reached a higher diversity than the mown plots, grazing at the intensity chosen appears to be nearer to the level of population reduction at which maximum diversity occurs.

Huston's model suggests that changes in diversity are smaller if two disturbance levels are compared at a high rate of competitive displacement than in a situation nearer to the optimum level. However, the difference between mowing and grazing is much more marked at high productivity levels, than at the low productivity level of the unfertilized mown and grazed plots in 1981. This is shown by the drop in

species level after 1970. Before that year grazing was combined with approximately the same fertilization level. One cannot simply equate the impact of population reduction caused by cattle grazing before 1970 with that of sheep grazing in our experiments, but both prevent potentially dominant species from reaching dominance, while mowing usually does so rather well at the lower productivity level, but not at the higher level. This results in a larger difference in diversity between the two disturbance types at high productivity levels. This may partly be explained by assuming that differences between the disturbances are not equivalent at different levels of production. In addition, however, it may be hypothesized that the shape of Huston's response surface should be more like a table mountain (Fig. 6). Then, at a certain rate of competitive displacement, above a critical level of disturbance impact causing strong reduction of the dominants, a 'plateau' is reached where changes in disturbance impact level only cause smaller, though still important, changes in diversity.

For specific situations, the general tendencies in the model must be qualified also in other ways. The rate of competitive displacement is determined mostly by actual growth and mortality rates of the dominant species, but as has been argued already the structure of the dominant plant species plays a role, too. Which species will be dominant at a specific productivity level, strongly depends on ratios of available nutrients (Thurston, 1969; Rorison, 1971; Tilman, 1982) and management practices (e.g., Wells, 1971).

The relation between grazing and population reduction in a specific case is influenced by the palatability of the dominants. In fact, at one grazing intensity, impact of population reduction may differ drastically for the different plant species in the vegetation; again, it is the reduction of dominants that is primarily important. In the model, this means that palatability of the dominants determines the size of the plateau of the table mountain, if grazing is practiced. A further complication may emerge, when after introduction of grazing another, unpalatable, species takes over and becomes dominant, again reducing diversity (McNaughton, 1977); for such catastrophic situations, the model is not suitable (Denslow, 1980; Peet et al., 1983).

The degree of population reduction caused by mowing depends both on the season of mowing and

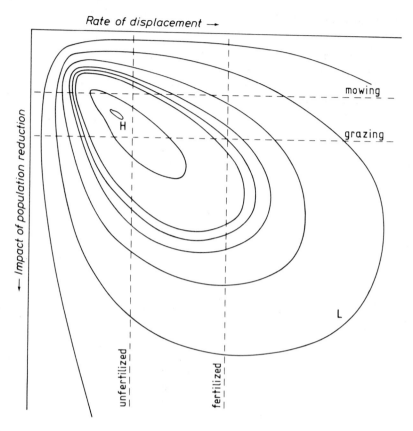

Fig. 6. Hypothesized response surface of diversity in relation to rate of competitive displacement and impact of population reduction, adapted from Huston (1979). Hypothetical positions of the fertilized and unfertilized plots on the *x*-axis and of mowing and grazing on the *y*-axis are indicated.

its frequency. Mowing once a year in different seasons results in different amounts of production harvested, but also reproduction of many species is affected differently (Wells, 1971). Also, structural aspects of the plants such as location of the meristem and tiller production will affect this relation.

Finally, the element of constancy, one of the decisive parameters in van Leeuwen's (1966, 1970) model, must be mentioned. Not only the frequency, but also the constancy over the years of population reduction affects diversity (Whittaker, 1977b; Peet *et al.*, 1983). The model predicts higher diversity with higher constancy. According to Abugov (1982), this applies only in situations with low disturbance rates. A useful test might involve the effect of alternate mowing in early summer and autumn.

The important aspect of Huston's model is, we think, that suppression of competitive exclusion does not imply the need for extensive niche differ-

entiation for coexistence, but rather implies that species with broadly overlapping niches may co-exist provided population reduction is severe enough (niche-sharing).

Management of chalk grasslands

The aims of acquisition and management of chalk grassland reserves, as for most nature reserves in general (Harper, 1971; Westhoff, 1971; van der Maarel, 1971, 1980; Bakker, 1979) are primarily to preserve or enhance the existing or former diversity. As many of the examples cited show, species composition is rather sensitive to management practices (e.g., Wells, 1969; Grubb, 1976). Moreover, there is an enormous diversity of chalk grasslands themselves, both within a region and over Western Europe as a whole (Willems, 1982). Therefore, such aims must be clearly specified with reference to the local or regional situation as well as in the context of the diversity of such grasslands throughout Europe.

For management choices, an optimal strategy must be carefully chosen (Swartzmann & Singh, 1974; Singh, 1976). Useful general insights may be gained by cautious application of diversity models, especially Huston's (1979) model, but these models generally are not applicable to specific situations. The relation between the rate of competitive exclusion and productivity level, and that between the impact of population reduction and a grazing or mowing regime are in particular not straightforward. These depend heavily on palatability and structure of the species involved and in the case of grazing on the species and breed of grazers used. Species density will generally reach some kind of saturation, but pattern diversity may increase yet further. Elucidation of the relation between these two kinds of diversity would certainly be rewarding and significantly contribute to an optimal management strategy.

References

Abugov, R., 1982. Species diversity and phasing of disturbance. Ecology 63: 289–293.

Alatalo, R. V., 1981. Problems in the measurements of evenness in ecology. Oikos 37: 199–204.

Al-Mufti, M. M. et al., 1977. A quantitative analysis of shoot phenology in herbaceous vegetation. J. Ecol. 65: 759–791.

Auclair, A. N. & Goff, F. G., 1971. Diversity relations of upland forests in the western Great Lakes area. Am. Nat. 105: 499–528.

Bakker, P. A., 1979. Vegetation science and nature conservation. In: M. J. A. Werger (ed.), The Study of Vegetation, pp. 247–288. Junk, The Hague.

Barkman, J. J., Doing, H. & Segal, S., 1964. Kritische Bemerkungen und Vorschläge zur quantitativen Vegetationsanalyse. Acta Bot. Neerl. 13: 394–419.

Boeijen, J. H. & Leeuwen, C. J. van, 1980. Een productie-oecologisch onderzoek aan enige kalkgraslanden in Zuid-Limburg. Internal Report Inst. Syst. Botany, Utrecht.

Boyle, J. R. & Voigt, G. K., 1973. Biological weathering of silicate minerals - implications for tree nutrition and soil genesis. Plant & Soil 38: 191–201.

Denslow, J. S., 1980. Patterns of plant species diversity during succession under different disturbance regimes. Oecologia (Berlin) 46: 18–21.

Grime, J. P., 1973. Control of species density in herbaceous vegetation. J. Environ. Manage. 1: 151–167.

Grime, J. P., 1974. Vegetation classification by reference to strategies. Nature (London) 250: 26–31.

Grime, J. P., 1979. Plant Strategies and Vegetation Processes. Wiley, Chichester.

Grubb, P. J., 1976. A theoretical background to the conservation of ecologically distinct groups of annuals and biennials in the chalk grassland ecosystem. Biol. Conserv. 10: 53–76.

Grubb, P. J., 1977. The maintenance of species-richness in plant communities: the importance of the regeneration niche. Biol. Rev. 52: 107–145.

Harper, J. L., 1971. Grazing, fertilizers and pesticides in the management of grasslands. In: E. Duffey & A. S. Watt (eds.), Symp. Br. Ecol. Soc. 11: 15–31.

Hope-Simpson, J. F., 1940. Studies of the vegetation of the English Chalk VI. Late stages in succession leading to chalk grassland. J. Ecol. 28: 386–402.

Huston, M., 1979. A general hypothesis of species diversity. Am. Nat. 113: 81–101.

Jackson, M. L. & Sherman, G. D., 1953. Chemical weathering of minerals in soils. Adv. Agron. 5: 219–318.

Jackson, T. A. & Voigt, G. K., 1971. Biochemical weathering of calcium-bearing minerals by rhizosphere micro-organisms, and its influence on calcium accumulation in trees. Plant & Soil 35: 655–658.

Leeuwen, C. G. van, 1966. A relation theoretical approach to pattern and process in vegetation. Wentia 15: 25–46.

Leeuwen, C. G. van, 1970. Raum-zeitliche Beziehungen in der Vegetation. In: R. Tüxen (ed.), Gesellschaftsmorphologie. Symp. Ber. Int. Ver. Vegetationskunde Rinteln 1966: 62–68.

Londo, G., 1983. Natuurbeheer en natuurbehoud van kalkgraslanden. Publ. Nat. Hist. Gen. Limburg 33 (1/2): 9–13.

Maarel, E. van der, 1971. Plant species diversity in relation to management. In: E. Duffey & A. S. Watt (eds.), Symp. Br. Ecol. Soc. 11: 45–63.

Maarel, E. van der, 1980. Towards an ecological theory of nature management. Verh. Ges. Oekologie (Freising-Weihenstephan 1979) 8: 13–24.

May, R. M., 1973. Stability and Complexity in Model Ecosystems. Princeton Univ. Press, Princeton.

McArthur, R. M. & Wilson, E. O., 1967. The Theory of Island Biogeography. Princeton Univ. Press, Princeton.

McNaughton, S. J., 1977. Diversity and stability of ecological communities: a comment on the role of empiricism in ecology. Am. Nat. 111: 515–525.

McNaughton, S. J., 1979. Grazing as an optimization process: grass-ungulate relationships in the Serengeti. Am. Nat. 113: 691–703.

Meijden, R. van der, Weeda, E. J., Adema, F. A. C. B. & Joncheere, G. J. de, 1983. Heukels-van der Meijden Flora van Nederland. Wolters-Noordhoff, Groningen.

Newman, E. I., 1982. Niche separation and species diversity in terrestrial vegetation. In: E. I. Newman (ed.), The Plant Community as a Working Mechanism. Spec. Publ. Br. Ecol. Soc. 1: 61–77.

Odum, E. P., 1969. The strategy of ecosystem development. Science 164: 262–270.

Odum, E. P., 1975. Diversity as a function of energy flow. In: W. H. van Dobben & R. H. Lowe-McConnell (eds.), Unifying Concepts in Ecology, pp. 11–14. Junk, The Hague.

Peet, R. K., 1974. The measurement of species diversity. Ann. Rev. Ecol. Syst. 5: 285–307.

Peet, R. K., 1978. Forest vegetation of the Colorado Front Range: patterns of species diversity. Vegetatio 37: 65–78.

Peet, R. K., Glenn-Lewin, D. C. & Wolf, J. W., 1983. Prediction of man's impact on plant species diversity: a challenge for vegetation science. In: W. Holzner, M. J. A. Werger & I.Ikusima (eds.), Man's Impact on Vegetation, pp. 41–54. Junk, The Hague.

Rorison, I. H.,1971. The use of nutrients in the control of the floristic composition of grassland. In: E. Duffey & A. S. Watt (eds.), Symp. Br. Ecol. Soc. 11: 65–77.

Singh, J. S., 1976. Structure and function of tropical grassland vegetation of India. Polish Ecol. Stud. 2 (2): 17–24.

Sousa, W. P., 1979. Disturbance in marine intertidal boulder fields: the non-equilibrium maintenance of species diversity. Ecology 60: 1225–1239.

Swartzman, G. L. & Singh, J. S., 1974. A dynamic programming approach to optimal grazing strategies using a successional model for a tropical grassland. J. Appl. Ecol. 11: 537–548.

Thurston, J. M., 1969. The effect of liming and fertilizers on the botanical composition of permanent grassland, and on the yield of hay. In: I. H. Rorison (ed.), Symp. Br. Ecol. Soc. 9: 3–10.

Tilman, D., 1982. Resource competition and community structure. Monographs in Population Biology 17. Princeton Univ. Press, Princeton.

Toai, T. V. & Linscott, D. L., 1979. Phytotoxic effects of decaying quackgrass (Agropyron repens) residues. Weed Science 27: 595–598.

Verkaar, H. J., Schenkeveld, A. J. & Klashorst, M. P. van de, 1983. On the ecology of short-lived forbs in chalk grasslands: dispersal of seeds. New Phytol. 95: 335–344.

Voskuilen, K. J. G.,1975. De invloed van bemesting en maaibeheer op de productie en op de vegetatiekundige samenstelling van een kalkgrasland in het Gerendal, Zuid-Limburg. Internal Report Inst. Syst. Botany, Utrecht.

Wallace, J. M. & Elliott, L. F., 1979. Phytotoxins from anaerobically decomposing wheat straw. Soil Biol. Biochem. 11: 325–330.

Wells, T. C. E., 1969. Botanical aspects of conservation management of chalk grasslands. Biol. Conserv. 2: 36–44.

Wells, T. C. E.,1971. A comparison of the effects of sheep grazing and mechanical cutting on the structure and botanical composition of chalk grassland. In: E. Duffey & A. S. Watt (eds.), Symp. Br. Ecol. Soc. 11: 497–515.

Wells, T. C. E., 1980. Management options for lowland grassland. In: I. H. Rorison & R. Hunt (eds.), Amenity Grassland, pp. 175–195. Wiley, Chichester.

Westhoff, V., 1970. Choice and management of nature reserves in the Netherlands. Bull. Jard. Bot. Nat. Belg. 41: 231–245.

Westhoff, V., 1971. The dynamic structure of plant communities in relation to the objectives of conservation. In: E. Duffey & A. S. Watt (eds.), Symp. Br. Ecol. Soc. 11: 3–14.

Westhoff, V., Bakker, P. A., Leeuwen, C. G. van & Voo, E. E. van der, 1970. Wilde Planten I. Ver. Behoud Natuurmonumenten, Amsterdam.

Whittaker, R. H., 1965. Dominance and diversity in land plant communities. Science 147: 250–260.

Whittaker, R. H., 1969. Evolution of diversity in plant communities. Brookhaven Symp. Biol. 22: 178–196.

Whittaker, R. H., 1972. Evolution and measurement of diversity. Taxon 21: 213–251.

Whittaker, R. H., 1977a. Evolution of species diversity in land communities. Evol. Biol. 10: 1–67.

Whittaker, R. H., 1977b. Animal effects on plant species diversity. In: R. Tüxen (ed.), Vegetation und Fauna. Ber. Int. Symp. Int. Ver. Vegetationskunde, Rinteln 1976: 409–425.

Willems, J. H., 1973. Limestone grassland vegetation in the central part of the French Jura, south of Champagnole (dept. Jura). Proc. Kon. Ned. Ac. Wet. Ser. C, 76: 231–244.

Willems, J. H.,1978. Phytosociological and ecological notes on chalk grasslands of southern England. Vegetatio 37: 141–150.

Willem, J. H., 1979. Experiments on the relation between species diversity and above-ground plant biomass in chalk grasslands. Acta Bot. Neerl. 28: 235.

Willems, J. H., 1980. An experimental approach to the study of species diversity and above-ground biomass in chalk grassland. Proc. Kon. Ned. Ac. Wet. Ser. C, 83: 279–306.

Willems, J. H., 1982. Phytosociological and geographical survey of Mesobromion communities in Western Europe. Vegetatio 48: 227–240.

Willems, J. H., 1983. Species composition and above-ground phytomass in chalk grassland with different management. Vegetatio 52: 171–180.

Zeevalking, H. J. & Fresco, L. F. M., 1977. Rabbit grazing and diversity in a dune area. Vegetatio 35: 193–196.

Accepted 14.2.1984.